SOLVING PROBLEMS
IN CHEMISTRY

SOLVING PROBLEMS
in CHEMISTRY
Second Edition

with emphasis on STOICHIOMETRY and EQUILIBRIUM

and applications in

Agriculture & Marine Sciences
Biological & Medical Sciences
Industrial Chemistry
Environmental Sciences

ROD O'CONNOR

Professor and Director of First Year Chemistry Programs
Texas A&M University

CHARLES MICKEY

Assistant Professor of Chemistry
Moody College of Marine Sciences

ALTON HASSELL

Assistant Program Director, First Year Chemistry
Texas A&M University

HARPER & ROW, PUBLISHERS

New York, Hargerstown, San Francisco, London

SOLVING PROBLEMS IN CHEMISTRY with emphasis on stoichiometry and equilibrium, Second Edition

Copyright © 1977 by Rod O'Connor, Charles Mickey, and Alton Hassell

ISBN 0-06-044882-2

DEDICATION

To our wives, Shirley O'Connor, Betty Mickey, and Pat Hassell, and to our excellent typists, Carla and Gary Wesson.

They sometimes thought _we_ were the "problems".

Contents

- PREFACE -

This is not a textbook of chemistry. It is an introduction to methods and skills used in solving problem situations in some of the more important areas of chemistry and is designed to be used in conjunction with a textbook or other supplementary materials. Its purpose is to provide additional help and practice for those seeking competency or proficiency in chemical problem solving or for those desiring the experience and challenge provided by problem situations.

It is not essential that all Units be studied or that they be followed in sequence. You many wish to concentrate your efforts on those Units whose topics represent areas of some difficulty to you, so that extra practice and more examples are useful, or on Units in which you have some particular interest in achieving greater proficiency.

Each Unit begins with a brief summary of necessary background information, referenced to suitable textbook sections or single-topic booklets which provide theoretical and factual background material. Your instructor may wish to suggest other references.

Following this introduction are brief statements of learning objectives at two levels, competency and proficiency. The distinction between these levels is basically that we would consider a competency level to be roughly equivalent to that expected of a C-grade student in a typical university general chemistry course, while the proficiency level corresponds to that of A and B student work. You may find that your own interests and needs are adequately satisfied for some areas by the competency level work, while the more challenging proficiency level may be more appropriate in other areas. (All proficiency level material is identified by an asterisk [*].)

Immediately following the statements of objectives are Pre-Test questions, at both levels. Successful completion of the competency level questions offers you a choice of proceeding on to the proficiency level, working on appropriate competency level Exercises or Relevant Problems, or terminating -- if "competency" (ability to handle simpler problem situations) is sufficient for your needs. If you have difficulties with the competency level Pre-Test, you should study the "Methods" work at that level. Successful completion of the proficiency level Pre-Test, or the final Self-Test questions based on the proficiency level "Methods" work, offers you the choices of terminating or proceeding on to the more challenging Relevant Problems in areas of your own interest.

The Relevant Problems are arranged in four categories: (A) Agriculture and Marine Sciences, (B) Biological and Medical Sciences, (I) Industrial Chemistry, and (E) Environmental Sciences. Problem solving may thus be practiced in an area related to personal interests or, by working problems from more than one Set, as exercises in examining a broad scope of chemical applications.

We hope you will find this book useful in expanding you capabilities to handle problems in chemistry and that the Relevant Problems will illustrate for you some of the many applications of chemistry to other fields. Your comments and suggestions would be most welcome.

Rod O'Connor
Charles Mickey
Alton Hassell

Texas A&M University
July 1977

SOLVING PROBLEMS
IN CHEMISTRY

INTRODUCTION: PROBLEM SOLVING

Some people seem to have an intuitive grasp of how to approach a problem to obtain a rapid and efficient solution. Although there is little doubt that some are more gifted than others in this respect, it is a talent that can be acquired. Proficiency in this skill, as in all skills, improves with practice.

When a mechanically inept customer takes his car to a mechanic with the complaint that it is getting poor gas mileage and makes a "funny rumbling sound," this poses a problem for the mechanic. Although the talented mechanic might not express it this way, his approach to this problem involves certain distinguishable phases:

1. Identification of what he wants to find out.
2. Recognition of what information is given that might be useful.
3. Identification of other information needed, but not given (i.e., information that must be obtained from memory and from other sources).
4. Selection of a method of attack on the problem.
5. Manipulation of information and tools to achieve a solution.
6. Checking that the solution is probably satisfactory.

For the mechanic, these steps might involve:

1. Finding out what is causing the poor gas mileage and unusual engine noise.
2. Given: poor mileage + "funny rumbling sound."
3. Other information: memory of previous experience with similar problems + application of various suitable test instruments and other pertinent observations.
4. Selecting parts to be replaced, adjustments to be made, and tools to be used.
5. Fixing the engine.
6. Test driving.

This system is a general one for solving all kinds of problems. It is the basis for most scientific research, and it is the key to solving all sorts of mathematical problems. The more problems you solve, the easier it becomes, until you, like the competent mechanic, can solve problems without formally thinking out each step of the method. Until the stage of "intuitive" proficiency is reached, however, it is worthwhile to proceed slowly and systematically through the various phases of problem solving.

Let us illustrate the approach by a "case study" of three problems.

==

PROBLEM 1

A dozen assorted ball bearings weighs 180 grams. What is the average mass of a bearing in this assortment?

SOLUTION

Step 1: Find: *mass per bearing* (in grams)

Step 2: Given: *180 grams per dozen* (of bearings)

Step 3: Needed, not given: *How many items per dozen?*

Answer (from memory): 12 per dozen

Step 4: Method of attack:

Divide mass by number to find average mass of each. Use unity factor method.

Step 5: Manipulation:

$$\frac{180 \text{ grams}}{\text{dozen (bearings)}} \times \frac{1 \text{ dozen}}{12} = \frac{180 \text{ grams}}{12 \text{ (bearings)}} = 15 \text{ grams (per bearing)}$$

Step 6: Check:

a. Dimensions canceled properly.

b. Math check:

15 grams x 12 = 180 grams

==

PROBLEM 2

A manufacturing concern plans to distribute nut/bolt/washer sets. Each set will consist of one 18.5 g nut, one 31.0 g bolt, and two 5.50 g washers. The sets will be sold in "1-dozen set" packages. What will be the net mass of each package?

SOLUTION

Step 1: Find: *net mass of a "1-dozen" package of the sets*

Step 2: Given: *1 nut/set @ 18.5 g/nut; 1 bolt/set @ 31.0 g/bolt; 2 washers/set @ 5.50 g/washer; 1 dozen per package.*

Step 3: Needed, not given:

How many items per dozen?

Answer (from memory): 12 per dozen

What does "net mass" mean?

Answer (from memory): mass of package *contents* (i.e., 1 dozen sets)

Step 4: Method of attack:

Find the mass of each set and multiply by the number of sets per package. Use unity factor method where appropriate.

(Note: Often there is no unique method of attack. Here, for example, we might have decided to find the mass of 1 dozen nuts, 1 dozen bolts, and 2 dozen washers and then added these together.)

2

Step 5: Manipulation:

$$\left[\frac{18.5\text{ g}}{1\text{ nut}} \times \frac{1\text{ nut}}{1\text{ set}}\right] + \left[\frac{31.0\text{ g}}{1\text{ bolt}} \times \frac{1\text{ bolt}}{1\text{ set}}\right] + \left[\frac{5.50\text{ g}}{1\text{ washer}} \times \frac{2\text{ washers}}{1\text{ set}}\right]$$

$$= \frac{[18.5 + 31.0 + (2)(5.50)]\text{ g}}{1\text{ set}} = \frac{60.5\text{ g}}{1\text{ set}}$$

$$\frac{60.5\text{ g}}{1\text{ set}} \times \frac{1\text{ dozen sets}}{1\text{ pkg}} \times \frac{12}{1\text{ dozen}} = \frac{(60.5 \times 12)\text{ g}}{1\text{ pkg}}$$

Estimate

$$6.05 \times 10^1 \times 1.2 \times 10^1 \simeq 6 \times 1 \times 10^2 \ (\sim 600)$$

Solve

$$\frac{(6.05 \times 1.2 \times 10^2)\text{ g}}{1\text{ pkg}} = 7.26 \times 10^2 \text{ g/pkg} \quad (726\text{ g/pkg})$$

Step 6: Check:

    a. Dimensions canceled properly.

    b. Arithmetic rechecked for accuracy.

==

* PROBLEM 3

    What is the maximum mass of nut/bolt/washer sets, as described in Problem 2, that could be prepared from 8.00 kg of nuts, 18.2 kg of bolts, and 12.2 kg of washers?

SOLUTION

Step 1: Find: *maximum mass of sets from given masses of components*

Step 2: Given: *8.00 kg of nuts, 18.2 kg of bolts, 12.2 kg of washers*

Step 3: Needed, not given:

    Composition of a set

    Answer: (Given in Problem 2)

    1 nut + 1 bolt + 2 washer

    Masses of components

    Answer: (Given in Problem 2)

    18.5 g/nut, 31.0 g/bolt, 5.50 g/washer

* Proficiency Level

Mass of a set

Answer: (Found during solution to Problem 2)

60.5 g/set

Step 4: Method of attack:

The word "maximum" in the problem suggests that there is some limitation other than just the sum of the masses of the components given. It must be recognized that we can no longer make sets when we run out of any one component. Thus, the *maximum* mass of sets will correspond to the mass obtainable from the limiting component (i.e., the <u>smallest</u> of the masses of sets obtainable from the masses of each component available). Since we are dealing with masses (rather than numbers), some appropriate equivalents would be useful in constructing unity factors.

$$18.5 \text{ g nuts} = 60.5 \text{ g sets}$$
$$31.0 \text{ g bolts} = 60.5 \text{ g sets}$$
$$(2 \times 5.50) \text{ g washers} = 60.5 \text{ g sets}$$

Then we must calculate the mass of sets obtainable by using up all of each component available and select the smallest of these values as the limit on the mass of sets obtainable.

Step 5: Manipulation:

From nuts

$$\frac{60.5 \text{ g sets}}{18.5 \text{ g nuts}} \times \frac{8.00 \text{ kg nuts}}{1} = 26.2 \text{ kg sets}$$

From bolts

$$\frac{60.5 \text{ g sets}}{31.0 \text{ g bolts}} \times \frac{18.2 \text{ kg bolts}}{1} = 35.5 \text{ kg sets}$$

From washers

$$\frac{60.5 \text{ g sets}}{11.0 \text{ g washers}} \times \frac{12.2 \text{ kg washers}}{1} = 67.1 \text{ kg sets}$$

Since our calculations show that we shall run out of nuts when 26.2 kg of sets have been constructed, the maximum mass of sets obtainable is 26.2 kg.

Step 6: Check:
   a. Dimensions cancel.
   b. Arithmetic rechecked for accuracy.
   c. Logic rechecked.

==

Each new problem poses its own unique challenges, and there is no "magic formula" for problem solving. The stepwise approach suggested should prove useful in establishing a systematic technique. Facility will improve with practice.

STOICHIOMETRY

UNIT 1 : CHEMICAL FORMULAS

A chemical formula uses a combination of element symbols and subscript numbers to represent a compound or polyatomic grouping.

sodium chloride	methane	nitrate ion
NaCl	CH_4	NO_3^-

Three general types of formulas are of considerable utility in chemistry. Empirical formulas, indicating only the simplest whole number ratio of combined atoms, are most useful for simple inorganic compounds such as potassium bromide (KBr) or magnesium sulfate ($MgSO_4$). Such formulas reveal little information about more complex species. Carbon, for example, forms a tremendous variety of covalent compounds (organic compounds) and empirical formulas provide no real description of most organic compounds. Such different species as benzene and acetylene, for example, are both represented by CH, while a number of complex sugars have the empirical formula CH_2O. (Does this seem appropriate for a "carbo-hydrate"?)

Molecular formulas indicate the actual number of each kind of atom in a unique chemical combination. Such formulas thus distinguish between benzene (C_6H_6) and acetylene (C_2H_2). Molecular formulas, such as $C_6H_{12}O_6$ for glucose, are sufficient for many purposes, including stoichiometric calculations.

Structural formulas of various types show how atoms are connected in molecules or polyatomic ions. These formulas may be relatively simple, such as $CH_3CH_2CH_3$ for propane, or - in expanded forms - they may offer more sophisticated descriptions of structure, ranging from two-dimensional formulations to representations of three-dimensional characteristics. To write structural formulas, we need to know a fair amount about chemical bonding and some common conventions. The experimental evidence required and its interpretation may be quite complex.

Empirical and molecular formulas can be determined from relatively simple experimental data, using rather straightforward mathematical techniques. Since more knowledge and experience are required for interpreting structural information, we shall reserve this topic for more detailed texts.

a compressed
structural
formula for
ethyl alcohol

C_2H_5OH

an expanded
structural
formula for
ethyl alcohol

H H
| |
H-C-C-O-H
| |
H H

a 3-dimensional
structural formula
for α-D-glucose

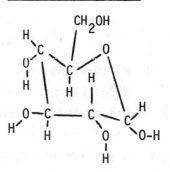

- References -

1. O'Connor, Rod. 1977. Fundamentals of Chemistry, second edition. New York: Harper & Row (Units 3, 4, 6, 7, 28, and Excursion 5)

2. Brown, Theodore, and H. Eugene LeMay, Jr. 1977. Chemistry: The Central Science. Englewood Cliffs, N.J.: Prentice-Hall (Chapters 3, 8, 9, 24)

3. Masterton, William, and Emil Slowinski. 1977. Chemical Principles, fourth edition. Philadelphia: W.B. Saunders (Chapters 3, 8, 10)

4. Nebergall, William, F.C. Schmidt, and H.F. Holtzclaw, Jr. 1976. College Chemistry, fifth edition. Lexington, MA: D.C. Heath (Chapters 1, 4, 6, 27)

5. Kieffer, W.F. 1963. The Mole Concept in Chemistry. New York: Van Nostrand/Reinhold

OBJECTIVES:

(1) Given the percentage composition of a compound, be able to calculate its empirical formula.

(2) Given the formula weight and the empirical formula of a compound, be able to calculate its molecular formula.

(3) Given appropriate analytical data for a binary (two-element) compound, be able to calculate its empirical formula.

*(4) Given appropriate analytical data for a multi-element compound, be able to calculate its empirical formula.

*(5) Given the empirical formula of a volatile compound and vapor density data, be able to calculate the molecular formula.

*Proficiency Level

PRE-TEST:

Necessary Atomic Weights:

carbon	(C)	12.01	nitrogen	(N)	14.01
copper	(Cu)	63.54	oxygen	(O)	16.00
hydrogen	(H)	1.008	sulfur	(S)	32.06

(1) A red-orange dye called Alizarin is an organic compound consisting of 70.02% carbon, 3.36% hydrogen, and 26.64% oxygen (each analysis being accurate to $\pm$ 0.01%). What is the empirical formula for Alizarin? _____

(2) The formula weight of Alizarin is 240.2. What is its molecular formula? _____

(3) A sulfide of copper was treated with acid in such a way as to convert essentially all of the sulfur to hydrogen sulfide (H_2S). If a 0.8144 g sample of the copper salt produced 0.1744 g of hydrogen sulfide, what was the empirical formula of the original salt? _____

*(4) Combustion of a 0.3082 g sample of hexamethylenediamine, a compound used in the production of Nylon-66, formed 0.7003 g of carbon dioxide and 0.3821 g of water. A separate nitrogen assay, using 1.270 g of the compound, produced 0.3723 g of ammonia (NH_3). What is the empirical formula of the original compound? _____

*(5) The vapor density of a sample of hexamethylenediamine, corrected to Standard Temperature and Pressure (STP), was reported as 5.19 g liter^{-1}. Using the information that 22.4 liters of gas, under ideal conditions, contains 1.00 mole of the gas at STP, calculate the molecular formula of this compound. _____

Answers and Directions:

(1) $C_7H_4O_2$, (2) $C_{14}H_8O_4$, (3) Cu_2S. If all are correct, go on to *(4) and *(5) or to RELEVANT PROBLEMS, UNIT 1. If you missed any, study METHODS, sections 1.1 through 1.3.
*(4) C_3H_8N, *(5) $C_6H_{16}N_2$. If both are correct, go on to RELEVANT PROBLEMS, UNIT 1. If you missed either, study METHODS, sections 1.4 and 1.5.

*Proficiency Level

METHODS

1.1 Empirical Formula from Percentage Composition

Each chemical element has its own atomic weight (Table 1.1), which may be thought of as the average mass of an atom of the element, expressed in units of <u>amu</u> (atomic mass units). The empirical formula of a compound shows the relative <u>number</u> of atoms, while the percentage composition of a compound shows the <u>mass</u> relationships. If we think of the percentage of an element in a compound as "amu's of element per 100 amu of compound," we can use such data with atomic weights to obtain the relative numbers of atoms.

Table 1.1 Atomic Weights of Some Common Elements
(to four significant figures)

Symbol	Name	Atomic Weight	Symbol	Name	Atomic Weight
Al	aluminum	26.98	H	hydrogen	1.008
As	arsenic	74.92	I	iodine	126.9
B	boron	10.81	Li	lithium	6.939
Ba	barium	137.3	Mg	magnesium	24.31
Br	bromine	79.91	Mn	manganese	54.94
C	carbon	12.01	N	nitrogen	14.01
Ca	calcium	40.08	Na	sodium	22.99
Cl	chlorine	35.45	O	oxygen	16.00
Cu	copper	63.54	P	phosphorus	30.97
Cr	chromium	52.00	S	sulfur	32.06
F	fluorine	19.00	Si	silicon	28.09
Fe	iron	55.85	Zn	zinc	65.37

(See Appendix A for complete listing.)

For example, let's consider a simple compound of carbon and hydrogen, propane (81.72% C, 18.28% H). Using unity factors we can determine an atomic ratio:

$$\frac{81.72 \text{ amu(C)}}{100 \text{ amu(cpd)}} \times \frac{100 \text{ amu(cpd)}}{18.28 \text{ amu(H)}} \times \frac{1 \text{ atom(C)}}{12.01 \text{ amu (C)}} \times \frac{1.008 \text{ amu(H)}}{1 \text{ atom(H)}} =$$

$$0.3752 \text{ atom(C)/atom(H)}$$

We could express this as an empirical formula, $C_{0.3752}H_1$, but conventionally we use only whole numbers in empirical formulas. What is the simplest number which can be multiplied by both 0.3752 and 1 to yield whole numbers? If we note that 0.3752 is about 3/8 (Table 1.2), then it is apparent that we should multiply both subscripts by 8:

$$C_{(3/8 \times 8)}H_{(1 \times 8)} = C_3H_8$$

In general, such a procedure will work for relatively simple compounds. With the more complex compounds it is usually more profitable to omit calculation of an empirical formula, if an accurate molecular weight is available, and determine the molecular formula directly.

Note that it does not matter which element is selected for "comparison". Had we chosen to calculate atoms(H)/atom(C) for propane, for example, we would have obtained $C_1H_{2.67}$ and, noting that 2.67 is about $2\frac{2}{3}$ (8/3), whole numbers would have been obtained by multiplying by 3:

$$C_{(1 \times 3)}H_{(8/3 \times 3)} = C_3H_8$$

Table 1.2 Some Decimal/Fraction Equivalents

decimal	fraction	decimal	fraction	decimal	fraction
$0.08\overline{33}$	1/12	$0.36\overline{36}$	4/11	0.625	5/8
$0.09\overline{09}$	1/11	0.375	3/8	$0.63\overline{63}$	7/11
$0.\overline{11}$	1/9	0.4	2/5	$0.6\overline{6}$	2/3
0.125	1/8	$0.41\overline{66}$	5/12	0.7143	5/7
0.1429	1/7	0.4286	3/7	$0.72\overline{72}$	8/11
$0.16\overline{6}$	1/6	$0.4\overline{4}$	4/9	0.75	3/4
$0.18\overline{18}$	2/11	$0.45\overline{45}$	5/11	$0.7\overline{7}$	7/9
0.2	1/5	0.5	1/2	0.8	4/5
$0.2\overline{2}$	2/9	$0.54\overline{54}$	6/11	$0.81\overline{81}$	9/11
0.25	1/4	$0.5\overline{5}$	5/9	$0.83\overline{3}$	5/6
$0.27\overline{27}$	3/11	0.5714	4/7	0.8571	6/7
0.2857	2/7	$0.58\overline{33}$	7/12	0.875	7/8
$0.3\overline{3}$	1/3	0.6	3/5	$0.8\overline{8}$	8/9

(A bar above a number, or pair of numbers, indicates repetitive digits, e.g., $1.3\overline{3} = 1.33333\ldots$)

EXAMPLE 1

Lactic acid, formed in the body during muscle activity, consists of 40.00%C, 6.71%H, and 53.29%O. What is its empirical formula?

SOLUTION:

Since the compound contains three elements, we'll have to find atomic ratios twice, using the same "comparison" element. Let's try hydrogen:

1. $\dfrac{40.00 \text{ amu(C)}}{100 \text{ amu(cpd)}}$ x $\dfrac{100 \text{ amu(cpd)}}{6.71 \text{ amu(H)}}$ x $\dfrac{1 \text{ atom(C)}}{12.01 \text{ amu(C)}}$ x $\dfrac{1.008 \text{ amu(H)}}{1 \text{ atom(H)}}$ =

$$0.500 \text{ atom(C)/atom(H)}$$

2. $\dfrac{53.29 \text{ amu(O)}}{100 \text{ amu(cpd)}}$ x $\dfrac{100 \text{ amu(cpd)}}{6.71 \text{ amu(H)}}$ x $\dfrac{1 \text{ atom(O)}}{16.00 \text{ amu(O)}}$ x $\dfrac{1.008 \text{ amu(H)}}{1 \text{ atom(H)}}$ =

$$0.500 \text{ atom(O)/atom(H)}$$

Hence, the empirical formula for lactic acid is $C_{0.5}H_1O_{0.5}$ or, in conventional whole number form,

$$C_{(0.5 \times 2)}H_{(1 \times 2)}O_{(0.5 \times 2)} = CH_2O$$

Note that subscripts of unity (1) are understood, not written.

- -

EXERCISE 1

Phenylpyruvic acid is produced in the body in abnormal amounts as the result of a molecular disease called phenylketonuria, which causes irreversible brain damage. Phenylpyruvic acid is composed of 65.85%C, 4.91%H, and 29.24%O. What is its empirical formula?

(answer, page 17)

- -

1.2 Molecular Formula from Formula Weight and Empirical Formula

The formula weight of a compound is the sum of all atomic weights of its component atoms, as expressed by its molecular formula. For example, for glucose, $C_6H_{12}O_6$, the formula weight is determined by:

C_6 (six carbons) : 6 x 12.01 =   72.06

H_{12} (twelve hydrogens) : 12 x 1.008 = 12.096

O_6 (six oxygens) : 6 x 16.00 =   <u>96.00</u>

                                 180.2   (rounded to four figures)

An <u>empirical</u> formula weight can be calculated in the same way. The molecular formula of a compound can then be determined by multiplying the subscript numbers in the empirical formula by the ratio of formula weight to empirical formula weight, since this shows us how many times the empirical formula is contained in the molecular formula.

EXAMPLE 2

The formula weight of lactic acid (Example 1) is 90.08. What is its molecular formula?

SOLUTION:

From Example 1, the empirical formula is CH_2O, for which the empirical formula weight is:

C: 1 x 12.01 =   12.01
H_2: 2 x 1.008 =   2.016
O: 1 x 16.00 =   16.00
                         30.03   (4-place)

then:

$$\frac{\text{form. wt.}}{\text{emp. form. wt.}} = \frac{90.08}{30.03} = 3$$

So the molecular formula is given by:

$$C_{(1 \times 3)}H_{(2 \times 3)}O_{(1 \times 3)} = \underline{C_3H_6O_3}$$

===

EXERCISE 2

Ethylene glycol, the compound most commonly employed in antifreeze preparations for automobile radiators, has the empirical formula CH_3O. Its molecular weight is 62.07. What is its molecular formula?

(answer, page 18)

===

1.3 Empirical Formula from Analytical Data

Quantitative chemical analysis can provide data useful in determining an empirical formula. In the simplest case, that of a binary (two-element) compound, it is necessary only to convert a weighed sample of the compound completely to a new compound of either of the original elements and to obtain the weight of the new compound, whose chemical formula is known. Using these weights and appropriate unity factors, it is easy to find the percentage composition, from which the empirical formula can be determined (section 1.1). It is only necessary to recognize that there is a direct conversion between the mass units of amu and grams (as described in Unit 3), so that atomic or formula weight ratios can be expressed using any mass units.

EXAMPLE 3

Calomel, a chloride of mercury, is used in medicine as a stimulant to bile secretion. A 0.7844 g sample of calomel was treated with sulfuric acid so that the chloride was converted completely to 0.1214 g of HCl. What is the empirical formula of calomel? [The atomic weight of mercury is 200.6.]

SOLUTION:

(to convert to%)⟶

$$\%Cl = \frac{0.1214 \text{ g(HCl)}}{0.7844 \text{ g(cpd)}} \times \frac{35.45 \text{ g(Cl)}}{36.46 \text{ g(HCl)}} \times \frac{100}{1} = 15.05\%$$

Since this is a binary compound, %Hg = 100.00% - 15.05% = 84.95%

Then, using the method of Section 1.1,

$$\frac{84.95 \text{ amu(Hg)}}{100 \text{ amu(cpd)}} \times \frac{100 \text{ amu(cpd)}}{15.05 \text{ amu(Cl)}} \times \frac{1 \text{ atom(Hg)}}{200.6 \text{ amu(Hg)}} \times \frac{35.45 \text{ amu(Cl)}}{1 \text{ atom(Cl)}} =$$

$$1.00 \text{ atom(Hg)/atom(Cl)}$$

so the empirical formula is HgCl.

[Note: The molecular formula of mercury(I) chloride is Hg_2Cl_2.]

EXERCISE 3

The "octane rating" of gasoline is based on the use of a reference compound, isooctane, assigned an octane rating of 100. When 0.2351 g of isooctane, a hydrocarbon (compound of carbon and hydrogen), was burned in a special combustion chamber, it was converted completely to water and carbon dioxide. The CO_2 formed was trapped and weighed as $BaCO_3$. If 3.2488 g of $BaCO_3$ was obtained, what is the empirical formula of isooctane?

(answer, page 18)

Extra Practice

EXERCISE 4

Oxides of nitrogen are among the byproducts of operation of internal combustion engines. One of these oxides consists of 30.45%N and 69.55%O. What is its empirical formula?

(answer, page 18)

EXERCISE 5

The formula weight of the nitrogen oxide described in EXERCISE 4 is 92.02. What is its molecular formula?

(answer, page 18)

EXERCISE 6

TEFLON is made from a fluorocarbon (compound of carbon and fluorine) starting material, perfluoroethylene. Analysis of the starting material was done by converting all the fluorine from a 0.3802 g sample to 0.5936 g of CaF_2. What is the empirical formula of perfluoroethylene?

(answer, page 19)

==

At this point you should try the <u>competency</u> level Self-Test questions on pages 16 and 17.

==

*1.4 Empirical Formula from Analytical Data

The empirical formula of a compound expresses the simplest whole-number ratios of combined atoms. Weight data from chemical analyses can easily be used to calculate the weight ratios of combined elements in a compound. All that we require, then, is some way of converting weight ratios to atom ratios. The key to such a conversion is the atomic weight, which may be used to form weight/atom unity factors having the dimensions of atomic mass units per atom ($amu\ atom^{-1}$).

One of the easier ways of determining an empirical formula of a multi-element compound involves three steps. First, the percentage of each element in the compound is determined from analytical data and appropriate unity factors. This allows us to express a weight-ratio formula in terms of atomic mass units of each element. Second, atomic weight unity factors are used to convert the weight-ratio formula to an atom-ratio formula. Finally, atom-ratio figures are divided by the smallest of the figures and then, if necessary, converted to whole numbers by use of appropriate decimal/fraction equivalents (Table 1.2).

*EXAMPLE 4

Xylocaine, a local anesthetic which has largely replaced novocaine in dentistry, is a compound of carbon, hydrogen, nitrogen, and oxygen. Combustion of a 0.4817 g sample of xylocaine produced 1.2665 g of CO_2 and 0.4073 g of H_2O. A separate nitrogen assay, using 0.3933 g of xylocaine, formed 0.0572 g of NH_3. What is the empirical formula of xylocaine?

<u>SOLUTION:</u>

1. % Composition

$$\%C = \frac{1.2665\ g\ (CO_2)}{0.4817\ g\ (cpd)} \times \frac{12.01\ g\ (C)}{44.01\ g\ (CO_2)} \times 100 = 71.75\%$$

$$\%H = \frac{0.4073\ g\ (H_2O)}{0.4817\ g\ (cpd)} \times \frac{2.016\ g\ (H)}{18.02\ g\ (H_2O)} \times 100 = 9.46\%$$

$$\%N = \frac{0.0572\ g\ (NH_3)}{0.3933\ g\ (cpd)} \times \frac{14.01\ g\ (N)}{17.03\ g\ (NH_3)} \times 100 = 11.96\%$$

$\%O = 100.00\% - (71.75\% + 9.46\% + 11.96\%) = 6.83\%$

So, the weight-ratio formula may be expressed as:

71.75 amu(C): 9.46 amu(H): 11.96 amu(N): 6.83 amu(O)

*Proficiency Level

2. Conversion to Atom-Ratios

$$\frac{71.75 \text{ amu(C)}}{12.01 \text{ amu atom}^{-1}} : \frac{9.46 \text{ amu(H)}}{1.008 \text{ amu atom}^{-1}} : \frac{11.96 \text{ amu(N)}}{14.01 \text{ amu atom}^{-1}} : \frac{6.83 \text{ amu(N)}}{16.00 \text{ amu atom}^{-1}} =$$

5.974 atom(C): 9.385 atom(H): 0.8537 atom(N): 0.4269 atom(O)

3. Obtaining Whole-Number Ratios

$$\frac{5.974 \text{ atom(C)}}{0.4269} : \frac{9.385 \text{ atom(H)}}{0.4269} : \frac{0.8537 \text{ atom(N)}}{0.4269} : \frac{0.4269 \text{ atom(O)}}{0.4269}$$

13.99 atom C: 21.98 atom(H): 2.000 atom(N): 1.000 atom(O)

These are close enough to whole numbers to give the empirical formula, $C_{14}H_{22}N_2O$ (This is also the molecular formula.)

---

*EXERCISE 7

Ethyl butyrate, which has the odor of pineapple, is used in preparing artificial flavoring agents. Combustion of 0.5309 g of ethyl butyrate produced 1.2069 g of CO_2 and 0.4916 g of water. Analysis showed that the only elements in the compound are carbon, hydrogen, and oxygen. What is the empirical formula of ethyl butyrate?

(answer, page 19)

---

*1.5 Molecular Formula from Empirical Formula and Vapor Density

The ratio of formula weight to empirical formula weight tells us how many times the empirical formula is contained in the molecular formula, i.e., the number by which we must multiply the subscripts in the empirical formula in order to obtain the molecular formula. There are many ways of determining formula weights. If the compound is volatile, its formula weight can be determined rather accurately from a measurement of vapor density. At STP (Standard Temperature and Pressure, 273°K, 760 torr), 22.4 liters of any gas, assuming ideal behavior, contains a mass in grams equal to the formula weight of the gaseous compound. The number of molecules in this quantity of the gas is 6.023 x 10^{23}, one mole (Unit 3), so 22.4 liters is sometimes referred to as the standard molar volume of a gas. It is important to note that gas volumes and gas densities must be compared at the same temperature and pressure and that the 22.4 value applies at STP.

For our present purposes, until we explore the mole concept in more detail, it is sufficient to note that the formula weight of a gaseous substance ("vapor") can be determined by multiplying the vapor density (in g liter^{-1} at STP) by 22.4 liters mole^{-1}.

*EXAMPLE 5

Paraldehyde, a sedative and hypnotic drug, has the empirical formula C_2H_4O. Its vapor density, corrected to STP, is 5.90 g liter^{-1}.   What is its molecular formula?

SOLUTION:

formula wt. = $\dfrac{5.90 \text{ g}}{\text{liter}}$ x $\dfrac{22.4 \text{ liters}}{\text{mole}}$ = 132 (g mole^{-1})

emp. form. wt., for C_2H_4O:

C_2: 2 x 12.01 = 24.02

H_4: 4 x 1.008 = 4.032

O : 1 x 16.00 = 16.00

$\overline{\hspace{2cm} 44.05}$

$\dfrac{\text{form. wt.}}{\text{emp. form. wt.}}$ = $\dfrac{132}{44.05}$ = 3.00

molecular formula = $C_{(2 \times 3)}H_{(4 \times 3)}O_{(1 \times 3)}$ = $C_6H_{12}O_3$

- -

*EXERCISE 8

The vapor density, corrected to STP, of ethyl butyrate (EXERCISE 7) is 5.19 g liter^{-1}. What is the molecular formula of this compound?

(answer, page 19)

- -

Extra Practice

*EXERCISE 9

Peroxyacetyl nitrate, a potent lachrymator (tear gas), has been detected in signifi-cant concentrations in photochemical smogs.   Complete combustion of 0.2818 g of this com-pound formed 0.2049 g of CO_2 and 0.0629 g of H_2O.   A separate nitrogen assay, using 0.3704 g of the compound, formed 0.0521 g of NH_3.   The only other element in the compound is oxygen. What is the empirical formula of peroxyacetyl nitrate?

(answer, page 20)

*EXERCISE 10

The vapor density of peroxyacetyl nitrate (EXERCISE 9), corrected to STP, is 5.40 g liter^{-1}.   What is its molecular formula?

(answer, page 20)

- -

SELF-TEST (UNIT 1)   [answers, page 20]

1.1 A chemical soil additive sold under the trade name NUTRIROOT consists of 35.00%N, 5.03%H, and 59.96%O.   What is its empirical formula?

1.2. Nicotine, the toxic alkaloid of the tobacco plant, has a molecular weight of 162.2. Its empirical formula is C_5H_7N. What is its molecular formula?

1.3. "Fool's Gold" is actually a compound of iron and sulfur, called iron pyrite. When a 0.6814 g sample of the pyrite was roasted in air, the sulfur was converted completely to 0.7276 g of SO_2. What is the empirical formula of the pyrite?

- -

If you completed Self-Test questions 1.1 - 1.3 correctly, you may go on to proficiency level work, try the Relevant Problems (Unit 1), or stop here. If not, you should consult your instructor for suggestions of further study aids.

- -

*1.4. Hexamethylenetetramine was used for the treatment of kidney and bladder infections prior to the development of modern antibiotic drugs. Combustion of 0.2897 g of this compound produced 0.5456 g of CO_2 and 0.2234 g of H_2O. Separate nitrogen assay of a 0.3801 g sample yielded 0.1848 g of ammonia. What is the empirical formula of this compound?

*1.5. The vapor density, corrected to STP, of hexamethylenetetramine (Question 1.4) is 6.26 g liter^{-1}. What is its molecular formula?

= =

If you completed Self-Test questions 1.4 and 1.5 correctly, you may go on to the Relevant Problems for Unit 1. If not, you should consult your instructor for suggestions of further study aids.

- -

ANSWERS to EXERCISES, Unit 1

1. ($C_9H_8O_3$) Solution:

$$\frac{65.85 \text{ amu(C)}}{100 \text{ amu(cpd)}} \times \frac{100 \text{ amu(cpd)}}{4.91 \text{ amu(H)}} \times \frac{1 \text{ atom(C)}}{12.01 \text{ amu(C)}} \times \frac{1.008 \text{ amu(H)}}{1 \text{ atom(H)}} =$$

$$1.126 \text{ atom(C)/atom(H)}$$

$$\frac{29.24 \text{ amu(O)}}{100 \text{ amu(cpd)}} \times \frac{100 \text{ amu(cpd)}}{4.91 \text{ amu(H)}} \times \frac{1 \text{ atom(O)}}{16.00 \text{ amu(O)}} \times \frac{1.008 \text{ amu(H)}}{1 \text{ atom (H)}} =$$

$$0.375 \text{ atom(O)/atom(H)}$$

from Table 1.2, $1.126 = 1\frac{1}{8} = 9/8$, $0.375 = 3/8$

$$C_{(9/8 \times 8)} H_{(1 \times 8)} O_{(3/8 \times 8)} = \underline{C_9H_8O_3}$$

2. $(C_2H_6O_2)$ Solution:

    empirical formula weight for CH_3O:

      C: 1 x 12.01   =   12.01

      H_3: 3 x 1.008 =   3.024

      O: 1 x 16.00   =   16.00

                           31.03

$$\frac{\text{form. wt.}}{\text{emp. form. wt.}} = \frac{62.07}{31.03} = 2$$

$$C_{(1 \times 2)}H_{(3 \times 2)}O_{(1 \times 2)} = \underline{C_2H_6O_2}$$

3. (C_4H_9) Solution:

$$\%C = \frac{3.2488 \text{ g } (BaCO_3)}{0.2351 \text{ g (cpd)}} \times \frac{12.01 \text{ g (C)}}{197.3 \text{ g } (BaCO_3)} \times 100 = 84.12\%$$

$\%H = 100.00\% - 84.12\% = 15.88\%$

$$\frac{84.12 \text{ amu(C)}}{100 \text{ amu(cpd)}} \times \frac{100 \text{ amu(cpd)}}{15.88 \text{ amu(H)}} \times \frac{1 \text{ atom(C)}}{12.01 \text{ amu(C)}} \times \frac{1.008 \text{ amu(H)}}{1 \text{ atom(H)}} =$$

$$0.4446 \text{ atom(C)/atom(H)}$$

from Table 1.2, $0.4446 \simeq 4/9$, so:

$$C_{(4/9 \times 9)}H_{(1 \times 9)} = \underline{C_4H_9}$$

4. (NO_2) Solution:

$$\frac{30.45 \text{ amu(N)}}{100 \text{ amu(cpd)}} \times \frac{100 \text{ amu(cpd)}}{69.55 \text{ amu(O)}} \times \frac{1 \text{ atom(N)}}{14.01 \text{ amu(N)}} \times \frac{16.00 \text{ amu(O)}}{1 \text{ atom(O)}} =$$

$$0.500 \text{ atom(N)/atom(O)}$$

$$N_{(0.5 \times 2)}O_{(1 \times 2)} = \underline{NO_2}$$

5. (N_2O_4) Solution:

    empirical formula weight for NO_2:

      N: 1 x 14.01   =   14.01

      O_2: 2 x 16.00 =   32.00

                          46.01

$$\frac{\text{form. wt.}}{\text{emp. form. wt.}} = \frac{92.02}{46.01} = 2$$

$$N_{(1 \times 2)}O_{(2 \times 2)} = \underline{N_2O_4}$$

6. (CF_2) Solution:

$$\%F = \frac{0.5936 \text{ g } (CaF_2)}{0.3802 \text{ g } (cpd)} \times \frac{38.00 \text{ g } (F)}{78.08 \text{ g } (CaF_2)} \times 100 = 75.98\%$$

$\%C = 100.00\% - 75.98\% = 24.02\%$

$$\frac{24.02 \text{ amu}(C)}{100 \text{ amu}(cpd)} \times \frac{100 \text{ amu}(cpd)}{75.98 \text{ amu}(F)} \times \frac{1 \text{ atom}(C)}{12.01 \text{ amu}(C)} \times \frac{19.00 \text{ amu}(F)}{1 \text{ atom }(F)} =$$

$$0.500 \text{ atom}(C)/\text{atom}(F)$$

$$C_{(0.5 \times 2)} F_{(1 \times 2)} = \underline{CF_2}$$

*7. (C_3H_6O) Solution:

$$\%C = \frac{1.2069 \text{ g } (CO_2)}{0.5309 \text{ g } (cpd)} \times \frac{12.01 \text{ g } (C)}{44.01 \text{ g } (CO_2)} \times 100 = 62.04\%$$

$$\%H = \frac{0.4916 \text{ g } (H_2O)}{0.5309 \text{ g } (cpd)} \times \frac{2.016 \text{ g } (H)}{18.02 \text{ g } (H_2O)} \times 100 = 10.41\%$$

$\%O = 100.00\% - (62.04\% + 10.41\%) = 27.55\%$

$$\frac{62.04 \text{ amu}(C)}{12.01 \text{ amu atom}^{-1}} : \frac{10.41 \text{ amu}(H)}{1.008 \text{ amu atom}^{-1}} : \frac{27.55 \text{ amu}(O)}{16.00 \text{ amu atom}^{-1}}$$

$$\frac{5.166 \text{ atom}(C)}{1.722} : \frac{10.33 \text{ atom}(H)}{1.722} : \frac{1.722 \text{ atom}(O)}{1.722}$$

$3.00 \text{ atom}(C) : 6.00 \text{ atom}(H) : 1 \text{ atom}(O)$

*8. ($C_6H_{12}O_2$) Solution:

$$\text{form. wt.} = \frac{5.19 \text{ g}}{\text{liter}} \times \frac{22.4 \text{ liters}}{\text{mole}} = 116 \text{ (g mole}^{-1})$$

emp. form. wt., C_3H_6O:

C_3: 3 x 12.01 = 36.03

H_6: 6 x 1.008 = 6.048

O: 1 x 16.00 = 16.00

58.08 (4-place)

$$\frac{\text{form. wt.}}{\text{emp. form. wt.}} = \frac{116}{58.08} = 2.00$$

$$\text{mol. formula} = C_{(3 \times 2)} H_{(6 \times 2)} O_{(1 \times 2)} = \underline{C_6H_{12}O_2}$$

*9. ($C_2H_3NO_5$) <u>Solution</u>:

$$\%C = \frac{0.2049 \text{ g } (CO_2)}{0.2818 \text{ g } (cpd)} \times \frac{12.01 \text{ g } (C)}{44.01 \text{ g } (CO_2)} \times 100 = 19.84\%$$

$$\%H = \frac{0.0629 \text{ g } (H_2O)}{0.2818 \text{ g } (cpd)} \times \frac{2.016 \text{ g } (H)}{18.02 \text{ g } (H_2O)} \times 100 = 2.50\%$$

$$\%N = \frac{0.0521 \text{ g } (NH_3)}{0.3704 \text{ g } (cpd)} \times \frac{14.01 \text{ g } (N)}{17.03 \text{ g } (NH_3)} \times 100 = 11.57\%$$

$\%O = 100.00\% - (19.84\% + 2.50\% + 11.57\%) = 66.09\%$

$$\frac{19.84 \text{ amu}(C)}{12.01 \text{ amu atom}^{-1}} : \frac{2.50 \text{ amu}(H)}{1.008 \text{ amu atom}^{-1}} : \frac{11.57 \text{ amu}(N)}{14.01 \text{ amu atom}^{-1}} : \frac{66.09 \text{ amu}(O)}{16.00 \text{ amu atom}^{-1}}$$

$$\frac{1.652 \text{ atom}(C)}{0.8258} : \frac{2.48 \text{ atom}(H)}{0.8258} : \frac{0.8258 \text{ atom}(N)}{0.8258} : \frac{4.131 \text{ atom}(O)}{0.8258}$$

2.00 atom(C): 3.00 atom(H): 1 atom(N): 5.00 atom(O)

*10. ($C_2H_3NO_5$) <u>Solution</u>:

$$\text{form. wt.} = \frac{5.40 \text{ g}}{\text{liter}} \times \frac{22.4 \text{ liters}}{\text{mole}} = 121 \text{ (g mole}^{-1})$$

emp. form. wt., for $C_2H_3NO_5$ (EXERCISE 9)

C_2: 2 x 12.01 = 24.02

H_3: 3 x 1.008 = 3.024

N : 1 x 14.01 = 14.01

O_5: 5 x 16.00 = <u>80.00</u>

121.05

$\dfrac{\text{form. wt.}}{\text{emp. form. wt.}} = \dfrac{121}{121} = 1$, so the molecular formula is the same as the empirical formula.

= =

<u>ANSWERS to SELF-TEST, Unit 1</u>

1.1. $N_2H_4O_3$ [This is ammonium nitrate, NH_4NO_3.]

1.2. $C_{10}H_{14}N_2$

1.3. FeS_2

*1.4. $C_3H_6N_2$

*1.5. $C_6H_{12}N_4$

A.1. Plant growth requires large quantities of "fixed" nitrogen, most of which is incorporated into proteins. A very efficient way to replenish soil nitrogen is through the
application of chemical fertilizers. Urea, a good vehicle for fixed nitrogen, is an
organic compound consisting of 20.00% carbon, 6.71% hydrogen, 46.65% nitrogen, and
26.64% oxygen. What is the empirical formula of urea?

A.2. At various geological times, shallow marine basins may be isolated by tectonic activity
which restricts circulation and facilitates evaporation. As the evaporation prog-
resses, the dissolved materials precipitate, forming a wide variety of compounds of
commercial value. Extensive marine evaporite deposits are found in the United States.
One of the major deposits, anhydrite, is an inorganic compound consisting of 29.44%
calcium, 23.55% sulfur, and 47.01% oxygen. What is the empirical formula of anhydrite?

A.3. Mascagnite, an inorganic fertilizer, provides the plant nutrients nitrogen and sulfur.
By analysis, a sample of mascagnite was found to contain 21.20% nitrogen, 6.10% hydrogen, 24.27% sulfur, and 48.43% oxygen. What is the empirical formula of mascagnite?

A.4. Diazinon, an organophosphorus insecticide, can be used to protect crops shortly before
harvest without leaving harmful residues, if properly applied. This insecticide kills
insects by inactivating enzymes essential to an insect's nervous system. Diazinon
consists of 47.36% carbon, 6.95% hydrogen, 9.20% nitrogen, 15.77% oxygen, 10.18%
phosphorus, and 10.54% sulfur. What is the empirical formula of Diazinon?

A.5. An agricultural fungicide called Captan is an organic compound consisting of 35.96%
carbon, 2.69% hydrogen, 35.5% chlorine, 4.67% nitrogen, 10.65% oxygen, and 10.67%
sulfur. What is the empirical formula for Captan?

A.6. Caproic acid, an organic compound, is obtained by hydrolysis of glyceryl esters found
in goat fat. Its empirical formula is C_3H_6O and its molecular weight is 116.16. What
is its molecular formula?

*A.7. Hexalure is a powerful synthetic insect attractant. It was used successfully in rid-
ding California's San Joaquin Valley of the pink bollworm moth. Combustion of 0.4065 g
of Hexalure produced 1.1399 g of CO_2 and 0.4407 g of water. Analysis of the compound showed that the only elements in the compound are carbon, hydrogen, and oxygen.
What is its empirical formula?

*A.8. To control the cotton bollworm moth, traps were baited with the attractant Hexalure
(problem A.7) to reveal the spread of the pest. Once located, the infested area was
saturated with sterile male moths that still retained their sexual aggressiveness.

*Proficiency Level

The vapor density of <u>Hexalure</u>, corrected to STP, is 12.61 g liter^{-1}. What is its molecular formula?

*A.9. Prostaglandins are under intensive investigation with respect to their hormonal properties in relation to female reproductive physiology. The natural product, 5-<u>trans</u>-PGA$_2$, was isolated from the marine Coelenterate, <u>Plexaura</u> <u>homomalla</u>. Combustion of 0.6535 g of 5-<u>trans</u>-PGA$_2$ produced 1.7201 g of CO_2 and 0.5280 g of water. Analysis of this lipid showed that the only elements in the compound are carbon, hydrogen, and oxygen. What is its empirical formula?

*A.10. The vapor density, corrected to STP, of 5-<u>trans</u>-PGA$_2$ (problem A.9) is 14.9 g liter^{-1}. What is its molecular formula?

<u>ANSWERS</u>: (A.1.) CH_4N_2O [$H_2N\overset{\overset{\text{O}}{\|}}{C}NH_2$], (A.2.) $CaSO_4$, (A.3.) $N_2H_8SO_4$ [more commonly known as ammonium sulfate, $(NH_4)_2SO_4$], (A.4.) $C_{12}H_{21}N_2O_3PS$, (A.5.) $C_9H_8Cl_3NO_2S$, (A.6.) $C_6H_{12}O_2$, (A.7.) $C_9H_{17}O$, (A.8.) $C_{18}H_{34}O_2$, (A.9.) $C_{10}H_{15}O_2$, (A.10.) $C_{20}H_{30}O_4$

RELEVANT PROBLEMS

Unit 1: *Chemical Formulas*

Set B: *Biological & Medical Sciences*

B.1. Dimethyl sulfide is spontaneously evolved from the red and green algae, <u>Polysiphonia</u> <u>fastigiata</u> and <u>Enteromorpha</u> <u>intestinalis</u>, when these sea plants are exposed to air. This organo sulfur compound consists of 38.66% carbon, 9.73% hydrogen, and 51.61% sulfur. What is its empirical formula?

B.2. The formation of dimethyl sulfide (problem B.1.) results from the enzymatic decomposition of methionine which is fairly widespread in both benthic and planktonic marine plants. Methionine consists of 40.25% carbon, 7.43% hydrogen, 21.45% oxygen, 9.39% nitrogen, and 21.49% sulfur. What is the empirical formula of methionine?

B.3. <u>Isoniazid</u>, was first studied by chemists and biologists in 1950. It is one of our serendipitous drugs that is credited with remarkable therapeutic action against tuberculosis. <u>Isoniazid</u> is an organic compound consisting of 52.54% carbon, 5.15% hydrogen, 30.64% nitrogen, and 11.67% oxygen. What is the empirical formula of <u>Isoniazid</u>?

B.4. The α-keratins are fibrous, insoluble proteins of animals derived from ectodermal cells. They include the structural protein elements of skin as well as the biological derivatives of ectoderm, such as hair, wool, scales, feathers, nails, horns, and silk. The α-keratins are rich in the amino acid cystine which consists of 29.99% carbon, 5.03% hydrogen, 26.63% oxygen, 11.66% nitrogen, and 26.69% sulfur. What is the empirical formula of cystine?

B.5. The cystine rich α-keratins contain many disulfide bridges formed by the oxidation of cysteine, an amino acid often referred to as the "reduced form" of cystine. The molecular weight of cystine (problem B.4) is 240.3. What is its molecular formula?

*B.6. Poisonous pufferfishes, family Tetrodontidae, are among the earliest described toxic marine organisms. Their ovaries and liver contain high concentrations of tetrodotoxin. The pharmacology of tetrodotoxin has become a valuable tool in neurophysiology research. Combustion of 0.4560 g of tetrodotoxin produced 0.7204 g of CO_2 and 0.2457 g of H_2O. A separate nitrogen assay using 0.5650 g of tetrodotoxin produced 0.0864 g of ammonia (NH_3). The only other element in the compound is oxygen. What is the empirical formula of tetrodotoxin? [This is also its molecular formula.]

*B.7. First isolated in 1820, caffeine is now an important commercial chemical. The pharmaceutical industry markets caffeine products ranging from "stay-awake" tablets to multi-component pain relief preparations. Caffeine, when pure, occurs as long, silky white needles. Chemical analysis shows that it is a compound of carbon, hydrogen, nitrogen, and oxygen. In a particular analysis, combustion of a 0.5120 g sample of caffeine formed 0.9284 g of CO_2 and 0.2373 g of H_2O. In a separate determination, the nitrogen of a 1.485 g caffeine sample was converted to 0.5208 g of ammonia (NH_3). What is the empirical formula of caffeine?

*B.8. Caffeine is a diuretic and a stimulant of the central nervous system. It is reported to be quite effective in the relief of minor fatigue, neuralgia, and headache resulting from eyestrain. A pure sample of caffeine, which sublimes at 178°C, was found to have a vapor density of 8.67 mg ml^{-1}, corrected to Standard Temperature and Pressure. What is the molecular formula of caffeine?

*B.9. The alkaloid content of the tobacco leaf averages between four and six percent. Of these organic bases, the principal one is nicotine. A liquid in the pure form, nicotine is a powerful and rapid poison. Various extracts and mixtures of nicotine are used as agricultural insecticides, such as BLACK LEAF - 40 (a 40% aqueous solution of nicotine sulfate). Combustion of a 0.4221 g of nicotine formed 1.145 g of CO_2 and 0.3285 g of H_2O. A separate nitrogen assay, using 1.450 g of nicotine, formed 0.3044 g of ammonia. What is the empirical formula of nicotine?

*Proficiency Level

*B.10. Many of the physiological effects of tobacco are due to its nicotine content. The drug has profound effects on the autonomic nerves and on the circulatory and digestive systems. Inhalation of the smoke from a single cigarette causes detectable vasoconstriction lasting up to sixty minutes, with a consequent increase in blood pressure. Nicotine, which boils at 124°C at a pressure of 17 torr, has a vapor density of 7.25 mg ml^{-1}, corrected to STP. What is the molecular formula of nicotine?

ANSWERS:

(B.1.) C_2H_6S, (B.2.) $C_5H_{11}NO_2S$, (B.3.) $C_6H_7N_3O$, (B.4.) $C_3H_6NO_2S$, (B.5.) $C_6H_{12}N_2O_4S_2$, (B.6.) $C_{12}H_{20}N_3O_8$, (B.7.) $C_4H_5N_2O$, (B.8.) $C_8H_{10}N_4O_2$, (B.9.) C_5H_7N, (B.10.) $C_{10}H_{14}N_2$

RELEVANT PROBLEMS

Unit 1: Chemical Formulas

Set I: Industrial Chemistry

I.1. About 80% of all rubber used in 1973 was synthetic, made from petroleum derivative raw materials. The quality control laboratory of a certain rubber company performs routine analysis on the raw materials used to manufacture their products. The combustion of a 0.6543 g sample of one raw material produced 2.130 g of carbon dioxide and 0.6538 g of water. Determine the empirical formula of this raw material, a hydrocarbon.

I.2. The oxo process, developed in the 1930's is now used to produce intermediates for synthetic lubricating oils and plasticizers. The process produces long-chain alcohols by the addition of carbon monoxide and hydrogen to olefins. What is the empirical formula of the olefin that is 14.37% H and 85.63% C?

I.3. If the formula weight of the olefin in problem I.2. is 70, what is the molecular formula?

I.4. Catalytic hydrogenation was discovered by Paul Sabatier in the 1890's when he attempted to make $Ni(C_2H_4)_4$. The compound that he actually made was a gas, 79.88% C and 20.11% H. What was the empirical formula?

- UNIT 1 -

I.5. During World War II, the "Fisher-Tropsch" reaction was developed in Germany to provide gasoline from coal. The reaction uses hydrogen gas with carbon monoxide under catalytic conditions to produce hydrocarbons which boil in the gasoline range. A 2.374 g sample of one product, on complete combustion, gave 7.317 g of CO_2 and 3.368 g of H_2O. What is the empirical formula?

I.6. If the product of the "Fisher-Tropsch" reaction described in problem I.5. has a formula weight of 114, what is the molecular formula?

*I.7. The raw material described in problem I.1 boils at -4.41°C and has a vapor density of 2.42 g liter^{-1}, corrected to STP. Assuming the vapor behaves as an ideal gas, determine the molecular formula of this compound.

*I.8. The gas made by Sabatier (problem I.4) had a vapor density of 1.34 g liter^{-1} at STP. Assuming ideal gas behavior, what was the molecular formula?

*I.9. Tetraethyl lead has been for many years an important antiknock agent in gasoline, although environmental concerns may soon cause its discontinuance. Tetraethyl lead is made by mixing a lead-sodium alloy with a chlorinated hydrocarbon in a high pressure, high temperature reaction vessel. The crude product is washed and distilled for purification. Analysis of the chlorinated hydrocarbon shows it to contain 54.95% chlorine. Combustion of a 0.2808 g sample of the compound formed 0.3831 g of of CO_2 and 0.1961 g of H_2O. The formula weight of this compound is 64.51. What is its molecular formula?

*I.10. Once a thriving industry, whaling now finds only a limited market for its products. "Whale oil" (spermaceti wax), however, is used in certain speciality ointments and cosmetics and in the manufacture of more luxurious candles. "Whale oil" is mainly a waxy ester, cetyl palmitate. Analysis shows this compound to consist entirely of carbon, hydrogen, and oxygen. Combustion of a 2.3836 g sample of cetyl palmitate produced 6.9364 g of CO_2 and 2.9311 g of H_2O. The molecular and empirical formulas of the compound are the same. What is the molecular formula of cetyl palmitate?

ANSWERS:
(I.1.) C_2H_3, (I.2.) CH_2, (I.3.) C_5H_{10}, (I.4.) CH_3, (I.5.) C_4H_9, (I.6.) C_8H_{18}, (I.7.) C_4H_6, (I.8.) C_2H_6, (I.9.) C_2H_5Cl, (I.10.) $C_{31}H_{64}O_2$

---

*Proficiency Level

E.1.   One of the man-made pollutants detected in the atmosphere in recent years contains 7.81% C and 92.19% Cl. This compound has been used as a dry-cleaning agent and in certain types of fire extinguishers, although its toxicity has led to major restrictions on its use within recent years. If its molecular formula is the same as its empirical formula, what is the molecular formula of this compound?

E.2.   California has set air-quality standards for several air pollutants. When a 0.7328 g sample of one of these pollutants, a hydrocarbon, was completely burned in oxygen, 0.9415 g of H_2O and 2.300 g CO_2 were produced. What is the empirical formula?

E.3.   The formula weight of the air pollutant mentioned in problem E.2 is 28. What is the molecular formula of this compound, a raw material for over two hundred commercial products?

E.4.   Benzo(α)pyrene is a known carcinogen that has been found in coal and cigarette smoke. The products from a complete combustion of a 5.789 g sample were 2.481 g of H_2O and 20.20 g of CO_2. What is the empirical formula?

E.5.   The formula weight of benzo(α)pyrene (problem E.4) is 252. What is the molecular formula?

E.6.   The food dye "Red 2", (amaranth) has recently been identified as a possible carcinogen and declared unsafe for human consumption. This dye has 39.87% C, 1.84% H, 23.90% O, 6.98% N, 11.45% Na and 15.96% S. What is the molecular formula if it is the same as the empirical formula?

*E.7. Once thought to be the ideal insecticide, p,p'-dichlorodiphenyltrichloroethane (DDT) has proven to be harmful to the planet's ecology when used in massive quantities, and its sale has been discontinued in most areas. Combustion of a 0.5883 g sample of DDT formed 1.0226 g of CO_2 and 0.1346 g of H_2O. A separate assay showed the compound to consist of 50.01% chlorine. What is the empirical formula of DDT?

*E.8. The vapor density of DDT (problem E.7), corrected to STP, is 15.8 mg ml^{-1}. What is the molecular formula of DDT?

*E.9. "Lindane" is the trade name for an insecticide which has been used successfully against certain DDT-resistant insects. This compound consists only of carbon, hydrogen, and chlorine, with the chlorine accounting for 73.14% of the weight. Combustion of a 0.7357 g sample of Lindane formed 0.6680 g of CO_2 and 0.1367 g of H_2O. What is its empirical formula?

*Proficiency Level

*E.10. The vapor density of Lindane (problem E.9), corrected to STP, is 13.0 mg ml^{-1}. What is the molecular formula?

ANSWERS:

(E.1.) CCl_4, (E.2.) CH_2, (E.3.) C_2H_4, (E.4.) C_5H_3, (E.5.) $C_{20}H_{12}$, (E.6.) $C_{20}H_{11}N_3O_9S_3Na_3$, (E.7.) $C_{14}H_9Cl_5$, (E.8.) $C_{14}H_9Cl_5$, (E.9.) $CHCl$, (E.10.) $C_6H_6Cl_6$

STOICHIOMETRY

UNIT 2: <u>BALANCING EQUATIONS</u>

A mathematician may say "two times three equals six", but it is more convenient for him to express this as a brief equation:

$$2 \times 3 = 6$$

In the same way, it is more cumbersome for the chemist to say that "two formula units of sodium hydroxide react with one of sulfuric acid to form two formula units of water and one of sodium sulfate" than it is to express this as a balanced chemical equation:

$$2\ NaOH + H_2SO_4 \rightarrow 2\ H_2O + Na_2SO_4$$

A balanced chemical equation indicates the exact ratio in which chemicals combine and products are formed. As such, the balanced equation is a most useful way of looking at the quantitative aspects of chemical processes.

Two features are required for a balanced equation. Since atoms can neither be created nor destroyed by chemical change, a balanced equation must show the same number of each element symbol on both sides of the equation. Electrical charge is also conserved during chemical change, so a balanced equation must have the same <u>net</u> charge for both sides of the equation.

Simpler chemical equations can be balanced rather quickly by trial-and-error use of numerical coefficients until both requirements for "balancing" are met. More complicated equations, particularly those describing electron-transfer (oxidation-reduction) processes, are often quite time consuming to balance by a trial-and-error approach. For such cases, a systematic method of balancing is worth learning.

- References -

1. O'Connor, Rod. 1977. <u>Fundamentals of Chemistry</u>, second edition. New York: Harper & Row (Units 3, 6, 19, 22)

2. Brown, Theodore, and H. Eugene LeMay, Jr. 1977. <u>Chemistry: The Central Science</u>. Englewood Cliffs, N.J.: Prentice-Hall (Chapters 3, 8, 12, 16, 19)

3. Masterton, William, and Emil Slowinski. 1977. <u>Chemical Principles</u>, fourth edition. Philadelphia: W. B. Saunders (Chapters 3, 18, 20, 22)

4. Nebergall, William, F. C. Schmidt, and H. F. Holtzclaw, Jr.. 1976. <u>College Chemistry</u>, fifth edition. Lexington, MA: D. C. Heath (Chapters 2, 14, 15, 16, 18)

5. Moeller, Therald, and Rod O'Connor. 1972. <u>Ions in Aqueous Systems</u>. New York: McGraw-Hill (Chapter 2)

OBJECTIVES:

(1) Given a chemical equation showing the complete symbols and formulas for reactants and products, be able to supply proper numerical coefficients to balance the equation.

(2) Given a net equation, showing the principal species involved in a chemical change, be able to supply proper numerical coefficients to balance the equation.

*(3) Given the net equation for an oxidation-reduction reaction in acidic or neutral solution, except for H^+ and H_2O, be able to use a systematic method for balancing the equation.

*(4) Given the net equation, except for OH^- and H_2O, for an oxidation-reduction process in alkaline solution, be able to complete and balance the equation.

PRE-TEST:

(1) Hexamethylenetetramine, one of the older drugs used in treating kidney infections, was believed to function in the kidney by releasing formaldehyde, a chemical toxic to most bacteria, according to the equation:

$$C_6H_{12}N_4 + H_2O \rightarrow H_2CO + NH_3$$

Supply the proper numerical coefficients to balance this equation.

(2) The action of hexamethylenetetramine as an antibiotic was dependent on the maintenance of a fairly acidic solution in the kidney to minimize the recombination of formaldehyde with ammonia. A dihydrogen phosphate salt was given to convert the ammonia formed to ammonium ion:

$$NH_3 + H_2PO_4^- \rightarrow NH_4^+ + PO_4^{3-}$$

Supply proper coefficients to balance this net equation.

*(3) Aqua regia ("Royal Water"), a mixture of nitric and hydrochloric acids, is one of the few reagents capable of dissolving metallic gold. Complete and balance the equation for this process:

$$Au + NO_3^- + Cl^- \rightarrow [\ AuCl_4\ ]^- + NO_2$$

*(4) A certain drain-cleaner is composed of a mixture of sodium hydroxide granules and aluminum turnings. When treated with water the mixture reacts to generate gaseous hydrogen, an aid in loosening the debris in a clogged drain. Complete and balance the equation for this process:

$$Al + OH^- \rightarrow [\ Al(OH)_4\ ]^- + H_2$$

Answers and Directions:

(1) $C_6H_{12}N_4 + \underline{6}\ H_2O \rightarrow \underline{6}\ H_2CO + \underline{4}\ NH_3$

(2) $\underline{2}\ NH_3 + H_2PO_4^- \rightarrow \underline{2}\ NH_4^+ + PO_4^{3-}$

If both are correct, go on to questions *(3) and *(4), or terminate if minimal competency is sufficient. If you missed any, study METHODS sections 2.1 and 2.2.

*(3) $Au + \underline{3} \ NO_3^- + \underline{4} \ Cl^- + \underline{6H}^+ \rightarrow [\ AuCl_4 \ ]^- + \underline{3} \ NO_2 + \underline{3H_2O}$

*(4) $\underline{2} \ Al + \underline{2} \ OH^- + \underline{6H_2O} \rightarrow \underline{2} \ [\ Al(OH)_4 \ ]^- + \underline{3} \ H_2$

If both are correct, go on to RELEVANT PROBLEMS, UNIT 2. If you missed any, study METHODS sections 2.3 and 2.4.

METHODS

2.1. Balancing a "Complete Formula" Equation

In simple chemical equations, such as those describing acid/base reactions or pre-cipitations by ion recombination, it is usually rather easy to determine appropriate numerical coefficients for balancing. When the formulas of all reactants and products are known, careful attention to subscripts will generally give clues to necessary coefficients.

A useful procedure is to first balance the species having the largest subscript, ex-cept for subscripts in polyatomic ions (e.g., PO_4^{3-}). If that doesn't immediately reveal all necessary coefficients, try multiplying all coefficients by the next largest subscript.

EXAMPLE 1

Some commercial baking powders contain a mixture of sodium bicarbonate (baking soda) and calcium dihydrogen phosphate. When the powder is moistened, carbon dioxide gas is liberated according to the equation:

$$NaHCO_3 + Ca(H_2PO_4)_2 \rightarrow Na_2HPO_4 + CaHPO_4 + CO_2 + H_2O$$

Supply the proper coefficients to balance this equation (and, thus, to show the ratio of components in the baking powder).

SOLUTION:

Neglecting subscripts in polyatomic ions (HCO_3^-, $H_2PO_4^-$, HPO_4^{2-}), the largest sub-script is 2 (in Na_2HPO_4 or in $Ca(H_2PO_4)_2$). The latter is changed, so we'll try a coefficient of 2 to balance the Na symbols:

$$\underline{2} \ NaHCO_3 + Ca(H_2PO_4)_2 \rightarrow Na_2HPO_4 + CaHPO_4 + CO_2 + H_2O$$

This suggests that we must also place a 2 in front of $\underline{CO_2}$ (because of $\underline{2} \ NaH\underline{C}O_3$) to balance the C symbols:

$$\underline{2} \ NaH\underline{C}O_3 + Ca(H_2PO_4)_2 \rightarrow Na_2HPO_4 + CaHPO_4 + \underline{2} \ \underline{C}O_2 + H_2O$$

Inspection then shows that Na, Ca, C, and P symbols are balanced, but there is a total of 6 H and 14 O in reactants and 4 H and 13 O in products. The difference is 2 H and 1 O, which may be accounted for by placing a coefficient of 2 for the H_2O, giving the final balanced equation:

$$2 \; NaHCO_3 + Ca(H_2PO_4)_2 \;\rightarrow\; Na_2HPO_4 + CaHPO_4 + 2 \; CO_2 + 2 \; H_2O$$

EXERCISE 1

Lead arsenate has found extensive use as an insecticide for certain fruit trees. It may be prepared from lead nitrate and sodium arsenate, according to the equation:

$$Pb(NO_3)_{2(aq)} + Na_3AsO_{4(aq)} \;\rightarrow\; Pb_3(AsO_4)_{2(s)} + NaNO_{3(aq)}$$

Balance this equation, thus establishing the ratio of reactants required for this preparation.

(answer, page 38)

2.2. Balancing "Net" Equations

By definition, a "net" equation shows only the symbols and formulas of species <u>actually participating</u> in a chemical change. Species which remain unchanged ("spectators") by the process are omitted. For example, the reaction between aqueous nitric acid and aqueous sodium hydroxide may be represented, in a "complete" formula equation as:

$$HNO_{3(aq)} + NaOH_{(aq)} \;\rightarrow\; NaNO_{3(aq)} + H_2O_{(\ell)}$$

However, nitric acid, sodium hydroxide, and sodium nitrate exist in aqueous solution primarily as ions, as indicated by

$$H^+_{(aq)} + NO^-_{3(aq)} + Na^+_{(aq)} + OH^-_{(aq)} \;\rightarrow\; Na^+_{(aq)} + NO^-_{3(aq)} + H_2O_{(\ell)}$$

Thus, sodium and nitrate ions remain unchanged by the process and we can represent the <u>net</u> change by an equation omitting these "spectator ions":

$$H^+_{(aq)} + OH^-_{(aq)} \;\rightarrow\; H_2O_{(\ell)}$$

Both "complete formula" and "net" equations are useful in describing chemical reactions and the one most convenient for the needs of a particular problem may be selected. By convention, "net" equations retain complete formulas for all solids (even ionic solids such as $CaSO_4$), liquids, gases, and weakly-ionized dissolved species (such as acetic acid, CH_3CO_2H).

Balancing a "net" equation is often easier than balancing one containing all complete formulas. It is critical to note that both atom balance <u>and charge balance</u> must be shown. That is, in the balanced equation the net charge (algebraic sum of <u>all</u> charges) must be the same for both reactants and products.

EXAMPLE 2

Artists often prepare decorative aluminum articles by etching with hydrochloric acid. The areas to be protected from the acid are painted and the acid then attacks any remaining exposed aluminum, according to the net equation:

$$Al_{(s)} + H^+_{(aq)} \rightarrow Al^{3+}_{(aq)} + H_{2(g)}$$

Balance this equation.

SOLUTION:

At first glance, it might appear that the equation could be balanced simply by placing a coefficient of 2 in front of the H^+ symbol:

WRONG $\quad Al_{(s)} + 2 H^+_{(aq)} \rightarrow Al^{3+}_{(aq)} + H_{2(g)}$

This does balance all symbols ("atom balance"), but there remains a charge imbalance (2+ for reactants and 3+ for products).

We may ask ourselves how best to achieve charge balance. A useful first try would consider looking for the smallest number containing both 2+ and 3+. The answer, 6+, could be obtained by multiplying the 2+ group by 3 and the 3+ symbol by 2:

STILL NOT BALANCED $\quad Al_{(s)} + 6 H^+_{(aq)} \rightarrow 2 Al^{3+}_{(aq)} + H_{2(g)}$

This achieves charge balance, but leaves an atom imbalance. Fortunately, the species affected are electrically neutral, so changing their coefficients will not disturb charge balance:

BALANCED $\quad \underline{2} Al_{(s)} + \underline{6} H^+_{(aq)} \rightarrow \underline{2} Al^{3+}_{(aq)} + \underline{3} H_{2(g)}$

Every "net" equation poses its own unique problem and no "routine method" is universally applicable. It is necessary to try, as systematically as possible, to choose coefficients which result in both charge balance and atom balance. It is frequently advantageous to balance charge first.

==

EXERCISE 2

Copper sulfate ("blue vitriol") is used as an additive to swimming pools to kill algae. It may be prepared by the action of hot sulfuric acid on copper:

$$Cu_{(s)} + H^+_{(aq)} + HSO_4^-{}_{(aq)} \rightarrow Cu^{2+}_{(aq)} + SO_{2(g)} + H_2O_{(\ell)}$$

Balance this equation.

(answer, page 38)

==

Extra Practice
EXERCISE 3

Many steel structures, such as the Golden Gate Bridge, are protected from corrosion by a coating of "red lead", Pb_3O_4, often covered by a more decorative paint. Balance the equation for the preparation of "red lead":

$$Pb + O_2 \rightarrow Pb_3O_4$$

(answer, page 38)

EXERCISE 4

Certain antacid preparations contain a mixture of sodium bicarbonate and citric acid. When added to water, the preparations "fizz" by release of carbon dioxide gas. Balance the net equation for this process:

$$H_3C_6H_5O_7(aq) + HCO_3^-(aq) \rightarrow C_6H_5O_7^{3-}(aq) + CO_2(g) + H_2O(\ell)$$

(answer, page 38)

===

At this point you should try the <u>competency</u> level Self-Test questions on page 37.

===

*2.3. Balancing Oxidation-Reduction Equations (Neutral or Acidic Conditions)

The selection of appropriate numerical coefficients to achieve both charge balance and atom (symbol) balance for equations representing electron-transfer processes is seldom easy by a trial-and-error approach. Since so many important chemical reactions involve oxidation and reduction, it is worthwhile to learn some systematic method of balancing these equations.

Several methods are available, but one of the most general uses two equations for "half-reactions", one involving oxidation and the other reduction. The advantages of this method include its systematic approach, which can be learned easily with practice, and its versatility for a wide variety of reactions, such as those of complex organic compounds for which oxidation number assignments (used in some alternative procedures) are difficult to determine. In addition, the half-reaction equations are often useful in describing electrode processes in electrochemical cells (Units 8 and 14).

It should be emphasized that balancing an equation is essentially a bookkeeping procedure and the various steps do not imply anything about the real stages of a reaction. The following sequence is suggested:

†*Since an aqueous solution provides H_2O, H^+, or OH^- (Sect. 2.4), the use of such species is compatible with chemical reality.*

step 1-Convert the net equation to two equations for plausible half-reactions.

step 2-In each new equation, balance all symbols except <u>H</u> and <u>O</u>.

†step 3-Use H_2O to balance <u>O</u> symbols.

†step 4-Use H^+ to balance <u>H</u> symbols.

step 5-Use e^- (electrons) to balance charge.

step 6-Multiply as necessary, maintaining balance, to equalize e^- numbers in the two equations.

*Proficiency Level

step 7-Add equations, eliminating duplications.

step 8-Check charge balance and atom balance, and be sure
coefficients are in lowest terms.

*EXAMPLE 3

   To illustrate use of this stepwise approach, let's see how we could balance the
equation for the reaction between methanol and dichromate ion in acidic solution to form
formic acid and chromium(III) ion. We can formulate the process, except for H^+ and H_2O, by:

$$CH_3OH + Cr_2O_7^{2-} \rightarrow HCO_2H + Cr^{3+}$$

Step 1-The most plausible half-reactions would involve changes
in the carbon and chromium species, respectively:

$$CH_3OH \rightarrow HCO_2H$$
$$Cr_2O_7^{2-} \rightarrow Cr^{3+}$$

Step 2-Balancing symbols other than H and O involves only the
second equation:

$$Cr_2O_7^{2-} \rightarrow \underline{2}\ Cr^{3+}$$

Step 3-Using H_2O to balance O symbols:

$$\underline{H_2O} + CH_3OH \rightarrow HCO_2H$$
$$Cr_2O_7^{2-} \rightarrow 2\ Cr^{3+} + \underline{7H_2O}$$

Step 4-Using H^+ to balance H symbols:

$$H_2O + CH_3OH \rightarrow HCO_2H + \underline{4H^+}$$
$$\underline{14H^+} + Cr_2O_7^{2-} \rightarrow 2\ Cr^{3+} + 7H_2O$$

Step 5-Using e^- to balance charge:

$$H_2O + CH_3OH \rightarrow HCO_2H + \underline{4H}^{\textcircled{+}} + \underline{4e}^{\textcircled{-}}$$
$$[zero] = [\ (4+) + (4-)\ ]$$
$$\underline{14H}^{\textcircled{+}} + Cr_2O_7^{\textcircled{2-}} + \underline{6e}^{\textcircled{-}} \rightarrow 2Cr^{\textcircled{3+}} + 7H_2O$$
$$[(14+) + (2-) + (6-)] = [2(3+)]$$

Step 6-To equalize electron numbers, noting that the smallest
number containing both 4 and 6 is 12, we multiply the
first equation by 3 ($3 \times 4e^- = 12e^-$) and the second by
2 ($2 \times 6e^- = 12e^-$):

$$3H_2O + 3CH_3OH \rightarrow 3HCO_2H + 12H^+ + \underline{12e^-}$$
$$28H^+ + 2Cr_2O_7^{2-} + \underline{12e^-} \rightarrow 4Cr^{3+} + 14H_2O$$

Step 7-To add the equations, while eliminating duplications, we
note that duplication of H_2O, H^+, and electrons suggests

subtraction from both reactants and products of $\underline{3H_2O}$, $\underline{12H^+}$, and $\underline{12e^-}$, leaving:

$$16 \ H^+ + 3 \ CH_3OH + 2 \ Cr_2O_7{}^{2-} \rightarrow 3 \ HCO_2H + 4 \ Cr^{3+} + 11 \ H_2O$$

<u>Step 8</u>-Checking both charge and atom balance:

$$[(16+) + 2(2-)] = 4(3+)$$

$$16\underline{H} + 12\underline{H} = 6\underline{H} + 22\underline{H}; \quad 3\underline{C} = 3\underline{C}; \quad 3\underline{O} + 14\underline{O} = 6\underline{O} + 11\underline{O}; \quad 4\underline{Cr} = 4\underline{Cr}$$

- -

*EXERCISE 5

Solutions of potassium permanganate are sometimes used as topical antiseptic agents, particularly in veterinary medicine. Such solutions, like all chemical solutions, should never be pipetted by mouth. Permanganate ion is not only poisonous, but also a powerful oxidizing agent capable of generating chlorine from the hydrochloric acid in the stomach:

$$MnO_4{}^- + Cl^- \rightarrow Mn^{2+} + Cl_2$$

Complete (add H^+ or H_2O as needed) and balance this equation.

(answer, page 39)

- -

*2.4. Balancing Oxidation-Reduction Equations (in Alkaline Solution)

The characteristic of an alkaline solution which distinguishes it from those which are classified as neutral or acidic is an excess of hydroxide ion in the alkaline solution. Thus, equations for reactions in an aqueous alkaline medium may show OH^- as a reactant or product, but not H^+. Many chemical species vary in composition between acidic and alkaline solutions, so proper formulations must be used. In addition, many species undergo different reactions under acidic or alkaline conditions.

A knowledge of proper formulations and the ability to predict products under varying conditions must await a more detailed study of chemistry. We will content ourselves at this point with simply learning how to balance an equation in which all reactants and products are specified, except for H_2O and OH^-.

Although there are alternative methods available, it is expedient to use the same steps for balancing an equation in alkaline solution that we used for the earlier cases (Section 2.3), with two additional steps:

<u>Step 9</u>-Convert any H^+ to H_2O and add <u>that</u> <u>number</u> of OH^- to the other side of the equation.

<u>Step 10</u>-Eliminate any duplication of H_2O by combination or subtraction.

We may think of Step 9, if you like, as adding sufficient OH^- to neutralize one side ($H^+ + OH^- \rightarrow H_2O$) and make the other alkaline, by an equivalent addition.

*EXAMPLE 4

The reaction between iodide and permanganate ions in alkaline solution is formulated, except for H_2O and OH^-, as:

$$I^- + MnO_4^- \rightarrow I_2 + MnO_2$$

If we follow the first 8 Steps outlined in Section 2.3, we obtain:

$$6I^- + 2MnO_4^- + 8H^+ \rightarrow 3I_2 + 2MnO_2 + 4H_2O$$

Then, conversion to a formulation consistent with alkaline conditions involves:

Step 9-Converting $\underline{8H^+}$ to $\underline{8H_2O}$ and adding $\underline{8OH^-}$ to the other side:

$$6I^- + 2MnO_4^- + 8H_2O \rightarrow 3I_2 + 2MnO_2 + 4H_2O + 8OH^-$$

Step 10-Eliminating duplication of H_2O by subtracting $\underline{4H_2O}$ from each side:

$$6I^- + 2MnO_4^- + 4H_2O \rightarrow 3I_2 + 2MnO_2 + 8OH^-$$

===

*EXERCISE 6

Gold may be recovered from lowgrade ores or mine "tailings" by a treatment with cyanide solution in the presence of air:

$$Au + CN^- + O_2 \rightarrow [Au(CN)_2]^- + OH^-$$

After which the dicyanoaurate(I) complex is reduced with powdered zinc:

$$2 [Au(CN)_2]^- + Zn \rightarrow 2Au + [Zn(CN)_4]^{2-}$$

Complete and balance the equation for the air oxidation of gold in contact with cyanide solution.

(answer, page 39)

===

Extra Practice

*EXERCISE 7

One of the principal mercury ores is cinnabar, containing the red-orange mercury(II) sulfide. When samples of the ore are to be dissolved and assayed for mercury content it is necessary to use aqua regia, since this sulfide is insoluble even in hot concentrated nitric acid. Complete and balance the equation for the reaction (in acid):

$$HgS_{(s)} + Cl^-_{(aq)} + NO^-_{3(aq)} \rightarrow [HgCl_4]^{2-}_{(aq)} + S_{(s)} + NO_{(g)}$$

(answer, page 40)

EXERCISE 8

Nitrates from excess agricultural fertilizers have become serious water pollutants in many areas. The direct quantitative determination of nitrate is difficult, so one method of assaying water samples for nitrate contamination uses the reduction of nitrate by zinc to ammonia, followed by a direct quantitative determination of the ammonia produced. Complete and balance the equation for the reduction (in alkaline solution):

$$NO^-_3 + Zn \rightarrow NH_3 + [Zn(OH)_4]^{2-}$$

(answer, page 40)

==

SELF-TEST (UNIT 2)   [answers, page 41]

2.1. One of the steps in the "blast furnace" process for recovering iron from its ores uses carbon monoxide gas formed by the incomplete oxidation of coke. Balance the equation for the reaction of this gas with iron(III) oxide:

$$CO_{(g)} + Fe_2O_{3(s)} \rightarrow Fe_{(\ell)} + CO_{2(g)}$$

2.2. One of the more common silver ores is argentite, containing silver sulfide. Silver may be obtained from this ore by a two step process, first using cyanide to dissolve the silver sulfide and then using zinc to liberate free silver. Balance the net equations for this process:

$$Ag_2S_{(s)} + CN^-_{(aq)} \rightarrow [Ag(CN)_2]^-_{(aq)} + S^{2-}_{(aq)}$$

$$[Ag(CN)_2]^-_{(aq)} + Zn_{(s)} \rightarrow Ag_{(s)} + [Zn(CN)_4]^{2-}_{(aq)}$$

- -

If you completed Self-Test questions 2.1 and 2.2 correctly, you may go on to the proficiency level, try the Relevant Problems (Unit 2), or stop here. If not, you should consult your instructor for suggestions of further study aids.

- -

*2.3. Iodic acid, for the preparation of sodium or potassium iodate, may be prepared by oxidizing iodine with concentrated nitric acid. Being relatively insoluble in the reaction mixture, iodic acid settles out as a white solid. Complete and balance the equation:

$$I_2 + H^+ + NO^-_3 \rightarrow HIO_3 + NO_2$$

*Proficiency Level

*2.4. "Unsaturated" in reference to an organic compound (as in "polyunsaturated vege-table oils") indicates the presence in the molecule of one or more carbon-carbon double bonds. A quick laboratory test (the Baeyer test) for unsaturation, if other easily oxidized groups are absent, uses an alkaline solution of potassium permanganate. A positive test depends on the change from the purple color of MnO_4^- to the brown color of MnO_2. Complete and balance:

$$CH_3CH_2CH_2\overset{H}{\underset{}{C}}=CH_2 + MnO_4^- \rightarrow CH_3CH_2CH_2\overset{H}{\underset{OH}{C}}-CH_2OH + MnO_2$$

_ _

If you completed Self-Test questions 2.3 and 2.4 correctly, you may go on to the Relevant Problems for Unit 2. If not, you should consult your instructor for suggestions of further study aids.

_ _

ANSWERS TO EXERCISES, Unit 2

1. $[3Pb(NO_3)_2 + 2Na_3AsO_4 \rightarrow Pb_3(AsO_4)_2 + 6NaNO_3]$ Solution:

    Neglecting subscripts in polyatomic ions (PO_4^{3-} and NO_3^-), the largest sub-scripts is $\underline{3}$ (in Na_3AsO_4 and in $Pb_3(AsO_4)_2$). If we simply balance the Na symbols by placing a 3 in front of $NaNO_3$, we wouldn't have balanced the Pb symbols or AsO_4^{3-} formulas.

    Proper coefficients are selected by noting that the $Pb_3(AsO_4)_2$ requires $\underline{3}$ $Pb(NO_3)_2$ and the two arsenate units, $Pb_3(AsO_4)_2$, require $\underline{2}$ Na_3AsO_4. Then we can see that $\underline{2}$ Na_3AsO_4 will require $\underline{6}$ $NaNO_3$. (The same conclusion is reached from $\underline{3}$ $Pb(NO_3)_2$.

2. $[Cu_{(s)} + 3H^+_{(aq)} + HSO_4^-_{(aq)} \rightarrow Cu^{2+}_{(aq)} + SO_{2(g)} + 2H_2O_{(\ell)}]$ Solution:

    Trying first for charge balance, we note that $\underline{3}H^+ + HSO_4^-$ will give a net 2+ to equal the charge of the Cu^{2+} in products. Then the total of 4 H symbols ($\underline{3}H^+ + \underline{HSO_4^-}$) suggests a coefficient of 2 for H_2O. This also balances the oxygen symbols.

3. $[3Pb + 2O_2 \rightarrow Pb_3O_4]$ Solution:

    The 4 oxygen in Pb_3O_4 require $\underline{2}$ O_2 and the three lead in Pb_3O_4 require $\underline{3}$ Pb.

4. $[H_3C_6H_5O_7 + 3HCO_3^- \rightarrow C_6H_5O_7^{3-} + 3CO_2 + 3H_2O]$ Solution:

    Balancing charge first requires $3HCO_3^-$ to balance the 3- charge of $C_6H_5O_7^{3-}$.

Since the $(C_6H_5O_7)$ unit remains intact, the carbon balance is attained by a coefficient of 3 for CO_2. Both hydrogen and oxygen are balanced by addition of a 3 in front of H_2O.

- -

*5. $[2MnO_4^- + 10Cl^- + 16H^+ \rightarrow 2Mn^{2+} + 5Cl_2 + 8H_2O]$ <u>Solution:</u>

Step 1- $MnO_4^- \rightarrow Mn^{2+}$

$Cl^- \rightarrow Cl_2$

Step 2- $MnO_4^- \rightarrow Mn^{2+}$

$\underline{2} Cl^- \rightarrow Cl_2$

Step 3- $MnO_4^- \rightarrow Mn^{2+} + \underline{4H_2O}$

$2Cl^- \rightarrow Cl_2$

Step 4- $\underline{8H^+} + MnO_4^- \rightarrow Mn^{2+} + \underline{4H_2O}$

$2Cl^- \rightarrow Cl_2$

Step 5- $8H^{\oplus} + MnO_4^{\ominus} + \underline{5e^{\ominus}} \rightarrow Mn^{2+} + 4H_2O$

$[(8+) + (1-) + (5-)] = (2+)$

$2Cl^{\ominus} \rightarrow Cl_2 + \underline{2e^{\ominus}}$

Step 6- (smallest multiple of 2 and 5 is <u>10</u>.)

$16H^+ + 2MnO_4^- + \underline{10e^-} \rightarrow 2Mn^{2+} + 8H_2O$

$10Cl^- \rightarrow 5Cl_2 + \underline{10e^-}$

Step 7- Addition, eliminating duplication of $\underline{10e^-}$:

(see answer)

Step 8- Check charge and atom balance.

*6. $[4Au + 8CN^- + O_2 + 2H_2O \rightarrow 4[Au(CN)_2]^- + 4OH^-]$ <u>Solution:</u>

Step 1- $Au + CN^- \rightarrow [Au(CN)_2]^-$

$O_2 \rightarrow OH^-$

Step 2- $Au + \underline{2CN^-} \rightarrow [Au(CN)_2]^-$
(1st equation only)

Step 3- $O_2 \rightarrow \underline{O}H^- + H_2\underline{O}$
(2nd equation only)

Step 4- $\underline{3H^+} + O_2 \rightarrow O\underline{H}^- + \underline{H_2}O$
(2nd equation only)

Step 5- $Au + 2CN^- \rightarrow [Au(CN)_2]^- + e^-$

$$[(2-)] = [(1-) + (1-)]$$

$$3H^+ + O_2 + 4e^- \rightarrow OH^- + H_2O$$

Step 6- $4Au + 8CN^- \rightarrow 4[Au(CN)_2]^- + \underline{4e^-}$

$$3H^+ + O_2 + \underline{4e^-} \rightarrow OH^- + H_2O$$

Step 7- $4Au + 8CN^- + 3H^+ + O_2 \rightarrow 4[Au(CN)_2]^- + OH^- + H_2O$

Step 8- CHECK

Step 9- $(+3OH^- \rightarrow 3H_2O)$ $\qquad\qquad$ $(+3OH^-)$

$$4Au + 8CN^- + 3H_2O + O_2 \rightarrow 4[Au(CN)_2]^- + 4OH^- + H_2O$$

Step 10-(Subtracting one H_2O gives answer.)

*7. $[3HgS + 12Cl^- + 2NO_3^- + 8H^+ \rightarrow 3[HgCl_4]^{2-} + 3S + 2NO + 4H_2O]$ Solution:

Step 1- $HgS + Cl^- \rightarrow [HgCl_4]^{2-} + S$

$$NO_3^- \rightarrow NO$$

Step 2- $HgS + \underline{4Cl^-} \rightarrow [HgCl_4]^{2-} + S$
(1st equation only)

Step 3- $NO_3^- \rightarrow N\underline{O} + \underline{2H_2O}$
(2nd equation only)

Step 4- $\underline{4H^+} + NO_3^- \rightarrow NO + \underline{2H_2O}$
(2nd equation only)

Step 5- $HgS + 4Cl^- \rightarrow [HgCl_4]^{2-} + S + 2e^-$

$$[(4-)] = [(1-) + (3-)]$$

$$4H^+ + NO_3^- + 3e^- \rightarrow NO + 2H_2O$$

$$[(4+) + (1-) + (3-)] = [zero]$$

Step 6- $(\underline{3} \times 2e^- = \underline{2} \times 3e^-)$

$$3HgS + 12Cl^- \rightarrow 3[HgCl_4]^{2-} + 3S + \underline{6e^-}$$

$$8H^+ + 2NO_3^- + \underline{6e^-} \rightarrow 2NO + 4H_2O$$

Step 7- (Combine, subtracting $6e^-$, to get answer)

Step 8- CHECK

*8. $[NO_3^- + 4Zn + 7OH^- + 6H_2O \rightarrow 4[Zn(OH)_4]^{2-} + NH_3]$ Solution:

Steps 1-8- $NO_3^- + 4Zn + 13H_2O \rightarrow 4[Zn(OH)_4]^{2-} + NH_3 + 7H^+$
$\qquad\qquad (+7OH^-)\downarrow \qquad\qquad\qquad\qquad (+7OH^- \rightarrow 7H_2O)\downarrow$

Step 9- $NO_3^- + 4Zn + 7OH^- + 13H_2O \rightarrow [Zn(OH)_4]^{2-} + NH_3 + 7H_2O$

40

Step 10-(Subtracting $7H_2O$ gives the answer.)

= =

ANSWERS to SELF-TEST, Unit 2

2.1. $3CO_{(g)} + Fe_2O_{3(s)} \rightarrow 2Fe_{(\ell)} + 3CO_{2(g)}$

2.2. $Ag_2S_{(s)} + 4CN^-_{(aq)} \rightarrow 2[Ag(CN)_2]^-_{(aq)} + S^{2-}_{(aq)}$

$2[Ag(CN)_2]^-_{(aq)} + Zn_{(s)} \rightarrow 2Ag_{(s)} + [Zn(CN)_4]^{2-}_{(aq)}$

*2.3. $I_2 + 10H^+ + 10NO_3^- \rightarrow 2HIO_3 + 10NO_2 + 4H_2O$

*2.4.
$$3CH_3CH_2CH_2\overset{\overset{\displaystyle H}{|}}{C}=CH_2 + 2MnO_4^- + 4H_2O \rightarrow 2OH^- + 2MnO_2 + 3CH_3CH_2CH_2\overset{\overset{\displaystyle H}{|}}{\underset{\underset{\displaystyle OH}{|}}{C}}-CH_2OH$$

A.1. Although the percentage of phosphorus in plant material is relatively low, it is an essential component of plants. The phosphate from <u>fluorapatite</u>, the principal phosphate mineral, is relatively unavailable to plants because of the low solubility of the salt. However, <u>fluorapatite</u> may be treated with sulfuric acid to produce the more soluble "superphosphate". Balance the equation for this conversion:

$$Ca_{10}(PO_4)_6F_2 + H_2SO_4 + H_2O \rightarrow HF + Ca(H_2PO_4)_2 \cdot H_2O + CaSO_4$$

A.2. Nitrification is a chemical process of enzymatic oxidation occurring with two groups of soil bacteria, <u>Nitrosomonas</u> and <u>Nitrobacter</u>. <u>Nitrosomonas</u> cause the trasformation of ammonium salts to nitrite,

$$NH_4^+ + O_2 \rightarrow NO_2^- + H_2O + H^+$$

whereas <u>Nitrobacter</u> mediate the oxidation of nitrite to nitrate:

$$NO_2^- + O_2 \rightarrow NO_3^-$$

This microbial transformation of ammonium salts to nitrates maximizes the assimilation of nitrogen by plants. Provide the proper coefficients to balance these equations.

A.3. In some areas of low rainfall, soils may become too alkaline due to the presence of basic salts such as sodium carbonate. These soils may be treated with iron sulfate, which will release sulfuric acid on hydrolysis:

$$Fe^{3+} + SO_4^{2-} + H_2O \rightarrow Fe(OH)_3 + H^+ + HSO_4^-$$

Balance this net equation.

A.4. The ionic composition of sea water results from a balance between the rate at which dissolved inorganic matter is added to the ocean from the land and the atmosphere, and the rate at which it is removed from the sea by incorporation into sediments or marine organisms, or by being returned to the atmosphere. For example, the equilibrium between <u>kaolinite</u> and <u>chlorite</u>, two clay minerals, is an important factor in maintaining the magnesium content of the sea water at a constant level:

$$Al_2Si_2O_5(OH)_4 + SiO_2 + Mg^{2+} + H_2O \rightarrow Mg_5Al_2Si_3O_{10}(OH)_8 + H^+$$

Supply the proper numerical coefficients to balance this equation.

A.5. Surface sea waters are generally quite low in silicon content because of the utilization of this element in the shells and skeletons of many marine organisms. The weathering of crystal rock, such as potassium feldspar provides a source of soluble silica in sea water. The overall reaction is:

$$KAlSi_3O_8 + H^+ + H_2O \rightarrow Al_2Si_2O_5(OH)_4 + H_4SiO_4 + K^+$$

Supply the proper coefficients to balance this equation.

*A.6. When land has been submerged for some time under sea water the sulfates are reduced to sulfides. Soil reclaimed from such marshlands and used for citrus groves develops high acidity detrimental to plant growth. Moreover, H_2S released by increased acidity is very toxic to citrus roots. Soils may be tested for potential acid-sulfate formation by a peroxide test illustrated by the equation:

$$FeS_2 + H_2O_2 \rightarrow Fe^{3+} + SO_4^{2-}$$

Supply the proper coefficients to balance this equation.

*A.7. Bacteria in marine sediments play a significant role in replenishing the oxygen of the atmosphere and in limiting the accumulation of organic sediments. These bacteria are buried several centimeters below the ocean floor, with limited access to free oxygen for respiration. In this environment they must use nitrates and sulfates to oxidize organic compounds. Supply the necessary coefficients to balance this equation, in which $C_6H_{12}O_6$ represents a typical organic compound.

$$C_6H_{12}O_6 + NO_3^- + H^+ \rightarrow CO_2 + N_2 + H_2O$$

*A.8. There are several ways by which nitrogen may be lost from soils in the gaseous form. For example, nitrites in a slightly acid soil will evolve gaseous nitrogen when they are in contact with ammonium salts or simple organic nitrogen compounds, such as urea. The following reaction is suggestive of what occurs with urea:

$$H^+ + NO_2^- + CO(NH_2)_2 \rightarrow CO_2 + N_2 + H_2O$$

Supply the necessary coefficients to balance this equation.

*A.9. The four major forms of sulfur in soils and fertilizers include sulfides, sulfates, organic sulfur compounds, and elemental sulfur. Microorganisms mediate reduction of sulfur in calcium sulfate deposits to produce elemental sulfur which is interspersed in the pores of the limestone product. Supply the proper coefficients to balance this equation for a process in slightly alkaline conditions:

$$Ca^{2+} + SO_4^{2-} + C_6H_{12}O_6 \rightarrow CaCO_3 + S + CO_2$$

*A.10. In marine environments having a dissolved oxygen level below 1-2 mg liter^{-1}, anaerobic microbial decomposition of marine phytoplankton takes precedence over aerobic decomposition. In a highly simplified form, the overall reaction for the decay of the marine phytoplankton can be represented by the equation:

$$C_{106}H_{263}N_{16}PO_{110} + SO_4^{2-} \rightarrow CO_2 + S^{2-} + NH_3 + HPO_4^{2-}$$

*Proficiency Level

Supply the proper coefficients to balance this equation, for the reaction in alkaline system.

ANSWERS:

(A.1.) $Ca_{10}(PO_4)_6F_2 + 7H_2SO_4 + 3H_2O \rightarrow 2HF + 3Ca(H_2PO_4)_2 \cdot H_2O + 7CaSO_4$

(A.2.)(a) $2NH_4^+ + 3O_2 \rightarrow 2NO_2^- + 2H_2O + 4H^+$ (b) $2NO_2^- + O_2 \rightarrow 2NO_3^-$

(A.3.) $Fe^{3+} + SO_4^{2-} + 3H_2O \rightarrow Fe(OH)_3 + 2H^+ + HSO_4^-$

(A.4.) $Al_2Si_2O_5(OH)_4 + SiO_2 + 5Mg^{2+} + 7H_2O \rightarrow Mg_5Al_2Si_2O_{10}(OH)_8 + 10H^+$

(A.5.) $2KAlSi_3O_8 + 2H^+ + 9H_2O \rightarrow Al_2Si_2O_5(OH)_4 + 4H_4SiO_4 + 2K^+$

(A.6.) $2FeS_2 + 15H_2O_2 \rightarrow 2Fe^{3+} + 2H^+ + 4SO_4^{2-} + 14H_2O$

(A.7.) $5C_6H_{12}O_6 + 24H^+ + 24NO_3^- \rightarrow 30CO_2 + 12N_2 + 42H_2O$

(A.8.) $2H^+ + 2NO_2^- + CO(NH_2)_2 \rightarrow CO_2 + 3H_2O + 2N_2$

(A.9.) $4Ca^{2+} + 4SO_4^{2-} + C_6H_{12}O_6 \rightarrow 4CaCO_3 + 4S + 2CO_2 + 6H_2O$

(A.10.) $C_{106}H_{263}N_{16}PO_{110} + 53SO_4^{2-} + 2OH^- \rightarrow 106CO_2 + 53S^{2-} + 16NH_3 + 108H_2O + HPO_4^{2-}$

RELEVANT PROBLEMS

Unit 2: *Balancing Equations*

Set B: *Biological & Medical Sciences*

B.1. Gastric hyperacidity refers to the secretion in the stomach of excess hydrochloric acid beyond that required for normal digestive processes. While prolonged or repetitive hyperacidity may be symptomatic of a severe disorder, occasional hyperacidity is not unusual and can be treated with any of a number of nonprescription "antacids". One of these is a mixture containing magnesium oxide and magnesium carbonate. Balance the equation for the reaction formulated as:

$Mg_2O(CO_3) + HCl \rightarrow MgCl_2 + H_2O + CO_2$

B.2. Most organic, and some inorganic, wastes are degraded in natural waters by oxidation processes. As the dissolved oxygen supply is depleted fish and other aquatic life may die. One of the methods used to determine dissolved oxygen in water is the Winkler method. The fundamental reactions in this method are represented as follows:

$MnCl_2 + NaOH \rightarrow Mn(OH)_2 + NaCl$

$Mn(OH)_2 + O_2 + H_2O \rightarrow Mn(OH)_4$

$$Mn(OH)_4 + HCl \rightarrow MnCl_4 + H_2O$$

$$MnCl_4 + KI \rightarrow MnCl_2 + KCl + I_2$$

$$Na_2S_2O_3 + I_2 \rightarrow Na_2S_4O_6 + NaI$$

Supply the necessary coefficients to balance these equations.

B.3. Carbon monoxide is a non-irritating gas, without color, taste, or odor, produced by the incomplete oxidation of combustible carbon-containing material. It is responsible for a large number of suicidal and accidental deaths. Inhalation of air containing only 1% of this gas may prove fatal within 10 to 20 minutes. Iodine pentoxide is used for detection and determination of carbon monoxide. Supply the necessary coefficients to balance the equation for the reaction involved:

$$CO + I_2O_5 \rightarrow CO_2 + I_2$$

B.4. Most municipal water plants must add an antibacterial agent to drinking water supplies. Although increasing attention is being given to the use of iodine for this purpose, the majority of water purification systems now add chlorine. One of the disadvantages of chlorination is the taste and odor of highly chlorinated water. This may be eliminated at the point of use by charcoal "filters". The charcoal "filters" used to remove chlorine for a more palatable drinking water do not actually filter out the chlorine. Rather, the chlorine reacts with the charcoal. Supply the coefficients necessary to balance the equation for this reaction:

$$Cl_2 + C + H_2O \rightarrow CO_2 + H^+ + Cl^-$$

B.5. For prolonged storage or long distance piping, ammonia may be added to chlorinated water. The NH_2Cl and $NHCl_2$ formed remain dissolved better than chlorine itself. The chloramine (NH_2Cl) and dichloramine ($NHCl_2$) formed when ammonia reacts with chlorine are useful because of their stability in water, even though they are less effective than chlorine itself as disinfectants. Balance the equations for the reactions producing these chloramines:

$$NH_3 + Cl_2 \rightarrow NH_2Cl + NH_4^+ + Cl^-$$

$$NH_3 + Cl_2 \rightarrow NHCl_2 + NH_4^+ + Cl^-$$

*B.6. An emergency treatment for oxalic acid poisoning uses a dilute solution of potassium permanganate as a stomach flush. The reaction is slow unless a trace of manganese(II) ion is added as a catalyst. To avoid an excess of permanganate, which would also be poisonous, the color of the fluid flushed from the stomach is carefully monitored and the process is terminated when the purple color of excess permanganate is ob-

*Proficiency Level

served. Note that Mn^{2+} appears in the equation for this reaction as a product, not a reactant. Thus, this is an example of an autocatalytic reaction (one producing its own catalyst). Supply the necessary coefficients to balance this equation:

$$H_2C_2O_4 + MnO_4^- + H^+ \rightarrow CO_2 + Mn^{2+} + H_2O$$

*B.7. Denitrification, the mechanism by which fixed nitrogen is returned to the atmosphere, is an important biological process. Moreover, denitrification is also useful in advanced water treatment for the removal of nutrient nitrate. The water is treated with a minimum amount of methanol under anaerobic conditions, and nitrogen gas is evolved according to the equation:

$$CH_3OH + NO_3^- + H^+ \rightarrow CO_2 + N_2 + H_2O$$

Supply proper coefficients to balance this net equation.

*B.8. Autotrophic bacteria are not dependent upon organic matter for growth, and may live in a completely inorganic medium. They use carbon dioxide as a carbon source and their energy source depends on the species of bacteria. An example of autotrophic bacteria is <u>Gallionella</u>. These bacteria grow in a medium consisting of ammonium chloride, phosphates, mineral salts, carbon dioxide, and solid iron(II) sulfide. Complete and balance the equation that represents their energy-yielding reaction:

$$FeS + O_2 \rightarrow Fe(OH)_3 + SO_4^{2-} \qquad \text{(in alkaline conditions)}$$

*B.9. Nutrients for algae, primarily nitrates and phosphates, may wash into the water from adjacent land areas. The phytoplankton thriving on this "inorganic broth" serve as rich fodder for fishes, after performing the essential task of helping to replenish atmospheric oxygen. Although photosynthetic phytoplankton are complex organisms, not single chemical compounds, their elemental composition may be represented by what might be called their empirical formula. This, in turn can be used in an equation representing their formation from inorganic nutrients and indicating something about their role in the CO_2 - to - O_2 exchange process. Supply the necessary coefficients to balance this equation:

$$CO_2 + H_2O + NO_3^- + H_2PO_4^- + H^+ \rightarrow C_{106}H_{263}N_{16}PO_{110} + O_2$$

*B.10. Medically, iodine is administered internally for hyperthyroidism and as an antidote for alkaloidal poisons. Topically it is used as a counterirritant, bactericide and fungicide. The iodine content of medicinal preparations can be assayed by a reductive titration. Balance the equation for the reduction of iodine by thiosulfate ion in alkaline solution:

$$S_2O_3^{2-} + I_2 \rightarrow SO_4^{2-} + I^-$$

ANSWERS:

(B.1.) $Mg_2O(CO_3) + 4HCl \rightarrow 2MgCl_2 + 2H_2O + CO_2$

(B.2.) (a) $MnCl_2 + 2NaOH \rightarrow Mn(OH)_2 + 2NaCl$ (b) $2Mn(OH)_2 + O_2 + 2H_2O \rightarrow 2Mn(OH)_4$

   (c) $Mn(OH)_4 + 4HCl \rightarrow MnCl_4 + 4H_2O$ (d) $MnCl_4 + 2KI \rightarrow MnCl_2 + 2KCl + I_2$

   (e) $2Na_2S_2O_3 + I_2 \rightarrow Na_2S_4O_6 + 2NaI$

(B.3.) $5CO + I_2O_5 \rightarrow 5CO_2 + I_2$

(B.4.) $2Cl_2 + C + 2H_2O \rightarrow CO_2 + 4H^+ + 4Cl^-$

(B.5.) (a) $2NH_3 + Cl_2 \rightarrow NH_2Cl + NH_4^+ + Cl^-$ (b) $3NH_3 + 2Cl_2 \rightarrow NHCl_2 + 2NH_4^+ + 2Cl^-$

(B.6.) $5H_2C_2O_4 + 2MnO_4^- + 6H^+ \rightarrow 10\ CO_2 + 2Mn^{2+} + 8H_2O$

(B.7.) $5CH_3OH + 6NO_3^- + 6H^+ \rightarrow 5CO_2 + 3N_2 + 13H_2O$

(B.8.) $4FeS + 9\ O_2 + 8\ OH^- + 2H_2O \rightarrow 4Fe(OH)_3 + 4SO_4^{2-}$

(B.9.) $106CO_2 + 122H_2O + 16NO_3^- + H_2PO_4^- + 17H^+ \rightarrow C_{106}H_{263}N_{16}PO_{110} + 138\ O_2$

(B.10.) $10\ OH^- + S_2O_3^{2-} + 4I_2 \rightarrow 2SO_4^{2-} + 8I^- + 5H_2O$

RELEVANT
PROBLEMS

Unit 2: *Balancing Equations*

Set I: *Industrial Chemistry*

I.1. The commercial grades of nitric acid produced range from 60 to 72 percent HNO_3. The acid is produced according to the following reactions. The nitric oxide formed in the last step is recycled back through the process. Balance each of the equations by providing the proper coefficients.

   a. $4NH_3 + 5O_2 \rightarrow 4NO + 6H_2O$

   b. $2NO + O_2 \rightarrow 2NO_2$

   c. $6NO_2 + 2H_2O \rightarrow 4HNO_3 + 2NO$

I.2. Hydrogen cyanide is used as a fumigant and as a raw material in the production of other compounds. Hydrogen cyanide can be produced commercially by the following process. Balance the equation by providing the proper coefficients.

   $NH_3 + O_2 + CH_4 \rightarrow HCN + H_2O$

I.3. The "Fisher-Tropsch" reaction mentioned in Relevant Problem I.5 (Unit 1), can be illustrated by the following equation. Balance the equation by supplying the proper coefficients.

   $H_2 + CO \xrightarrow{\text{(catalyst)}} C_8H_{18} + H_2O$

I.4. Yellow phosphorus is produced by reducing phosphate rock (calcium phosphate) with coke (carbon) in the presence of sand. The process is carried out in an electric furnace and may be represented by the following equation. Provide the proper coefficients to balance this net equation.

$$(PO_4)^{3-} + C + SiO_2 \rightarrow SiO_3^{2-} + CO + P_4$$

I.5. The primary products of the chrome industry are sodium chromate (Na_2CrO_4) and sodium dichromate ($Na_2Cr_2O_7$). These compounds may be converted to other chromium-containing chemicals on demand. The net process for the preparation of sodium chromate is described by the following equation. Supply the coefficients necessary to balance this equation.

$$FeCr_2O_4 + Na_2CO_3 + O_2 \rightarrow Na_2CrO_4 + Fe_2O_3 + CO_2$$

*I.6. The assay of hydrogen peroxide in commercial products and in relatively pure aqueous solutions free from organic materials, which react with permanganate, is performed by weighing the sample into dilute sulfuric acid and titrating with standardized potassium permanganate solution, according to the following equation. Provide the coefficients necessary to balance this equation.

$$H_2SO_4 + KMnO_4 + H_2O_2 \rightarrow H_2O + O_2 + MnSO_4 + K_2SO_4$$

*I.7. Iron and steel products are of critical importance to the economy of any industrial society, yet many tons of iron are lost each year from corrosion in moist air. Balance the equation for one of the steps in iron corrosion:

$$Fe + O_2 + H_2O \rightarrow Fe(OH)_2 + H_2$$

*I.8. The first step in the industrial synthesis of potassium permanganate is a fused salt oxidation as shown in the following equation. Balance this equation.

$$MnO_2 + KOH + O_2 \rightarrow K_2MnO_4 + H_2O$$

*I.9. Aniline is an important intermediate in the manufacture of certain dyes and rubber additives. One source of aniline is the reduction of nitrobenzene by iron in an acidic aqueous mixture. Complete and balance the equation for this process.

$$C_6H_5NO_2 + Fe \rightarrow C_6H_5NH_2 + Fe_3O_4$$

*I.10. The leather part of a pair of "chrome-tanned" shoes will outwear two or three pairs of "vegetable-tanned" shoes. The "chrome liquor" for the tanning process is made by the following reaction. Balance this equation by supplying the correct coefficients.

$$Na_2Cr_2O_7 + SO_2 + H_2O \rightarrow Cr(OH)SO_4 + Na_2SO_4$$

*Proficiency Level

ANSWERS:

(I.1.) (a) $4NH_3 + 5 O_2 \rightarrow 4NO + 6H_2O$ (b) $2NO + O_2 \rightarrow 2NO_2$

(c) $3NO_2 + H_2O \rightarrow 2HNO_3 + NO$

(I.2.) $2NH_3 + 3 O_2 + 2CH_4 \rightarrow 2HCN + 6H_2O$

(I.3.) $17H_2 + 8CO \rightarrow C_8H_{18} + 8H_2O$

(I.4.) $4PO^{3-} + 10 C + 6SiO_2 \rightarrow 6SiO_3^{2-} + 10 CO + P_4$

(I.5.) $4FeCr_2O_4 + 8Na_2CO_3 + 7 O_2 \rightarrow 8Na_2CrO_4 + 2Fe_2O_3 + 8CO_2$

(I.6.) $2KMnO_4 + 3H_2SO_4 + 5H_2O_2 \rightarrow 5O_2 + 2MnSO_4 + 8H_2O + K_2SO_4$

(I.7.) $3Fe + O_2 + 4H_2O \rightarrow 3Fe(OH)_2 + H_2$

(I.8.) $2MnO_2 + 4KOH + O_2 \rightarrow 2K_2MnO_4 + 2H_2O$

(I.9.) $4C_6H_5NO_2 + 9Fe + 4H_2O \rightarrow 4C_6H_5NH_2 + 3Fe_3O_4$

(I.10.) $Na_2Cr_2O_7 + 3SO_2 + H_2O \rightarrow 2Cr(OH)SO_4 + Na_2SO_4$

RELEVANT
PROBLEMS

Unit 2: Balancing Equations

Set E: Environmental Sciences

E.1. Because of its ability to dissolve or disperse so many different substances, water
rarely exists in a pure form. Water may be purified by a number of methods depending upon the use for the water and the nature of the impurities present. Impurities
may be dissolved solids such as soluble mineral salts, suspended material such as
clay or sand, dissolved gases such as carbon dioxide, oxygen or sulfur dioxide, or
bacteria, plant spores, pollen and other types of organic material. Unfortunately,
rivers and other bodies of water are frequently used as dumping places for sewage
and industrial wastes, resulting in the pollution of the water and the ultimate
destruction of the fish and aquatic plant life. The major water purification problem,
however, is to provide safe drinking water. The water supply for a city may be
purified by aeration, chlorination, filtration, or sedimentation and coagulation, or
by any combination of these purification steps. In the process of sedimentation and
coagulation, the water is made to flow into shallow concrete basins where suspended
materials start to settle out. To hasten this process, aluminum sulfate and lime are
added to the water and a sticky, jelly-like precipitate of aluminum hydroxide is
formed. As this material slowly settles, it catches and carries bacteria and sediment
to the bottom of the tank. The chemical reaction involved in the formation of the

coagulating agent, $Al(OH)_3$, is represented by the following equation. Provide the coefficients necessary to balance this equation.

$$Al_2(SO_4)_{3(aq)} + Ca(OH)_{2(aq)} \rightarrow Al(OH)_{3(s)} + CaSO_{4(s)}$$

E.2. Tannery wastes contain sodium and calcium sulfides which are oxygen consumers. The sodium sulfide may be oxidized to sulfur and sodium hydroxide by atmospheric air through catalytic action of an iron(II)-iron(III) redox system. One of the five main reactions of this system is listed below. Balance this reaction by providing the correct coefficients.

$$FeSO_4 + Ca(OH)_2 + H_2O + O_2 \rightarrow CaSO_4 + Fe(OH)_3$$

E.3. When coal containing sulfuritic materials, such as iron pyrite (FeS_2), is exposed to oxygen from the atmosphere in the presence of sufficient water, sulfuric acid is formed. If this material is then transported in sufficient quantity by surface waters into the natural drainage system, the resulting flow may in many cases be of such an acidic level that is unsatisfactory for supporting plant and animal life. Since the rate at which the acid is produced is directly proportional to the amount of exposed sulfuritic material, the problem is generally aggravated by the opening of new surfaces unless special precautions are instituted. In the past, new mining areas have been developed and existing mining operations expanded without any consideration of the pollution aspects of mine drainage. Unlike most industrial wastes that cease when the plant closes down, acid mine drainage will continue indefinitely even after coal mining operations have been suspended. Thus, unless protective measures are taken, the problem can only grow in severity until water resources throughout the mining region are in jeopardy. The oxidation of iron pyrite and the subsequent hydrolysis of its oxidation products may be represented by the following equation. Supply the coefficients necessary to balance the equation.

$$FeS_2 + H_2O + O_2 \rightarrow Fe_2O_3 + H_2SO_4$$

E.4. Another method of treating tannery wastes (problem E.2.) is with flue gases from the combustion of low-grade coal. The flue gases will contain carbon dioxide and sulfur dioxide in the appropriate proportions for the process described in the following net equation. Balance this equation by supplying the correct coefficients.

$$S^{2-} + H_2O + CO_2 + SO_2 \rightarrow HCO_3^- + S$$

*E.5. Cyanides are toxic to most mammals. Wastes from many electroplating processes contain cyanides. Cyanides can be oxidized by several reagents but chlorine or hypochlorites are most commonly used. Balance the following net equation which describes

*Proficiency Level

the oxidation of cyanide by chlorine.

$$CN^- + Cl_2 + OH^- \rightarrow N_2 + CO_3^{2-} + Cl^- + H_2O$$

*E.6. As a method to test for sulfur dioxide in air, Elbert Weaver suggested an aspirator to pull large volumes of air through water. The water would react with the sulfur dioxide to form a solution of sulfurous acid. The acid would then be titrated by potassium permanganate. Balance the equation for this method by supplying the proper coefficients.

$$H_2SO_3 + KMnO_4 \rightarrow K_2SO_4 + MnSO_4 + H_2SO_4 + H_2O$$

(Note: This method works well only when large quantities of SO_2 are present.)

*E.7. Sulfur dioxide and hydrogen sulfide are major air pollutants from coal stacks, refinery gas, or coke-oven gas. One method of removing these noxious gases from the exhaust also produces sulfur for commercial use. The gases, in proper volume ratio, are passed over a catalyst at 250°C. Balance the following equation that describes the recovery of sulfur from H_2S and SO_2.

$$H_2S + SO_2 \rightarrow S + H_2O$$

*E.8. Many plants obtain nitrogen from the soil in the form of nitrate (NO_3^-), to which it has been previously converted by soil bacteria from organic forms of nitrogen entering the soil as animal and plant remains. Plants use nitrate to manufacture organic nitrogen compounds, such as proteins. In addition to being used by higher plants, the nitrate nitrogen of the soil may be used by microorganisms, be lost in drainage, or escape from the nitrogen cycle in some gaseous form. The chemical reduction of nitrate nitrogen to gaseous compounds or elemental nitrogen is called denitrification. Nitrates in solution may evolve gaseous nitrogen by reaction with certain ammonium salts, with simple amines, or urea, and even with some non-nitrogenous sulfur compounds and carbohydrates. This mode of denitrification is enhanced by poor drainage and lack of aeration. The following reaction suggests what might occur to nitrate in contact with a carbohydrate in moist, alkaline soil.

$$C_6H_{12}O_6 + KNO_3 \rightarrow CO_2 + H_2O + N_2 + KOH$$

Supply the coefficients necessary to balance the preceding equation.

*E.9. Any type of water pollution produces economic, esthetic and health problems. Few of the complex chemical compounds that form the wastes of industrial processes are removed by existing waste water treatment plants and their ultimate effects, both on stream life and on human health, are still undetermined. In addition, domestic wastes containing a variety of organic compounds are added to the streams, rivers, and lakes. Fresh flowing water has a remarkable ability to absorb many organic wastes, oxidizing them to useful or innocuous substances. Part of this purification is entirely mechanical, resulting from the motion of the water. This motion stirs up the waste matter, dissolving some and breaking larger aggregates into particles which either settle to the bottom or are oxidized. Bodies of water can "metabolize" wastes, just as a living organism does. The water absorbs oxygen from the air and from water plants, which release it during the process of photosynthesis. The dissolved oxygen then oxidizes the organic wastes so that nothing remains except carbon dioxide, water, and small quantities of ash. If a body of water is overloaded with wastes it will suffer from "indigestion", i.e., it cannot oxidize any more wastes until it absorbs more oxygen. The "metabolic" oxidation of carbohydrate wastes, found in streams, rivers and lakes, can be represented by the following equation. Supply the coefficients necessary to balance this equation.

$$C_6H_{10}O_{5\,(aq)} + O_{2\,(aq)} + OH^- \rightarrow HCO_{3\,(g)}^- + H_2O_{(\ell)}$$

*E.10. A possible cause of both air and water contamination is the industrial process of producing methyl bromide from methanol. Methyl bromide is a poisonous gas which requires a closed refrigerated system to prevent air pollution. The sulfuric acid requires a liquid waste disposal system. Balance the equation that describes this production process by supplying the proper coefficients.

$$CH_3OH + Br_2 + S \rightarrow CH_3Br + H_2SO_4 + H_2O$$

ANSWERS:

(E.1.) $Al_2(SO_4)_3 + 3Ca(OH)_2 \rightarrow 2Al(OH)_3 + 3CaSO_4$

(E.2.) $4FeSO_4 + 4Ca(OH)_2 + 2H_2O + O_2 \rightarrow 4CaSO_4 + 4Fe(OH)_3$

(E.3.) $4FeS_2 + 8H_2O + 15\ O_2 \rightarrow 2Fe_2O_3 + 8H_2SO_4$

(E.4.) $2S^{2-} + 2H_2O + 4CO_2 + SO_2 \rightarrow 4HCO_3^- + 3S$

(E.5.) $2CN^- + 5Cl_2 + 12\ OH^- \rightarrow N_2 + 2CO_3^{2-} + 10\ Cl^- + 6H_2O$

(E.6.) $5H_2SO_3 + 2KMnO_4 \rightarrow 2H_2SO_4 + 2MnSO_4 + K_2SO_4 + 3H_2O$

(E.7.) $2H_2S + SO_2 \rightarrow 3S + 2H_2O$

(E.8.) $5C_6H_{12}O_6 + 24KNO_3 \rightarrow 30\ CO_2 + 12N_2 + 24KOH + 18H_2O$

(E.9.) $C_6H_{10}O_5 + 6\ O_2 + 6\ OH^- \rightarrow 6HCO_3^- + 5H_2O$

(E.10) $6CH_3OH + 3Br_2 + S \rightarrow 6CH_3Br + H_2SO_4 + 2H_2O$

UNIT 3: THE MOLE CONCEPT

Sometimes it is convenient to count the things we're dealing with and sometimes it is more convenient to measure them some other way, perhaps by mass or volume. Consider gravel, for example. If we were out on a river bank throwing gravel at a floating can, we might want to count the pieces of gravel to keep track of our accuracy. But if we wanted to order some gravel for a new driveway, we wouldn't ask for 3.6×10^5 pieces of gravel. (If we did, the hauling company might be a bit surprised!) Rather, we would order "6 cubic meters" or "4 metric tons" or some similar unit, knowing (or hoping) that the quantity specified by volume or mass would contain the right number of pieces of gravel to do the job. The point is, that when numbers of things get so large that counting becomes more trouble than it's worth, it is easier to make our direct measurements by mass or volume.

Of course, we don't have to count things in units of one. We have a number of other useful counting units, such as the dozen (12) or the thousand (10^3). In dealing with very large numbers of things it would be handy to have some analogous counting unit. The number of atoms in even a small sample of material is very large (much larger than the number of pebbles in a metric ton of gravel). A counting unit useful in dealing with chemical processes would have to contain a very large number of unit particles. The mole has been defined for just such a purpose:

1 MOLE CONTAINS 6.023 X 10²³ UNIT PARTICLES

Why such an odd number? Remember that our direct measurement of very large numbers is more convenient by mass or volume than by actual counting. In dealing with chemical things we have some handy mass digits to work with - the atomic weights of the elements (which can be combined to make formula weights of compounds or other polyatomic species). Thus, as a matter of convenience, we define a mole as the number of unit particles in one gram formula weight (or gram atomic weight) of any substance. The actual number of unit particles in a mole is then determined experimentally to be 6.023×10^{23}. In this case, then, we base our counting unit on convenient masses rather than on convenient numbers (as in the dozen or the thousand).

It is important to recognize that the mole itself is based on the formula weight (or atomic weight) expressed in grams. Related units such as the millimole (10^{-3} mole, 6.023×10^{20} unit particles) or the metric ton-mole (formula weight expressed in metric tons, 6.023×10^{29} unit particles) are useful in a number of applications, but these express different quantities of unit particles.

The use of the mole concept enables us to produce a variety of very useful unity factors for dealing with the quantitative aspects of chemical measurement or chemical change. The general form of these unity factors will be:

$$\frac{1 \text{ mole}(X)}{[\text{form. wt.}] \text{ g}(X)} \qquad \text{or} \qquad \frac{[\text{form. wt.}] \text{ g}(X)}{1 \text{ mole}(X)}$$

For example, we can write such unity factors as:

$$\frac{1 \text{ mole (Na)}}{23.0 \text{ g (Na)}} \qquad \frac{2.0 \text{ g (H}_2)}{1 \text{ mole (H}_2)} \qquad \frac{1 \text{ mole (Na}_2\text{CO}_3)}{106 \text{ g (Na}_2\text{CO}_3)}$$

We will find that unity factors of this type are invaluable for the calculation of quantities involved in chemical processes.

-References -

1. O'Connor, Rod. 1977. Fundamentals of Chemistry, second edition. New York: Harper & Row (Unit 4)
2. Brown, Theodore, and H. Eugene LeMay, Jr. 1977. Chemistry: The Central Science. Englewood Cliffs, N. J.: Prentice-Hall (Chapter 3)
3. Masterton, William, and Emil Slowinski. 1977. Chemical Principles, fourth edition. Philadelphia: W. B. Saunders (Chapters 2,3)
4. Nebergall, William, F. C. Schmidt, and H. F. Holtzclaw, Jr. 1976. College Chemistry, fifth edition. Lexington, MA: D. C. Heath (Chapter 2)
5. Kiefer, W. F. 1963. The Mole Concept in Chemistry. New York: Van Nostrand/Reinhold

---

OBJECTIVES:

(1) Given the number of moles, or the mass, of a pure substance and its formula (or atomic) weight, be able to calculate the number of grams or moles, respectively, represented.

(2) Given the chemical formula of a compound and the number of moles, or the mass, of a sample, be able to calculate the number of grams or moles, respectively, represented (using tabulated atomic weights).

*(3) Given a description of a mixture, in terms of mass composition, be able to determine the number of moles of any specified component (using tabulated atomic weights).

- -

PRE-TEST:

Necessary Atomic Weights:			
H (1.008)	N (14.01)	F (19.00)	S (32.06)
O (16.00)	Cl (35.45)	C (12.01)	

*Proficiency Level

(1) Special electrochemical cells have been developed in space vehicle research programs for operation at either very high or very low temperatures, outside the temperature range of normal cells.

    (a) A cell capable of operating at $500^\circ C$ uses a magnesium bar as one electrode and mixed copper oxides as the other. If 32 g of Cu_2O (form. wt. 143) is required, to how many moles does this correspond?

    _____

    (b) If the magnesium electrode is converted to 0.45 mole of magnesium oxide (form. wt. 40.3) during cell operation, how many grams of MgO are formed? _____

(2) Early refrigeration units used sulfur dioxide gas, but this noxious material was replaced many years ago by the odorless, nontoxic freons, such as Freon-12, CCl_2F_2. The freons may, in turn, soon be replaced because of the potential environmental problems with respect to possible freon damage to the ozone in the upper atmosphere.

    (a) How many moles of SO_2 were used in a refrigeration compressor charged with 156 g of the gas? _____

    (b) How many grams of Freon-12 would correspond to the number of moles of SO_2 used in question 2a? _____

*(3) With a potential shortage of natural gas, the U.S. has reconsidered possible utilization of some earlier fuel gas mixtures. One of these, "producer gas", is a mixture of carbon monoxide, nitrogen, and hydrogen made by passing a hot blast of moist air through a bed of heated coke or anthracite coal. A particular sample of "producer gas" was analyzed and found to contain (by weight) 36% CO, 60% N_2, and 4.0% H_2. How many moles of each gas would be found in 5.00 metric tons of this mixture? _____

Answers and Directions:

    (1) (a) 0.22   (b) 18 g   (2) (a) 2.44   (b) 295 g

If all are correct, go on to question *3. If you missed any, study METHODS, sections 3.1 and 3.2.

*(3) 6.4×10^4 mole CO, 1.1×10^5 mole N_2, 9.9×10^4 mole H_2

If all are correct, go on to RELEVANT PROBLEMS, Unit 3. If you missed any, study METHODS, section 3.3.

---

*Proficiency Level

METHODS

3.1 Mole/Mass Conversions (Formula Weight Known)

If we know the formula weight (or atomic weight) of a pure substance and wish to convert moles to grams, or grams to moles, the problem is quite simple. It involves only an understanding of the mole concept and an easy exercise in the use of mole/mass unity factors:

$$no.\ grams = no.\ moles\ x\ \frac{(form.\ wt.)\ g}{1\ mole}$$

$$no.\ moles = no.\ grams\ x\ \frac{1\ mole}{(form.\ wt.)\ g}$$

EXAMPLE 1

How many moles of sodium bicarbonate ($NaHCO_3$, form. wt. 84.0) are contained in a 2.00 lb (907.2 g) box of baking soda?

SOLUTION:

$$\frac{907.2\ g\ (NaHCO_3)}{1}\ x\ \frac{1\ mole\ (NaHCO_3)}{84.0\ g\ (NaHCO_3)} = \underline{10.8\ mole}$$

EXERCISE 1

In developing an improved procedure for the hydrogenation of vegetable oils to produce shortenings for kitchen use, a research chemist utilized 5.00 moles of the oil (formula weight 568) and 3.2 g of a platinum catalyst (atomic weight 195). How many grams of the oil were used? How many moles of platinum?

(answer, page 60)

3.2 Mole/Mass Conversions (Formula Weight from Formula)

If we need a formula weight and have only the formula of the chemical substance, it is a simple task to add up all the component atomic weights. Then the mole/mass conversion step is the same as that described in Section 3.1.

All we need to remember is to multiply a subscript by the atomic weight of the element and then sum the results (Section 1.2). For example, the formula weight of glucose ($C_6H_{12}O_6$) is found by:

C_6: 6 x 12.01 = 72.06
H_{12}: 12 x 1.008 = 12.096
O_6: 6 x 16.00 = 96.00
form. wt.: 180.2 (rounded to 4 digits)

EXAMPLE 2

Mercury fulminate, $Hg(CNO)_2$, is a very shock-sensitive explosive used in "blasting caps". How many grams of this compound would correspond to 0.500 mole?

SOLUTION:

First we must find the formula weight of mercury fulminate, using tabulated atomic weights (Appendix A):

$$Hg: \quad 1 \times 200.6 \quad = \quad 200.6$$
$$C_2: \quad 2 \times 12.01 \quad = \quad 24.02$$
$$N_2: \quad 2 \times 14.01 \quad = \quad 28.02$$
$$O_2: \quad 2 \times 16.00 \quad = \quad \underline{32.00}$$
$$\text{form. wt.:} \quad 284.6$$

Then we can use a mass/mole unity factor:

$$\frac{0.500 \text{ mole}}{1} \quad x \quad \frac{284.6 \text{ g}}{1 \text{ mole}} = \underline{142 \text{ g}}$$

==

EXERCISE 2

The preparation of mercury fulminate (EXAMPLE 2) is a dangerous procedure. Not only is the product highly explosive, but one must work with liquid mercury (hazardous vapors) and concentrated nitric acid (very corrosive). In a test procedure, 25 moles of nitric acid were used and 207 g of mercury fulminate were isolated. How many grams of HNO_3 were used? How many moles of $Hg(CNO)_2$ were isolated?

(answer, page 61)

==

Extra Practice

EXERCISE 3

The liquid propellant rockets used to launch the Gemini satellites utilized a mixture of hydrazines oxidized by dinitrogen tetroxide, N_2O_4. A test engine was designed to require 18 moles of hydrazine (N_2H_4, form. wt. 32.0). How many grams of hydrazine were required?

(answer, page 61)

EXERCISE 4

A hydrazine/N_2O_4 mixture ignites spontaneously. Such systems are called "hypergolic". How many moles of N_2O_4 would be required for a hypergolic propellant system using 150 g of N_2O_4?

(answer, page 61)

==

At this point you should try the competency level Self-Test questions on page 59.

*3.3 Mole/Mass Conversions Involving Mixtures

The utility of mole/mass unity factors (Section 3.1) can be applied to a wide variety of situations which require some conversion between mass and number of moles. Although each situation requires its own analysis, at some stage of the calculations a mole/mass unity factor will prove useful.

A general approach requires first an identification of what answer is required. Then we must survey any information given to find items needed and to determine additional necessary information not given (e.g., a formula weight). Having accumulated the data required, we must assemble an appropriate calculation "set-up". This is the stage at which the mole/mass unity factor will prove valuable. All that then remains is the arithmetic.

*EXAMPLE 3

A well-known brand of "iodized" salt contains 0.010% potassium iodide (by weight). How many moles of KI are found in a typical 26 oz (737 g) box of this salt?

SOLUTION:

To find: moles of KI

Information available: %KI, total mass

Information needed, not given: form. wt. of KI

(use atomic weights, Appendix A)

form. wt. = 39.1 + 126.9 = 166

Set-up (using unity factors):

$$\frac{737 \text{ g (salt)}}{1} \times \frac{0.010 \text{ g (KI)}}{100 \text{ g (salt)}} \times \frac{1 \text{ mole (KI)}}{166 \text{ g (KI)}}$$

Arithmetic:

$$\frac{737 \times 10^{-2}}{10^2 \times 166} = \underline{4.44 \times 10^{-4}} \text{ mole}$$

*EXERCISE 5

The venom of the common honeybee is an aqueous mixture of many different compounds. Among these is histamine (whose physiological activity is counteracted by "antihistamine" drugs), to the extent of about 0.013% by weight. An "average" bee sting injects into the victim about 35 mg of the aqueous mixture. How many moles of histamine, $(C_5H_7N_2)NH_2$, are

*Proficiency Level

injected in the "average" bee sting?

(answer, page 60)

===

Extra Practice

*EXERCISE 6

Analysis of the air over a congested metropolitan area by a pollution control center revealed a concentration of 2.4×10^{-6} mole of SO_2 per liter of air. Under the conditions of the analysis, each mole of SO_2 gas corresponds to 26 liters of SO_2. The maximum prolonged exposure level of SO_2 is set at 20 ppm (parts per million), by volume. Was the air being analyzed above allowed limits of SO_2?

(answer, page 61)

===

SELF-TEST (UNIT 3) [answers, page 61]

3.1. One of the first propellant gases to be used in aerosol cans was nitrous oxide, perhaps better known as "laughing gas". How many grams of nitrous oxide (N_2O, form. wt. 44) are contained in an aerosol can freshly charged with 5.6 moles of the gas?

3.2. Laughing gas can be prepared by the cautious heating of ammonium nitrate. Incautious heating can result in a violent explosion. If a laughing gas preparation is planned to use 190 g of NH_4NO_3, how many moles of the salt does this represent?

- -

If you completed Self-Test questions 3.1 and 3.2 correctly, you may go on to the proficiency level, try the Relevant Problems (Unit 3), or stop here. If not, you should consult your instructor for suggestions of further study aids.

- -

*3.3 The controlled decompostion of 120 g of ammonium nitrate yielded 54 g of water and a quantity of nitrous oxide ("laughing gas") equivalent in moles to half the number of moles of water formed. What mass of N_2O was produced? Was any other product formed, within the limits of the product weights reported?

- -

If you completed question 3.3 correctly, you may go on to the Relevant Problems for Unit 3. If not, you should consult your instructor for suggestions of further study aids.

- -

ANSWERS *to* EXERCISES, *Unit 3*

1. (2840 g of oil, 0.016 mole of platinum) <u>Solution</u>:

    $$\frac{5.00 \text{ moles (oil)}}{1} \quad x \quad \frac{568 \text{ g (oil)}}{1 \text{ mole (oil)}}$$

    $$\frac{3.2 \text{ g (Pt)}}{1} \qquad \frac{1 \text{ mole (Pt)}}{195 \text{ g (Pt)}}$$

2. (1580 g HNO_3, 0.727 mole $Hg(CNO)_2$) <u>Solution</u>:

    <u>formula weights</u> ----

            HNO_3 $Hg(CNO)_2$

    H: 1 x 1.008 = 1.008   (see EXAMPLE 2)

    N: 1 x 14.01 = 14.01   284.6

    O_3: 3 x 16.00 = <u>48.00</u>

                       63.0

    <u>conversions</u> ----

            HNO_3 $Hg(CNO)_2$

    $$\frac{25 \text{ moles}(HNO_3)}{1} \quad x \quad \frac{63.0 \text{ g}(HNO_3)}{1 \text{ mole}(HNO_3)} \quad \Big| \quad \frac{207 \text{ g}}{1} \quad x \quad \frac{1 \text{ mole}}{284.6 \text{ g}}$$

3. (576 g) <u>Solution</u>:

    $$\frac{18 \text{ moles}(N_2H_4)}{1} \quad x \quad \frac{32.0 \text{ g}(N_2H_4)}{1 \text{ mole}(N_2H_4)}$$

4. (1.63 moles) <u>Solution</u>:

        N_2: 2 x 14.01 = 28.02

        O_4: 4 x 16.00 = <u>64.00</u>

            form. wt.: 92.0

    $$\frac{150 \text{ g}(N_2O_4)}{1} \quad x \quad \frac{1 \text{ mole}(N_2O_4)}{92.0 \text{ g}(N_2O_4)}$$

---

*5. (4.1×10^{-8} mole) <u>Solution</u>:

    To find: moles of histamine (h)

    Information available: %(h), total mass

    Information needed, not given: form. wt.

              (from atomic wts., 111)

Set-up (using unity factors):

$$\frac{35 \times 10^{-3}\ g(venom)}{1} \times \frac{0.013\ g(h)}{100\ g(venom)} \times \frac{1\ mole(h)}{111\ g(h)}$$

*6. (yes) Solution:

To find: ppm SO_2 (to compare with 20 ppm limit)

Information available: definition of ppm, moles of SO_2 per liter air, volume per mole of SO_2

Information needed, not given: none

Set-up (using unity factors and definition of ppm):

$$\frac{2.4 \times 10^{-6}\ mole\ (SO_2)}{1\ liter\ (air)} \times \frac{26\ liters\ (SO_2)}{1\ mole\ (SO_2)} \times \frac{10^6\ liters}{1\ million\ liters}$$

(answer, 62 ppm exceeds 20 ppm limit)

- -

ANSWERS to SELF-TESTS, Unit 3

3.1. 246 g

3.2 2.38 moles

*3.3. 66 g (no other product, 66 + 54 = 120)

A.1. Most nitrogen fertilizers are ammonia based. In the Haber process, atmospheric nitrogen is catalytically combined with hydrogen to form ammonia. The ammonia can be applied directly or it can be employed as a raw material for the manufacture of urea, nitrates or other nitrogen compounds. If 2.27×10^5 g of ammonia (NH_3, form. wt. 17.03) is required for a test plot of wheat, to how many moles does this correspond?

A.2. The marine coelenterate _Palythoa mammillosa_, is a colonial invertebrate resembling small sea anemones. Widely distributed in shallow tropical waters, _Palythoa_ often overgrows and smothers other organisms. The _Palythoa_ species produce a powerful toxin, palytoxin, which is one of the most potent marine toxins known and the most powerful nonprotein toxin known in nature. Calculate the number of grams in 3.0 moles of palytoxin ($C_{145}H_{264}N_4O_{78}$).

A.3. The compound 2,4-dichlorophenoxyacetic acid (2,4-D) is a plant hormone with herbicidal properties. However, recent studies have indicated that the use of 2,4-D has caused a significant increase in the insect and pathogen pests on corn. The use of 2,4-D increases the protein levels in corn plants and this may have favored the growth of pests. In one test plot 2,4-D ($C_8H_6Cl_2O_3$) was applied at the rate of 0.55 kg per hectare. Calculate the number of moles of 2,4-D used on a 3.0 hectare test plot.

A.4. "Red tides", the sporadic and unpredictable appearance of red colored organisms over large stretches of ocean in the temperate zones, have been known and feared since antiquity. Such outbreaks result in massive fish kills which generally last a few weeks. The causative agent, saxitoxin, was isolated as a metabolite of an axenic culture of _Gonyaulax catenella_. In test animals the LD_{50} for saxitoxin, $C_{10}H_{15}N_7O_3 \cdot 2HCl$, was eight micrograms per kilogram of body weight. Calculate the number of moles in 8.0 micrograms of saxitoxin.

A.5. The best measurement of aeration status of a soil is the oxygen diffusion rate. This determines the rate at which oxygen can be replenished if it is used by respiring plant roots or replaced by water. Root growth stops when the oxygen diffusion rate drops to about 6.25×10^{-9} mole cm^{-2} min^{-1}. How many _grams_ of oxygen does this represent per square centimeter for a 1.0 minute interval?

*A.6. Each fertilizing material, whether it is a single compound or a complete ready-to-apply mixture, must carry a guarantee as to its content of nutrient elements. The

*Proficiency Level

simplest form of guarantee is a mere statement of the relative amounts of N, P, and K or N, P_2O_5, and K_2O. Thus, a 12-24-12 fertilizer contains 12 percent total nitrogen, 24 percent available P_2O_5, and 12 percent water soluble K_2O. How many moles of each ingredient would be found in a 110 kilograms of this fertilizer?

*A.7. Inorganic phosphate exists in the sea almost entirely as one of the orthophosphate ions. The distribution of the several forms of phosphorus in the sea is generally controlled by biological and physical agencies that are similar to those which influence the marine chemistry of nitrogen. One kilogram of "average" sea water contains about 7.0×10^{-5} grams of dissolved orthophosphate. At 20°C and a pH of 8.0, 87% of the phosphate occurs as HPO_4^{2-}, 12% as PO_4^{3-}, and 1.0% of $H_2PO_4^{-}$. How many moles of each form would be found in 5.0 kilograms of "average" sea water?

*A.8. The most abundant anion dissolved in sea water is the chloride ion (Cl^-). Its concentration is 19.344 g kg^{-1} of water of salinity 35‰. The volume of the world's oceans is about 1.5×10^{21} liters. How many moles of chlorine (Cl_2) are potentially available from the oceans? (Assume that the density of the "average" sea water is 1.0 g ml^{-1}.)

*A.9. Most of the dissolved inorganic species in sea water are electrolytes, although neutral species such as H_3BO_3 and dissolved gases are included in the category of inorganic species. Competitive complexation with the major anions (e.g., Cl^-, OH^-, SO_4^{2-}, HCO_3^-, PO_4^{3-}) is the main factor controlling the nature of the inorganic species present in sea water. One type of interaction between sea water cations and anions is purely electrostatic as found in ion-pair aggregates (e.g., $CaPO_4^-$). It is now believed that 99.6% of the PO_4^{3-} species (problem A.7) in sea water forms ion-pairs with Ca^{2+} and Mg^{2+} cations. How many moles of PO_4^{3-} remain as the free anion? (Assume that the volume of the world's oceans is 1.5×10^{21} liters; the density is 1.0 g ml^{-1}.)

*A.10. The production and use of phosphates for agriculture, industry, and the home have increased enormously over the past thirty years. The mined phosphate, primarily $Ca_3(PO_4)_2$, is "consumed" by conversion to non-recoverable compounds. Therefore, phosphate for fertilizer may become one of our long term resource problems. About 4.0×10^{11} moles of calcium phosphate are mined annually. Approximately 80.0% of this phosphate is converted to fertilizer. How many kilograms of $Ca_3(PO_4)_2$ are converted to fertilizer?

ANSWERS:

(A.1.) 1.33×10^4 moles, (A.2.) 9.9×10^3 g, (A.3.) 7.5 moles, (A.4) 2.3×10^{-8} mole, (A.5.) 2.0×10^{-7} g, (A.6.) 9.4×10^2 mole[N], 1.9×10^2 mole[P_2O_5], 1.4×10^2 mole[K_2O], (A.7.) 3.2×10^{-6} mole[HPO_4^{2-}], 4.4×10^{-7} mole[PO_4^{3-}], 3.7×10^{-8} mole[$H_2PO_4^{-}$], (A.8.) 4.1×10^{20} moles, (A.9.) 5.3×10^{11} moles, (A.10.) 9.9×10^{10} kg

RELEVANT
PROBLEMS

Unit 3: *The Mole Concept*

Set B: *Biological & Marine Sciences*

B.1. Aluminum hydroxide, $Al(OH)_3$, is used in pharmacy as a gel and sold under such names as Amphojel, Vanogel, and Creamalin. The commercial preparation Amphojel is used as a gastric antacid in treating peptic ulcer. An average dose of Amphojel contains 1.4 g of aluminum hydroxide (formula weight, 77.99). Calculate the number of moles of aluminum hydroxide in this dosage.

B.2. Some chemical substances emitted by animals may attract the opposite sex, determine odor trails to be followed, mark territory, identify friends, and raise alarm in case of danger. The remarkable potency of such chemical messengers, called phero-mones, was demonstrated when a male gypsy moth was called from a distance of a quarter of a mile by one nanogram (10^{-9} g) of the sex attractant methyl eugenol ($C_{11}H_{14}O_2$). How many moles of methyl eugenol are represented by 1.0×10^{-9} gram?

B.3. One of the most effective remedies developed in primitive cultures was cinchona, an antimalarial first used as a brew of the powdered bark of the Peruvian cinchona tree. The principal alkaloid in the bark was shown to be quinine, $C_{20}H_{24}N_2O_2$. Quinine is still widely used in the form of the dihydrochloride ($C_{20}H_{24}N_2O_2 \cdot 2HCl$) as an anti-pyretic and analgesic for cattle. If a veterinarian writes a prescription for twelve 10 g capsules of this salt, how many moles of the compound are required to fill the prescription?

B.4. Glucose is produced by plants but it is also a most essential animal body chemical. The oxidation of glucose supplies the energy for muscle and gland activity and body heat. Glucose ($C_6H_{12}O_6$), occurs naturally in fruits and other parts of plants. Calculate the number of grams in 9.0 moles of glucose.

B.5. Aspirin is used extensively in medicine as an antipyretic (substance that will re-duce or prevent fever) and analgesic (substance that will reduce or prevent pain). It is the most widely used synthetic drug because it is inexpensive, relatively safe, and easily available. An average oral dose of acetylsalicylic acid ($C_9H_8O_4$) is 0.65 g. Calculate the number of moles of acetylsalicylic acid in this dose.

*B.6. The venom of the honeybee is an aqueous mixture of several different compounds. Among these is the pheromone isopentyl acetate, to the extent of about 0.0080% by weight. This pheromone is discharged together with the venom when a bee stings an enemy, thus attracting other bees to sting near the same spot. An "average" bee sting injects into the victim about 35 mg of the venom. How many moles of isopentyl acetate, $C_7H_{14}O_2$, are released with the "average" bee sting?

*B.7. Hemoglobin is a complex protein molecule that has the ability to combine with atmospheric oxygen, forming oxyhemoglobin. This occurs in the capillaries of the lungs, and the oxygen is then transported in the arterial blood to the tissues, where it is released as needed. The percent of iron found in hemoglobin obtained from the red blood corpuscles of most animal species is 0.33%. How many moles of hemoglobin (molecular weight = 68,000) would contain 900 g of iron?

*B.8. When a muscle is stimulated, it requires energy to perform its work. Needed energy can be provided quickly by the hydrolysis of so-called "high energy phosphate" bonds, actually P-O-P anhydride links. These P-O-P links are found in the compound adenosine triphosphate (ATP). Average mammalian skeletal muscle, at rest, contains 375 mg of ATP per 100 g of tissue. How many moles of ATP ($C_{10}H_{16}P_3N_5O_{13}$) are available in a muscle section weighing 1.2 kg?

*B.9. Biological diseases are generally susceptible to specific chemical therapeusis. In recent years, a variety of specific antibacterial drugs have been developed. Dequalinium chloride (BAQD 10), a bacteriostat, has a solubility of 1.0 gram in 200 ml of water. Calculate the number of moles of dequalinium chloride, $C_{30}H_{40}Cl_2N_4$, that will dissolve in 1.0 liter of water.

*B.10. The antimetabolite 6-mercaptopurine (6 MP) has been used in the maintenance therapy of leukemia in children and is the active metabolite of the immunosuppressive drug, azathioprine. A local pharmacy has a stock bottle of 6 MP labeled as a 1.0% solution. The solution, originally had a volume of 500 ml. How many moles of 6-mercaptopurine ($C_5H_4N_4S$) were required to prepare the solution? (Assume that the density of the solution is 1.0 g ml^{-1}.)

---

*Proficiency Level

ANSWERS:

(B.1.) 1.8×10^{-2} mole, (B.2.) 5.6×10^{-12} mole, (B.3.) 0.30 mole, (B.4.) 1.6×10^{3} g,

(B.5.) 3.6×10^{-3} mole, (B.6.) 2.2×10^{-8} mole, (B.7.) 4.0 moles, (B.8.) 8.9×10^{-3} mole,

(B.9.) 9.5×10^{-3} mole, (B.10.) 3.3×10^{-2} mole

RELEVANT
PROBLEMS

Unit 3: The Mole Concept

Set I: Industrial Chemistry

I.1. Butadiene is used in making styrene-butadiene rubber, polybutadiene and neoprene. The end products that involve butadiene are tires and other rubber and plastic products. The U.S. butadiene production was about 1.59×10^{9} g in 1976. How many moles of butadiene (formula weight 54) were produced?

I.2. Only four elements are recovered from the sea in commercially significant amounts: chlorine, sodium, magnesium and bromine. The sodium concentration is about 0.70 moles per kilogram of seawater. How many grams of sodium (atomic weight 23.0) are in a kilogram of seawater?

I.3. The chrome tanning solution (problem I.10, Unit 2) contains 5 lbs (2270 g) of $Na_2Cr_2O_7 \cdot 2H_2O$ (formula weight 298.0) for 400 gallons (1514 ℓ) of solution. How many moles of $Na_2Cr_2O_7 \cdot 2H_2O$ are needed to make 400 gal of this solution?

I.4. It requires 25.0 g of potassium ferricyanide ($K_3Fe(CN)_6$) per liter to make a black-and-white bleach preparation. This solution is used in the development of high-quality color prints. How many moles of $K_3Fe(CN)_6$ are required for one liter of this solution?

I.5. "Alkylation" processes are exothermic and are fundamentally similar to polymerization except that only part of the hydrocarbon charging stock need be unsaturated. The product, "alkylate", is a mixture of saturated, stable, highly branched gasoline components with desirable high octane ratings. Continuing improvements in technology make HF-catalyzed "alkylation" more attractive to oil refiners for production of high octane gasolines. Simplified processing in newer units reduces investments and operating costs, thus the high quality anhydrous HF at lower cost provides added incentive for refiners to consider this route to higher quality gasolines. A new petroleum refinery produces 3000 barrels of high octane gasoline per day. The chemical engineer in charge of the operation has developed a process that utilizes 68.1 grams of catalytic HF per barrel of "alkylate". Determine the number of moles of hydrogen fluoride needed to supply the operation for seven days.

I.6. For the past 20 years the petrochemical industry has been able to utilize plentiful, low-cost natural gas and "natural gas liquids" as its primary feedstocks. Now this era is ending. The consumption of natural gas is exceeding the rate of discovery of new domestic reserves and the price of gas is rising. The price of "natural gas liquids" will rise accordingly. As a result of the decreasing supply and increasing price of "natural gas liquids", the future expansion of olefin manufacturing facilities will be based almost exclusively on heavy oil feedstocks. Facilities using heavy oil feedstocks accounted for only 12% of all the ethylene produced in 1970. This type of production is expected to exceed 50% by 1980. It is technically possible for a petrochemical complex using heavy liquids as feedstocks to operate independently from a refinery. It would include a simple crude oil distilling unit from which the feedstock liquids for the olefins plant, e.g., ethylene, would be produced. The additional heavy fractions produced plus "energy products" generated in the olefins plant would be sold. Economics, however, favor petrochemical complexes which are in close proximity to refineries producing a full range of petrochemicals. Perhaps the most important petrochemical is ethylene. The demand for ethylene, $H_2C=CH_2$, is increasing at an annual rate of 10 percent from the output of 7.26×10^{12} grams in 1970. Determine the number of moles of ethylene produced from heavy oil feedstocks in 1970.

*I.7. Dimethyl sulfoxide, DMSO, is used as an industrial solvent because of its solvating properties and its ability to disperse charged solutes. This versatile compound is prepared by the air-oxidation of dimethyl sulfide in the presence of nitrogen oxides, according to the reaction:

$$2 \; CH_3-S-CH_3 + O_2 \xrightarrow{NO_2 \; (catalyst)} 2 \; CH_3-\underset{O}{\overset{\|}{S}}-CH_3$$

The compatibility of DMSO with acidic and alkaline solutions permits the formulation of strippers (paint and adhesives) utilizing the cleaning action of both materials, thus increasing the rate of stripping. This application of DMSO is becoming increasingly important in the cleaning and reclaiming of coated metal parts. A novel combination of 9 parts DMSO and 1 part concentrated nitric acid (by volume) at 125°C will rapidly remove even the toughest epoxy coatings and adhesives from surfaces such as aluminum, magnesium and stainless steel. Determine the number of moles of DMSO required to prepare 25 liters of this stripper. (The density of DMSO is 1.1 g/ml).

*Proficiency Level

*I.8. Industrially, copper chromites are supplied in a high oxidation state and, as such, they are hydrogenation catalysts. Upon reduction, copper chromites become dehydrogenation catalysts. Barium is frequently incorporated in some copper chromite complexes as a stabilizing agent. However, unstabilized copper chromite catalysts are often used successfully for special hydrogenations and are generally preferred for dehydrogenations. The industrial copper chromites are manufactured to several different formulations, including both powdered and tableted catalysts for batch and fixed bed processes. A special copper chromite catalyst has an apparent bulk density of 73 lbs ft^{-3}, a surface area of 30 m^2 g^{-1} and a pore volume of 0.69 cm^3 g^{-1}. Determine the number of moles of chromium in 2.3 kg of this catalyst, which was formulated as 40% CuO and 60% Cr_2O_3.

*I.9. Nitroparaffins are used as intermediates to other products and as solvents. The industrial preparation of nitroparaffins is a continuous process using propane and nitric acid as raw materials. The typical distribution of products is 10% (by weight) nitromethane (CH_3NO_2), 25% nitroethane ($CH_3CH_2NO_2$), 25% 1-nitropropane ($CH_3CH_2CH_2NO_2$), and 40% 2-nitropropane ($CH_3CH(NO_2)CH_3$). if 2000 moles of 2-nitropropane are made, what is the total number of grams of nitroparaffins produced?

*I.10. Baking powders as leavening agents are generally used in place of baking soda alone (sodium bicarbonate) because of the unpleasant taste of the Na_2CO_3 that is formed from thermal decomposition of sodium bicarbonate. An acid such as anhydrous monocalcium phosphate is added to baking powder to completely decompose the sodium bicarbonate. Baking powder contains about 28% sodium bicarbonate ($NaHCO_3$), enough acid and 20%-40% corn starch or flour filler. The filler is added to improve distribution in the dough and to prevent advance reaction between the acid and the bicarbonate. How many moles of $NaHCO_3$ are in a 200 g package of baking powder?

ANSWERS:

(I.1.) 2.9×10^7 moles, (I.2.) 16 g, (I.3.) 7.62 moles, (I.4.) 7.60×10^{-2} mole, (I.5.) 7.15×10^4 moles, (I.6.) 3.11×10^{10} moles, (I.7.) ~320 moles, (I.8.) 18 moles, (I.9.) 4.5×10^5 g, (I.10.) 0.67 mole

---

E.1. Lead poisoning can lead to convulsions, delirium, coma, severe and irreversible brain damage, blindness, paralysis, mental retardation, and death. Unlike many other pollutants, lead is a cumulative poison. Most lead entering the human body does so through contaminated food. This lead, which enters the stomach, is rather inefficiently absorbed by the body, and only about 5 to 10% of the lead actually enters the blood stream. The total daily intake of lead in the food and drink of an individual American is typically 3.0×10^{-4} g. How many moles of lead (atomic weight 207) are in the daily intake of one American?

E.2. The compound 1,1,1-trichloro-2,2-bis(p-chlorophenyl)ethane is commonly known as DDT. It is a good insecticide but is also toxic to human and animal life. Nature can handle many potential pollutants if they are present in small quantities, but DDT is persistent in the environment. The sediments from the river near one plant were about 1% DDT. If the formula weight of DDT is 355, how many moles correspond to the 2.0×10^{6} g of DDT that has been dredged from this river so far?

E.3. A change in the temperature of a stream can completely upset the ecological balance of that stream. Many fish, for instance, can live only in a range of a few degrees of the optimum temperature. Environmentalists are often concerned with the amount of heat delivered to a stream by an industrial concern, and with the amount of water used by the facility. One industrial process to produce methylamine uses 927 moles of water (formula weight 18.0) per gram of product. How many grams of water are used for each gram of methylamine produced?

E.4. Tetramethyl lead [$(CH_3)_4Pb$] has been used as a gasoline antiknock agent. Some 70% to 80% of the lead from this compound is exhausted into the atmosphere as small particles of poisonous lead compounds. How many moles of tetramethyl lead correspond to 8.17×10^{9} grams produced in a typical year.

E.5. In 1937, 2-naphthylamine ($C_{10}H_7NH_2$) was shown to be carcinogenic. It is an optical bleaching agent and is used as a dye intermediate. Doses of 0.5 g per day caused bladder cancer in dogs. How many moles were in a dose?

E.6. Air pollution by man is seemingly a minor problem on a global scale. However, the balance of nature is a delicate one and the increasing addition of atmospheric contaminants, when coupled with other ecological damage, could prove disastrous. At best, we are faced with very serious localized problems in metropolitan and industrial areas where the addition rate of air contaminants exceeds the combined rates of their dissipation and removal by natural systems. Sulfur dioxide emissions, from the burning of sulfur-containing fuels and the smelting of sulfide ores, were estimated at 3.2×10^{13} g in the United States in 1972. How many <u>moles</u> of SO_2 does this represent?

*E.7. Few of the substances we think of as air pollutants are original to man. Natural processes supply oxides of sulfur (e.g., from volcanoes), particulate matter (e.g., from dust storms), hydrocarbons (e.g., from decomposition of organic matter), and other undesirable species in truly enormous quantities. Nature, however, has ways of removing and using natural contaminants. Oxides of sulfur and nitrogen, for example, washed from the air by rain, supply necessary inorganic chemicals to the soil. Sulfur dioxide in the atmosphere slowly combines with oxygen to form SO_3, a process catalyzed by sunlight or dust particles. The SO_3 combines with moisture to form sulfuric acid. A concentration exceeding 0.40 ppm of SO_3 in air is injurious to most sensitive plants, causing significant decrease in growth rates. How many moles of SO_3 are in 1.00 km^3 of air containing 0.40 ppm SO_3 (by weight), if the air density is 1.34 g ml^{-1}?

*E.8. Many organic, and some inorganic, wastes are degraded in natural waters by oxidation processes. In some cases, these involve direct reaction of chemical wastes with molecular oxygen dissolved in the water. More commonly, aerobic bacteria in the water degrade the chemicals while utilizing dissolved oxygen. In either case, oxygen is consumed and, if waste degradation demands abnormally high amounts of O_2, fish and other aquatic life cannot survive in the oxygen-depleted waters. The biological oxygen demand (BOD) is defined as:

$$BOD = \frac{\Delta (no. \ mg \ O_2)}{no. \ liters \ water} \qquad (in \ ppm \ O_2)$$

and is determined by mixing a measured sample of polluted water with a known volume of oxygen-saturated pure water. The total O_2 concentration in the mixture is measured just after mixing and again after five days incubation at 20°C. It is the <u>change</u> in O_2 concentration (Δ) over this period that is used to calculate the BOD. A sewage sample was found to have a BOD of 165 ppm. How many <u>moles</u> of oxygen are required over a 5 day period by a sewage volume of 25,000 liters?

*E.9. If 28% of the required oxygen is converted to CO_2 during the waste degradation (problem E.8), how many moles of CO_2 are formed over a 5 day period? How many grams?

*E.10. Many industrial plants have developed methods to clean up the discharged waste from their process. One plant uses a <u>S</u>pent <u>A</u>cid <u>R</u>ecovery unit. The original waste water was 40% by weight H_2SO_4. After the SAR unit, the sulfuric acid was only 1.0%. How-

*Proficiency Level

ever, this sulfate concentration was still high enough to inhibit the normal bio-logical treatment process. If the waste water was discharged at 2.75×10^7 g per hour, how many moles of sulfuric acid were discharged in the original waste water in a week? How many moles of sulfuric acid were in the treated water in a week, if only sulfuric acid was removed by the SAR unit?

ANSWERS:

(E.1.) 1.4×10^{-6} mole, (E.2.) 5.6×10^3 moles, (E.3.) 1.67×10^4 g, (E.4.) 3.06×10^7 moles, (E.5.) 3×10^{-3} mole, (E.6.) 5.0×10^{11} moles, (E.7.) 6.7×10^6 moles, (E.8.) 1.3×10^2 moles, (E.9.) 3.6×10^1 moles, 1.6×10^3 g, (E.10.) 1.9×10^7 moles, 2.9×10^5 moles

---

STOICHIOMETRY

UNIT 4: MASS/MASS PROBLEMS

All chemical reactions are governed by the law of mass conservation, i.e., there is no detectable loss of mass during a chemical change. This means that the masses of reaction products formed are quantitatively related to those of reactants consumed. The mole concept (Unit 3) permits us to establish this relationship, provided that we have some way of determining the mole ratios involved.

A balanced chemical equation is one way of showing the mole relationships among reaction products and reactants. Consider for example the decomposition of ammonium nitrate to nitrous oxide and water, as expressed by:

$$NH_4NO_3 \rightarrow N_2O + 2H_2O$$

This equation tells us a number of things about the reaction. All of the nitrogen atoms of an ammonium nitrate unit are converted to N_2O, along with one-third of the NH_4NO_3 oxygen. All of the hydrogen and two-thirds of the oxygen of an ammonium nitrate unit are converted to water. In terms of numbers of species:

one(1) NH_4NO_3 unit forms one(1) N_2O and two(2) H_2O molecules

one dozen(12) NH_4NO_3 units would form one dozen(12) N_2O and two dozen(24) H_2O molecules

one thousand(10^3) NH_4NO_3 units would form one thousand(10^3) N_2O and two thousand (2×10^3) H_2O molecules

one MOLE (6×10^{23}) NH_4NO_3 units would form one MOLE (6×10^{23}) N_2O and two MOLES ($2 \times 6 \times 10^{23}$) H_2O molecules

Remember that counting very large numbers of things is generally less convenient than a weighing operation, or some other method of determining total quantity. From our definition of the MOLE (Unit 3) and calculation of formula weights, we can write:

$$1 \text{ mole } (NH_4NO_3) = 80 \text{ g } (NH_4NO_3)$$

$$1 \text{ mole } (N_2O) = 44 \text{ g } (N_2O)$$

$$1 \text{ mole } (H_2O) = 18 \text{ g } (H_2O)$$

Then, if 80 g of NH_4NO_3 were decomposed according to the chemical equation given, we would expect to form 44 g of N_2O and 36 g (2 moles) of H_2O. Since the mole ratios are always the same for this reaction, we could easily calculate the mass of any one species from the measured mass of any other, using the mole/mass relationships involved.

Mass/mass stoichiometry, the calculation of mass changes on the basis of established mole ratios, is a very useful procedure. It enables the chemist to calculate the quantities of reactants needed to make a given amount of a new compound. It provides the molecular biologist with a quantitative way of looking at the chemical processes of living organisms. It allows the engineer to scale up a pilot process for a profitable industrial operation.

Chemical reactions are not always simple and mass/mass calculations based on a single balanced equation sometimes represent idealized situations. In such cases, mass/mass stoichiometry still provides useful information, the <u>theoretical</u> quantities of chemical substances involved. When coupled with experimentally-determined data, these theoretical calculations reveal the "efficiency" of chemical processes.

- References -

1. O'Connor, Rod. 1977. <u>Fundamentals</u> <u>of</u> <u>Chemistry</u>, second edition. New York: Harper & Row (Unit 4)
2. Brown, Theodore, and H. Eugene LeMay, Jr. 1977. <u>Chemistry</u>: <u>The</u> <u>Central</u> <u>Science</u>. Englewood Cliffs, N.J.: Prentice-Hall (Chapter 3)
3. Masterton, William, and Emil Slowinski. 1977. <u>Chemical</u> <u>Principles</u>, fourth edition. Philadelphia: W.B. Saunders (Chapter 3)
4. Nebergall, William, F.C. Schmidt, and H.F. Holtzclaw, Jr. 1976. <u>College</u> <u>Chemistry</u>, fifth edition. Lexington, MA: D.C. Heath (Chapter 2)
5. Nash, Leonard K. 1966. <u>Stoichiometry</u>. Reading, MA: Addison-Wesley

---

OBJECTIVES:

(1) *Given the balanced equation for a chemical process and the mass of any one chemical substance involved, be able to calculate the theoretical mass of any other component of the reaction system.*

(2) *Given the balanced chemical equation, the mass of the limiting reagent, and the actual yield of a product, be able to calculate the percentage yield (efficiency) of the conversion.*

*(3) *Given a balanced chemical equation, the percentage yield (efficiency) for a particular process, and the mass of a product desired, be able to calculate the minimum mass required for a specified reactant.*

*(4) *Given information from which appropriate mole ratios and mass data can be determined, be able to calculate the mass of a component of the reaction system.*

PRE-TEST:

Necessary Atomic Weights:		
H (1.008)	O (16.00)	S (32.06)
C (12.01)	Na (22.99)	Ca (40.08)
N (14.01)	P (30.97)	Cu (63.55)

*Proficiency Level

73

(1) "Baking powder" consists of a mixture of "baking soda" (sodium bicarbonate, $NaHCO_3$) and some solid acid so that the moistened mixture will react to liberate CO_2 gas. When sodium dihydrogen phosphate is used in the mixture, the reaction of the moistened powder may be formulated as:

$$NaHCO_{3(aq)} + NaH_2PO_{4(aq)} \rightarrow Na_2HPO_{4(aq)} + H_2O_{(\ell)} + CO_{2(g)}$$

What mass of sodium dihydrogen phosphate must be used to consume 168 g of "baking soda", assuming 100% efficiency? _____

(2) Calcium cyanide powder is sometimes used by beekeepers to destroy a colony of diseased bees. This powder must be stored in a sealed container, since moist air decomposes it, according to the equation:

$$Ca(CN)_{2(s)} + 2H_2O_{(\ell)} \rightarrow Ca(OH)_{2(s)} + 2HCN_{(g)}$$

A new can of calcium cyanide was accidentally left open in a storage room where the relative humidity was high. When it was discovered, the can was sent to a laboratory for analysis. A 5.00 g sample of the homogenized solid was found to contain 0.83 g of calcium hydroxide. What percentage of the original calcium cyanide had been decomposed by moisture? _____

*(3) Methyl salicylate ("oil of wintergreen") is sold in dilute solution as a flavoring agent and in a grease base as "external aspirin". The compound is prepared from methanol and salicylic acid:

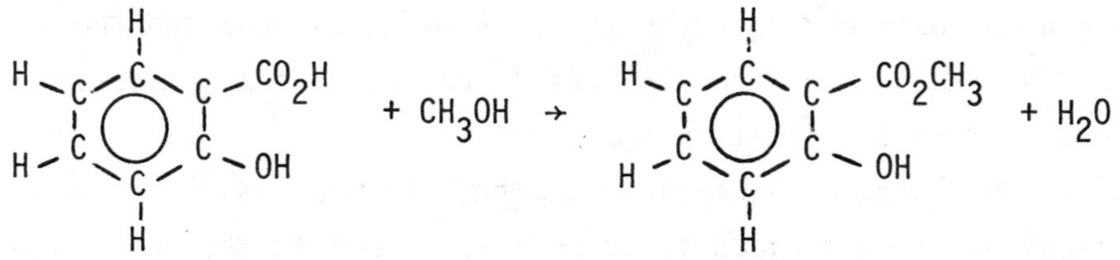

Under typical reaction conditions, the synthesis is 72% efficient when equimolar amounts of reactants are used. What is the minimum mass of salicylic acid $(C_7H_6O_3)$ required for production of 8.0 oz (227 g) of methyl salicylate? _____

*(4) A low-grade copper ore contains 8.62% by weight Cu_2S. How many metric tons of copper could be obtained from 250 metric tons of the ore by a conversion process which is 78% efficient in production of metallic copper? _____

Answers and Directions:

(1) 240 g, (2) 21%

*If both are correct, go on to questions *3 and *4. If you missed any, study*

METHODS, sections 4.1 and 4.2.

*(3) 286 g, *(4) 13 metric tons

If both are correct, go on to the RELEVANT PROBLEMS, Unit 4. If you missed any, study METHODS, sections 4.3 and 4.4.

METHODS

4.1 Mass/Mass Problems (Assuming 100% Efficiency)

There are several ways of approaching mass/mass stoichiometry problems, but one of the easiest uses a simple proportion method derived from the balanced chemical equation and appropriate formula weights. We may think of this as a series of steps:

1. Identify the two chemicals in the equation for which the mass/mass relation-ships are needed.
2. Below each of their formulas, write the corresponding formula weights multiplied by the coefficients of the respective formulas, as shown in the balanced equation.
3. Above each of their formulas, write the known mass and a symbol for the unknown mass (e.g., w), respectively.
4. Equate the two ratios established by "above" and "below" quantities and solve for the unknown term.

For example, suppose we wished to know the mass of sodium hydroxide required for neutraliza-tion of 116 g of sulfuric acid, according to the equation:

$$2NaOH + H_2SO_4 \rightarrow Na_2SO_4 + 2H_2O$$

Using the "steps" described, we would proceed as follows:

Step 1 - $\boxed{2NaOH}$ + $\boxed{H_2SO_4}$ → Na_2SO_4 + $2H_2O$

Step 2 - $\boxed{2NaOH}$ $\boxed{H_2SO_4}$

(2 moles x 40.0 g mole^{-1}) (1 mole x 98.1 g mole^{-1})

Step 3 -   w   116 g

   2NaOH H_2SO_4

   2 x 40.0 g   98.1 g

Step 4 - $\dfrac{w}{2 \times 40.0 \text{ g}} = \dfrac{116 \text{ g}}{98.1 \text{ g}}$

from which, $w = \dfrac{2 \times 40.0 \text{ g} \times 116 \text{ g}}{98.1 \text{ g}} = \underline{94.6 \text{ g}}$

EXAMPLE 1

Chlorine for use in water purification systems may be obtained from industries using the electrolytic decompostion of sea water, for which the chemical change may be represented by:

$$2NaCl_{(aq)} + 2H_2O_{(\ell)} \rightarrow 2NaOH_{(aq)} + H_2(g) + Cl_2(g)$$

What mass of sodium chloride would be consumed for the production of 25 metric tons of chlorine, assuming 100% efficiency.

SOLUTION:

$$\boxed{2NaCl} \quad + \quad \dots \quad \rightarrow \quad \dots \quad + \quad \boxed{Cl_2}$$

(2 moles x 58.4 g mole^{-1})   (1 mole x 70.9 g mole^{-1})

$$\frac{w}{2 \times 58.4\ g} = \frac{25\ metric\ tons}{70.9\ g}$$

$$w = \frac{25\ metric\ tons \times 2 \times 58.4\ g}{70.9\ g} = \underline{41\ metric\ tons}$$

EXERCISE 1

Both iron and chromium for use in making chrome steel can be obtained by the reduction of chromite ore by coke:

$$FeCr_2O_4 + 4C \rightarrow Fe + 2Cr + 4CO$$

How many metric tons of coke (C) must be used to produce 37 metric tons of chromium by this process, assuming 100% efficiency?

(answer, page 83)

4.2 Calculation of Percentage Yield (Efficiency)

A balanced chemical equation is something of an idealized formulation for a real chemical process. In the sense that the equation represents an exact relationship among the species actually involved in the chemical reaction formulated and a conformity with the law of mass conservation, the equation is completely valid. However, many reactions do not proceed to completion, that is, net change may cease while appreciable amounts of original reactants remain. This case will be discussed in a study of chemical equilibrium, beginning with Unit 10. In other cases, some or all of the original reactants may interact in various ways so that a single equation represents only one of a number of competing reactions. Of course, there is a simpler reason why real processes may give us less than 100% of the expected product. The stages in handling, transferring, isolating, or purifying desired products may involve losses which may reflect imperfect techniques or intrinsic characteristics of the process. The isolation of an "insoluble" product by filtration, for example, always involves some loss since the product is unlikely to be "completely insoluble" in the solvent being employed.

All of these circumstances suggest that the calculation of a product amount expected on the basis of mass/mass stoichiometry will give a number larger than that actually found in a real process. Thus, simple stoichiometry allows us to calculate only the theoretical yield of a product. The actual yield for any particular process must be determined experimentally.

A comparison between actual and theoretical yields provides a measure of the efficiency of a chemical process. Such a comparison is normally expressed as a "percentage yield":

$$\% \ yield = \frac{actual \ yield}{theoretical \ yield} \times 100\%$$

The yield of a chemical process can often be improved by finding more favorable reaction conditions. One method often employed, when economically feasible, is to add an excess of one or more reactants, thus improving the conversion of some other reactant, typically a more expensive chemical, to the desired product. Of course, the formation of a product stops when all of any necessary reactant is consumed. When excess of some reactant is used, the necessary chemical whose complete consumption halts product formation is called the limiting reagent. It is this reactant which must be used in calculating the theoretical yield.

EXAMPLE 2

Nitroglycerin, more properly called glyceryl trinitrate, is a powerful explosive, unstable to heat or shock. It is prepared under carefully controlled conditions from glycerol, nitric acid, and catalytic amounts of sulfuric acid. The net process may be represented by:

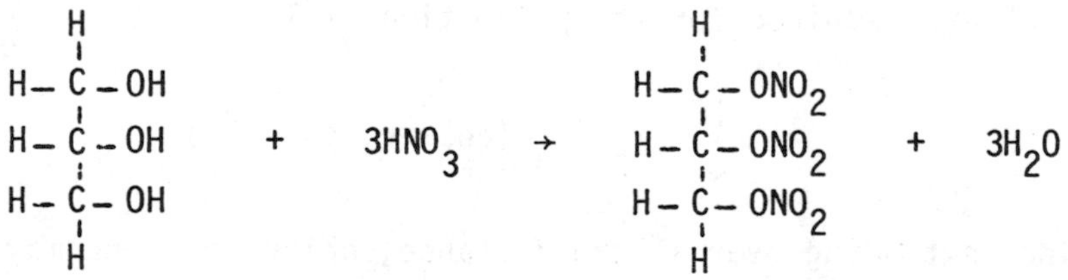

In a test preparation, 25 g of glycerol ($C_3H_8O_3$) was treated with excess nitric acid and 53 g of glyceryl trinitrate ($C_3H_5N_3O_9$) was isolated. What was the percentage yield?

SOLUTION:

$$\boxed{\underset{\text{(1 mole x 92 g mole}^{-1})}{\overset{25 \text{ g}}{C_3H_8O_3}}} + \dots \rightarrow \boxed{\underset{\text{(1 mole x 227 g mole}^{-1})}{\overset{t}{C_3H_5N_3O_9}}}$$

$$\frac{25 \text{ g}}{92 \text{ g}} = \frac{t}{227 \text{ g}}$$

$$t = \frac{227 \text{ g x } 25 \text{ g}}{92 \text{ g}} = 62 \text{ g}$$

(theoretical yield)

$$\% \text{ yield} = \frac{53 \text{ g}}{62 \text{ g}} \times 100\% = \underline{85\%}$$

= =

EXERCISE 2

Ethyl formate, an artificial flavoring agent having the characteristic odor of rum, may be prepared by the reaction of formic acid with ethanol:

$$HCO_2H + C_2H_5OH \rightarrow HCO_2C_2H_5 + H_2O$$

If a mixture of 75 g of formic acid with a slight excess of ethanol resulted in the formation of 81 g of ethyl formate, what was the percentage yield for the process?

(answer, page 84)

= =

Extra Practice

EXERCISE 3

"Stannous fluoride" [tin(II) fluoride] is a common toothpaste additive as a source of fluoride ion for retarding tooth decay. It may be manufactured from tin and anhydrous hydrogen fluoride:

$$Sn + 2HF \rightarrow SnF_2 + H_2$$

How many grams of HF are required for the production of 1.0 oz (28.4 g) of "stannous fluoride"?

(answer, page 84)

EXERCISE 4

For water lines extending over a long distance, chlorine alone may not provide adequate protection against bacteria because of a slow loss of chlorine from the water. In such cases, the municipal water plant may also add a small amount of ammonia, some of which reacts with chlorine to form trichloramine, an antibacterial agent having a long term

78

stability in water. In a test run, 10 mg of chlorine and a slight excess of ammonia were mixed in 1.00 liter of water. The trichloramine, formed by:

$$3Cl_2 + NH_3 \rightarrow NCl_3 + 3HCl,$$

was measured to be 2.2 mg. What was the percentage yield?

==

At this point you should try the _competency_ level Self-Test questions on page 82.

==

*4.3 Reactant-Requirement Calculation

Since most real chemical processes are less than 100% efficient, it is necessary either to settle for less than the theoretical maximum amount of product or to increase _reactant_ amounts as required. If it is desired to prepare a specific quantity of some chemical, then the theoretical quantities of reactants, as calculated from stoichiometric considerations, must be increased to compensate for the experimentally-determined in-efficiency.

Consider for example the synthesis of ethyl formate, a compound having the odor of rum, from formic acid and ethanol under conditions found to give a product yield of 67%. If a fixed quantity of ethyl formate, let's say 50 kg, is to be manufactured, then stoi-chiometry suggests that 50 kg at 67% yield would correspond to (50 x 100/67) kg at 100% efficiency. Since direct stoichiometric calculation assumes 100% efficiency, we would then have to calculate the quantities of formic acid and ethanol needed to make (50 x 100/67) kg of the product.

*EXAMPLE 3

Older type wooden matches used a phosphorus sulfide as the "fuel component" of the match head. The sulfide is manufactured by heating a mixture of sulfur and red phosphorus:

$$4 P + 3 S \rightarrow P_4S_3$$

In a typical industrial operation, using a stoichiometric ratio of reactants, the product was isolated in 82% yield. What mass of phosphorus would be required for the production of 18 metric tons of P_4S_3 by this process?

SOLUTION:

We may calculate either the theoretical mass of phosphorus required to prepare 18 metric tons of P_4S_3 and "correct" this to "real" conditions by the ratio 100/82, or the theoretical mass of P_4S_3 (18 metric tons x 100/82). We'll illustrate the latter

*Proficiency Level

approach:

$$\text{theor. yld.} \quad P_4S_3 = \frac{18 \text{ metric tons}}{1} \times \frac{100}{82} = 22 \text{ metric tons}$$

Then mole ratios from the balanced equation and mole/mass unity factors can be used to complete the calculation:

$$\frac{22 \text{ metric tons}(P_4S_3)}{1} \times \frac{1 \text{ metric ton-mole}(P_4S_3)}{220 \text{ metric tons}(P_4S_3)} \times \frac{4 \text{ metric ton-moles}(P)}{1 \text{ metric ton-mole}(P_4S_3)} \times$$

$$\frac{31 \text{ metric tons}(P)}{1 \text{ metric ton-mole}(P)} = \underline{12 \text{ metric tons}}$$

[metric tons(P_4S_3) → metric ton-moles(P_4S_3) → metric ton-moles(P) → metric tons(P)]

==

*EXERCISE 5

"Chloral hydrate", $Cl_3CCH(OH)_2$, is a powerful sleep-producing drug. When mixed with whiskey, this compound forms the infamous "Mickey Finn" cocktail once used in waterfront bars in San Francisco to "Shanghai" sailors for unexpected long voyages. The product is formed by the reaction of water with trichloroethanal:

```
    Cl H                          Cl H
    \ |                           \ |
 Cl-C-C=O  +  H2O            Cl-C-C-OH
    /                             / |
    Cl                           Cl OH
```

What mass of trichloroethanal is needed by a process which is 73% efficient to produce 8.0 oz (227 g) of "chloral hydrate"?

(answer, page 84)

==

*4.4 Mass Calculations From Process Descriptions

Mass/mass stoichiometry does not necessarily require a balanced chemical equation. Any information which provides the necessary mole ratio and mass data can be used to make the appropriate calculations. Chemical names, for example, may tell us the necessary formulas from which mole ratios might be found. If we were considering a process such as the conversion of "copper(I) sulfide to copper", we could write Cu_2S → $2Cu$ to show that the ratio 2 moles(Cu)/1 mole(Cu_2S) could be used.

Each situation may be unique in the way information is presented, but a general problem-solving approach offers a systematic way of analyzing the problem.

*EXAMPLE 4

Experimental projects for water purification by reverse osmosis, the forcing of pure water through a semipermeable membrane from some aqueous solution, require membranes able to withstand high pressures. Natural membranes are not strong enough, but a synthetic membrane of great strength can be made by precipitating gelatinous $Cu_2[Fe(CN)_6]$ on un- glazed porcelain. How much copper(II) sulfate would be required for preparation of 38 kg of the gelatinous salt by a process which is 88% efficient?

SOLUTION:

To find: mass of copper(II) sulfate needed

Information available: mass of product desired, % efficiency

Information needed, not given: formula of copper(II) sulfate (from memory, $CuSO_4$), mole ratio, formula weights

Set-up (using unity factors): (C-F used to represent $Cu_2[Fe(CN)_6]$)

$$\frac{38\ kg(C\text{-}F)}{1} \times \frac{100}{88} \times \frac{1\ kg\text{-}mole(C\text{-}F)}{339\ kg(C\text{-}F)} \times \frac{2\ kg\text{-}mole(CuSO_4)}{1\ kg\text{-}mole(C\text{-}F)} \times \frac{160\ kg(CuSO_4)}{1\ kg\text{-}mole(CuSO_4)}$$

Arithmetic:

$$\frac{38 \times 100 \times 2 \times 160}{88 \times 339} = 41\ kg(CuSO_4)$$

===

*EXERCISE 6

Ethylene glycol, $HOCH_2CH_2OH$, for use as an automobile "antifreeze", can be prepared by bubbling ethylene gas ($H_2C{=}CH_2$) through a solution of potassium permanganate in aqueous potassium hydroxide. Each mole of ethylene consumed uses 2/3 mole of permanganate. Both ethylene and KOH are used in excess since they may be recovered and a typical process gives a glycol production of 93%, based on permanganate consumption. How many kilograms of potassium permanganate are required for the preparation of 100 kg of ethylene glycol?

(answer, page 85)

===

Extra Practice

*EXERCISE 7

Silicon carbide, commonly known as "carborundum", is one of the hardest substances known, finding extensive use in the manufacture of abrasives. The compound is made by heating a mixture of sand and coke in an electric furnace, in simplified formulation:

$$SiO_2 + 3C \rightarrow SiC + 2CO$$

When excess sand is used, 87% of the coke is typically converted to silicon carbide. How many metric tons of coke must be used to produce 250 metric tons of "carborundum"?

(answer, page 85)

EXERCISE 8

Most silicates can be dissolved by hydrofluoric acid, hence the utility of this re-agent for the etching of glass. With aqueous hydrofluoric acid, a mixture of soluble products is formed, but the solution process always requires a minimum of four moles of HF per mole of silicon. What is the minimum mass, in grams, of HF required to remove 15% of the silicon by etching from 150 g of a Pyrex glass, which analysis shows to contain 81% SiO_2 by weight?

(answer, page 85)

= =

SELF-TEST (UNIT 4) [answers, page 85]

4.1. An alloy, called Misch metal, of some of the rare earths found extensive military use in the preparation of "tracer bullets" as early as World War I. Such bullets form a brilliant light pattern as the alloy begins to burn, allowing rapid-fire guns to correct their firing trajectories in night actions. Misch metal is 50% cerium by weight and the combustion of heated cerium may be represented by:

$$4\ Ce + 3\ O_2 \rightarrow 2\ Ce_2O_3$$

What mass of the oxide would have been formed during a battle in which enough "tracer bullets" were fired to have burned 38 kg of cerium?

4.2. Until recently, sodium nitrite was used extensively in the meat packing industry as an antioxidant additive to preserve the red color associated with "fresh" meat. Its use has been largely abandoned as the result of evidence that nitrite may react with certain amines in the stomach to produce carcinogenic (cancer-causing) chemicals called nitrosoamines. In a laboratory test situation, the reaction of excess diethyl amine with 20 g of sodium nitrite, in acidic solution, formed 27 g of N-nitrosodiethyl amine, according to the equation:

$$(C_2H_5)_2NH + NaNO_2 + HCl \rightarrow (C_2H_5)_2NNO + NaCl + H_2O$$

What percentage of the sodium nitrite was converted to the nitrosoamine?

- -

If you completed Self-Test questions 4.1 and 4.2 correctly, you may go on to the proficiency level, try the RELEVANT PROBLEMS (Unit 4), or stop here. If not, you should consult your instructor for suggestions of further study aids.

---

*4.3. Carbon tetrachloride has been used for many years as a non-flammable "cleaning fluid", although it is rapidly being replaced by other substances whose vapors are less hazardous to inhale. Carbon tetrachloride, along with chloroform and other products, is formed by the gas-phase reaction of methane with chlorine. We can write a single equation for its preparation as:

$$CH_4 + 4Cl_2 \rightarrow CCl_4 + 4HCl$$

In a typical plant run, using excess chlorine, 72% of the original methane was converted to CCl_4. How much methane would be needed to produce 180 metric tons of CCl_4 by this process?

*4.4. Silver utensils used for stirring "scrambled eggs" tarnish rapidly by the formation of dark Ag_2S from reaction of the silver with sulfur-containing amino acids in the egg proteins. The tarnish can be removed by various "silver cleaning" agents, but repetitive tarnishing and cleaning of silverplated utensils can remove the relatively thin silver "plate". Metallic silver is, of course, inert to proper "cleaning agents". What percentage of the silver in a 95 g spoon would be lost by the removal of 14 g of Ag_2S tarnish over a period of extensive use and cleaning?

If you completed Self-Test questions 4.3 and 4.4 correctly, you may go on to the RELEVANT PROBLEMS for Unit 4. If not, you should consult your instructor for suggestions of further study aids.

ANSWERS to EXERCISES, Unit 4

1. (17 metric tons) Solution:

$$.... + \boxed{4C}^{\,w} \quad .... + \boxed{2Cr}^{\,\text{37 metric tons}} + ...$$
$$(4 \text{ moles} \times 12.0 \text{ g mole}^{-1}) \quad (2 \text{ moles} \times 52.0 \text{ g mole}^{-1})$$

$$\frac{w}{4 \times 12.0} = \frac{37 \text{ metric tons}}{2 \times 52.0}$$

$$w = \frac{4 \times 12.0 \times 37 \text{ metric tons}}{2 \times 52.0}$$

*Proficiency Level

2. (67%) <u>Solution</u>:

$$\begin{array}{ccccccc} \overset{75\ g}{\boxed{HCO_2H}} & + & \ldots & \rightarrow & \overset{t}{\boxed{HCO_2C_2H_5}} & + & \ldots \\ (1\ mole \times 46\ g\ mole^{-1}) & & & & (1\ mole \times 74\ g\ mole^{-1}) & & \end{array}$$

$$\frac{75\ g}{46\ g} = \frac{t}{74\ g}$$

$$t = \frac{74\ g \times 75\ g}{46\ g} = 121\ g$$

$$\%\ yield = \frac{81\ g}{121\ g} \times 100\%$$

3. (7.25 g) <u>Solution</u>:

$$\begin{array}{ccccccc} \ldots & + & \overset{w}{\boxed{2HF}} & \rightarrow & \overset{28.4\ g}{\boxed{SnF_2}} & + & \ldots \\ & & (2\ moles \times 20.0\ g\ mole^{-1}) & & (1\ mole \times 156.7\ g\ mole^{-1}) & & \end{array}$$

$$\frac{w}{2 \times 20.0\ g} = \frac{28.4\ g}{156.7\ g}$$

$$w = \frac{2 \times 20.0 \times 28.4\ g}{156.7}$$

4. (39%) <u>Solution</u>:

$$\begin{array}{ccccccc} \overset{10\ mg}{\boxed{3\ Cl_2}} & + & \ldots & \rightarrow & \overset{t}{\boxed{NCl_3}} & + & \ldots \\ (3\ moles \times 70.9\ g\ mole^{-1}) & & & & (1\ mole \times 120.4\ g\ mole^{-1}) & & \end{array}$$

$$\frac{10\ mg}{3 \times 70.9\ g} = \frac{t}{120.4\ g}$$

$$t = \frac{120.4 \times 10\ mg}{3 \times 70.9} = 5.7\ mg$$

$$\%\ yield = \frac{2.2\ mg}{5.7\ mg} \times 100\%$$

==

*5. (277 g) <u>Solution</u>:

(Using tc to represent trichlorethanal and ch to represent "chloral hydrate")

$$theor.\ yield\ (ch) = \frac{227\ g}{1} \times \frac{100}{73} = 311\ g(ch)$$

$$\frac{311\ g(ch)}{1} \times \frac{1\ mole(ch)}{165\ g(ch)} \times \frac{1\ mole(tc)}{1\ mole(ch)} \times \frac{147\ g(tc)}{1\ mole(tc)}$$

*6. (183 kg) <u>Solution</u>:

To find: mass of potassium permanganate needed

Information available: formula of ethylene glycol, mole ratio of ethylene and potassium permanganate, % efficiency

Information needed, not given: formula of potassium permanganate (from memory $KMnO_4$), mole ratio of ethylene to ethylene glycol (from carbon balance, 1:1) formula weights

Set-up (Using unity factors):

(using E to represent ethylene and G to represent the glycol)

$$\frac{100.0 \text{ kg(G)}}{1} \times \frac{100}{93} \times \frac{1 \text{ mole(G)}}{62 \text{ g (G)}} \times \frac{1 \text{ mole(E)}}{1 \text{ mole(G)}} \times \frac{2 \text{ moles}(KMnO_4)}{3 \text{ moles(E)}} \times \frac{158 \text{ g }(KMnO_4)}{1 \text{ mole}(KMnO_4)}$$

*7. (258 metric tons) <u>Solution</u>:

$$\text{theor. yield (SiC)} = \frac{250 \text{ metric tons}}{1} \times \frac{100}{87} = 287 \text{ metric tons}$$

$$\frac{287 \text{ metric tons(SiC)}}{1} \times \frac{1 \text{ mole(SiC)}}{40.1 \text{ g(SiC)}} \times \frac{3 \text{ moles(C)}}{1 \text{ mole(SiC)}} \times \frac{12.0 \text{ g(C)}}{1 \text{ mole(C)}}$$

*8. (24 g) <u>Solution</u>:

To find: minimum mass of HF needed

Information available: HF/Si mole ratio, % SiO_2 in sample, % Si to be removed, mass of sample

Information needed, not given: mole ratio Si/SiO_2 (from Si balance, 1:1) formula weights

Set-up (using unity factors):

$$\frac{150 \text{ g(glass)}}{1} \times \frac{81 \text{ g}(SiO_2)}{100 \text{ g(glass)}} \times \frac{1 \text{ mole}(SiO_2)}{60.1 \text{ g}(SiO_2)} \times \frac{1 \text{ mole(Si)}}{1 \text{ mole}(SiO_2)} \times \frac{15}{100} \times \frac{4 \text{ moles(HF)}}{1 \text{ mole(Si)}} \times \frac{20.0 \text{ g(HF)}}{1 \text{ mole(HF)}}$$

[g(glass) → g(SiO_2) → moles(SiO_2) → moles(Si) → moles used → moles(HF) → g(HF)]

= =

<u>ANSWERS to SELF-TEST, Unit 4</u>

4.1.   45 kg

4.2.   90%

*4.3.   26 metric tons

*4.4.   13%

A.1. Frequently, elemental sulfur is used for overcoming soil deficiencies of this element. The sulfur is first converted to sulfuric acid (H_2SO_4). This is largely initiated by <u>thiobacillus thioxidans,</u> a sulfur-oxidizing microbe that appears to be universally present in soils. This process is described by the following equations:

$$S + O_2 \xrightarrow{\text{bacteria}} SO_2$$

$$2H_2O + O_2 + 2SO_2 \rightarrow 2H_2SO_4$$

For each 45.5 kilograms of elemental sulfur applied to the soil, how many kilograms of sulfuric acid could be produced?

A.2. In regions of naturally acidic soils, found in most of the high-rainfall regions of the earth where sulfur-deficiency is also likely to develop, oxidation of applied sulfur (problem A.1) adds to the soil acidity, often with adverse results for agriculture. One way to circumvent this problem is to add pulverized limestone ($CaCO_3$) to neutralize the sulfuric acid resulting from the oxidation of elemental sulfur.

$$(CaCO_3 + H_2SO_4 \rightarrow CaSO_4 + H_2O + CO_2)$$

How many kilograms of sulfuric acid can be neutralized by 142.0 kilograms of pulverized limestone?

A.3. The growth of marine plants is of great importance since plants provide the basis of the marine food chain which culminates in fish and marine mammals. During the photosynthetic process the plants remove dissolved carbon dioxide and nutrients from the water and, using solar energy, convert them to phytoplankton protoplasm as represented in the overall equation:

$$106\ CO_2 + 16\ NO_3^- + 122\ H_2O + 17\ H^+ + H_2PO_4^- \xrightarrow{\text{(light)}} C_{106}H_{263}N_{16}PO_{110} + 138\ O_2$$

In a system containing 90.0 g of carbon dioxide, 75.0 g of nitrate ion (NO_3^-), 0.50 g of dihydrogen phosphate ion ($H_2PO_4^-$), and excess acid, 15.0 g of oxygen was produced. What was the percentage yield of oxygen?

A.4. Typical farm crops are adversely affected in several ways by soil compaction: the growth of the roots is curtailed; the absorption of nutrients and water is greatly reduced; and the formation of compounds toxic to many plants is favored. The anaerobic decomposition of glucose occurs as follows:

$$C_6H_{12}O_6 \rightarrow 3CO_2 + 3CH_4$$

86

If the bacterially mediated reduction of 150 g of glucose, in a highly compacted soil, resulted in the formation of 35.0 g of methane, what was the percentage yield for the process?

A.5. Some components of deep-sea sediments result from the formation of solid matter in the sea by inorganic reactions. Barite ($BaSO_4$), the principal sulfate in deep-sea sediments, is an evaporite and is formed in coastal areas where there is a restricted inflow of sea water, favorable prevailing winds, and extreme heat to induce evaporation. Barite nodules are formed as follows:

$$Ba^{2+} + SO_4^{2-} \rightarrow BaSO_4$$

In a selected coastal system, the water contained 2800 mg of SO_4^{2-} ions per kilogram of water and a slight excess of Ba^{2+} ions. The barite formed was measured as 5654 mg per kilogram of water. What was the percentage yield?

*A.6. The "lithogenous" components of deep-sea sediments are formed by land erosion, submarine volcanoes or submarine weathering processes. The reaction of minerals with sea water is probably the most important way by which chemical breakdown occurs. The submarine weathering of the feldspar mineral, <u>orthoclase</u>, is represented by the equation:

$$2KAlSi_3O_8 + 8H_2O \rightarrow Al_2O_3 \cdot 3H_2O + 2K^+ + 10\,H^+ + 6SiO_3^{2-}$$

What mass of <u>orthoclase</u> is needed by a process which is 85% efficient to produce 454 g of aluminum oxide trihydrate?

*A.7. <u>Chloropicrin</u>, CCl_3NO_2, is an insecticide which has been used successfully against several varieties of insects that attack cereals and grains. This compound is easily formed by the reaction of nitromethane with chlorine:

$$CH_3NO_2 + 3Cl_2 \rightarrow CCl_3NO_2 + 3HCl$$

When excess chlorine is used, 91% of the nitromethane is typically converted to chloropicrin. How many kilograms of nitromethane must be used to produce 12 kilograms of chloropicrin?

*A.8. Soil acidity and the substandard nutritional conditions that accompany it result from a deficiency of exchangeable metal cations. The quantity of these absorbed cations controls the percentage of base saturation and thereby determines the hydrogen ion concentration in the soil solution. Calcium and magnesium are the two cations best suited for controlling soil acidity. A representative commercial oxide of lime, quicklime, contains 77% calcium oxide. How many kilograms of hydrochloric acid, from the soil solution, could be neutralized by the contents of a 110 kilo-

*Proficiency Level

gram bag of this quicklime if the neutralization reaction has an efficiency of 85% in soil conditions?

*A.9. The composition of sea water has been a challenge to chemists since Antoine Lavoisier made his first analyses. As the world's population increases more people are becoming increasingly interested in the sea and its resources. Currently, four elements are being recovered from the sea commercially: sodium, magnesium, chlorine, and bromine. Magnesium ions are precipitated from sea water with calcium hydroxide. The precipitate is acidulated with hydrochloric acid and the resulting magnesium chloride is electrolyzed to produce magnesium metal. What is the minimum mass, in kilograms, of magnesium chloride required to electrolytically produce 1000 kilograms of magnesium at 89% efficiency?

*A.10. The average bromine content of sea water, the world's most important source for this element, is approximately 73 grams per metric ton. Bromine is extracted from sea water according to the overall reaction:

$$2Br^- + Cl_2 \rightarrow 2Cl^- + Br_2$$

When excess chlorine is used, 96% of the bromide ion is typically converted to elemental bromine. How many kilograms of bromide ion must be consumed under these conditions to produce 50 kilograms of bromine? How many metric tons of sea water would be processed?

ANSWERS:

(A.1.) 139 kg, (A.2.) 139 kg, (A.3.) 66%, (A.4.) 87.4%, (A.5.) 83.1%, (A.6.) 2.0×10^3 g, (A.7.) 4.9 kg, (A.8.) 94 kg, (A.9.) 4.4×10^3 kg, (A.10.) 52 kg [Br^-], approx. 710 metric tons [sea water]

RELEVANT PROBLEMS

Unit 4: *Mass/Mass Problems*

Set B: *Biological & Medical Sciences*

B.1. Certain bacteria are able to generate the energy required for formation of ATP (adenosine triphosphate) from oxidation of inorganic raw materials without using light energy. One type of chemosynthetic bacteria can oxidize sulfur to bisulfate according to the net equation:

$$2S + 3O_2 + 2H_2O \xrightarrow{\text{bacteria}} 2H^+ + 2HSO_4^- + \text{Energy}$$

What mass of sulfur would be required for the chemosynthesis of 12.5 g of bisulfate?

B.2. One type of chemosynthetic bacteria can oxidize nitrite (NO_2^-) to nitrate (NO_3^-); another converts ammonium ion (NH_4^+) to nitrite. Although the role of these organisms in the overall energy economy of nature is slight, they perform essential functions in the nitrogen cycles of living systems. The nitrite bacteria oxidize ammonium ion to nitrite according to the net equation:

$$2NH_4^+ + 3O_2 \xrightarrow{\text{bacteria}} 2NO_2^- + 2H_2O + 4H^+ + \text{Energy}$$

If a mixture of 110 g of ammonium ion with an excess of oxygen resulted in the bacterially mediated formation of 271.5 g of nitrite ion, what was the percentage yield for the process?

B.3. Amphojel, (aluminum hydroxide), may be used as a protective against gastrointestinal disturbances when administered together with poorly tolerated drugs, such as acetylsalicylic acid. When used as a gastric antacid the Amphojel functions by neutralizing the hydrochloric acid in the patient's gastric juice. How many grams of hydrochloric acid can be neutralized by a typical oral dose of 2.4 g of aluminum hydroxide? Assume the reaction to be approximated by the equation:

$$Al(OH)_3 + 3HCl \rightarrow AlCl_3 + 3H_2O$$

B.4. Aspirin, the world's largest selling analgesic, is synthesized by acetylation of salicylic acid with acetic anhydride according to the equation:

$$C_7H_6O_3 + (CH_3CO)_2O \rightarrow C_9H_8O_4 + CH_3CO_2H$$

How many kilograms of salicylic acid would be required to produce a typical one-day U.S. supply (1.76×10^5 kg) of aspirin? (Note: Aspirin really is sold at the rate of 1.76×10^5 kg day^{-1} in the United States).

B.5. Esters occur extensively in nature. Many of the odors and flavors characteristic of fruits and flowers are the result of esters found in the plant products. The natural fats, one of the three major classes of foodstuffs, are high molecular weight esters. Several esters are used medicinally. For example, glyceryl trinitrate ("nitroglycerine") is used to dilate coronary arteries and lower blood pressure. The reaction for preparation of glyceryl trinitrate may be formulated as:

$$C_3H_5(OH)_3 + 3HNO_3 \xrightarrow[\text{[cold]}]{(H_2SO_4 \text{ catalyst})} C_3H_5(NO_3)_3 + 3H_2O$$

In a pilot run, 40 g of glycerol was treated with excess nitric acid and 64.5 g of glyceryl trinitrate was isolated. What was the percentage yield?

89

*B.6. The uptake of dissolved minerals in the soil by plants is the principal path by which animals eventually obtain their mineral requirements. However, when the naturally occurring salts are insoluble or are not widely distributed, the mineral must be provided in a useable form as a food supplement. For example, to reduce tooth decay, fluoride ions (F^-) are added to drinking water in the form of soluble sodium fluoride, NaF. This compound is made commercially by fusing a mixture of cryolite and sodium hydroxide:

$$2Na_3AlF_6 + 8NaOH \rightarrow 2NaAlO_2 + 12NaF + 4H_2O$$

When excess sodium hydroxide is used, 78.0% of the cryolite is converted to sodium fluoride. How many metric tons of cryolite must be used to produce 150 metric tons of sodium fluoride for water fluoridation?

*B.7. During periods of vigorous exercise the rate of oxygen supply in muscle tissue may not keep pace with demand. In such cases, pyruvic acid is biochemically reduced to lactic acid. The accumulation of lactic acid in the muscle produces pain, fatigue, and stiffness. After periods of heavy breathing, which provides additional oxygen, the reaction is reversed, changing lactic acid back to pyruvic acid:

$$2CH_3\text{-}CH\text{-}C{\overset{O}{\underset{OH}{}}}_{OH} + O_2 \rightarrow 2CH_3\text{-}C\text{-}C{\overset{O}{}}_{OH} + 2H_2O$$

In a typical skeletal muscle, having excess oxygen available, 95% of the lactic acid is converted to pyruvic acid. How much lactic acid is needed to produce 20 mg of pyruvic acid by this process?

*B.8. In the human body some of the amino acids that are not used for tissue building are decomposed to ammonia, carbon dioxide and water, at the same time producing energy. Ammonia, a toxic by-product of the deamination of amino acids, is removed from the body, predominately in the form of urea. The formation of urea takes place in the liver by a series of reactions that may be represented by the net equation:

$$2NH_3 + CO_2 \xrightarrow{\text{(enzymes)}} NH_2\text{-}\overset{O}{\overset{\|}{C}}\text{-}NH_2 + H_2O$$

What mass of ammonia is used by a process that is 87% efficient to produce 30 g of urea? (Note: 30 g is the average amount excreted per day.)

*B.9. Many inorganic salts are essential to proper growth and metabolism of the body. Some salts regulate the irritability of nerve and muscle cells, the beating of the heart, and the proper osmotic pressure of the cells. Iron salts are necessary for the formation of hemoglobin, calcium and phosphate salts regulate the formation of bones and teeth, and iodine salts maintain the proper functioning of the thyroid gland. Potassium iodide, which may be used as an expectorant or a source of essen-

*Proficiency Level

tial iodine, may be prepared according to the equation:

$$HI + KHCO_3 \rightarrow KI + H_2O + CO_2$$

What mass of hydrogen iodide is required, in a process that is 95% efficient, to produce 2500 kg of potassium iodide for therapeutic uses?

*B.10.   One of the most widely distributed organic acids in the plant kingdom is a mono-hydroxytricarboxylic acid, citric acid. Found in highest concentration in citrus fruits, it is present also in many other fruits, such as raspberries, cranberries, strawberries, and pineapple. Magnesium citrate, mixed with sodium bicarbonate and sugar, is sold as a mild laxative. When excess magnesium carbonate is used, 83% of the citric acid is recovered as magnesium citrate, according to the equation:

$$2H_3C_6H_5O_7 + 3MgCO_3 \rightarrow Mg_3(C_6H_5O_7)_2 + 3H_2O + 3CO_2$$

How much citric acid is needed to produce 12 kg of magnesium citrate?

ANSWERS:

(B.1.) 4.13 g,   (B.2.) 96.8%,   (B.3.) 3.4 g,   (B.4.) 1.35×10^5 kg,   (B.5.) 65%,   (B.6.) metric tons,   (B.7.) 22 mg,   (B.8.) 20 g,   (B.9.) 2.0×10^3 kg,   (B.10.) 12 kg

---

RELEVANT PROBLEMS

Unit 4: Mass/Mass Problems

Set I: Industrial Chemistry

I.1.   Fluorine is used directly or combined with hydrogen, metals, or certain nonmetals to form useful inorganic fluorides. Many of these compounds are, in turn, used for organic fluorinations, including the production of fluorocarbons. The inorganic fluorides are generally classified as "hard", "moderate", or "soft" fluorinating reagents. During fluorinations, "hard" fluorinating reagents (e.g., F_2, ClF_3, AgF_2, CoF_3, PbF_4 and CeF_4) oxidize the elements to their highest oxidation states or, in organic compounds, cause fragmentations and polymerizations. "Moderate" fluorinating reagents (e.g., HgF_2, SbF_5, SbF_3 and CaF_2) are selective and fluorinate only certain groups. These reagents have limited capability for the substitution of fluorine for halogens or hydrogen. However, these reagents are very useful in the fluorination of halocarbons, especially those in which the more polar chlorides are present. "Soft" fluorinating reagents are those which, under normal conditions, do not cause fragmentation of the functional groups, do not saturate double bonds, and do not oxidize metals to their highest oxidation states. Synthetic chemical industries have appreciable investemnts in fluorination technology. One of the "moderate" fluorinating reagent, antimony pentafluoride (SbF_5), is prepared industrially in an aluminum apparatus according to the equation

$$SbCl_5 + 5HF \rightarrow 5HCl + SbF_5$$

How much antimony pentachloride is needed to prepare 1.00 kg of antimony pentafluoride?

I.2. Silica ("SiO_2") will fuse and react with metal oxides. The products of such reactions are glasses, noncrystalline supercooled liquids of variable composition. Soft glass, made by fusing silica, sodium carbonate (as a source of Na_2O), and calcium carbonate (as a source of CaO), is characterized by a high coefficient of thermal expansion and shatters when subjected to thermal shock. Heat-resistant glasses, such as Pyrex and Kimax, are made with aluminum oxide, Al_2O_3, and boron trioxide, B_2O_3. These glasses have very low coefficients of thermal expansion, and are used for high-temperature work. "Soda-lime glass" (soft glass), formulated as $Na_2O \cdot CaO \cdot 6SiO_2$, is used for the manufacture of inexpensive bottles. The glass is prepared by fusing sodium carbonate, calcium carbonate and "silicon dioxide", according to the following equation:

$$Na_2CO_3 + CaCO_3 + 6SiO_2 \xrightarrow{\text{fusion}} Na_2O \cdot CaO \cdot 6SiO_2 + 2CO_2$$

Determine how many kilograms of "silicon dioxide" ("SiO_2") would be used to produce enough glass to make 20,000 bottles, each weighing 350 g.

I.3. Acetylene is distributed as a bottled fuel gas, chiefly for use in oxyacetylene torches used in welding and cutting metals. The most important commercial use for acetylene, however, is in the synthesis of complicated organic compounds such as neoprene rubber and vinyl resins. There are two standard methods for the production of acetylene. The older method employs the reaction of calcium carbide with water, according to the following equation:

$$CaC_2 + 2H_2O \rightarrow HC \equiv CH + Ca(OH)_2$$

Alternately, acetylene is produced commercially by the thermal decomposition (cracking) of certain hydrocarbons such as methane, ethane, propane and butane. The methane is converted according to the following equation:

$$2CH_4 \xrightarrow{\text{(catalyst)}} HC \equiv CH + 3H_2$$

How many metric tons of methane are required to make 8300 metric tons of acetylene if the conversion efficiency is 100%?

I.4. Hydroquinone is used as a dye intermediate and polymerization inhibitor. It is added to fats and oils as an antioxidant or preservative, but its principal use is as a photographic developer. Acetylene is one of the raw materials in the production of hydroquinone, as shown in the following equation:

$$2HC\equiv CH + 3CO + H_2O \xrightarrow{\text{(heat, pressure, catalyst)}} \underset{(C_6H_6O_2)}{\text{[benzene ring with OH at top and OH at bottom]}} + CO_2$$

What is the % yield if 51 kilograms of hydroquinone are made from 28 kilograms of acetylene?

I.5. One step in producing pictures from a piece of exposed film is to "develop" the film (reduce the light-activated silver halide crystals) by reaction with an organic reducing agent such as hydroquinone (HO-⟨◯⟩-OH), as shown in the following equation:

$$C_6H_6O_2 + 2AgBr + 2OH^- \rightarrow C_6H_4O_2 + 2Ag + 2H_2O + 2Br^-$$

What is the % efficiency if 12 mg of silver bromide crystals produce 3.1 mg of $C_6H_4O_2$?

*I.6. Sodium hydroxide, NaOH, is frequently called caustic soda in commerce, and its solutions are sometimes referred to as lye. Sodium hydroxide converts some types of animal and vegetable matter into soluble materials by chemical action. It is described as a very "caustic" substance because of its destructive effect on skin, hair, and wool. Most commercial sodium hydroxide is produced as a by-product from the electrolysis of sodium chloride solutions, according to the equation:

$$2NaCl + 2H_2O \xrightarrow{D.C.} 2NaOH + Cl_2 + H_2$$

The electrolytic preparation of 908 kg of sodium hydroxide typically requires 2500 kilowatt hours of electricity. However, approximately 794 kg of chlorine and 248 cubic meters of hydrogen are simultaneously produced. How many kilograms of caustic soda could be obtained by electrolyzing a brine containing 2.9×10^3 kg of sodium chloride, in a cell that has a 95% conversion efficiency? How many kilograms each of chlorine and hydrogen are produced simultaneously?

*I.7. Acrylonitrile's main use is in polyacrylonitrile fibers. Dyeable Orlon is made by polymerizing acrylonitrile with modifying constituents. The end uses of acrylic

*Proficiency Level

fibers include sweaters, women's coats, men's winter suiting, carpets and blankets. One method of making acrylonitrile is from nitric oxide and an excess of propylene, as shown in the following equation:

$$4C_3H_6 + 6NO \rightarrow 4C_3H_3N + N_2 + 6H_2O$$

At a conversion rate of 80%, how much nitric oxide is needed to make 387 metric tons of acrylonitrite?

*I.8. Sulfuric acid is such a versatile compound to the economy that its production and consumption are used as economic indicators. Most of the sulfuric acid produced in the United States today is made by the contact process, which is steadily replacing the older lead chamber process. Both processes can use either "brimstone" (sulfur) or pyrites as alternative raw materials, with the choice dependent primarily on price. Because of the recent concern over air pollution, industrial by-product gases are becoming increasingly important sources of sulfur dioxide for the production of sulfuric acid. Regardless of the raw material being used, impurities which might combine with the catalyst and "poison" it must be removed from the sulfur dioxide. The purified sulfur dioxide is mixed with air and passed through heated iron pipes which contain the catalyst, usually vanadium pentoxide, V_2O_5. This close "contact" of the sulfur dioxide and the catalyst gives the <u>contact</u> <u>process</u> its name. While sulfur dioxide and oxygen of the air are both adsorbed on the surface of the catalyst, they react to form sulfur trioxide. The gaseous sulfur trioxide does not dissolve in or react readily with pure water. Consequently, it is absorbed in 97% sulfuric acid, in which it is readily soluble. Subsequently, the sulfur trioxide combines with 3% water and forms 100% sulfuric acid. Modern sulfuric acid plants vary in daily capacity from 100 metric tons of 100% acid to 1,000 metric tons of 100% acid. How many metric tons of pure iron pyrite (FeS_2) would be needed to produce 700 metric tons of pure sulfuric acid, assuming 90% conversion efficiency?

*I.9. Tetraethyl lead [$Pb(C_2H_5)_4$] has been the principal antiknock compound for gasoline. However, because of the harmful lead compounds that are exhausted from gasoline engines, research is attempting to replace "leaded" gasolines. Tetraethyl lead is made by the action of ethyl chloride (C_2H_5Cl) on a lead-sodium alloy. How many metric tons of ethyl chloride are needed to produce 2.5×10^5 metric tons of tetraethyl lead, if the efficiency is 75%?

*I.10. Another use for tetraethyl lead is in the production of ethyl mercuric chloride [C_2H_5HgCl], better known as <u>Ceresan</u>. It is an organic fungicide used in the treatment of seeds and as a general garden spray. <u>Ceresan</u> is also one of the two best antimildewing agents for paints. It is made by treating tetraethyl lead with mercury(II) chloride [$HgCl_2$]. How many metric tons of $HgCl_2$ are needed to react with 2.5×10^5 metric tons of tetraethyl lead, assuming that transfer of all ethyl groups is 90% efficient?

ANSWERS:

(I.1.) 1.38 kg, (I.2.) 5.27×10^3 kg, (I.3.) 1.0×10^4 metric tons, (I.4.) 86%, (I.5.) 90%, (I.6.) 1.9×10^3 kg NaOH, 1.7×10^3 kg Cl_2, 4.7×10^1 kg H_2, (I.7.) ~410 metric tons, (I.8.) ~480 metric tons, (I.9.) 2.7×10^5 metric tons, (I.10.) 9.3×10^5 metric tons

---

RELEVANT PROBLEMS

Unit 4: Mass/Mass Problems

Set E: Environmental Sciences

E.1. A network of pipelines linking principal refining areas with the heaviest marketing centers helps cut down the distribution cost for the flood of gasoline and other fuel demands. Prior to the "energy crunch", the federal and state gasoline taxes added up to about 38 percent of the cost of the fuel. National surveys show that improved refining, formulation, and distribution still make gasoline one of the best bargains in the American family budget, but not necessarily a bargain in the "ecological budget". An "import" car has a gas tank that holds 37.9 liters (about 10 gallons). Octane (C_8H_{18}), a major constituent of gasoline, has a density of 0.70 g ml^{-1}. What mass of carbon monoxide would be emitted into the atmosphere by the incomplete combustion of one tank of pure octane, according to the equation:

$$C_8H_{18} + 11 \ O_2 \rightarrow 5 \ CO_2 + 3 \ CO + 9 \ H_2O$$

E.2. High concentrations of sulfuric acid from atmospheric pollution are capable of attacking a wide variety of building materials. Especially susceptible are such carbonate-containing substances as marble, limestone, roofing slate, and mortar. The carbonates in these materials are converted to sulfates which are moderately water-soluble. The material becomes pitted and weakened mechanically as the soluble sulfates are leached away by rainwater. The conversion reaction is:

$$CaCO_3 + H_2SO_4 \rightarrow CaSO_4 + CO_2 + H_2O$$

(limestone)

How many grams of sulfuric acid are needed to remove 11.3 g of limestone, if the conversion is 100%?

E.3. One proposed method of determining the concentration of certain atmospheric "oxidants" is to use a solution containing starch, a known amount of $Na_2S_2O_3$, and excess KI. As air is passed through the solution, the atmospheric "oxidants" will change the I^- to I_2 as shown in the equation where O_3 represents the "oxidants".

$$O_3 + 2KI + H_2O \rightarrow O_2 + 2KOH + I_2$$

As long as $Na_2S_2O_3$ is present the following subsequent reaction will occur.

$$I_2 + 2Na_2S_2O_3 \rightarrow 2NaI + Na_2S_4O_6$$

When the $Na_2S_2O_3$ is depleted the I_2 will cause the starch solution to turn blue. By measuring the amount of air that passes through before the solution turns blue, the concentration of "oxidants" in the air can be measured. (Any reducing gases such as SO_2 or H_2S will interfere with this test.) How many milligrams of ozone (O_3) will it take to use up 175 milligrams of $Na_2S_2O_3$?

E.4. For every 1000 gallons (3784 ℓ) of gasoline burned in a typical automobile engine, 2300 lbs (1.04×10^3 kg) of CO are emitted. Octane (C_8H_{18}) has a density of 0.70 g ml^{-1} and is a major component of gasoline. What is the percent conversion to CO if we assume the gasoline is all octane and is burned according to the following equation:

$$C_8H_{18} + 11 O_2 \rightarrow 5 CO_2 + 3 CO + 9 H_2O ?$$

E.5. Complete combustion of octane yields CO_2 and H_2O by the reaction:

$$2 C_8H_{18} + 25 O_2 \rightarrow 16 CO_2 + 18 H_2O$$

What is the percent yield if 2650 g of octane produced 5110 g of carbon dioxide?

*E.6. The removal of SO_2 from flue gases is receiving much of the attention in the battle against sulfur oxides pollution. No method has yet proved satisfactory enough to find widespread use. One process, hopefully of value in power plants, involves the injection of limestone into the combustion zone of the furnace. The limestone reacts with SO_2 according to the equation:

$$2 CaCO_3 + 2 SO_2 + O_2 \rightarrow 2 CaSO_4 + 2 CO_2$$

If the percent yield is 88%, how much SO_2 was removed if 245 kg of $CaSO_4$ was produced?

*Proficiency Level

*E.7. One of the waste water pollutants from the manufacture of coke, steel, and ferro-alloys is phenol. Ozone is able to oxidize phenol to oxalic acid by the following equation:

$$C_6H_5OH + 11\ O_3 \rightarrow 3(CO_2H)_2 + 11\ O_2$$

How much ozone is needed to produce 3.07 g of oxalic acid if the process is 93.5% efficient?

*E.8. According to the National Industrial Pollution Control Council, nature is responsible for about two-thirds of the sulfur emitted into the atmosphere. The major natural sources are volcanoes. Man's input is mostly in the form of sulfur dioxide resulting from combustion of sulfur-containing fuels (coal or oil) by power plants. Smelting sulfide ores and some chemical operations, such as sulfuric acid manufacture, also produce modest amounts of sulfur dioxide. Photochemical reactions can convert sulfur dioxide to sulfur trioxide, which rapidly combines with water vapor in the atmosphere to form acid mists. In the absence of rain over an extended period, such mists can accelerate the deterioration of paints, metals, building materials and textiles. Both sulfur dioxide and sulfur trioxide are soluble in water and can be washed from the air by rainfall. This is part of nature's design, since sulfur is essential to plant nutrition. However, large amounts can be harmful if the soil is low in alkali. In areas of localized high concentrations of SO_2 and SO_3, the rains may contain harmful concentrations of sulfurous and sulfuric acids. The estimated U.S. sulfur dioxide emissions for 1972 totaled 3.5×10^7 metric tons. How many metric tons of sulfuric acid (H_2SO_4), assuming complete conversion, may have rained down on the United States as a result of this atmospheric pollution?

*E.9. Contaminated water can be a menace to life. Dysentery and typhoid fever are endemic in many parts of the underdeveloped world where primitive sanitary facilities permit sewage to enter drinking water sources. In the industrialized nations, careful treatment of drinking water is routine and waterborne diseases have been practically eliminated. Purifying sewage before disposal is an essential step in preventing the pollution of our water supplies. In a typical sewage treatment process, the raw sewage is first put through a comminutor, a device that pulverizes the coarse debris. It is then mixed with bacteria-rich water in an aerator. As propellor blades agitate the mixture, the bacteria transform organic matter into harmless by-products. Heavy sludge is removed in a settling basin, after which the water is filtered through a bed of sand. At this point the water is a clear liquid, which is chlorinated to destroy any remaining bacteria. The water is now pure enough to drink and is usually piped into streams where it flows back to the natural water supplies of the earth. Cellulose waste in sewage is oxidized quantitatively by aerobic bacteria in sewage treatment plants to CO_2 and H_2O. How much oxygen is consumed for the degradation of 1.00 metric ton of cellulose waste, represented by $(C_6H_{10}O_5)_n$, by a process of 85% efficiency?

*E.10. Phosphates in sewage, primarily from detergents and excess agricultural fertilizers, are not biodegraded and their accumulation in natural waters has been linked to increasing growth of algae. In an attempt to reduce phosphate pollution, detergent manufacturers have investigated alternatives such as NTA, $[N(CH_2CO_2H)_3]$, (nitrilotriacetic acid). Evidence of the toxicity of NTA to animal life resulted in a ban on its use in 1970. One biodegradation sequence of NTA can be formulated as producing nitrate ions, carbon dioxide, hydrogen ions and water. What mass of nitrate, another water contaminant, would be formed by this degradation per metric ton of NTA consumed assuming 92% efficiency?

ANSWERS:

(E.1.) 2.0×10^4 g, (E.2.) 11.1 g, (E.3.) 26.6 mg, (E.4.) 53%, (E.5.) 62.5%, (E.6.) ~130 kg, (E.7.) 6.42 g, (E.8.) 5.4×10^7 metric tons, (E.9.) 1.4 metric tons, (E.10.) 0.30 metric ton

---

STOICHIOMETRY

UNIT 5: GAS VOLUME PROBLEMS

Stoichiometric calculations involving <u>volumes</u> of gaseous products or reactants can be made in basically the same way as those in which quantities are measured by mass (Unit 4). Two additional factors, however, must be considered.

The first is based on <u>Avogadro's</u>[1] <u>law</u>: *equal volumes of gases at the same temperature and pressure contain equal numbers of gaseous particles.* Since the <u>MOLE</u> is a convenient unit particle counting number for chemical substances (Unit 3), it is necessary, considering Avogadro's law, only to determine experimentally the volume of a mole of <u>one</u> gas (e.g., H_2, 2.016 g) at a known temperature and pressure in order to know the volume of a mole of <u>any</u> gas at that temperature and pressure. It would be particularly useful to have some standard conditions for such a <u>molar volume</u>. These have been defined for gases as Standard Temperature and Pressure (<u>STP</u>), $273°K$[2] and 760 torr[3]. Under <u>these</u> conditions, the volume of one mole of any gas (assuming "ideal" behavior) is 22.4 liters. Thus, we have a new, and very useful, set of unity factors:

$$\text{at STP} \quad \frac{1.00 \text{ mole(gas)}}{22.4 \text{ liters}} \quad \text{and} \quad \frac{22.4 \text{ liters}}{1.00 \text{ mole(gas)}}$$

The second factor which must be considered in dealing with gases is that gas volumes, unlike masses, vary appreciably with changes in pressure or temperature. If we are to work with mole/volume unity factors, we do not want to be limited to volumes at $273°K$ and 760 torr. Real chemical processes occur over a wide range of temperatures and pressures. There are several ways of "correcting" volumes for pressure or temperature variation, i.e., of calculating volume changes. For our purposes, a convenient "formula" is derived from a combination of <u>Boyle's law</u>[4] and <u>Charles' law</u>[5]:

$$\frac{P_1 V_1}{T_1} = \frac{P_2 V_2}{T_2}$$

If we plan to use the Standard Molar Volume in stoichiometric calculations, then the "corrected volume" for other conditions, "P_{exp} and T_{exp}", is derived from:

---

[1] *Amedeo Avogadro, a physics professor at the University of Turin, suggested this idea in 1811. It was subsequently rejected by John Dalton because it required "polyatomic molecules" and was not generally accepted by the scientific community until 1858, by which time the evidence was overwhelming.*

[2] $°K = °C + 273°$

[3] 760 torr = 760mm$_{Hg}$ = 1.00 atm

[4] Robert Boyle (1662): $P_1 V_1 = P_2 V_2$

[5] Jacques Charles (1787): $P_1/T_1 = P_2/T_2$

(Note that a capital T denotes °K.)

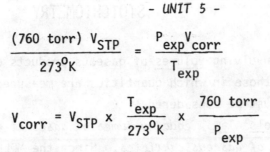

$$\frac{(760 \text{ torr}) V_{STP}}{273^{0}K} = \frac{P_{exp} V_{corr}}{T_{exp}}$$

$$V_{corr} = V_{STP} \times \frac{T_{exp}}{273^{0}K} \times \frac{760 \text{ torr}}{P_{exp}}$$

This formula is quite useful in gas volume stoichiometry. More general approaches should be employed for other types of gas volume correction problems not involving STP conditions.

The use of mole/volume unity factors and gas volume "corrections" permits us to expand our applications of stoichiometry to a new range of processes.

- References -

1. O'Connor, Rod. 1977. Fundamentals of Chemistry, second edition. New York: Harper & Row (Unit 10)

2. Brown, Theodore, and H. Eugene LeMay, Jr. 1977. Chemistry: The Central Science. Englewood Cliffs, N.J.: Prentice-Hall (Chapter 5)

3. Masterton, William, and Emil Slowinski. 1977. Chemical Principles, fourth edition. Philadelphia: W.B. Saunders (Chapters 5, 17)

4. Nebergall, William, F.C. Schmidt, and H.F. Holtzclaw, Jr. 1976. College Chemistry, fifth edition. Lexington, MA: D.C. Heath (Chapters 10, 24, 25)

5. Parsonage, N.G. 1966. The Gaseous State. Elmsford, N.Y.: Pergamon

---

OBJECTIVES:

(1) *Given the balanced equation for a chemical process, be able to make appropriate stoichiometric calculations involving gas volumes under Standard Conditions (STP)*

(2) *Given the balanced equation for a chemical process, be able to make appropriate stoichiometric calculations involving gas volumes, including those of gas mixtures, under conditions other than STP.*

*(3) *Given a description of a chemical process with information from which appropriate mole ratios and mass (and/or volume) data can be determined, be able to calculate specified gas volumes under given conditions of temperature and pressure.*

PRE-TEST:

Necessary Atomic Weights:	
H (1.008)	N (14.01)
O (16.00)	Bi(209.0)

*Proficiency Level

(1) "Laughing gas", nitrous oxide, can be prepared by the carefully controlled heating of ammonium nitrate, according to the equation:

$$NH_4NO_{3(s)} \rightarrow N_2O_{(g)} + 2H_2O_{(g)}$$

How many liters of nitrous oxide (at STP) could be produced by the decomposition of 24 g of ammonium nitrate? _____

(2) Ammonium nitrate may also decompose explosively, as in the famous Texas City disaster of 1947 in which a ship loaded with NH_4NO_3 fertilizer exploded. The detonation of ammonium nitrate may be shown as:

$$2NH_4NO_{3(s)} \rightarrow 2N_{2(g)} + O_{2(g)} + 4H_2O_{(g)}$$

How many <u>total</u> liters of gases would be formed by the explosive decomposition of 1.00 lb (454 g) of ammonium nitrate at 819°C and 640 torr? _____

*(3) Metallic bismuth is used in the manufacture of various low-melting alloys, such as those used in automatic sprinkler systems. The free metal may be prepared from <u>bismite</u> ore, containing Bi_2O_3, by heating the ore with coke and draining off the low-melting bismuth as a liquid. For each mole of Bi_2O_3 processed by this method, three moles of carbon monoxide gas are released. What volume, in liters, of this toxic by-product are formed at 546°C and 700 torr for each ton (907 kg) of Bi_2O_3 processed? _____

Answers and Directions:

(1) 6.7 liters   (2) 2.11 x 10^3 liters

*If both are correct, go on to question *3. If you missed either, study METHODS, sections 5.1 and 5.2.*

*(3) 4.26 x 10^5 liters

If your answer was correct, go on to RELEVANT PROBLEMS (Unit 4). If you missed it, study METHODS, section 5.3.

METHODS

5.1 Gas Stoichiometry at STP

Gas volume stoichiometry problems may be treated by the same "proportion method" we developed for mass/mass problems. When volumes under Standard Conditions (273°K, 760 torr) are to be calculated, the only change from the mass/mass method is to substitute the <u>molar volume</u>, 22.4 liters, for the <u>formula weight</u> for any gaseous substance whose volume is involved in the calculation. All other aspects of the method remain the same, except that any masses used, including formula weights must be expressed in units of grams.[6]

EXAMPLE 1

When a wooden match containing P_4S_3 in the match head is ignited, a white smoke of P_4O_{10} and gaseous SO_2 are released. Calculate the volume of SO_2 (at STP) that could be formed from the complete combustion of 0.25 g P_4S_3, according to the equation:

$$P_4S_{3(s)} + 8O_{2(g)} \rightarrow P_4O_{10(s)} + 3SO_{2(g)}$$

SOLUTION: (following the general method of Unit 4, Competency Level)

$$
\underset{(1 \text{ mole} \times 220 \text{ g mole}^{-1})}{\overset{0.25 \text{ g}}{\boxed{P_4S_{3(s)}}}} + \ldots \rightarrow \ldots + \underset{(3 \text{ moles} \times 22.4 \text{ liters mole}^{-1})}{\overset{V_{STP}}{\boxed{3\ SO_{2(g)}}}}
$$

(@STP)

$$\frac{0.25 \text{ g}}{220 \text{ g}} = \frac{V_{STP}}{(3 \times 22.4) \text{ liters}}$$

$$V_{STP} = \frac{(3 \times 22.4) \text{ liters} \times 0.25 \text{ g}}{220 \text{ g}} = 0.076 \text{ liters}$$

$$(76 \text{ ml})$$

EXERCISE 1

Calcium carbide was used in miners' lamps before the much safer battery-operated lamps were available. A regulated drip of water onto the calcium carbide liberated acetylene gas which was ignited to produce a bright white flame. The acetylene formation is described by:

$$CaC_{2(s)} + 2H_2O_{(\ell)} \rightarrow Ca(OH)_{2(s)} + C_2H_{2(g)}$$

How many grams of water must be added to excess calcium carbide to generate 15 liters of acetylene (measured at STP)?

(answer, page 107)

5.2 Gas Mixtures and "Non-STP" Conditions

When more than one gas is involved in a stoichiometric calculation, as in a case where the total volume of different gaseous products is to be determined, the combined

----

[6]This permits us to equate ratios of mass/molar mass and volume/molar volume.

molar volumes of the individual gases can be used to form the denominator of a volume/molar volume ratio. The calculation then proceeds along the lines of the case of a single gas as described in Section 5.1.

The use of the value "22.4 liters" for the molar volume applies to specific conditions, corresponding to a temperature of $273^{\circ}K(0^{\circ}C)$ and a pressure of 760 torr. Since the number of particles in a given volume of gas depends on the temperature and pressure, volume/molar volume ratios are meaningful only under comparable temperature and pressure conditions. A simple way of treating gas stoichiometry problems including conditions other than those of STP follows a two-step procedure whose sequence depends on the nature of the problem.

a. If a volume at other than STP is to be <u>calculated</u>:

<u>Step 1</u> - Calculate V_{STP} by the methods for simple gas stoichiometry (EXAMPLE 1)

<u>Step 2</u> - "Correct" the volume to the conditions given.

$$\frac{760 \text{ torr} \times V_{STP}}{273^{\circ}} = \frac{P_{exp} \times V_{corr}}{T_{exp}} \qquad \text{(page 99)}$$

b. If a volume measured at other than STP is to be <u>used</u> in calculating another quantity:

<u>Step 1</u> - "Correct" the given volume to STP

$$\frac{760 \text{ torr} \times V_{STP}}{273^{\circ}} = \frac{P_{exp} \times V_{exp}}{T_{exp}}$$

<u>Step 2</u> - Use the V_{STP} in the stoichiometric calculation (EXERCISE 1)

We will illustrate only case <u>a</u>, since the principles are the same in both cases. For an example of case <u>b</u>, see EXERCISE 4.

EXAMPLE 2

Most children's "Chemistry Sets" contain a simple alcohol burner, often using methanol for fuel. The complete combustion of methanol is given by the equation:

$$2 \text{ CH}_3\text{OH}_{(\ell)} + 3 \text{ O}_{2(g)} \rightarrow 2 \text{ CO}_{2(g)} + 4 \text{ H}_2\text{O}_{(g)}$$

What total gas volume would result, at $400^{\circ}C$ and 680 torr, from the complete combustion of 35 g of methanol?

SOLUTION:

Step 1 -

$$\underset{(2 \text{ moles} \times 32.0 \text{ g mole}^{-1})}{\underset{35 \text{ g}}{\boxed{2 \text{ CH}_3\text{OH}}}} + \quad .... \quad \rightarrow \quad \underset{(6 \text{ moles} \times 22.4 \text{ liters mole}^{-1})}{\overset{V_{STP}}{\boxed{2 \text{ CO}_{2(g)} + 4 \text{ H}_2\text{O}_{(g)}}}}$$

$$[2 + 4]$$

103

$$V_{STP} = \frac{(6 \times 22.4) \text{ liters} \times 35 \text{ g}}{2 \times 32.0 \text{ g}} = 73.5 \text{ liters}$$

Step 2 - $(400^{\circ}C + 273^{\circ} = 673^{\circ}K)$

$$V_{corr} = 73.5 \text{ liters} \times \frac{673^{\circ}}{273^{\circ}} \times \frac{760 \text{ torr}}{680 \text{ torr}} = \underline{202 \text{ liters}}$$

EXERCISE 2

The explosion of nitroglycerine (glyceryl trinitrate) generates a large amount of heat and produces an enormous gas volume from a small sample of the explosive. The reaction may be represented in simplefied form as:

$$4 \, C_3H_5(ONO_2)_3{}_{(\ell)} \rightarrow 12 \, CO_2{}_{(g)} + 10 \, H_2O_{(g)} + 6 \, N_2{}_{(g)} + O_2{}_{(g)}$$

What total gas volume, in liters at $1000^{\circ}C$ and 900 torr, would result from the explosive decomposition of 1.00 oz (28.4 g) of nitroglycerine?

(answer, page 108)

Extra Practice

EXERCISE 3

When hydrocarbon fuels, such as gasoline, burn in an internal combustion engine, part of the carbon is converted to carbon monoxide because of the incomplete oxidation of the fuel. A simplified equation for the incomplete combustion of octane can be used to illustrate a typical situation:

$$C_8H_{18} + 11 \, O_2 \rightarrow 3 \, CO + 5 \, CO_2 + 9 \, H_2O$$

How many liters of carbon monoxide (measured at STP) would result from the burning of 1.00 gallon (2800 g) of octane according to the equation given?

(answer, page 108)

EXERCISE 4

Many years ago nitric acid, used in the manufacture of nitrate fertilizers and explosives, was made almost exclusively from imported "Chile saltpeter" ($NaNO_3$). This salt was, therefore, of tremendous importance to the economy of Chile. Alternative methods of producing nitric acid were devised prior to World War II and by 1940 "Saltpeter" was used for less than 10% of the U.S. production of nitric acid. One of the more efficient methods of making nitric acid starts with the gas phase oxidation of ammonia to nitric oxide (NO). The nitric oxide is, in turn, oxidized by air to nitrogen dioxide, which is absorbed by water to form nitric acid. The three stage process may be summarized by a single equation showing a simplified relationship between the original reactants and the final product:

$$NH_{3(g)} + 2\,O_{2(g)} \rightarrow HNO_{3(aq)} + H_2O_{(\ell)}$$

Assuming 100% efficiency, how many grams of nitric acid could be formed from a reaction mixture of 1.00 liter of $NH_{3(g)}$ and 2.00 liters of $O_{2(g)}$ at 200°C and 1200 torr?

(answer, page 108)

===

At this point you should try the competency level Self-Test questions on page 107.

===

*5.3 Gas Stoichiometry from Process Descriptions

Stoichiometric problems involving gas volumes are not essentially different from those dealing only with masses. Mole/molar volume unity factors (e.g., 1 mole(gas)/22.4 liters @ STP) can be employed in the same way as mole/mass unity factors. If conditions other than STP are involved, temperature and pressure ratios can be included in the total set-up. We must remember that any use of 22.4 liters mole^{-1} implies a particular set of conditions, 273°K(0°C) and 760 torr(1.00 atm).

Recognition of how a gas volume varies with temperature and pressure changes (directly with temperature, in °K, and inversely with pressure) must be included in the problem analysis leading to a proper set-up. A useful ploy is to include appropriate temperature ratio and pressure ratio "correction" terms as an initial or final part of the set-up, labelling the conditions applying to the remaining segment with T,P notations and the pressure and temperature ratios segment with the T,P "correction" applied. This is illustrated in the SOLUTION to EXAMPLE 3.

*EXAMPLE 3

Ammonia, for use as a soil additive or as a starting material for the manufacture of other nitrogen compounds, may be prepared by the Haber process. Nitrogen and hydrogen gases in the mole ratio $1N_2:3H_2$ are combined in the presence of a catalyst under conditions of high temperature and pressure. Two moles of ammonia are formed for each mole of nitrogen consumed. The yield of ammonia is limited by the conditions of equilibrium (Part II of this text) so that the process is appreciably less than 100% efficient. Assuming a 61% conversion efficiency, what mass (in grams) of ammonia would be formed from an initial mixture of 15 liters of nitrogen and 45 liters of hydrogen, measured at the reaction conditions of 500°C and 500 atm?

SOLUTION:

To find: mass of NH_3 formed

Information available: reactant quantities and conditions, % yield, mole ratios

Information needed, not given: molar volume (from memory, 22.4 liters per mole @ STP),

T and P-ratios, formula weight of NH_3

Set-up (Using unity factors):

(Since N_2 and H_2 were mixed in a volume ratio equivalent to the stoichiometric ratio for reaction, we could compute the product yield from either gas volume given. We'll use that of nitrogen.)

$$\left[\frac{15 \text{ liters}(N_2)}{1} \times \frac{273°}{773°} \times \frac{500 \text{ atm}}{1.00 \text{ atm}}\right] \times \left[\frac{1 \text{ mole}(N_2)}{22.4 \text{ liters}(N_2)}\right] \times \frac{2 \text{ moles}(NH_3)}{1 \text{ mole}(N_2)} \times \frac{17.0 \text{ g}(NH_3)}{1 \text{ mole}(NH_3)} \times \frac{61}{100}$$

["correcting" from 500°C(773°K), 500 atm to STP]   [@ STP]   ["efficiency factor"]

$$= 2.5 \times 10^3 \text{ g}$$

----

*EXERCISE 5

Pentaerythritol tetranitrate (PETN) is an explosive used in making "shaped charges", as in the World War II antitank grenades. When the compound is formed with a binder into a cone, its detonation characteristics are such as to direct most of the explosive force towards the base of the cone. Product temperatures of several thousand degrees, coupled with an enormous volume change (each mole of PETN releases 2 moles CO, 3 moles CO_2, 2 moles N_2, and 4 moles $H_2O_{(g)}$), result in a very high explosive force. What total gas volume at 3000°C and 800 torr would result from the detonation of 1.00 lb (454 g) of PETN [$C(CH_2ONO_2)_4$]?

(answer, page 109)

Extra Practice
*EXERCISE 6

Liquid hydrazine, N_2H_4, can be used as the fuel for a monopropellant rocket. The hydrazine from the fuel tank is injected into a propulsion chamber containing a decomposition catalyst. For each mole of liquid hydrazine decomposed, one mole of N_2 and two moles of H_2 are formed. The gaseous products expand as the temperature is raised by the heat of decomposition, providing an added rocket thrust as the heated gases are exhausted from the propulsion chamber. How many grams of hydrazine would be required to produce a total gaseous product volume of 250 liters at 700°C and 500 torr?

(answer, page 109)

SELF-TEST (UNIT 5) [answers, page 109]

5.1. One of the earlier gases developed as a chemical warfare agent is the very toxic cyanogen chloride, prepared by treating aqueous hydrocyanic acid with chlorine:

$$HCN_{(aq)} + Cl_{2(aq)} \rightarrow HCl_{(aq)} + CNCl_{(g)}$$

How many grams of hydrogen cyanide would be required for the production of 50 liters of cyanogen chloride (measured at STP), assuming 100% efficiency?

5.2. With the current interest in locating alternative sources of methane to supplement dwindling supplies of "natural gas", consideration is being given to the production of methane from coal. When steam is passed over a bed of heated coal, a mixture of carbon monoxide and hydrogen (with minor amounts of other gases) is formed. This mixture, called "water gas", was used as a fuel gas in the days of the "gaslight era", but it is much more hazardous for home use than methane. It is possible to combine carbon monxide and hydrogen, using a suitable catalyst, to form methane:

$$CO_{(g)} + 3\ H_{2(g)} \rightarrow CH_{4(g)} + H_2O_{(g)}$$

What total volume of reactant gases, in the stoichiometric ratio of 1 CO: $3H_2$, at $500^{o}C$ and 76,000 torr would be required for the production of 50 g of methane, assuming complete conversion?

- -

If you completed Self-Test questions 5.1 and 5.2 correctly, you may go on the proficiency level, try the RELEVANT PROBLEMS (Unit 5), or stop here. If not, you should consult your instructor for suggestions of further study aids.

- -

*5.3. Copper statues, or other copper objects, exposed to moist air gradually acquire a "patina" of malachite green, formulated as $Cu_2CO_3(OH)_2$. The reaction consumes one mole each of carbon dioxide, oxygen, and water per mole of the salt formed. How many grams of copper would be converted to malachite green by reaction with 75 liters of CO_2 (measured at $47^{o}C$ and 700 torr) and equivalent amounts of O_2 and H_2O?

- -

If you answered this question correctly, you may go on to the RELEVANT PROBLEMS for Unit 5. If not, you should consult your instructor for suggestions of further study aids.

- -

ANSWERS to EXERCISES, Unit 5

1. (24 g) Solution:

$$\cdots + \boxed{2H_2O_{(\ell)}} \rightarrow \cdots + \boxed{\overset{\text{15 liters}}{C_2H_{2(g)}}}$$

(2 moles x 18.0 g mole^{-1}) (1 mole x 22.4 liters mole^{-1})

$$\frac{w}{(2 \times 18.0 \text{ g})} = \frac{15 \text{ liters}}{22.4 \text{ liters}}$$

$$w = \frac{(2 \times 18.0) \text{ g} \times 15 \text{ liters}}{22.4 \text{ liters}}$$

2. (80 liters) Solution:

Step 1 -

$$\overset{\text{28.4 g}}{\boxed{4C_3H_5(ONO_2)_{3(\ell)}}} \rightarrow \overset{V_{STP}}{\boxed{12CO_{2(g)} + 10H_2O_{(g)} + 6N_{2(g)} + O_{2(g)}}}$$

(4 moles x 227 g mole^{-1}) (29 moles x 22.4 liters mole^{-1})

[12+10+6+1]

$$V_{STP} = \frac{(29 \times 22.4) \text{ liters} \times 28.4 \text{ g}}{(4 \times 227) \text{ g}} = 20.3 \text{ liters}$$

Step 2 - (1000°C + 273° = 1273°K)

$$V_{corr} = 20.3 \text{ liters} \times \frac{1273°}{273°} \times \frac{760 \text{ torr}}{900 \text{ torr}}$$

3. (1650 liters) Solution:

$$\overset{\text{2800 g}}{\boxed{C_8H_{18}}} + \cdots \rightarrow \overset{V_{STP}}{\boxed{3 CO_{(g)}}} + \cdots$$

(1 mole x 114.2 g mole^{-1}) (3 moles x 22.4 liters mole^{-1})

$$V_{STP} = \frac{(3 \times 22.4) \text{ liters} \times 2800 \text{ g}}{114.2 \text{ g}}$$

4. (2.57 g) Solution:

(Since the volume ratio of gaseous reactants is the same as the mole ratio expressed by the balanced equation, we can work with a single, combined, volume/molar volume term.)

Step 1 - $V_{STP} = 3.00 \text{ liters} \times \dfrac{273°}{473°} \times \dfrac{1200 \text{ torr}}{760 \text{ torr}} = 2.74 \text{ liters}$

Step 2 -   2.74 liters   w

$$\boxed{NH_{3(g)} + 2O_{2(g)}} \quad \rightarrow \quad \boxed{HNO_3}$$

(3 moles x 22.4 liters mole^{-1})   (1 mole x 63.0 g mole^{-1})

[1+2]

$$w = \frac{63.0 \text{ g} \times 2.74 \text{ liters}}{(3 \times 22.4) \text{ liters}}$$

===

*5.   (4.03 x 10^3 liters) Solution:

To find: total gas volume at 3000°C (3273°K) and 800 torr

Information available: mass and formula of PETN, mole ratio

Information needed, not given: molar volume (from memory, 22.4 liters per mole @ STP), T and P-ratios, formula weight of PETN

Set-up (using unity factors and combined gas volumes):

$$\left[\frac{454 \text{ g(PETN)}}{1} \times \frac{1 \text{ mole(PETN)}}{316 \text{ g(PETN)}} \times \frac{11 \text{ moles(gas)}}{1 \text{ mole(PETN)}} \times \frac{22.4 \text{ liters}}{1 \text{ mole(gas)}}\right] \times \left[\frac{3273°}{273°} \times \frac{760 \text{ torr}}{800 \text{ torr}}\right]$$

[combined gas volume of (2+3+2+4) moles @ STP]   [STP→3000°C, 800 torr]

*6.   (22 g) Solution:

$$\left[\frac{250 \text{ liters(gas)}}{1} \times \frac{273°}{973°} \times \frac{500 \text{ torr}}{760 \text{ torr}}\right]\left[\frac{1 \text{ mole(gas)}}{22.4 \text{ liters(gas)}}\right]\left[\frac{1 \text{ mole}(N_2H_4)}{3 \text{ moles(gas)}} \times \frac{32.0 \text{ g}(N_2H_4)}{1 \text{ mole}(N_2H_4)}\right]$$

[total gas volume @ 700°C,   [@ STP]   [1 N_2 + 2H_2]
500 torr → STP]

===

ANSWERS to SELF-TEST, Unit 5

5.1.   60 g

5.2.   7.9 liters

*5.3.   334 g

A.1. Gaseous losses of soil nitrogen may occur when nitrites in a slightly acidic soil
are brought in contact with simple organic nitrogen compounds, such as urea. This
type of gaseous loss is strictly chemical and does not require the presence of
microorganisms or adverse soil conditions. The change occurs as shown:

$$2H^+_{(aq)} + 2NO^-_{2(aq)} + NH_2\text{-}\underset{O}{\overset{}{C}}\text{-}NH_{2(aq)} \rightarrow CO_{2(g)} + 3H_2O_{(l)} + 2N_{2(g)}$$

How many kilograms of urea must be added to a soil solution containing excess
nitrite to generate 40 liters of nitrogen (measured at STP)?

A.2. During the bacterial decomposition of organic sulfur compounds, sulfides are formed
along with other incompletely oxidized substances such as elemental sulfur. The
elemental sulfur is subject to oxidation just as are the ammonium compounds formed
when nitrogenous materials are decomposed. The oxidation of sulfur may be illustrated by the equation:

$$2S_{(s)} + 3 O_{2(g)} + 2H_2O_{(l)} \rightarrow 2H^+_{(aq)} + 2HSO^-_{4(aq)}$$

What volume of oxygen, in liters at 95°C and 748 torr, would be required to oxidize
50 grams of elemental sulfur?

A.3. Bacterial organisms, such as some <u>Pseudomonas</u> species, which are able to grow in
anoxic (or nearly anoxic) sea water and marine sediments use nitrate ions as
alternative electron acceptors instead of oxygen in the oxidation of organic matter.
The overall reaction occurs as shown by the equation:

$$4NO^-_{3(aq)} + 2H_2O_{(l)} \rightarrow 2N_{2(g)} + 5 O_{2(g)} + 4 OH^-_{(aq)}$$

Assuming 100% efficiency, how many grams of nitrate ion would be consumed by formation of a product mixture of 2.0 liters of $N_{2(g)}$ and 5 liters of $O_{2(g)}$ at 20°C and
800 torr?

A.4. Prior to man's industrialization of the world, the oxygen demands of weathering
of the land were restored by photosynthesis. The drains of carbon dioxide from
the atmosphere as a result of soil weathering and photosynthesis were restored by
precipitation reactions in the ocean, chiefly by the precipitation of carbonates and
silicates. The net reaction for silicate precipitation may be formulated as:

$$Mg^{2+}_{(aq)} + SiO_{2(dispersed)} + 2HCO^-_{3(aq)} \rightarrow MgSiO_{3(s)} + 2CO_{2(g)} + H_2O_{(l)}$$

How many grams of magnesium silicate would be precipitated during the formation of
60 liters of carbon dioxide, measured at 35°C and 790 torr?

A.5. A major source of sulfur entering the atmosphere is hydrogen sulfide originating from the bacterial reduction of sulfate in anoxic marine basins and intertidal flats. In the atmosphere, the hydrogen sulfide is converted readily to sulfur dioxide by the following overall equation:

$$2H_2S_{(g)} + 3 O_{2(g)} \rightarrow 2SO_{2(g)} + 2H_2O_{(g)}$$

What total gas volume would result, at 500°C and 820 torr, from the complete atmospheric oxidation of 100 grams of hydrogen sulfide?

*A.6. Ammonia produced by the Haber process may be utilized as a fertilizer in three different ways. It may be liquified and applied directly; it may be dissolved in water, forming ammonium hydroxide which may be applied to the crop; or the ammonia may be used to synthesize ammonium salts that can be used as fertilizers. In the reaction of ammonia with sulfuric acid, two moles of ammonia and one mole of sulfuric acid are consumed per mole of the salt, ammonium sulfate, formed. How many grams of sulfuric acid would be converted to ammonium sulfate by reaction of 75 liters of ammonia (measured at 25°C and 800 torr)?

*A.7. Phosphorus is a very critical nutrient for the growth of plants in the marine environment. A deficiency of phosphorus may prevent other nutrients from being acquired by the plants. The phosphate level in the oceans appears to be controlled by the solubility of $Ca_3(PO_4)_2$, which is in turn a function of ocean acidity. Ocean acidity is controlled by the dissolved carbon dioxide concentration. For each mole of calcium phosphate dissolved by reacting with four moles of carbon dioxide and four moles of water, three moles of calcium ion, two moles of dihydrogen phosphate ion and four moles of bicarbonate ion are formed. Assuming 73% conversion efficiency, what volume (in liters) of carbon dioxide (measured at 25°C and 830 torr) would be required to form 3.0 kg of dihydrogen phosphate ion?

*A.8. "Diammonium phosphate" is a synthetic fertilizer that is becoming increasingly important. Its phosphorus as well as its nitrogen, both vital nutrients, are water soluble. Such compounds are very important where a high degree of water solubility is required. Assuming a 90% conversion efficiency, what volume (in liters) of ammonia measured at 100°C and 1100 torr would be required to form 1200 kg of "diammonium phosphate"? The equation for this reaction may be formulated as:

$$2NH_{3(g)} + H_3PO_{4(\ell)} \rightarrow (NH_4)_2HPO_{4(s)}$$

*A.9. Facultative bacteria in marine sediments prefer elemental oxygen for their respiratory processes, but under circumstances of inadequate aeration they can use the

*Proficiency Level

combined oxygen in sulfates as shown by the equation:

$$2CH_2O_{(s)} + H^+_{(aq)} + HSO^-_{4(aq)} \xrightarrow{\text{(bacteria)}} H_2S_{(g)} + 2H_2O_{(\ell)} + 2CO_{2(g)}$$

Assuming 81% conversion efficiency, what total volume (in liters) of gaseous products measured at 75°C and 900 torr would be formed by the anaerobic decomposition of 10.0 kg of organic compounds (represented by CH_2O)?

*A.10. Life requires nitrogen that has been "fixed" through combination with other elements. A part of the natural supply of fixed nitrogen is produced in the sea by a complex system of bacterial oxidation and reduction reactions. For every four moles of ammonia and three moles of oxygen produced, the microbe Azobacter consumes two moles of nitrogen and six moles of water. How many grams of nitrogen would be required to produce a total gaseous product volume of 300 liters at 500°C and 500 torr? (Assume both oxygen and ammonia to be "gaseous products".)

ANSWERS:

(A.1.) 0.054 kg, (A.2.) 72 liters, (A.3.) 11 g, (A.4.) ~120 g, (A.5.) 345 liters, (A.6.) ≈160 g, (A.7.) 1.9×10^3 liters, (A.8.) 4.3×10^5 liters, (A.9.) 9.8×10^3 liters, (A.10.) 24.9 g

---

RELEVANT
PROBLEMS

Unit 5: Gas Volume Problems

Set B: Biological & Medical Sciences

B.1. In photosynthesis, light energy from the sun is used by green plants to remove carbon dioxide and water from the environment. For each molecule pair of carbon dioxide and water removed, part of a molecule of carbohydrate is produced and one molecule of oxygen (O_2) is returned to the environment. The overall reaction for this process may be formulated as:

$$CO_{2(g)} + H_2O_{(\ell)} \rightarrow [CH_2O]_{(s)} + O_{2(g)}$$
$$\text{("carbohydrate")}$$

How many grams of water will be consumed by excess carbon dioxide in the photosynthetic production of 100 liters of oxygen (measured at STP)?

B.2. Three general stages are recognized in the photosynthetic conversion of carbon dioxide to carbohydrate. In the first stage, catalyzed by enzymes, water is oxidized to give "bound" electrons and oxygen. In the second stage, these "bound" electrons are excited to higher energy levels by the absorption of light by photosynthetic pigment molecules (chlorophylls). How many liters of oxygen (at STP) are released to the environment by a large tree during fixation of 412 g of CO_2 by photosynthesis?

B.3. The final stage of photosynthesis is called the dark reaction because it does not directly require light. The light is only necessary in the first steps for production of the reducing agent nicotinamide adenine dinucleotide phosphate (NADPH) and of the adenosine triphosphate (ATP) necessary for completing the synthesis reactions of the third stage. The third stage is characterized by the production of carbohydrate. Assuming quantitative conversion, what mass of carbohydrate could be formed during a photosynthetic process which releases 40 liters of oxygen (at STP)? [For the equation involved, see problem B.1.]

B.4. Carbon monoxide is a non-irritating gas, without color, taste, or odor, produced by the incomplete oxidation of combustible carbon-containing material. It is responsible for a major proporation of all suicidal and accidental deaths. Inhalation of air containing only 1% of this gas may prove fatal within 10 to 20 minutes, depending upon the degree of physical activity and the respiratory rate of the individual. An improperly adjusted stove formed carbon monoxide at 780°C and 810 torr according to the simplified equation:

$$2 C_{(s)} + O_{2(g)} \rightarrow 2 CO_{(g)}$$

How many grams of oxygen will be consumed by excess carbon in the formation of 1.3×10^3 liters of carbon monoxide?

B.5. The exhaust of an automobile contains approximately 7% carbon monoxide by volume. Since the absorption of this substance may be cumulative, toxic effects depend upon the total exposure, i.e., the concentration in the inspired air and the duration of exposure. Levels of the gas in the inspired air lower than one percent may cause slowly developing damage and eventuate in death several days after the last exposure. Carbon monoxide displaces oxygen from hemoglobin to produce carboxyhemoglobin:

$$O_2Hb_{(aq)} + CO_{(g)} \rightarrow COHb_{(aq)} + O_{2(g)}$$

How many liters of oxygen (at 37°C and 740 torr) will be displaced from hemoglobin (Hb) by 10 grams of carbon monoxide?

*B.6. Nitrous oxide, N_2O, is used as an inhalation anesthetic. It is a colorless gas with a slightly sweet odor and taste and is only moderately soluble in the blood. The compound does not react with hemoglobin but is simply transported as a dissolved

*Proficiency Level

species. The maximum concentration of nitrous oxide that can be safely administered for maintenance of anesthesia is 70 percent, by volume. Frequently, nitrous oxide is administered in conjunction with another drug such as halothane or ether because the N_2O alone is not potent enough for some patients. Nitrous oxide is prepared by the controlled thermal decomposition of ammonium nitrate:

$$NH_4NO_3(s) \xrightarrow{260°} N_2O(g) + 2H_2O(g)$$

Assuming 81% conversion efficiency, how many grams of ammonium nitrate would be needed to produce a total gaseous product volume of 30.0 liters at 260°C and 600 torr?

*B.7. Chlorine, a greenish-yellow gas, is the most widely used disinfectant employed for killing bacteria in water. Moreover, chlorine is a good virucide and amebicide. Human reaction to chlorine gas is characterized by skin, eye, and respiratory tract irritation. Inhalation of chlorine gas may cause cough, pulmonary edema, hypotension and death. Chlorine reacts with and destroys lung tissue. Chlorine is produced by electrolysis from fused chlorides or aqueous solutions of alkali metal chlorides according to the equation:

$$2NaCl_{(aq)} + 2H_2O \xrightarrow{\text{(electrolysis)}} 2NaOH_{(aq)} + Cl_{2(g)} + H_{2(g)}$$

What total volume (in liters) of product gases, at 75°C and 1000 torr, would be formed by the eletrolysis of 5.0 kilograms of sodium chloride, assuming complete conversion?

*B.8. A peptic ulcer, a common digestive disturbance, may form when the digestive action of hydrochloric acid and pepsin is exerted on the wall of the stomach. The acid first erodes the mucous membrane lining and eventually eats into the deeper muscle tissue. In advanced stages, the acid may cause the complete perforation of the stomach wall, allowing gastric juices to flow directly into the abdominal cavity. Without prompt surgical repair, perforation may cause death. The chief efforts in initial treatment of the simple peptic ulcer are directed toward reducing the hyper-acidity by exercising emotional control, selection of a proper diet, and rapid removal of excess acidity with antacids.

Precipitated chalk, $CaCO_3$, is a common component of antacids used to control acidity in the stomach. Its action can be formulated by the net equation:

$$2H^+_{(aq)} + CaCO_{3(s)} \longrightarrow H_2O_{(\ell)} + CO_{2(g)} + Ca^{2+}_{(aq)}$$

How many liters of carbon dioxide would be produced by a standard dose of 2.0 grams of calcium carbonate in a gastric juice containing excess acid, at 37°C and 750 torr?

*B.9. Proteins are complex organic molecules composed of large numbers of units known as amino acids. Plant proteins consumed by animals are converted to amino acids during digestion. These amino acids are then used in synthesizing animal tissues (proteins). Both growth and repair of animal tissues depend on intake of plant proteins. Not all of the amino acid molecules absorbed by the blood are used in cell protein synthesis. Some are broken down by a chemical activity of the liver, called oxidative deamination, into two parts. The carbon fragment forms intermediates that may be converted to carbohydrates and the nitrogen fragment is converted to ammonia, an intermediate in the formation of urea. For example, the oxidative deamination of alanine to pyruvic acid may be characterized by the equation:

$$CH_3CHCO_2^- \quad \xrightarrow{\text{(amino acid oxidase)}} \quad CH_3\overset{\overset{O}{\|}}{C}CO_2H + NH_3$$
$$\underset{NH_3}{\overset{+}{|}}$$

How many grams of pyruvic acid would be produced by this process to correspond to an amount of ammonia that would occupy a volume of 100 ml as a gas (measured at 37°C and 750 torr)?

*B.10. The large intestine serves principally to collect and dispose of materials that have not been previously digested. Undigested food passes from the small intestine to the large intestine by muscular movements similar to those in the esophagus. The material passing into the large intestine is a semisolid containing water, inorganic salts, bile salts, bile pigments, undigested food particles, unused digestive juices and epithelial tissues from the walls of the digestive tract. Chemical reactions in the large intestine are due to bacteria, which are present in large numbers. Certain of these bacteria ferment undigested carbohydrates into compounds which they can use for food and energy. The by-products of this fermentation are organic acids, such as acetic acid, and gases such as carbon dioxide and methane. These gases can cause inflation of the intestinal tract, producing a feeling of discomfort. Assuming a 75% fermentation efficiency, how many grams of glucose would be required to produce a total gaseous product of 2.5 liters at 37°C and 790 torr? The bacterial fermentation reaction is characterized by the equation:

$$C_6H_{12}O_6(s) \quad \xrightarrow{\text{(bacteria)}} \quad 2CH_3CO_2H_{(aq)} + CO_{2(g)} + CH_{4(g)}$$

ANSWERS:

(B.1.) 80.4 g, (B.2.) 210 liters, (B.3.) 54 g, (B.4.) 260 g, (B.5.) 9.3 liters, (B.6.) 18 g, (B.7.) 1.9×10^3 liters, (B.8.) 0.52 liter, (B.9.) 0.34 g, (B.10.) 12 g

RELEVANT
PROBLEMS

I.1. Sodium hypochlorite is employed as a disinfectant and deodorant in dairies, cream-
eries, water supplies, sewage disposal plants, and for household purposes. It is
also used as a bleach in laundries. As a bleaching agent, it is very useful on
cotton, linen, jute, artificial silk, and paper pulp. The equation for the most
common method of production is:

$$Cl_2 + 2NaOH \rightarrow NaCl + H_2O + NaOCl$$

What volume of chlorine (at STP) is needed to produce 438 kg of sodium hypochlorite?

I.2. Reforming catalysts have been used in order to make high-octane fuels. One type of
reaction dehydrogenates and cyclizes a straight-chain hydrocarbon. In the following
equation the raw material n-heptane has an octane number of 0. The product, toluene,
has an octane number of more than 100.

$$CH_3CH_2CH_2CH_2CH_2CH_2CH_3 \xrightarrow{\text{(catalyst)}} CH_3 + 4H_2$$

How many liters of hydrogen at STP are released in the production of 1.00 metric ton
of toluene?

I.3. Acrylic resins and plastics have wide use because of their clarity, brilliance, ease
of forming, and light weight. The applications include cockpit canopies, gun tur-
rets, spray shields, and boats. Emulsions of these resins are widely used as textile finishes, leather finishes, and paints. One of the intermediates in the production of these resins and plastics is acrylic acid. How many liters of total
gaseous reactants (at STP) are needed to produce 437 kg of acrylic acid?

$$C_2H_2{}_{(g)} + H_2O{}_{(g)} + CO{}_{(g)} \rightarrow C_2H_3COOH$$

I.4. The explosive properties of dynamite and nitroglycerine are due partly to the large
energy change associated with the decomposition of the nitrate esters, but more
important is the large volume change resulting from the formation of gaseous products. Such a volume change, occurring at a very rapid rate, in any confined space
produces an enormous sudden pressure. The explosive decomposition of nitroglycerine
is approximated by:

$$4\ C_3H_5(NO_3)_3{}_{(\ell)} \rightarrow 12\ CO_2{}_{(g)} + 10\ H_2O{}_{(g)} + 6\ N_2{}_{(g)} + O_2{}_{(g)}$$

116

The energy released by the explosion of relatively small amounts of this compound may produce temperatures as high as 3000°C. What change in volume would result from the explosive decomposition of 1.00 liter of liquid nitroglycerine (density 1.59 g ml^{-1}) to produce gaseous products at 2500°C in an open field where the atmospheric pressure was 700 torr?

I.5. What pressure, in torr, would result if the decomposition of 1.00 liter of liquid nitroglycerine occurred in a closed room having a volume of 300 m^3 (with original room air at 25°C and 700 torr), with the gaseous products at 2500°C? (Assume that room air was also heated to 2500°C.)

*I.6. Dichlorodifluoromethane, with the trade name of Freon-12, has been one of the more important of the commercial organic halides. Freon-12 was first produced as a refrigerant. It first appeared as an aerosol propellant in 1945. It was considered particularly desirable for this use because of its nonflammability, stability, lack of odor, and extremely low toxicity. Probably the first application of dichlorodifluoromethane as an aerosol propellant was in the "bug bombs" used by the military during World War II. In 1976, the formulations which employed the aerosol propellants included approximately 500 different products. In addition to the wide applications of dichlorodifluoromethane as an aerosol propellant, which are rapidly disappearing because of environmental concerns, it still finds extensive use in the refrigeration field. In general, chlorofluorocarbons account for 85% of the refrigerant market and in industrial refrigeration, where very low temperatures are required, dichlorodifluoromethane is preferred. A commercial grade of hydrogen fluoride, HF, was allowed to react with carbon tetrachloride in the presence of an antimony trifluoride catalyst:

$$CCl_4 + 2HF \xrightarrow{\text{SbF}_3} CF_2Cl_2 + 2HCl$$

The product, 2.1×10^5 ℓ of dichlorodifluoromethane, was obtained at 735 torr and 60°C. What was the volume of CF_2Cl_2 at STP? If the yield of Freon-12, based on HF used, was 80%, what volume of HF gas at 735 torr and 60°C was required?

*I.7. Solid carbon dioxide, "dry ice", is a unique industrial product of importance in refrigeration of ice cream, meats, and other foods. In addition to its refrigerating effect it has the added advantage that a carbon dioxide atmosphere reduces meat and food spoilage by bacteria. Carbon dioxide is also useful for neutralizing alkalies and has advantages over ordinary acids because it is easily shipped in the solid form, is noncorrosive, and is light in weight. Chemically it is equivalent to approximately 2.5 times its mass in sulfuric acid and about 5 times its mass in

*Proficiency Level

hydrochloric acid. With respect to food refrigeration, "dry ice" is primarily a transport refrigerant, more desirable than water ice because of its dryness, its low temperature, and the insulating and desiccating action of the gas evolved as it sublimes. About 454 kg of dry ice will chill a refrigerator car for a transcontinental rail trip without recharging, whereas the amount of water ice required is about 1800 kg initially, with two or three rechargings. Carbon dioxide used for the preparation of dry ice is usually obtained from one of three different sources: the combustion of carbon, the fermentation of sugar, or the extraction of CO_2 from the flue gases of other industrial operations. What volume of dry oxygen at 27° and 750 torr would be required to prepare enough carbon dioxide, by the combustion of carbon, to make a 100 kg block of "dry ice"?

*I.8.   "Dry ice" is produced by chilling and pressurizing gaseous carbon dioxide to form a liquid. The chilled liquid is allowed to vaporize through a nozzle where it forms a snow, because of the additional cooling achieved by the rapid expansion. The snow is formed inside a press where it is converted into 100 kg blocks under a high pressure. The blocks are quartered, and shipped as 25 cm cubes wrapped in kraft paper. Determine the approximate volume of dry carbon dioxide gas, at STP, that must be condensed to form a 100 kg block of dry ice.

*I.9.   Hydrogen cyanide is used as a fumigant and for the production of a number of other compounds. The reaction of HCN with either acetylene or ethylene oxide will produce acrylonitrile, although these are no longer the principal methods for making acrylonitrile. Hydrogen cyanide can be produced by the following high-temperature reaction.

$$2NH_3 + 3 O_2 + 2CH_4 \rightarrow 2HCN + 6H_2O$$

If all species are gases, what is the volume change in the production of 58 g of HCN, at 1000°C and 1.00 atm?

*I.10. Over 90% of trichloroethylene is used in metal degreasing. The rest is used in solvent extractions, such as making decaffeinated coffee. Most of the trichloroethylene is made by a two step process:

$$2Cl_{2(g)} + C_2H_{2(g)} \rightarrow C_2H_2Cl_{4(\ell)}$$
$$C_2H_2Cl_{4(\ell)} \rightarrow C_2HCl_{3(\ell)} + HCl_{(g)}$$

The second step is 90% efficient and the process is carried out at 300°C and one atmosphere pressure. The first step occurs at 50°C. How many liters each of chlorine and acetylene (measured at 50°C, 1.00 atm) are necessary to produce 158 ℓ of HCl as a by-product (measured at 300°C, 1.00 atm)?

ANSWERS:

(I.1.) 1.32×10^5 liters, (I.2.) 9.7×10^5 liters, (I.3.) 4.08×10^5 liters, (I.4.) 1.25×10^4 liters, (I.5.) 6.5×10^3 torr, (I.6.) 1.7×10^5 liters $[CF_2Cl_2]$, 5.3×10^5 liters [HF], (I.7.) 5.67×10^4 liters, (I.8.) 5.09×10^4 liters, (I.9.) ~110 liters, (I.10.) 198 liters $[Cl_2]$, 99 liters $[C_2H_2]$

RELEVANT
PROBLEMS

Unit 5: Gas Volume Problems

Set E: Environmental Sciences

E.1. Aerobic digestion is a process in which excess biological sludge is aerated for long periods of time, resulting in the oxidative destruction of cellular solids. If the bacterial cells are represented by the simplified chemical formula $C_5H_7NO_2$, then the cellular destruction through aerobic digestion is represented by:

$$C_5H_7NO_2 + 5\ O_2 \rightarrow 5CO_2 + 2H_2O + NH_3$$

How many liters of combined CO_2 and NH_3 are produced at STP by destruction of 2.5 g of cells?

E.2. A potential source of air pollution is the commercial production of mercury. Nearly all mercury is produced from the sulfide ore cinnabar (HgS) by means of roasting the ore in air. The reaction is:

$$HgS + O_2 \rightarrow Hg + SO_2$$

Although both the mercury and the SO_2 are intended for complete collection, some of both products may escape into the atmosphere. How much SO_2 gas is produced at STP when 43 metric tons of Hg are made?

E.3. Smelters account for about 15% of the sulfur oxide pollution, which is accompanied by a consumption of atmospheric oxygen. A reason for this is the natural occurrence of many useful elements in the form of sulfide ores. For example, sulfide ores are important sources of copper, zinc, mercury (problem E.2), and lead. A typical re-action is the production of lead oxide.

$$2PbS + 3\ O_2 \rightarrow 2PbO + 2SO_2$$

How many liters of oxygen (measured at STP) are required for production of 327 kg of PbO?

E.4. <u>Dacron</u> is a form of polyethylene terephthalate. An intermediate, terephthalic acid, is made by the following reaction:

$$C_6H_4(CH_3)_2 + 4HNO_3 \rightarrow C_6H_4(COOH)_2 + 4H_2O + 4NO$$

In one year 2.04×10^8 kg of terephthalic acid was manufactured by this process. The conditions are 149°C and 13.6 atmospheres pressure. How many liters of the pollutant, nitric oxide, were produced at reaction conditions during that year?

E.5. Between 1963 and 1970, carbon monoxide emissions from new cars were reduced by 65 percent from previous uncontrolled levels. The Clean Air Act Amendment of 1970 required that carbon monoxide emission must be 90 percent lower in 1975 and later models vehicles than they were in 1970 models. Most methods for reducing carbon monoxide tend to increase the amount of nitrogen oxides produced in the combustion process. Automobile manufactures have reduced carbon monoxide emissions by increasing the amount of air in air-fuel mixtures and are experimenting with carburetor refinements and other engine design changes and with such anti-pollution devices as catalytic converters to further limit those emissions. They are attacking the nitrogen oxides problem with catalytic converters, changes in compression ratio, and other methods. Studies show that although emission-reducing engineering changes such as lower engine compression ratios, leaner air-fuel mixtures, and exhaust recirculating do indeed reduce emissions, they also cut mileage and performance. As a result pollution <u>per gallon of gas</u> burned is still a major problem. An untuned car engine can produce 710 liters of exhaust per mile at a temperature of 27°. How many moles of carbon monoxide are delivered to the atmosphere on a 100 mile trip if the exhaust is 5.8% carbon monoxide, by volume? (Assume a pressure of 760 torr.)

*E.6. The problem carbon monoxide presents to humans is the strong bond it forms with the iron in hemoglobin. This can be shown by the following equation, in which HE represents a heme group.

$$CO + HE\text{-}FeO_2 \rightarrow HE\text{-}FeCO + O_2$$

By combining with hemoglobin, CO "starves" the body of oxygen. This produces listlessness and can lead to death. The reaction is reversible, so that pure oxygen can release the bound CO. In some cities where policemen stand directing traffic in the swirl of auto exhaust fumes, "oxygen breaks" are given them each half hour. What volume of pure oxygen (at 40°C and 800 torr) is needed to replace 23 mg of bound carbon monoxide?

*E.7. Hydrogen sulfide is one of the air pollutants from certain industrial processes. The paint industry has found that adding 5.0% by weight of mercury phenylacetate

[Hg(O$_2$CCH$_2$C$_6$H$_5$)$_2$] will prevent mildew. In H$_2$S-polluted air the mercury additive can be the cause of darkening of the paint, since mercury(II) ion reacts with the hydrogen sulfide to form black mercuric sulfide.

$$Hg^{2+} + H_2S \rightarrow HgS + 2H^+$$

What volume of H$_2$S at 25°C and 700 torr will react with all the mercury ion in 12 liters of paint (density 2.78 g ml^{-1})?

*E.8. The acute irritant effect of sulfur dioxide is related to its high solubility, which is 280 volume percent in water at 37°C. Accidental inhalation of high concentrations of SO$_2$ results in rapid solution in the mucous membranes of the upper airways, with the formation of the irritant compound H$_2$SO$_3$. The initial irritation causing bronchorrhea is rapidly followed by sloughing and a necrotizing bronchitis. If exposure is prolonged or heavy, the lower respiratory tract becomes exposed, with the development of an acute chemical pulmonary edema. Recovery may be complete. However, residual mucosal scars and submucosal thickening are not uncommon, particularly if the mucosal sloughs are deep and secondary infection is prominent. Chronic irritation from long-term exposure to low doses of sulfur dioxide, as in polluted city air, is very important in terms of community health, particularly in relation to the development of chronic obstructive lung disease. Twenty ppm (by volume) of sulfur dioxide constitutes the maximum allowable concentration for prolonged exposure in man. The "standard man" inhales about 2.0 x 10^4 liters of air during a 24 hour period. What volume of SO$_2$ would the "standard man" inhale during a 24 hour period if the atmospheric SO$_2$ pollution level is 20 ppm? How many moles of sulfurous acid could this produce?

*E.9. High temperature combustion is the major man-made source of nitric oxide (NO) and nitrogen dioxide (NO$_2$). The primary means for their control is preventive, that is, avoiding their formation rather than trying to recapture or neutralize them after formation. This involves carrying out combustion at the lowest possible temperatures and the lowest excess air consistent with complete oxidation. This is not always compatible with the needs of utility and industrial furnaces (so-called stationary sources) or of gasoline engines, since heat reduction is reflected in loss of power. As a result, other procedures are necessary to reduce emissions to acceptable levels. The primary obstacles to attaining satisfactory controls are the time and money required to complete the monumental task of replacing or modifying the vast number of furnaces and engines which were not designed to meet current air quality standards. Utility plants now under construction are being designed to reduce nitrogen oxide emissions to standards which have been established by the Environmental Protection

Agency (EPA). Flue gas recirculation and staged combustion are two methods being applied. Every thousand cars crawling through congested urban streets daily discharge, on the average, 114 kg of nitrogen dioxide into the air. What volume of atmospheric nitrogen (STP) is fixed (i.e., combined with oxygen) per thousand cars?

*E.10. Nitric oxide (NO) is produced simultaneously during the formation of nitrogen dioxide (NO_2). Upon exposure to oxygen in the atmosphere, the nitric oxide is photochemically oxidized to nitrogen dioxide. Assuming that 22.7 kg of nitric oxide are emitted daily for every thousand cars moving through congested streets, determine the volume (liters) of nitrogen dioxide (STP) that would be produced by its oxidation.

ANSWERS:

(E.1.) 3.0 liters, (E.2.) 4.8×10^6 liters, (E.3.) 4.92×10^4 liters, (E.4.) 1.25×10^{10} liters, (E.5.) ~170 moles, (E.6.) 20 ml, (E.7.) 94 liters, (E.8.) 400 ml, 1.8×10^{-2} mole, (E.9.) 2.78×10^4 liters, (E.10.) 1.69×10^4 liters

STOICHIOMETRY

UNIT 6: SOLUTION CHEMISTRY

Some of the most interesting of chemical phenomena occur in aqueous solutions. The chemical and biochemical degradation of dissolved wastes in rivers, lakes, and seas is of vital concern to environmentalists. The analysis of dissolved species in blood and urine provide the medical profession with valuable clues to health problems. Engineers play vital roles in the development of water purification systems or industrial plants producing solutions ranging from antiseptics to zoological preservatives.

The properties of aqueous solutions depend both on the nature of the dissolved species (solute) and the concentration of the solution (e.g., ratio of solute quantity to solution volume). The key to solution stoichiometry is the use of solution volume coupled with some convenient concentration expression. For chemical applications, molarity (symbolized as M) is a particularly useful concentration unit, as defined by:

$$M = \frac{no.\ moles\ (solute)}{no.\ liters\ (solution)}$$

This "working definition" permits us to write:

$$no.\ moles\ (solute) = M \times no.\ liters\ (solution)$$

or, for smaller scale applications:

$$no.\ mmols^1\ (solute) = M \times no.\ ml\ (solution)$$

The combination of solution volume and molarity will prove most useful in stoichiometric calculations based on chemical reactions in aqueous solutions.

- References -

1. O'Connor, Rod. 1977. Fundamentals of Chemistry, second edition. New York: Harper & Row (Units 13, 19, 22, and Excursion 2)

2. Brown, Theodore, and H. Eugene LeMay, Jr. 1977. Chemistry: The Central Science. Englewood Cliffs, N.J.: Prentice-Hall (Chapters 12, 16, 18)

3. Masterton, William, and Emil Slowinski. 1977. Chemical Principles, fourth edition. Philadelphia: W.B. Saunders (Chapters 12, 13, 20)

4. Nebergall, William, F.C. Schmidt, and H.F. Holtzclaw, Jr. 1976. College Chemistry, fifth edition. Lexington, MA: D.C. Heath (Chapters 13, 15, 16)

OBJECTIVES:

(1) *Given the chemical formula of the solute, the total solution volume, and either the molarity or the solute mass, be able to calculate the solute mass or molarity, respectively.*

---

^{1}1 mmol = 10^{-3} mole

(2) Given the balanced equation for a chemical process in aqueous solution, the molarity and volume of one reactant solution and either the molarity or the volume of another reactant solution, be able to calculate the volume or molarity of the second reactant solution.

*(3) Given the description of a chemical process involving aqueous solutions, including data from which necessary concentrations, volumes, and mole ratios can be determined, be able to make the stoichiometric calculation indicated.

PRE-TEST:

Necessary Atomic Weights:		
H (1.008)	O (16.00)	Na (22.99)
S (32.06)	Cl (35.45)	

(1) Sulfuric acid is one of the most versatile commercial chemicals, finding applications in such diverse industrial products as agrichemicals, detergents, and explosives. Two of the more familiar uses of the acid are in liquid drain cleaners and as the electrolyte in automobile batteries. A nationally-advertised liquid drain cleaner contains 175 g of H_2SO_4 per 100 ml of solution. What is its molarity? _____ The acid in an ordinary automobile battery is 5.00 $\underline{M}$ H_2SO_4. How many grams of sulfuric acid are contained in 500 ml of this solution? _____

(2) Sodium thiosulfate, photographer's "hypo", is used in the "fixing bath" for exposed film to wash out residual silver bromide as the soluble $[Ag(S_2O_3)_2]^{3-}$ complex. "Hypo" solutions may be checked periodically by a technique called titration. A measured volume of the thiosulfate solution is withdrawn and allowed to react with increments of a triiodide solution of known concentration, added from a calibrated measuring device called a burette. A little starch added to the thiosulfate before titration serves to indicate when reaction is complete by formation of a blue color when a tiny excess of triiodide is added. The net reaction may be formulated as:

$$2\ S_2O_3^{2-} + I_3^- \rightarrow S_4O_6^{2-} + 3\ I^-$$

In a particular test run, 25.0 ml of a thiosulfate solution required 18.2 ml of 0.156 $\underline{M}$ I_3^- for complete reaction. What was the molarity of the "hypo" solution tested? _____

*Proficiency Level

*(3) "Recycled" paper tends to be grey in color, but the fibers may be soaked in a chlorine bleach to "whiten" them. Before the fibers are then formed into paper, they must be washed with a sodium thiosulfate solution to remove excess chlorine. Each mole of thiosulfate reduces four moles of chlorine to chloride ion. A stock solution of sodium thiosulfate was prepared by using the "recipe" of 400 g of the pentahydrate salt ($Na_2S_2O_3 \cdot 5\ H_2O$) in sufficient water to form 1.00 liter of solution. What is the molarity of this solution? _____

How many liters of this solution would be required for the reduction of 20 lb (9072 g) of chlorine to chloride? _____

Answers and Directions:

(1) 17.8 $\underline{M}$, 245 g, (2) 0.227 $\underline{M}$

*If both are correct, go on to question *3. If you missed either, study METHODS, sections 6.1 and 6.2.*

*(3) 1.61 $\underline{M}$, 20 liters

If both answers are correct, go on to RELEVANT PROBLEMS, Unit 6. If you missed either answer, study METHODS, section 6.3.

METHODS

6.1 Molarity as a Concentration Unit

Since the working definition of molarity (symbolized by $\underline{M}$) is:

$$\underline{M} = \frac{\text{no. moles (solute)}}{\text{no. liters (solution)}}$$

it is a relatively simple procedure to use a given solution volume with molarity to find solute mass or with solute mass to find molarity. Both calculations are essentially mole/mass conversions as studied in Unit 3.

If we know the solution volume and the mass of solute used, then we may simply convert mass to moles, using the unity factor 1 mole/form. wt. (g), and "plug this in" the equation defining molarity. If, on the other hand, it is the solute mass that we wish to calculate, a reverse sequence must be used:

Step 1 - Find moles (solute) from

no. moles (solute) = $\underline{M}$ x no. liters (solution).

Step 2 - Convert moles to mass by using the unity factor:

$$\frac{\text{form. wt. (g)}}{\text{1 mole}}$$

EXAMPLE 1

Vinegar is a dilute aqueous solution of acetic acid, containing pigments and flavoring agents derived from the particular source of the vinegar (e.g., apples for "cider" vinegar or grapes for "wine" vinegar). Titration analysis of a commercial vinegar sample showed the solution to be 0.64 $\underline{M}$ in acetic acid. How many grams of CH_3CO_2H are present in 1.00 pt (473 ml) of this vinegar?

SOLUTION:

First, find the number of moles of acetic acid in 473 ml of the vinegar.

no. moles (solute) = $\underline{M}$ x no. liters (solution)

$$\text{no. moles}(CH_3CO_2H) = \frac{0.64 \text{ mole}(CH_3CO_2H)}{1 \text{ liter (vinegar)}} \times \frac{473 \text{ ml (vinegar)}}{1} \times \frac{1 \text{ liter}}{1000 \text{ ml}} =$$

$$0.30 \text{ mole}(CH_3CO_2H)$$

Then, using a mass/mole unity factor including the formula weight (60.1) of acetic acid, convert 0.30 mole to grams.

$$\frac{0.30 \text{ mole}(CH_3CO_2H)}{1} \times \frac{60.1 \text{ g}(CH_3CO_2H)}{1 \text{ mole}(CH_3CO_2H)} = \underline{18 \text{ g}}$$

EXERCISE 1

Narcotics agents often use chemical reagents for making quick field tests on substances suspected of containing illegal drugs. Preliminary test results may provide sufficient evidence for the arrest and detention of a suspect, but more extensive analyses are normally performed in forensic laboratories to supply sufficient information for court action. A variety of test reagents are available for indicating the possible presence of drugs of different types. Scheibler's reagent, for example, can be used to detect alkaloid drugs, such as those derived from the opium poppy. This reagent is prepared by dissolving 20 g of sodium tungstate, $Na_2WO_4 \cdot 2H_2O$, in sufficient 0.10 $\underline{M}$ sodium phosphate solution (containing a little nitric acid) to form 100 ml of final reagent solution. What is the molarity of tungstate in the final solution? How many grams of Na_3PO_4 are needed for preparation of 250 ml of the 0.10 $\underline{M}$ sodium phosphate solution?

(answers, page 132)

6.2 Solution Stoichiometry Using Molarities

The relationship between molarity, volume, and number of moles (or mmols, using ml volume units) provides us with a simple way of making mole ratio (or mmol ratio) calculations based on a balanced equation. The technique is similar to that developed for mass/mass problems (Unit 4), except that formula weights need not be determined.

Using the two species of interest from the balanced equation, the denominators for the ratio terms are simply the coefficients shown in the equation for those reagents. These numbers indicate reacting ratios by single formula units, by moles, by mmols, or by any other proportional units. Numerators for the respective ratios are expressed as moles (M x no. liters) or mmols (M x no. ml), whichever is most convenient, provided only that both numerators use the same units. The unknown quantity, either volume or molarity of one reagent, is then found by equating the two ratios formed.

EXAMPLE 2

Phosphoric acid may be prepared commercially by heating crushed phosphate rock with sand and coke in the presence of air and absorbing the resulting P_4O_{10} vapor in water. The aqueous acid finds applications in the production of fertilizers (e.g., $(NH_4)_3PO_4$), detergent additives, tile-cleaning preparations, insecticides, and a host of other chemicals. Quality-control laboratories continually monitor the phosphoric acid produced to be sure it meets announced specifications. A check on the concentration of the product solution can be made quickly by diluting a measured aliquot for titration with sodium hydroxide, according to the equation:

$$H_3PO_4 + 3NaOH \rightarrow Na_3PO_4 + 3H_2O$$

What was the concentration (molarity) of a diluted sample, 25.0 ml of which was exactly neutralized by 31.6 ml of 0.108 M NaOH?

SOLUTION:

(Using an approach similar to that developed in Unit 4)

(25.0 ml x C) ... H₃PO₄ ... 1 mmol ... + ... (31.6 ml x 0.108 M) ... 3 NaOH → .... 3 mmols

$$\frac{25.0 \text{ ml} \times C}{1 \text{ mmol}} = \frac{31.6 \text{ ml} \times 0.108 \text{ M}}{3 \text{ mmols}}$$

$$C = \frac{31.6 \text{ ml} \times 0.108 \text{ M}}{25.0 \text{ ml} \times 3} = 0.0455 \text{ M}$$

EXERCISE 2

Lactic acid is responsible for the sour taste of sour milk and sauerkraut. The reaction of this acid with "limewater", aqueous $Ca(OH)_2$, may be represented as:

$$Ca(OH)_{2(aq)} + 2CH_3\underset{OH}{CHCO_2H} \rightarrow 2H_2O + Ca(CH_3\underset{OH}{CHCO_2})_2$$

What volume of 0.125 $\underline{M}$ calcium hydroxide would be required to neutralize all of the lactic acid in 3.50 liters of its 0.650 $\underline{M}$ solution?

(answer, page 132)

- -

Extra Practice

EXERCISE 3

Many common laudry bleaches contain sodium or calcium hypochlorite as the active ingredient. CLOROX, for example, contains approximately 52 g of NaOCl per liter of solution. What is the molarity of this solution?

(answer, page 132)

EXERCISE 4

Butyric acid, $C_3H_7CO_2H$, is responsible for the rancid odor of spoiled butter. In food quality analysis, the butyric acid content of a butter sample can be determined by extracting the acid from a weighted sample of butter and titrating a solution of the extracted acid with sodium hydroxide:

$$NaOH + C_3H_7CO_2H \rightarrow H_2O + Na(C_3H_7CO_2)$$

In a particular analysis, a 25.0 ml aliquot of a butyric acid solution was neutralized by 17.6 ml of 0.118 $\underline{M}$ NaOH. What was the molarity of the butyric acid?

(answer, page 132)

At this point you should try the competency level Self-Test questions on page 130.

*6.3 Solution Concentrations and Volumes in Stoichiometry

The principal difference between solution stoichiometry and that applied in mass/mass or gas volume problems lies in the nature of the measurements made. Instead of weighing a pure substance or measuring a gas volume directly, we now determine a chemical quantity by measuring the volume of a liquid mixture (solution) and using that information, along with some concentration data, to find indirectly the amount of solute of interest.

The "working definition" of molarity is particularly useful in solving solution

*Proficiency Level

problems:

$$\underline{M} = \frac{\text{no. moles (solute)}}{\text{no. liters (solution)}} = \frac{\text{no. mmols (solute)}}{\text{no. ml (solution)}}$$

The choice between "moles" and "mmols" is purely one of convenience. The former is more useful when volumes are measured in liters, reserving "mmols" for smaller-scale situations. Solution concentrations and volumes are easily applied in the "unity factor method" of problem solving.

*EXAMPLE 3

For prolonged space flights, both food storage and waste removal pose significant problems. One of the possibilities being investigated is the synthesis of useable foods from simple chemical species. Formaldehyde (H_2CO), which could be prepared from CO_2, can be converted to a mixture of carbohydrates by heating the aqueous aldehyde with calcium hydroxide. In a particular set of experiments, a stock solution was prepared by dissolving 460 g of formaldehyde in sufficient water to form 10.0 liters of solution. From this stock, 90.0 ml aliquots were removed and mixed with 10.0 ml portions of 0.33 $\underline{M}$ Ca(OH)$_2$ for studies of the effect of heating time and temperature on carbohydrate formation. Under the most favorable conditions studied, 58% of the formaldehyde was converted to a mixture of carbohydrates, 3.2% of which was found to be glucose, $C_6H_{12}O_6$. What was the molarity of the stock solution used? How many liters of such a solution would be needed to produce 4.0 oz (113 g) of glucose under the best conditions found in these experiments?

SOLUTION:

Molarity of the stock reagent:

$$\underline{M} = \frac{460 \text{ g } (H_2CO)}{10.0 \text{ liters (solution)}} \times \frac{1 \text{ mole } (H_2CO)}{30.0 \text{ g } (H_2CO)} = \underline{1.53 \text{ M}}$$

Total stock reagent needed for 113 g of glucose:

(Note that "mass balance" would require 6 moles of H_2CO for each mole of $C_6H_{12}O_6$ under conditions of 100% conversion efficiency.)

$$\frac{113 \text{ g}(C_6H_{12}O_6)}{1} \times \frac{1 \text{ mole}(C_6H_{12}O_6)}{180 \text{ g}(C_6H_{12}O_6)} \times \frac{6 \text{ moles}(H_2CO)}{1 \text{ mole}(C_6H_{12}O_6)} \times \frac{100}{3.2} \times \frac{100}{58} \times \frac{1 \text{ liter(solution)}}{1.53 \text{ moles}(H_2CO)} =$$

[glucose "yield" corrections]   __132 liters__

(Common sense would suggest that "food syntheses" of this type would not be practical. It would be considerably simpler to transport the 113 g of glucose than to carry the chemicals needed to make it. Fortunately, some procedures investigated show greater promise.)

*EXERCISE 5

In the fermentation of sugars by yeast in aqueous solution, enzymes (complex organic catalysts) are involved in various steps of the conversion. The alcohol concentration of the fermentation mixture is limited by the fact that high concentrations of alcohol deactivate some of the necessary enzymes. Thus, alcoholic beverages, such as whiskey, which have a high alcohol content have to be made by distillation of more dilute fermentation solutions. The fermentation of natural mixtures, such as grape juice, is a relatively complex process. We can, however, gain some insights by considering the fermentation of pure glucose ($C_6H_{12}O_6$), which results in the formation of two moles each of ethanol (C_2H_5OH) and carbon dioxide per mole of glucose consumed. If the yield of ethanol is 85% in a particular glucose-fermentation method, how many grams of glucose would be needed to produce 250 ml of 1.3 $\underline{M}$ ethanol? (This is the concentration of ethanol in an "average" beer.)

(answer, page 133)

Extra Practice

*EXERCISE 6

One of the earliest methods of testing the "hardness" of water (i.e., its content of ions such as Ca^{2+}, which precipitate soaps) involved the titration of the water sample with a soap solution of known concentration. Since the mixture will not form a stable foam until all the insoluble calcium soaps have precipitated, the endpoint of the titration is determined by continuously shaking the mixture during addition of the soap solution until stable "soap suds" are observed, indicating that a slight excess of soap has been added. Two moles of soap are consumed per mole of Ca^{2+} in the water sample. How many milligrams[2] of Ca^{2+} are contained in 1.00 liter of a solution from which a 25.0 ml aliquot required 16.8 ml of a standardized soap titrant labeled "6.25×10^{-3} $\underline{M}$ soap", assuming all "hardness" is due to Ca^{2+} ?

(answer, page 133)

SELF-TEST (UNIT 6)   [answers, page 133]

6.1 A solution containing 360 g of potassium hydroxide (KOH) per liter is frequently

---

[2]A concentration expressed in mg per liter is commonly referred to as "parts per million" (ppm), a useful unit for very dilute solutions. Strictly speaking, of course, ppm refers to mg(solute) per $\underline{kg}$ (solution), but the density of a $\underline{dilute}$ aqueous solution is approximately 1.0 g ml^{-1} (i.e., a liter of the solution weighs about 1.0 kg).

used as a carbon dioxide-absorbant. What is the molarity of this solution? After use, a 250 ml portion of this solution was found to be 2.11 $\underline{M}$ in potassium carbonate (from $2KOH + CO_2 \rightarrow K_2CO_3 + H_2O$). How many grams of potassium carbonate had been formed in this solution?

6.2 Oxalic acid, a poisonous compound, is found in certain vegatables such as spinach and rhubarb (but in concentrations well below toxic limits). The manufacturers of a "spinach juice concentrate", however, routinely test their product by an oxalic acid analysis to be certain of avoiding any problems from an unexpectedly high concentration of this chemical. A titration with potassium permanganate is used for the oxalic acid assay, according to the net equation:

$$5H_2C_2O_4 + 2MnO_4^- + 6H^+ \rightarrow 10CO_2 + 2Mn^{2+} + 8H_2O$$

What is the molarity of an oxalic acid solution requiring 23.2 ml of 0.127 $\underline{M}$ permanganate for a 25.0 ml aliquot of the solution?

If you completed the Self-Test questions 6.1 and 6.2 correctly, you may go on to the proficiency level, try the RELEVANT PROBLEMS (Unit 6), or stop here. If not, you should consult your instructor for suggestions of further study aids.

----

*6.3 Consumer-Protection groups have long been concerned with "truth-in-labelling" requirements. One of their interests has been the use of mixed-fiber fabrics for clothing improperly labelled as "100% wool". A solution called Schweitzer's reagent can be used for testing presumably "woolen" fabrics. Prepared by bubbling air for about an hour through a mixture of 50 g of fine copper turnings and 300 ml of 15 $\underline{M}$ ammonia, this reagent will dissolve cotton, linen, silk; and some synthetic fibers. True wool fibers are left undissolved. The exact composition of the reagent cannot be determined from its "recipe", since some of the original ammonia is swept out by the air stream and a variety of complex copper species are formed. The original reagent prepared cannot, of course, be used repeatedly for fabric testing and a new mixture must be prepared when tests show the ammonia content to be below a specified limit. What percentage of the original ammonia in the "recipe" was lost by vaporization and use of a portion of the mixture from which a 1.00 ml aliquot, diluted to 50.0 ml, was neutralized by 33.7 ml of 0.178 $\underline{M}$ H_2SO_4? Assume that all ammonia (both free and in complexes) was neutralized and that each mole of sulfuric acid consumed two moles of ammonia.

*Proficiency Level

If you answered this Self-Test question correctly, you may go on to the RELEVANT PROBLEMS for Unit 6. If not, you should consult your instructor for suggestions of further study aids.

ANSWERS to EXERCISES, Unit 6

1. (0.61 $\underline{M}$, 4.1 g) Solution:

for sodium tungstate (formula weight 330)

$$\underline{M} = \frac{20 \text{ g}(Na_2WO_4 \cdot 2H_2O)}{0.100 \text{ liter (solution)}} \times \frac{1 \text{ mole}(Na_2WO_4 \cdot 2H_2O}{330 \text{ g}(Na_2WO_4 \cdot 2H_2O)}$$

for sodium phosphate (formula weight 164)

$$\text{no. moles}(Na_3PO_4) = \frac{0.10 \text{ mole}(Na_3PO_4)}{1 \text{ liter}} \times \frac{0.250 \text{ liter}}{1} = 0.025 \text{ mole}$$

$$\frac{0.025 \text{ mole}}{1} \times \frac{164 \text{ g}(Na_3PO_4)}{1 \text{ mole}}$$

2. (9.10 liters) Solution:

(V x 0.125 $\underline{M}$)   (3.50 liters x 0.650 $\underline{M}$)

$$\boxed{Ca(OH)_2}$$
1 mole

+

$$\boxed{\begin{array}{c} 2 \text{ CH}_3\text{CHCO}_2\text{H} \\ | \\ \text{OH} \end{array}}$$
2 moles

$$V = \frac{3.50 \text{ liters} \times 0.650 \underline{M}}{2 \times 0.125 \underline{M}}$$

3. (0.70 $\underline{M}$) Solution:

$$\frac{52 \text{ g}(NaOCl)}{1 \text{ liter (solution)}} \times \frac{1 \text{ mole}(NaOCl)}{74.4 \text{ g}(NaOCl)}$$

4. (0.083 $\underline{M}$) Solution:

(17.6 ml x 0.118 $\underline{M}$)   (25.0 ml x C)

$$\boxed{NaOH}$$
1 mmol

+

$$\boxed{C_3H_7CO_2H}$$
1 mmol

$$C = \frac{17.6 \text{ ml} \times 0.118 \underline{M}}{25.0 \text{ ml}}$$

- UNIT 6 -

*5. (34 g) <u>Solution</u>:

$$\frac{1.3 \text{ moles}(C_2H_5OH)}{1 \text{ liter(solution)}} \times \frac{0.250 \text{ liter(solution)}}{1} \times \frac{1 \text{ mole}(C_6H_{12}O_6)}{2 \text{ moles}(C_2H_5OH)} \times \frac{180 \text{ g}(C_6H_{12}O_6)}{1 \text{ mole}(C_6H_{12}O_6)} \times \frac{100}{85}$$

[molarity → moles ethanol → moles glucose → g of glucose (theor.) → actual g of glucose]

*6. (84.2 mg) <u>Solution</u>:

(Note that mmols and ml will be convenient units for this problem.)

$$\frac{6.25 \times 10^{-3} \text{ mmol(soap)}}{1 \text{ ml(titrant)}} \times \frac{16.8 \text{ ml(titrant)}}{25.0 \text{ ml("water")}} \times \frac{1 \text{ mmol}(Ca^{2+})}{2 \text{ mmol(soap)}} \times \frac{40.1 \text{ mg}(Ca^{2+})}{1 \text{ mmol}(Ca^{2+})} \times \frac{10^3 \text{ ml("water")}}{1}$$

==

<u>ANSWERS to SELF-TEST, Unit 6</u>

6.1. 6.42 <u>M</u>, 72.9 g

6.2. 0.295 <u>M</u>

*6.3. 20% (the mixture tested was 12.0 <u>M</u> in NH_3)

A.1. The components of sea water can be conveniently divided into two groups: major components (present in quantities greater than 1.0 mg ℓ^{-1}) and minor components (present in quantities less than 1.0 mg ℓ^{-1}). Major components are present in essentially the same proportions throughout the oceans of the world, and are independent of the total salt content, whereas the minor components vary with locality, depth, and runoff from land. The major cations, sodium, magnesium, calcium, potassium, and strontium, can be determined directly by atomic absorption spectroscopy. The standard solution used in this technique for determination of calcium is prepared by dissolving 2.497 g of dried calcium carbonate in a minimum volume of 3.0 $\underline{M}$ nitric acid and diluting to 1.0 liter. What is the molarity of Ca^{2+} ion in the final solution? How many grams of HNO_3 are needed for preparation of 300 ml of the 3.0 $\underline{M}$ nitric acid solution?

A.2. Selenium is widely distributed in nature but usually in concentrations of less than 500 ppm. It is frequently found in soils as selenide compounds, which are chemically analagous to sulfides. The element is concentrated by a number of plants, some of which cannot grow in soils containing no selenium minerals. These plants, called selenium "indicators", are useful in locating seleniferous soils. There are striking differences between the "indicator" plants and the non-indicators with respect to the amount of selenium they may absorb without injurious effects. The "indicators" may accumulate several thousand parts per million without showing signs of poisoning, whereas, the maximum for a typical food plant is approximately 300 parts per million. Selenium compounds are toxic to both man and other animals in small quantities, therefore the World Health Organization has recommended a limit of 0.05 mg $liter^{-1}$ for irrigation waters. The standard selenium solution used in the quantitative determination of selenium is prepared by dissolving 1.000 g of selenium in 80 ml of 7.85 $\underline{M}$ nitric acid and diluting to one liter. What is the molarity of selenium in the final solution? How many grams of HNO_3 are needed for preparation of 80 ml of the 7.85 $\underline{M}$ nitric acid solution?

A.3. "Sodic" soils contain sufficient sodium to interfere with the growth of most crop plants. Their exchangeable-sodium percentage is fifteen or more. The damaging effects on plants are due to the toxicity of the excess sodium ions, as well as the hydroxide ions usually found in such soil. The pH of "sodic" soils usually falls in the range 8.5 to 10.0. This extreme alkalinity, resulting from the sodium carbonate present, causes the soil surface to be discolored by the dispersed humus

134

carried upward by capillary water. "Sodic" soils are frequently located in small areas called "slick spots" surrounded by soils that are relatively productive. The high pH is largely due to the hydrolysis of sodium carbonate which occurs as follows:

$$Na_2CO_{3(aq)} + 2H_2O_{(\ell)} \rightarrow 2NaOH_{(aq)} + H_2O_{(\ell)} + CO_{2(g)}$$

What volume of 0.250 $\underline{M}$ sodium carbonate would be required to produce 5.40 liters of soil solution 0.80 $\underline{M}$ in sodium hydroxide, assuming 100% conversion under conditions of CO_2 removal?

A.4. Most of the dissolved constituents of sea water have had thousands of years to reach equilibrium. Nevertheless, thermodynamically unstable species, such as manganese(II), iodide, and arsenite (AsO_3^{3-}) ions, persist. Their occurrence can be attributed to a lack of reaction sites where equilibrium might be obtained. Manganese occurs in sea water largely as the divalent ion, part of which appears to be complexed with organic matter. Particulate manganese is only found in surface waters, and no evidence for the colloidal MnO_2 has been found in the water column, yet it occurs widely in ferromanganese nodules of the pelagic sediments. The widespread occurrence of tetravalent manganese as MnO_2 in the ferromanganese nodules suggest that the associated iron oxides provide the necessary catalytic sites for formation. The manganese(II) ion is readily converted to the tetravalent form by the reaction:

$$4\ OH^- + 2\ Mn^{2+} + O_2 \rightarrow 2\ MnO_2 + 2\ H_2O$$

In a particular reaction, a 30.0 ml aliquot of oxygenated sea water was allowed to react with 65.0 ml of 3.1×10^{-6} $\underline{M}$ NaOH. Assuming the NaOH only reacted with the manganese(II) ion in the sea water sample, what was the molarity of the manganese(II) ion?

A.5. As nutrients, silicates differ from the inorganic compounds of nitrogen and phosphorus in not being a universal requirement of living matter. Silicates are present, however, in large quantities in diatoms which dominate the phytoplankton in the cooler ocean waters. The silicate cycle corresponds in many ways with the cycle of phosphorus, silicate and phosphate being depleted or regenerated simultaneously. In the oceans, in general, the concentrations of silicate vary greatly in their proportion to the phosphate and nitrate present. Such variation arises from the fact that in different parts of the oceans the proportions of diatoms to other phytoplankton which do not require silicon differ greatly. In addition, the maximum possible solubility of amorphous silica in sea water may be limited by precipitation reactions of dissolved silica with other ions present in sea water. The formation of sepiolite occurs in this manner:

$$2Mg^{2+}_{(aq)} + 3SiO_{2\,(dispersed)} + 4\,OH^-_{(aq)} \rightarrow Mg_2Si_3O_{8(s)} + 2H_2O_{(\ell)}$$

What was the molar concentration of dispersed SiO_2 in an aliquot of sea water, 45.0 ml of which required 31.2 ml of 2.1×10^{-4} $\underline{M}$ NaOH for quantitative conversion of silica to sepiolite? (Assume excess Mg^{2+}.)

*A.6. The soil chemist specializes in the study of the vast array of elements and compounds present in the mineral and organic components of soil. A large amount of the work performed in the soils laboratory involves soil analysis to determine the presence and amounts of various chemical species and the acidity or alkalinity of the soil sample. Through such analyses the chemist is able to learn of soil needs and is then able to develop fertilizers and other chemical amendments to meet those needs. Soil water becomes acid by absorbing carbon dioxide from the air. The carbonic acid that is formed from the reaction of carbon dioxide from the air with soil water is a weak acid solution. A titration with sodium hydroxide is used for the carbonic acid assay, according to the equation:

$$NaOH_{(aq)} + H_2CO_{3\,(aq)} \rightarrow NaHCO_{3\,(aq)} + H_2O_{(\ell)}$$

How many milligrams of carbon dioxide are contained in 1.00 liter of a soil water carbonic acid solution from which a 30.0 ml aliquot required 18.4 ml of a standard sodium hydroxide titrant labeled 4.2×10^{-3} $\underline{M}$?

*A.7. Marine microorganisms, the "single-celled" organisms in the sea that fall within the microscopic range, form a large fraction of the marine biomass and, therefore, are responsible for a significant part of the chemical changes that occur in the marine environment. The "marine environment" is a collective term used to include water, sediment, air over the water, intertidal areas and the spray zones subject to sea salt spray. The marine microorganisms include two functional types: the primary producers (chemo- or photosynthetic), and the mineralizers or heterotrophic organisms. Some of the products formed by the primary producers include mono-saccharides and polysaccharides, and glycolic, oxalic, and pyruvic acids. In its simplified form, the production of glucose results from the combination of six moles of carbon dioxide with six moles of water. If the yield of glucose is 90% efficient for a particular marine algae, how many grams of carbon dioxide would be needed to produce 200 ml of a 0.15 $\underline{M}$ glucose solution in sea water?

*A.8. In sulfur-deficient areas crops generally have an appearance similar to that resulting from nitrogen deficiency or from lack of water, which markedly affects the amount of nitrogen available to crops. When sulfur or sulfate is applied to the soil, the plants turn dark green and the treated areas stand out prominently in the

*Proficiency Level

field. In cases of sulfur deficiency, the youngest leaves were the first to turn yellow, whereas the oldest leaves were the first when lack of nitrogen was a limiting factor in plant growth. Sulfur-deficient plants tend to have very slender and woody stems. Furthermore, the growth of roots is large in relation to that of other parts of the plant. In addition to losses by leaching and absorption by plants, sulfates are also depleted in soils by certain types of bacteria, viz., Desulfovibro and Desulfotomaculum. These organisms use the combined oxygen in the sulfate to oxidize organic materials. In the formation of two moles each of formic acid (HCO_2H) and water and one mole of sulfide (S^{2-}), two moles of methyl alcohol (CH_3OH) and one mole of sulfate are consumed. If the yield of a sulfide is 75% in a particular sulfate-reduction, how many grams of sulfate would be needed to produce 500 ml of 0.014 $\underline{M}$ sulfide?

*A.9. The chemistry of lead in the marine environment is highly dependent on the properties of the individual lead species. Like most other elements and compounds, lead enters the ocean through runoff water from the land and through the atmosphere. Before man interferred, rivers were the primary source of lead for the oceans. This lead was usually transported to the sea as insoluble forms attached to organic or inorganic particles or as soluble inorganic complexes. A substantial part of the 10^{11} grams of lead aerosols that are produced today will fallout over the ocean. Analyses of lead in the Greenland ice sheet have shown an increase in lead fallout of approximately 500-fold over the last 3000 years. A substantial part of this increase occurred after the introduction of tetraethyl lead in gasoline. Since lead is a cumulative poison its levels are continually monitored. The iodimetric chromate titration is useful for the volumetric determination of lead. One mole of chromate (CrO_4^{2-}) is consumed per mole of Pb^{2+} ion in the water sample. How many milligrams of Pb^{2+} are contained in 1.00 liter of a sample from which a 25.0 ml aliquot required 10.3 ml of a standardized chromate titrant labeled 5.4 x 10^{-4} $\underline{M}$?

*A.10. Boron, chlorine, cobalt, copper, iron, manganese, molybdenum, and zinc are among the important plant micronutrients. Many of these essential micronutrients function as components of plant enzymes. The chloride anion is used by most crop plants in larger amounts than any other micronutrient except iron. Chloride is found normally in soils in the range of 10-1000 ppm, usually as NaCl or KCl. Since most chlorides are soluble, the ion is readily redistributed through flowing or percolating ground waters. In well-drained humid regions, most chloride is leached from the soil, but in poorly drained soils its concentration may reach the point of salt toxicity to plants. The chloride ion concentration is frequently monitored by titrating the chloride ion in an aliquot of soil solution with silver nitrate. Potassium chromate is used as an indicator because when all the chloride ion is used-up by silver ion, the chromate will then react with the excess silver ion, forming a red-brown precipitate. One mole of silver ion is consumed per mole of Cl^- ion in the soil solution. How many milligrams of Cl^- are contained in 1.00 liter of a solution from which a 30.0 ml aliquot required 50.0 ml of a standardized silver nitrate titrant labeled 0.0120 <u>M</u>?

ANSWERS:

(A.1.) 0.025 <u>M</u>, 57 g, (A.2.) 0.013 <u>M</u>, 40 g, (A.3.) 8.6 liters, (A.4.) 3.4×10^{-6} <u>M</u>, (A.5.) 1.1×10^{-4} <u>M</u>, (A.6.) 114 mg, (A.7.) 8.8 g, (A.8.) 0.90 g, (A.9.) 46 mg, (A.10.) 710 mg

---

RELEVANT
PROBLEMS

Unit 6: Solution Chemistry

Set B: Biological & Medical Sciences

B.1. Calcium is one of the most abundant ions in the body. In the adult body there may be as much as two kilograms of calcium ion. The normal daily adult requirement is approximately 0.8 to 1.0 g of calcium. During periods of active growth, such as pregnancy, this requirement is increased to approximately 2.0 g daily. Calcium is abundant in many foods of the usual diet, but the greatest amount is derived from milk, cheese, and eggs. Most vegetables and meats are poor sources. The average concentration of Ca^{2+} ion in human blood serum is about 10 mg per 100 ml. What is the <u>molarity</u> of Ca^{2+} in this solution?

B.2. Calcium is chiefly absorbed in the duodenum, where its rate of absorption is controlled by vitamin D activity, the hydrogen ion concentration in the intestine, and other food factors in the diet. Calcium ion serves four major functions in the body: bone and tooth formation, aid to blood coagulation, maintenance of normal cell membrane permeability, and maintenance of normal neuromuscular function. Calcium ion is excreted by action of the kidneys, the liver, and the large intestine. The kidney threshold level is 1.5×10^{-3} <u>M</u>. When the blood calcium levels exceed this threshold, calcium is excreted in the urine. What mass of calcium is contained in 3.0 ℓ of blood that has reached the kidney threshold level?

B.3. The stomach is a pouchlike portion of the digestive tract where part of the digestive process occurs. In addition, the stomach serves as a temporary reservoir for food.

The cells that compose the inner stomach wall have specialized duties. They produce hydrochloric acid, protective mucus, and digestive enzymes such as pepsin and rennin. These products form the "gastric juice" which supports the chemical work of the stomach. The chemical action is aided by the muscular activity of the stomach. The walls contract and relax, mixing the "gastric juice" with the food and thus helping enzymatic digestion. A 75 ml sample of fresh gastric juice had a hydrochloric acid concentration of 0.17 $\underline{M}$. The reaction of this solution with sodium bicarbonate may be represented as:

$$H^+_{(aq)} + HCO_3^-{}_{(aq)} \rightarrow H_2O_{(\ell)} + CO_{2(g)}$$

What volume of 0.23 $\underline{M}$ sodium bicarbonate would be required to neutralize the hydrochloric acid in this sample?

B.4. The term ethyl alcohol "poisoning" should be used only with reference to acute alcoholism. Chronic alcoholism is associated with many functional and organic derangements, but these are, for the most part, due to concomitant dietary imbalances and cannot be ascribed merely to alcohol "poisoning". Absorption of alcohol occurs rapidly in the stomach and intestines so that it is detectable in the blood within minutes after ingestion, although peak levels are not usually reached until one to two hours later. Irrespective of the volume of alcohol intake, there is a fixed rate of metabolic oxidation of alcohol (about 10 ml per hour). Absorption of large amounts is not accompanied by an increased rate of utilization and must result in rising blood levels of unreacted alcohol. A blood alcohol (C_2H_5OH) concentration of 0.065 $\underline{M}$ causes obvious signs of intoxication in most humans. A concentration of 0.17 $\underline{M}$ is usually lethal. If a person has a total blood volume of 7.0 liters, what is the mass, in grams, of C_2H_5OH that corresponds to the difference between the intoxicating and lethal doses?

B.5. The metabolism of ingested foods produces chemical waste products that must be eliminated from the body in the process called excretion. In protein metabolism, nitrogenous waste products result from some of the chemical degradations of freed amino acids. Other such waste products occur as by-products of the synthesis of proteins. Nonprotein nitrogenous wastes include urea and uric acid. Any great accumulation of wastes in the tissues, especially "nonprotein nitrogen", causes rapid tissue poisoning, starvation, and eventually suffocation. Tissues filled with waste products cannot absorb either food or oxygen. Fever, convulsions, coma, and death are inevitable if nonprotein wastes do not leave the tissues. This "uremic syndrome" (toxic condition resulting from an accumulation of nitrogenous waste products in the blood) commonly occurs as a result of kidney malfunction. In clinical chemistry, urea in blood can be determined by titrating with standard hypobromite according to the equation:

$$H_2N\text{-}\overset{\overset{\displaystyle O}{\displaystyle \|}}{C}\text{-}NH_2 + 3\ OBr^- \rightarrow CO_2 + N_2 + 2\ H_2O + 3\ Br^-$$

In one case of uremic poisoning, a 25.0 ml aliquot of a patient's blood sample containing nitrogenous waste was oxidized by 18.5 ml of 0.125 $\underline{M}$ hypobromite. What was the molarity of urea in the blood?

*B.6. Milk is secreted by the mammary glands of a living animal. It is very complex in composition and liable to considerable variation in its composition. The relative proportions of its various components are dependent upon a number of factors such as diet, seasonal variations, type of animal, and climatic conditions. Chemically, milk is essentially a mixture of proteins, milk sugar (lactose) and salts - sodium chloride, potassium chloride, magnesium citrate, potassium phosphate and calcium citrate - dissolved in water, and holding in suspension fat in the form of finely divided globules. The calcium in a sample of milk was precipitated as the oxalate. The precipitate was subsequently dissolved in sulfuric acid and the oxalate was titrated with standardized potassium permanganate. Two moles of permanganate (MnO_4^-) were consumed for every five moles of calcium in the original sample. What mass of Ca^{2+} are contained in 1.00 liter of milk from which a 20.0 ml aliquot required 18.4 ml of a standardized permanganate titrant labeled 0.110 $\underline{M}$?

*B.7. The fundamental importance of milk as a "universal food" makes it a subject of considerable interest. Because it is so widely consumed and so readily polluted, it is intimately connected with the health of all people. One of the components of milk having nutrient value is the disaccharide lactose, which does not otherwise occur significantly in nature. When milk is kept at a temperature of 75 to 100 degrees, the bacteria in it multiply so rapidly that in ten to twelve hours the milk is sour. It is not practicable or desirable to keep these lactic-acid bacteria entirely out of milk, but if, as soon as it is drawn, the milk is cooled to a temperature of about 40 degrees and is kept cool the bacteria multiply very slowly and the milk will remain sweet much longer. The bacterial conversion of lactose to lactic acid gives sour milk its tart taste. The fermentation of lactose ($C_{12}H_{22}O_{11}$) in milk results in the formation of four moles of lactic acid $\left(\begin{matrix} CH_3CH\text{-}CO_2H \\ | \\ OH \end{matrix} \right)$ per mole of lactose consumed. If the yield of lactic acid is 78% in a particular sour milk sample, how many grams of lactose would be needed to produce 335 ml of 1.5 $\underline{M}$ lactic acid?

*B.8. The neuropathology of arsenic poisoning is complex because of the complicated metabolic effects of the generalized disorder that it produces. For example, there are severe gastrointestinal, hepatic, and renal lesions. Formerly, treatment of syphilis with arsenicals or the management of many illnesses with arsenic solutions was a common cause of arsenic intoxication. At present, most of the few cases that occur each year arise either from homicidal intent, or the accidental ingestion of arsenic in insecticides. A titration with potassium bromate solution may be used for arsenic assay. One mole of bromate is consumed for every three moles of arsenic(III) in a water sample. What mass of arsenic(III) is contained in 1.00 ℓ of a solution from which a 25.0 ml aliquot required 15.3 ml of a standardized bromate titrant labeled 8.30 x 10^{-2} $\underline{M}$?

*B.9. In many nutritional diseases, the nervous system is affected by a biochemical disorder which does not produce morphologic changes. In others, it is affected by indirect intervention of primary damage to another tissue or organ. The most important nutritional diseases which produce morphologic change in the nervous system are those due to deficiencies of B vitamins. In clinical practice, pellagra in man and blacktongue in dogs are diseases difficult to define, and probably represent a complex deficiency of more than one factor, although depletion of niacin seems to be the determining factor. The treatment of niacin ($C_6H_5NO_2$) with ammonia results in the formation of one mole of nicotinamide ($C_6H_6N_2O$) per mole of niacin consumed. If the conversion efficiency is 91%, how many grams of niacin would be needed to produce 150 ml of 1.4 $\underline{M}$ nicotinamide?

*B.10. Ethyl alcohol can produce intoxication, yet it is problematic whether this drug can produce any specific morphologic change in the brain or spinal cord. There is no change in acute alcoholic intoxication and the vague changes described in delirium tremens and alcoholic hallucinosis are not likely to be toxic effects of the alcohol. They are, rather, the result of complicating anoxia, a complicating liver disease, or a nutritional deficiency. The oxidation of ethyl alcohol (C_2H_5OH) with potassium dichromate, results in the formation of one mole of acetaldehye (CH_3CHO) for mole of ethyl alcohol consumed. If the yield of acetaldehyde is 75% in a particular oxidation, how many grams of ethyl alcohol would be needed to produce 500 ml of a 2.0 $\underline{M}$ acetaldehyde?

RELEVANT
PROBLEMS

Unit 6: Solution Chemistry

Set I: Industrial Chemistry

I.1. The principal component of most commercial "anti-freeze" solutions is 1,2-
dihydroxyethane, more commonly known as ethylene glycol ($HOCH_2CH_2OH$). Different
brands of antifreeze vary primarily in the nature and concentrations of additives,
such as rust retardants. A typical commercial operation will produce 50-100 million kilograms of ethylene glycol per year. A solution which is 60% ethylene glycol
and 40% water (by weight) will protect a radiator against freezing to temperatures
as low as -55°F. What is the molarity of ethylene glycol in a solution of 515 g of
ethylene glycol in 333 ml of solution?

I.2. A method has recently been demonstrated to develop, fix and rinse a color negative
in the darkroom. The initial black and white print produced is then converted to
a full color print with room lights on. The developer contains 2.0 g of Metol
(p-N-methylaminophenol, HO-⬡-N$\overset{H}{\underset{CH_3}{<}}$), 25.0 g of sodium sulfite (Na_2SO_3), 6.0 g
of hydroquinone (HO-⬡-OH), and 40.0 g of sodium carbonate monohydrate
($Na_2CO_3 \cdot H_2O$) in one liter of solution. What is the molarity of each compound?

I.3. The lead-acid storage battery is the most widely used of the secondary storage
cells. The cells of the battery produce electrical energy by electron transfer
in oxidation-reduction reactions. The cells of the storage battery, after partial
discharge, can be converted stoichiometrically back to the original chemical reactants. There are three active materials in each cell, the PbO_2 plates, the lead
(Pb) plates and the electrolyte, dilute sulfuric acid. The purity of the electrolyte is important because traces of certain impurities, such as acetates, will
damage the battery. A factory producing lead storage batteries for automobiles requires 3785 liters per day of 33.4% (by weight) sulfuric acid (specific gravity =
1.25 g/ml). What volume (liters) of 98% H_2SO_4 (specific gravity = 1.844 g/ml) is
required to prepare the daily supply of battery acid?

I.4. The concentration of the electrolyte in the automobile battery is usually determined by measuring the specific gravity of the solution. The important factor in

142

selecting the proper concentration of sulfuric acid is the conductivity of the solution. The conductivity of aqueous sulfuric acid is a maximum at about 30% (by weight) concentration. Since conductivity is also a function of temperature, the environment where the battery is to be used must also be considered when manufacturing the battery acid. The chemical change in a lead storage battery during discharge may be represented by:

$$Pb + PbO_2 + 2H_2SO_4 \rightarrow 2PbSO_4 + 2H_2O$$

A battery containing 500 ml of electrolyte solution was allowed to discharge to the point at which the specific gravity of the electrolyte was 1.11, corresponding to a sulfuric acid concentration of 15.6% (by weight). What was the molarity of H_2SO_4 in the final solution? What mass of lead was consumed during the discharge process?

I.5. Sodium chlorate is commonly used as a weed killer and in textile printing and dyeing. It is also used extensively in the production of chlorine dioxide, an important bleach for wood pulp. One step in producing ClO_2 is to electrolyze brine (6.4 $\underline{M}$ NaCl) to an equimolar solution of 3.2 $\underline{M}$ NaCl and 3.2 $\underline{M}$ $NaClO_3$. How many grams of sodium chlorate could be produced from 3.8 ℓ of brine?

I.6. Hydrazine has been used to a limited extent as the fuel in rocket propulsion systems. Other commerical uses for hydrazine include the production of blowing agents for making some plastics, such as polyurethane foam, and the use in boiler water to prevent boiler corrosion. One commercial method used for producing hydrazine is the reaction of aqueous sodium hypochlorite with ammonia.

$$2NH_3 + NaOCl \rightarrow N_2H_4 + NaCl + H_2O$$

If 43.4 g of hydrazine are produced from 734 ml of sodium hypochlorite solution, what was the molarity of the solution?

*I.7. One of the most important commercial uses for chlorine is in preparing bleaching agents, especially for cotton cloth. These are not used with silk or wool since the bleaching action destroys these animal fibers. The action of chlorine bleaches is due to the reaction of chlorine with water, forming the unstable hypochlorous acid

$$Cl_2 + H_2O \rightarrow HCl + HOCl$$

The subsequent decomposition of the hypochlorous acid, forming atomic oxygen, is what does the "bleaching".

$$HOCl \rightarrow HCl + [O]$$

*Proficiency Level

One of the most popular chlorine bleaches, sold under a variety of trade-names (e.g., Dazzle, Clorox) is an aqueous solution of sodium hypochlorite, NaOCl. Sodium hypochlorite can be prepared by several different methods. One of the most common is to add chlorine to an aqueous solution of sodium hydroxide:

$$2NaOH + Cl_2 \rightarrow NaCl + NaOCl + H_2O$$

Sodium hypochlorite solutions are very effective in removing fruit and other stains, and as an aid in laundering. They have the advantage over many other types of bleaches because the fabrics come out softer and rinse cleaner, and this insures better strength of the fiber and a more permanent white. Clorox is a 5.25% (by weight) solution of sodium hypochlorite, NaOCl (specific gravity = 1.035 g ml^{-1}). How much chlorine (in kg) would be needed to react with sodium hydroxide to prepare 8000 liters of Clorox? What is the molarity of sodium hypochlorite in Clorox?

*I.8. Except for the small amount imported from Chile, the world's supply of nitrates comes from nitric acid made in chemical plants. The last step in the production of nitric acid yields a solution 50-60% HNO$_3$ by weight. It takes three moles of nitrogen dioxide to produce two moles of nitric acid. How many grams of nitrogen dioxide are used in making 475 liters of 52.3% nitric acid solution, by weight (density 1.33 g ml^{-1})?

*I.9. It is necessary to start the production of aluminum metal with the purest possible alumina, because impurites in the latter tend to contaminate the metal. The chief ore used in producing aluminum is bauxite. The essential step in the purification of alumina from bauxite is to treat it under pressure in a hot aqueous solution of caustic soda. The product is in the form of Na$_3$AlO$_3$. What molarity of NaOH is needed to make 43 metric tons of Na$_3$AlO$_3$ from 3.45×10^5 ℓ of NaOH solution.

*I.10. Sodium cyanide is used in treating gold ore, in the case-hardening of steel, in electroplating, in organic reactions, in the preparation of hydrocyanic acid, and in making adiponitrile. A commercial method consists of neutralizing hydrocyanic acid (HCN) with caustic soda (NaOH). How many liters of 3.78 M NaOH will it take to neutralize 578 liters of 1.27 M HCN?

ANSWERS:

(I.1.) 24.9 M, (I.2.) 0.016 M, 0.198 M, 0.055M, 0.323 M, (I.3.) 1290 ℓ, (I.4.)1.77 M, 107 g, (I.5.) 1300 g, (I.6.) 1.85 M, (I.7.) 4.14×10^2 kg, 0.730 M, (I.8.) 3.62×10^5 g, (I.9.) 2.6 M, (I.10.) 194 ℓ

RELEVANT
PROBLEMS

E.1. Nitrogen comprises approximately four-fifths of the atmosphere. Although it is
relatively inert, at high temperatures or through various biological reactions it
can combine with oxygen to form several oxides. Natural causes include lightning,
volcanic eruptions, and decomposition of organic matter. This is part of nature's
design since nitrogen is essential to plant growth but cannot be utilized by most
green plants in its elemental form. The production of nitrogen oxides through natural causes is 10 times greater than from man-made sources. Nearly half of man's
input comes from so-called stationary sources, utility plants which generate electricity or steam and manufacturing operations which produce high-grade metals such
as stainless steel, chromium and nickel. Gasoline engines account for most of the
other half. While much of the nitrogen oxides are washed from the air by rainfall,
thus benefiting soil and plants, residues play a major role in photochemical smog,
a hazy mixture of eye, nose and throat irritants. In addition, rains from air
heavily polluted by nitrogen oxides, or sulfur oxides, may be appreciably acid-
contaminated. Rainwater analyses for some rains occurring in March 1973 in eastern
Tennessee revealed a nitric acid concentration of 2.2×10^{-6} g liter^{-1}. What is
the molarity, with respect to the nitric acid, of the rainwater?

E.2. The concentration of sulfuric acid in the same rain (problem E.1) was determined to
be 6.6×10^{-5} M. The amount of water which fell during this rain was measured as
6000 liters per acre. How many grams of sulfuric acid were deposited on each acre?

E.3. As a pure substance, water is odorless, colorless and tasteless. It is a compound
of great stability, a remarkable solvent, and a powerful source of chemical energy.
When frozen into a solid it expands, instead of contracting, and the lighter solid
floats on the more dense liquid. It can absorb and release more heat energy per
mole than most other substances. In many of its properties, such as its freezing
and boiling points, water is an oddity, an exception to the rules. It is the
structure and dipolar character of the water molecule which endows water with such
potency as a solvent and serves, indirectly, as the source of its other exceptional
properties. Each molecule of water in the atmosphere serves as a "bundle of heat
energy". It is the dynamic flow of water vapor in the atmosphere that gives rise
to global climates and local weather. An adequate water supply is necessary for all

145

plants and animals. Water dissolves and distributes such necessities of life as oxygen, carbon dioxide and salts. However, it can transmit pathogenic bacteria and toxic chemicals equally well. In the human body, water is essential for blood circulation, waste removal, and muscle response. San Mateo County lake water, which comes from the surrounding watershed and furnishes about 55% of the San Francisco water supply, requires the addition of chlorine gas (Cl_2) to kill harmful bacteria. The addition of 1.24×10^6 g of chlorine per day produces a concentration of approximately 1.7×10^{-5} M. What is the volume of water (liters) used from this source by San Francisco each day?

E.4.   The concentration of calcium ion in Sierra-Calaveras water, used by many San Francisco residents, is 2.4 mg $liter^{-1}$. How many grams of sodium carbonate (Na_2CO_3) would be required to remove the Ca^{2+} from 1.0×10^9 liters of this water? (This is about 45% of the daily water consumption in San Francisco).

$$[Ca^{2+}_{(aq)} + CO^{2-}_{3(aq)} \rightarrow CaCO_{3(s)}]$$

E.5.   Anyone who has been downwind of "sewage" understands one reason for treating it, besides the pollution of water and the possibility of the spread of disease. During the anerobic decay of organic matter in sewage, sulfate ions are reduced to foul smelling organic sulfur compounds or hydrogen sulfide. If a particular organic compound in sewage is oxidized by an equimolar amount of sulfate ion, what was the molarity of this organic compound in 475,000 liters of sewage requiring all the sulfate in 83,000 liters of 0.237 M SO_4^{2-} solution?

*E.6.   The wastes from electroplating plants may cause some special problems. Zinc and copper are typically electroplated from alkaline solutions containing sodium cyanide. Nickel and chromium, on the other hand, are usually plated from acidic solutions. If the waste electrolytes from the different processes are mixed, poisonous hydrogen cyanide may be released into the environment. The acidic conditions used in chromeplating also occur in a "preplating" conversion of chromate to the chromium(III) for eventual electrolytic reduction, for example:

$$7H^+ + 2CrO_4^{2-} + 3HSO_3^- \rightarrow 2Cr^{3+} + 3SO_4^{2-} + 5H_2O$$

In a pilot plant operation, 189 liters of bisulfite solution was required for reduction of 435 liters of 0.35 M chromate solution. What was the molarity of the HSO_3^-?

*E.7.   The waste cyanide-containing alkaline electrolyte solutions may be treated with chlorine or sodium hypochlorite to destroy the cyanide. Four moles of sodium hypo-

*Proficiency Level

146

chlorite are needed for each mole of cyanide ion oxidized. If the concentration of cyanide ions is 500 ppm in waste electrolyte (density: 1.01 g ml^{-1}) and 2.65×10^6 liters of this are treated each day, what molarity of NaOCl is needed if 1.60×10^4 liters of this solution is used?

*E.8. Calcium ions make water "hard". Calcium soaps precipitate as "bathtub ring" and insoluble salts form in pans where water is boiled. "Washing soda" (Na_2CO_3) has been used for removal of the calcium ions as $CaCO_3$ to "soften" water. How many liters of 1.87 $\underline{M}$ Na_2CO_3 will be required to remove the Ca^{2+} from 4.0×10^6 liter of 100 ppm Ca^{2+} solution (density: 0.998 g ml^{-1})?

*E.9. One method of waste disposal that has been used for several years is to pump the waste into deep wells for storage. A certain insecticide plant has a well that is 2600 feet deep into a dolomite layer which is very porous. The top of the well is sealed to prevent surface water contamination. The wastewater contains 2.0% (by weight) NaOH. How much 6.03 $\underline{M}$ H_2SO_4 would be required (in liters) to neutralize 487,000 liters of the waste solution (density: 1.17 g ml^{-1})?

*E.10. One method of detecting SO_2 in the atmosphere is to pull the air through water. The sulfur dioxide will react with water to form sulfurous acid. The acid solution is then titrated with potassium permanganate. Two moles of $KMnO_4$ are required to oxidize five moles of H_2SO_3. What volume of 0.15 $\underline{M}$ $KMnO_4$ will be needed to oxidize 11 ml of sulfurous acid solution resulting from absorption of all the SO_2 from 780 g of 130 ppm SO_2-contaminated air?

ANSWERS:

(E.1.) 3.5×10^{-8} $\underline{M}$, (E.2.) 39 g, (E.3.) 1.0×10^9 liters, (E.4.) 6.3×10^6 g, (E.5.) 4.1×10^{-2} $\underline{M}$, (E.6.) 1.2 $\underline{M}$, (E.7.) 12.8 $\underline{M}$, (E.8.) 5.3×10^3 liters, (E.9.) 2.4×10^4 liters, (E.10.) 4.2 ml

STOICHIOMETRY

UNIT 7: ENTHALPY PROBLEMS

Every chemical reaction is associated with an energy change. Of the various ways of classifying "energy", one of the simplest and most familiar is heat. Some chemical reactions, such as the burning of methane, evolve heat. Others require heat to convert reactants to products. Such reactions are said to be endothermic, while those releasing heat are termed exothermic.

Chemical thermodynamics is concerned with heat and other forms of energy associated with chemical processes. Of particular interest from a practical point of view are two quantities, heat (symbolized by q) and work (w). The distribution of energy from a reaction, such as the combustion of methane, between waste heat and useful work is of major concern in attempts to improve the efficiency of fuel utilization. We can alter conditions to vary the waste heat/useful work ratio of a chemical process, within certain limits. For example, the waste heat/work ratio for a typical electrical generator using steam produced by heat from burning methane is about 2:1 (i.e., only about 1/3 of the energy of combustion is actually useful). The same oxidation of methane can be performed in a "fuel cell" for direct production of electrical energy. In the latter case the waste heat/work ratio is reduced to about 1:5, a significant improvement in efficiency of fuel use.

There are both theoretical and practical limits on energy production and its conversion to work. Many of the "practical" limitations are best dealt with in the areas of engineering. We shall limit our present considerations to a single theoretical limitation, the enthalpy change for a chemical process. This quantity, symbolized as ΔH^1, can be measured experimentally by techniques of calorimetry or calculated from experimental values for appropriate chemical species. We shall restrict our calculations to ΔH values under a particular set of conditions, defined as the thermodynamic standard state conditions:

1. 25°C (298°K)   [note difference from STP used for gas problems]

2. all gases at 1.00 atm (760 torr)

3. all pure substances in their most stable forms

4. all dissolved species at unit activity (approx. 1 M)

Use of standard state quantities may be indicated by a superscript zero: e.g., ΔH^O indicates an enthalpy difference based on standard state values.

Enthalpy is referred to as a state function, which means that the enthalpy (heat content) change for a process depends only on the initial and final states and not, as with the waste heat/work ratio, on the particular pathway selected for the process. Thus, the ΔH^O for the combustion of methane, for example, is the same whether it is a "burning"

---

[1]H is the standard thermodynamic symbol for enthalpy (heat content) and the capital Greek delta (Δ) is used to indicate the difference between two quantities.

- UNIT 7 -

process or a "fuel cell" process. We can take advantage of the <u>state function</u> characteristic of ΔH^O values to calculate the standard enthalpy change for a chemical reaction from tabulated values of <u>standard heats of formation</u>, as given in Appendix F. The key relationship may be expressed as:

$$\Delta H^O_{reaction} = \Sigma(\Delta H^O_f \text{ of products}) - \Sigma(\Delta H^O_f \text{ of reactants})$$

Tabulated standard heats of formation (ΔH^O_f) are expressed (Appendix F) in kcal <u>per mole</u>, so the number of moles of any species must be taken into account in enthalpy calculations. Note that ΔH^O_f refers to <u>formation</u> from the elements, so ΔH^O_f for the stable free element is zero.

We shall find enthalpy calculations useful in determining standard heats of reaction or, when these are obtained from experimental data, in finding the standard heat of formation for some chemical species.

- References -

1. O'Connor, Rod. 1977. <u>Fundamentals of Chemistry</u>, second edition. New York: Harper & Row (Units 5, 12, 23)

2. Brown, Theodore, and H. Eugene LeMay, Jr. 1977. <u>Chemistry: The Central Science</u>. Englewood Cliffs, N.J.: Prentice-Hall (Chapters 4, 15, 19, 25)

3. Masterton, William, and Emil Slowinski. 1977. <u>Chemical Principles</u>, fourth edition. Philadelphia: W.B. Saunders (Chapters 4, 14, 23)

4. Nebergall, William, F.C. Schmidt, and H.F. Holtzclaw, Jr. 1976. <u>College Chemistry</u>, fifth edition. Lexington, MA: D.C. Heath (Chapters 8, 16, 20, 22)

5. Pimentel, George C. and Richard D. Spratley. 1969. <u>Understanding Chemical Thermodynamics</u>. San Francisco: Holden-Day.

OBJECTIVES:

(1) *Given the formulas and standard heats of formation for the reactants and products of a chemical process, be able to write the balanced equation and to use it in determining the standard heat of reaction for the process.*

*(2) *Given a description of a chemical process and necessary ΔH^O data, be able to calculate the ΔH^O_f for any one component of the reaction system, or the standard heat of reaction, as indicated.*

*Proficiency Level

PRE-TEST·

Necessary Thermodynamic Data:

compound	ΔH°_f (kcal mole^{-1})	compound	ΔH°_f (kcal mole^{-1})
$H_2O_{(\ell)}$	-68.3	$H_2O_{(g)}$	-57.8
$H_2SO_{4(\ell)}$	-194.6	$N_2H_{4(\ell)}$	$+12.1$
$NaOH_{(s)}$	-102.0	$N_2O_{4(g)}$	$+2.3$
$Na_2SO_{4(s)}$	-330.9		

(1) Contact with a concentrated acid or alkali in the laboratory or in the home (e.g., a liquid drain cleaner) can result in severe skin burns. Proper emergency treatment for such an accident requires immediate removal of the chemical by thorough washing (or, with a dry chemical such as quicklime, brushing off followed by washing). No attempt should ever be made to "neutralize" a concentrated acid or alkali on the skin. Such a procedure invariably adds a severe heat burn to the chemical damage, since neutralization reactions are typically quite exothermic. Write a balanced equation for the reaction of lye (NaOH) with sulfuric acid (H_2SO_4) to form liquid water and sodium sulfate (Na_2SO_4) and calculate the standard heat of reaction for the neutralization, in kcal per mole of sodium hydroxide.

*(2) The hypergolic (self-igniting) rocket propellant mixture of hydrazine and dinitrogen tetroxide generates "thrust" by expelling gases from the rocket chamber. A large gas volume results from the reaction ($3N_2H_{4(\ell)} + N_2O_{4(g)} \rightarrow 4N_{2(g)} + 2H_{2(g)} + 4H_2O_{(g)}$) and a further volume increase results from gas expansion caused by the large heat of reaction. How much heat would be generated (under thermodynamic standard state conditions) by the reaction of 10.0 lb (4536 g) of hydrazine (form. wt. 32.1) with a stoichiometric amount of N_2O_4?

Answers and Directions:

(1) $2\ NaOH + H_2SO_4 \rightarrow Na_2SO_4 + 2\ H_2O$, $\Delta H^\circ = -34.4$ kcal mole^{-1} (NaOH)

If you answered this question correctly, you may go on to the proficiency level. If you missed it, study METHODS, section 7.1.

*(2) -12.7×10^3 kcal

If you answered this question correctly, you may go on to the RELEVANT PROBLEMS, Unit 7. If you missed it, study METHODS, section 7.2.

METHODS

7.1 Standard Heats of Reaction

Since the balanced chemical equation (Unit 2) establishes the mole ratios of all

*Proficiency Level

species involved, it is useful in determining the net enthalpy change for a chemical process. Using lower case letters to represent numerical coefficients and capital letters to indicate chemical species, we may write a "generalized" equation:

$$aA + bB + cC + \ldots \rightarrow wW + xX + yY + \ldots$$

Then, since ΔH_f^o data is tabulated (Appendix F) in kilocalories per mole (kcal mole^{-1}), the net enthalpy change for a molar scale process will be:

$$\Delta H_{reaction}^o = [w(\Delta H_f^o \text{ for } W) + x(\Delta H_f^o \text{ for } X) + y(\Delta H_f^o \text{ for } Y) + \ldots]$$
$$- [a(\Delta H_f^o \text{ for } A) + b(\Delta H_f^o \text{ for } B) + c(\Delta H_f^o \text{ for } C) + \ldots]$$

We may choose to express the standard heat of reaction in terms of any one of the chemical substances involved. To do so, it is necessary only to divide the $\Delta H_{reaction}^o$ found by the coefficient of the particular chemical of interest, e.g.,

$$\frac{\Delta H_{reaction}^o}{a}$$ would give the units of kcal per mole of A [kcal mole^{-1} (A)].

Note that a negative sign for ΔH indicates heat evolved and a positive sign heat consumed.

If we wished to calculate a standard heat of reaction on any other scale, we would have to find the actual number of moles of the chemical specified (from a mass → mole conversion, Unit 3) and multiply that number by the "kcal per mole" of that chemical.

We shall work at this time only with molar scale processes to illustrate simple ΔH calculations.

EXAMPLE 1

A variety of artificial fuel gases have been prepared from coke or anthracite coal as possible alternatives to "natural gas" (methane). Among these are producer gas (30-40% CO, 2-5% H_2, 60% N_2) made by passing air through heated coke or coal and water gas (40-50% CO, 45-50% H_2, 3-7% CO_2, 4-5% N_2) made from coal and steam. Neither of these gases is very safe for home use because of the danger of carbon monoxide poisoning, but both have found a number of applications as industrial fuels. To compare the heating qualities of water gas and methane, calculate the standard heats of combustion for each of the following reactions, in kcal mole^{-1} of CO_2 formed, using ΔH_f^o data from Appendix F:

water gas combustion (idealized)

$$CO_{(g)} + H_{2(g)} + O_{2(g)} \rightarrow CO_{2(g)} + H_2O_{(g)}$$

methane combustion

$$CH_{4(g)} + 2\ O_{2(g)} \rightarrow CO_{2(g)} + 2\ H_2O_{(g)}$$

SOLUTION:

(using ΔH_f° data from Appendix F)

water gas

$$\Delta H^\circ_{reaction} = [\Delta H_f^\circ(CO_2) + \Delta H_f^\circ(H_2O)] - [\Delta H_f^\circ(CO) + \Delta H_f^\circ(H_2) + \Delta H_f^\circ(O_2)]$$

$$= [(-94.0) + (-57.8)] - [(-26.4) + (O) + (O)]$$

$$= \underline{-125.4 \text{ kcal mole}^{-1}} \ (CO_2)$$

methane

$$\Delta H^\circ_{reaction} = [\Delta H_f^\circ(CO_2) + \underline{2}\Delta H_f^\circ(H_2O)] - [\Delta H_f^\circ(CH_4) + \underline{2}\Delta H_f^\circ(O_2)]$$

$$= [(-94.0) + 2(-57.8)] - [(-17.9) + 2(O)]$$

$$= \underline{-191.7 \text{ kcal mole}^{-1}} \ (CO_2)$$

EXERCISE 1

Most explosives liberate large volumes of gaseous products and generate considerable amounts of heat. Calculate the standard enthalpy change for the explosive decomposition of nitroglycerine, after supplying the coefficient needed to balance the equation. Report the answer in kcal mole^{-1} (nitroglycerine).

$$\_ \ C_3H_5(ONO_2)_{3(\ell)} \rightarrow 12 \ CO_{2(g)} + 10 \ H_2O_{(g)} + 6 \ N_{2(g)} + O_{2(g)}$$

[ΔH_f° for nitroglycerine is -58.1 kcal mole^{-1}. For other values needed, see Appendix F.]

(answer, page 156)

===

Extra Practice

EXERCISE 2

The alkali metals (Li, Na, K, etc.) are used in a number of laboratory operations. Sodium for example, may be pressed into ribbon or wire by extruding the relatively soft metal through a hole in a steel cylinder. The ribbon or wire can be used to remove the last traces of moisture from an organic solvent, such as diethyl ether, which must be anhydrous (water-free) for some particular use. Hydrogen gas is liberated as the sodium reacts with the moisture. This reaction is quite exothermic and the heat may be sufficient, if enough reactants are present, to ignite the hydrogen formed. Thus, solvent drying with sodium requires that most of the water has been removed first by some other method to avoid a serious fire. Balance the equation for the reaction:

$$Na_{(s)} + H_2O_{(\ell)} \rightarrow H_{2(g)} + NaOH_{(s)}$$

then calculate the standard enthalpy change in kcal mole^{-1} (Na).

[For ΔH_f° data, see Appendix F.]   (answer, page 156)

= =

At this point you should try the <u>competency</u> level Self-Test question on page 155.

= =

*7.2 ΔH° - Values in Chemical Processes

Since enthalpy is a state function and standard heats of reaction are available in units of kcal mole^{-1} (Appendix F), we can use a "generalized" equation to see how enthalpy data can be obtained for any particular quantities of interest.

Consider the "generalized" equation, using lower case letters for coefficients of a balanced equation and capital letters to represent chemical species:

$$aA + bB + \ldots \rightarrow xX + yY + \ldots$$

On the basis of the state function character of enthalpy, we can write:

$$\Delta H_{reaction}^\circ = \Sigma (\Delta H_f^\circ \text{ of products}) - \Sigma (\Delta H_f^\circ \text{ of reactants})$$

or, in terms of our "generalized" equation:

$$\Delta H_{reaction}^\circ = [x\Delta H_f^\circ(X) + y\Delta H_f^\circ(Y) + \ldots] - [a\Delta H_f^\circ(A) + b\Delta H_f^\circ(B) + \ldots]$$

If we use the coefficients as the number of moles, since ΔH_f° values are tabulated as kcal mole^{-1}, the heat of reaction will be expressed as <u>kilocalories</u> for a molar-scale process.

To find enthalpy data on any other scale it is necessary only to apply the appropriate mass/mole unity factors. It is particularly convenient to calculate the enthalpy change for the molar-scale reaction first, in kilocalories per mole of the particular substance of interest. Any necessary mole-mass conversion can be included at this stage. The conversion to kcal mole^{-1} requires use of the number of moles of the substance for the molar-scale process, as expressed by the appropriate coefficient in the balanced equation. In our "generalized" case, for example, we could report the heat of reaction per mole of Y as:

$$\frac{\Delta H_{reaction}^\circ}{y} \quad [\text{kcal mole}^{-1} (Y)]$$

*EXAMPLE 2

Trinitrotoluene (TNT) is a well known explosive. Unlike nitroglycerine, TNT is not very shock sensitive and can, therefore, be used in military shells fired from guns by explosive charges. In fact, TNT is so difficult to explode that "booster" additives must be used in addition to a detonator chemical such as lead azide $Pb(N_3)_2$. The explosion of TNT may be represented as:

*Proficiency Level

$$2 \; C_7H_5(NO_2)_{3(s)} \rightarrow 7 \; C_{(s)} + 7 \; CO_{(g)} + 3 \; N_{2(g)} + 5 \; H_2O_{(g)}$$

How much heat would be generated by the detonation of 5.00 lb (2268 g) of TNT? (ΔH_f° for TNT is -87.1 kcal mole^{-1}. Other data needed will be found in Appendixes A and F. Assume Standard State conditions.)

SOLUTION:

(Using ΔH_f° given for TNT and other data as needed from Appendix F)

$\Delta H_{reaction}^\circ = [7\Delta H_f^\circ(C) + 7\Delta H_f^\circ(CO) + 3\Delta H_f^\circ(N_2) + 5\Delta H_f^\circ(H_2O_{(g)})] - [2\Delta H_f^\circ(TNT)]$

$\qquad = [7(zero) + 7(-26.4) + 3(zero) + 5(-57.8)] - [2(-87.1)]$

$\qquad = -299.6$ kcal

$$\frac{(-299.6) \, kcal}{2 \; moles(TNT)} \times \frac{1 \; mole(TNT)}{227 \; g \; (TNT)} \times \frac{2268 \; g \; (TNT)}{1} = \underline{-1500 \; kcal}$$

----

*EXERCISE 3

Internal combustion engines are inefficient fuel users. Hydrocarbon fuels, such as the mixture known as "gasoline", are only partially converted to their complete oxidation products. There are two results of this inefficiency. Exhaust gases contain incomplete combustion products, such as carbon monoxide, and less energy is generated than would have occurred for complete combustion. Write the balanced equation for the complete combustion of octane, C_8H_{18}, to CO_2 and H_2O and calculate the heat of combustion of 1.00 gal(2659 g) of octane. Calculate for contrast the heat of reaction of 1.00 gallon of octane for the incomplete combustion represented by:

$$C_8H_{18(\ell)} + 10 \; O_{2(g)} \rightarrow C_{(s)} + 3 \; CO_{(g)} + 4 \; CO_{2(g)} + 9 \; H_2O_{(g)}$$

[$\Delta H_f^\circ(C_8H_{18})$ = -49.8 kcal mole^{-1}. Use Appendix F for other necessary data. Assume Standard State conditions.]

(answer, page 156)

==

Extra Practice

*EXERCISE 4

With the long-delayed public recognition of the "energy crisis", much attention has been focused on relatively novel sources of power, such as "geothermal plants" like those now operating in northern California. It would seem at first glance that the tapping of underground steam should provide both economical and pollution-free energy production. Unfortunately, most such natural steam sources contain sufficient hydrogen sulfide to result in a significant (and unpleasant) odor for many miles downwind of a typical geothermal plant operation. It has been suggested that the hydrogen sulfide might be removed economically by using it to manufacture some commercially valuable sulfides, such as the yellow CdS used in certain pigments. If the heat of reaction between solid cadmium chloride and gaseous hydrogen sulfide (forming $CdS_{(s)}$ and $HCl_{(g)}$) is +19.4 kcal mole^{-1} (CdS), what is the standard heat of formation of solid CdS? [ΔH_f° for $CdCl_{2(s)}$ is -93.0 kcal mole^{-1}. Other necessary data are given in Appendix F.]

(answer, page 157)

==

SELF-TEST (UNIT 7) [answers, page 157]

7.1. At one time "triple-superphosphate of lime", $Ca(H_2PO_4)_2$, was one of the most commercially valuable soil additives. When supplied to alkaline soils, the "triple superphosphate" provided both an acidification action and phosphate needed as a plant nutrient. Calculate the standard heat of reaction, in kcal mole^{-1} [$Ca(H_2PO_4)_2$], for the manufacture of "triple superphosphate" from phosphoric acid and phosphate rock, after determining the coefficient needed to balance the equation:

$$Ca_3(PO_4)_{2(s)} + 4H_3PO_{4(aq)} \rightarrow \underline{}Ca(H_2PO_4)_{2(s)}$$

compound	ΔH_f° (kcal mole^{-1})
$H_3PO_{4(aq)}$	-306.2
$Ca_3(PO_4)_{2(s)}$	-986.2
$Ca(H_2PO_4)_{2(s)}$	-744.4

- -

If you completed this problem correctly, you may go on to the proficiency level, try the RELEVANT PROBLEMS (Unit 7), or stop here. If not, you should consult your instructor for suggestions of further study aids.

- -

*7.2. Environmentalists have long been concerned with the increasing concentrations of lead in the atmosphere, most of which comes from the lead additives used as "antiknock" agents in gasoline. In an attempt to eliminate "leaded" gasolines, a number of alternative "antiknock" compounds have been investigated. One of the simplest of these is benzene (C_6H_6), but the health hazards of benzene vapors argue against its extensive use in gasolines. Benzene, unlike tetraethyl lead, does burn without introducing new combustion residues into the atmosphere. How much heat would be generated by the combustion of 1.00 pt (416 g) of benzene, under conditions of complete oxidation to CO_2 and H_2O gases? [For ΔH_f° values, see Appendix F. Assume Standard State conditions.]

*Proficiency Level

If you completed this problem correctly, you may go on to the RELEVANT PROBLEMS (Unit 7). If not, you should consult your instructor for suggestions of further study aids.

ANSWERS to EXERCISES, Unit 7

1. [-368 kcal mole^{-1} (nitroglycerine)] <u>Solution</u>:

    To balance the equation, a coefficient of 4 is needed:

    $$\underline{4}\ C_3H_5(ONO_2)_{3(\ell)} \rightarrow 12\ CO_{2(g)} + 10\ H_2O_{(g)} + 6\ N_{2(g)} + O_{2(g)}$$

    For the equation as shown, corresponding to decomposition of <u>4 moles</u> of nitroglycerine:

    $$\Delta H^\circ_{reaction} = [12\Delta H^\circ_f(CO_2) + 10\Delta H^\circ_f(H_2O) + (zero)] - [4\Delta H^\circ_f\ (nitroglycerine)]$$
    $$= [12(-94.0) + 10(-57.8)] - [4(-58.1)] = -1474\ kcal$$

    To report this in kcal <u>per mole</u> of nitroglycerine:

    $$\Delta H^\circ = \frac{-1474\ kcal}{4\ moles(nitroglycerine)}$$

2. [-33.7 kcal mole^{-1} (Na)] <u>Solution</u>:

    $$\underline{2}\ Na_{(s)} + \underline{2}\ H_2O_{(\ell)} \rightarrow H_{2(g)} + \underline{2}\ NaOH_{(s)}$$

    $$\Delta H^\circ_{reaction} = [(zero) + 2\Delta H^\circ_f(NaOH)] - [(zero) + 2\Delta H^\circ_f(H_2O_{(\ell)})]$$
    $$= [2(-102.0)] - [2(-68.3)] = -67.4\ kcal$$

    per mole (Na): $\dfrac{-67.4\ kcal}{2\ moles\ (Na)}$

- -

*3. (complete: 2.85×10^4 kcal, incomplete: 2.16×10^4 kcal) <u>Solution</u>:

    complete combustion: $(2\ C_8H_{18} + 25\ O_2 \rightarrow 16\ CO_2 + 18\ H_2O)$

    $$\Delta H^\circ_{comb} = [16\Delta H^\circ_f(CO_2) + 18\Delta H^\circ_f(H_2O)] - [2\Delta H^\circ_f(C_8H_{18}) + 25\Delta H^\circ_f(O_2)]$$

    $$= [16(-94.0) + 18(-57.8)] - [2(-49.8) + 25(zero)]$$

    $$= -2444.8\ kcal$$

    $$\frac{(-2444.8\ kcal)}{2\ moles(C_8H_{18})} \times \frac{1\ mole(C_8H_{18})}{114.2\ g(C_8H_{18})} \times \frac{2659\ g(C_8H_{18})}{1}$$

    incomplete combustion: $(C_8H_{18} + 10\ O_2 \rightarrow C + 3\ CO + 4\ CO_2 + 9\ H_2O)$

    $$\Delta H^\circ_{reaction} = [\Delta H^\circ_f(C) + 3\Delta H^\circ_f(CO) + 4\Delta H^\circ_f(CO_2) + 9\Delta H^\circ_f(H_2O)] - [\Delta H^\circ_f(C_8H_{18})]$$

    $$= [(zero) + 3(-26.4) + 4(-94.0) + 9(-57.8)] - [-49.8]$$

    $$= -925.6\ kcal$$

$$\frac{(-925.6 \text{ kcal})}{1 \text{ mole}(C_8H_{18})} \times \frac{1 \text{ mole}(C_8H_{18})}{114.2 \text{ g}(C_8H_{18})} \times \frac{2659 \text{ g}(C_8H_{18})}{1}$$

*4. (-34.2 kcal mole^{-1}) <u>Solution</u>:

$$CdCl_{2(s)} + H_2S_{(g)} \rightarrow CdS_{(s)} + 2HCl_{(g)}$$

$$\Delta H^\circ_{reaction} = [\Delta H^\circ_f(CdS) + 2\Delta H^\circ_f(HCl)] - [\Delta H^\circ_f(CdCl_2) + \Delta H^\circ_f(H_2S)]$$

$$+ 19.4 = [\Delta H^\circ_f(CdS) + 2(-22.1)] - [(-93.0) + (-4.8)]$$

$$\Delta H^\circ_f(CdS) = (+19.4) + (44.2) - (93.0) - (4.8)$$

==

<u>ANSWERS to SELF-TEST, Unit 7</u>

7.1. -7.4 kcal mole^{-1} [Ca(H$_2$PO$_4$)$_2$]

*7.2. 4000 kcal

A.1. When the accumulation of organic matter in sea water is great, the dissolved oxygen in the water may become depleted, leading to a condition variously designated as anaerobic, anaeric or anoxic. Under such conditions, the oxidation of organic matter may continue by means of anaerobic bacterial processes in which nitrate-nitrite, sulfate-sulfite, and/or carbon dioxide serve as an oxygen source. The reduced products of these substances accumulate in the water in addition to the products of the oxidation of the organic matter. Together, these products modify the proportions of the components of the water in ways that differ from those characteristic of chemical oxidation in the presence of oxygen. A major denitrification reaction taking place in the absence of molecular oxygen may be represented by the equation:

$$NH_{3(g)} + HNO_{3(aq)} \xrightarrow{\text{bacteria}} N_{2(g)} + H_2O_{(\ell)}$$

Balance the equation for the reaction and then calculate the standard enthalpy change in kcal mole^{-1} (NH$_3$). [For HNO$_{3(aq)}$, $\Delta H_f^o = -49.4$ kcal mole^{-1}. See Appendix F for other data.]

A.2. Sugar cane is a coarse grass grown in tropical and semi-tropical areas for its stems, the juice of which is used for making sugar and syrup. It differs from ordinary sorghum (commonly called cane) in containing a higher percentage of sugar in its juices, and also in not producing seed in semi-tropical areas, and only sparingly in tropical areas. Sorghum produces an abundance of seed in a compact panicle at the top of the stalk. The sugar cane is used primarily for making sugar, while sorghum is used for making molasses. The plants of sugar cane vary in height from eight to fifteen feet. The stems are usually close-jointed and very leafy. Sugar cane was the first plant used for the commercial manufacture of sugar. The complete anaerobic decomposition of moist cane sugar, as in the rotting of sugar cane, occurs as follows:

$$H_2O_{(\ell)} + C_{12}H_{22}O_{11} \rightarrow CO_{2(g)} + CH_{4(g)}$$

Supply the coefficients necessary to balance the equation and then calculate the standard enthalpy change in kcal mole^{-1} (C$_{12}$H$_{22}$O$_{11}$).

Compound	ΔH_f^o (kcal mole^{-1})
$C_{12}H_{22}O_{11}$	−529.7
CO_2	− 94.0
CH_4	− 17.9

A.3. It has been suggested that the most efficient ways to increase the world's food supply is by treating land and crops with proper chemicals: fertilizers to enrich depleted soils; pesticides, dusts and sprays, to minimize the damage of insects, vermin and diseases; and poisons to control pests that damage or ruin stored grain. The type and amount of insecticide to use varies with season, climatic conditions and the type of insect. Stomach poisons are generally recommended for control of insects with chewing or biting mouth parts. Contact poisons are usually employed when the insect has sucking or piercing mouth parts, and where it is impractical to use a stomach poison. Fumigants can be used most successfully in enclosures where the gas produced can be confined for several hours at a time. Repellents are substances, such as lime, which are used to drive away insects when it is impossible to use any of the other types of insecticides. Carbon disulfide is a good fumigant for grain weevils and similar pests. Carbon disulfide is prepared by the reaction of a hydrocarbon, usually methane, with sulfur over silica gel, according to the equation:

$$CH_{4(g)} + S_{(g)} \xrightarrow[\text{(500-700°)}]{(Si_2O_3)} CS_{2(g)} + H_2S_{(g)}$$

Balance the equation for the reaction and then calculate the standard enthalpy change in kcal mole^{-1} (CS_2).

Compound	$\Delta H^°_f$ (kcal mole^{-1})
CH_4	-17.9
CS_2	+28.1
H_2S	- 4.8

A.4. Carbon disulfide volatilizes readily to form a poisonous vapor which is heavier than air (vapor density = 2.67 g ℓ^{-1}). On that account the corn or other material to be fumigated should be placed in a solid box, or, if in very large quantities, in a room that is reasonably air tight. The carbon disulfide should be placed on top of the material in open shallow vessels at the rate of 15 pounds per thousand cubic feet of space in the box or room. The box or room should be closed at once and kept closed for 24 hours. If it is a room, it should be throughly aired for two or more hours before any one enters it. It is necessary to keep in mind the explosive properties of carbon disulfide when it is being used as a fumigant. A lighted cigar, a match, or even a spark is sufficient to ignite it and cause an explosion. Carbon disulfide burns according to the equation:

$$CS_{2(g)} + O_{2(g)} \rightarrow CO_{2(g)} + SO_{2(g)}$$

Calculate the standard enthalpy change for the burning of carbon disulfide, after supplying the coefficients needed to balance the equation. Report the answer in kcal mole^{-1} (CS_2).

Compound	ΔH°_f (kcal mole^{-1})
CS_2	28.1
CO_2	-94.0
SO_2	-70.9

A.5. The principal causes of plant diseases are bacteria, viruses, and fungi. These organisms do not as a rule manufacture their own foods, and hence must take foods that plants have made. When the spore of a destructive fungus lodges on a plant under conditions favorable to its development, it sends out threadlike growths which pierce the epidermis of the plant. This filament continues developing and dividing within the plant, and may extend a long distance from the point of entry. Extending thus among and into the cells, the fungus feeds on the plant. The cells may break down and the plant wither as a result of this attack, or an abnormal growth may take place, producing the warts and galls so often seen on infected plants. A well known fungicide of long standing, the "Bordeaux Mixture", is made from five parts copper(II) sulfate, five parts calcium oxide and 400 parts water. The copper(II) sulfate used in the "Bordeaux Mixture" is formed commercially according to the equation:

$$CuO_{(s)} + H_2SO_{4(\ell)} \rightarrow CuSO_{4(s)} + H_2O_{(\ell)}$$

Calculate the standard enthalpy change in kcal mole^{-1} ($CuSO_4$).

Compound	ΔH°_f (kcal mole^{-1})
CuO	- 37.6
$CuSO_4$	-184.4
H_2O	- 68.3
H_2SO_4	-194.6

*A.6. The acetate unit, $CH_3CO_2^-$, is one of the most important intermediates in the metabolic cycles of many aerobic respiratory organisms. However, under the natural anaerobic conditions characteristic of most of the marine sediments, methane bacteria are probably the only important group that can degrade acetate. Apparently, methane formation does not occur in marine sediments until sulfate reduction is complete. This appears to happen because the sulfate-reducing bacteria compete favorably for hydrogen liberated during fermentation, or because the hydrogen sulfide, produced by

*Proficiency Level

sulfate reduction, is toxic for methane bacteria. Write the equation for the conversion of acetic acid to methane and carbon dioxide, and calculate the standard enthalpy change for the decomposition of 300 g of acetic acid. [Use Appendix F for necessary data.]

*A.7. The application of lime, CaO, to an acidic soil will cause a significant increase in the soil pH, i.e., the hydrogen ion concentration will decrease while the hydroxide ion concentration will increase. The liming of acid soils is a desirable practice because it increases the availability of several essential nutrients, such as calcium, magnesium, phosphates, and molybdates. At the same time, the availability of iron, manganese, and aluminum will decrease. This is a favorable reaction since under acidic conditions these species may be present in quantities toxic to many plants. Calculate the standard enthalpy change (in kcal) that occurs when 1020 g of calcium oxide added to a moist soil forms calcium hydroxide.

Compound	ΔH°_f (kcal mole^{-1})
$CaO_{(s)}$	-151.9
$H_2O_{(\ell)}$	- 68.3
$Ca(OH)_{2(s)}$	-235.8

A.8. Most "complete" fertilizers tend to form an acid residue when added to soils. This is primarily due to the presence of various nitrogen carriers, especially those which supply NH_4^+ or produce NH_4^+ when added to the soil. For example, when ammonium sulfate is added to the soil, some of the ammonium ions will be absorbed by the soil colloidal matter, displacing equivalent quantities of other cations. If metallic cations are displaced, their loss by leaching is enhanced. This ultimately may result in lowering of soil pH. Moreover, if hydrogen ions are released, sulfuric acid appears in the soil solution. The most important effect of ammonium ions occurs when they are "nitrified". Write the balanced equation for the O_2 oxidation of ammonium nitrate to nitric acid and water and calculate the standard enthalpy change (in kcal) for the oxidation of 400 g of ammonium nitrate (NH_4NO_3). [Use Appendix F for the necessary data.]

*A.9. The oceans provide a relatively constant chemical and physical environment for life. Marine organisms do not have to contend with large variations of temperature and amounts of nutrients that affect organisms on land. However, the oceans are becoming increasingly polluted as a result of man's interaction with his environment. Levels of pollution are increasing at an alarming rate in many estuaries and even in some coastal waters with limited circulation. The current pollution problems stem from dumping large quantities of waste materials, including dredge spoils, toxic and dangerous chemicals from the chemical industry, and domestic and industrial sewage, into the oceans. When sewage containing large quantities of organic compounds, phosphates, and nitrogen compounds are dumped into the ocean, bacterial oxidation of the waste organic content may cause depletion of the oxygen content of the water so that it will not support life. In the absence of oxygen the nitrogen compounds such as ammonia are subject to anaerobic oxidation by nitrate ions. If the heat of reaction between ammonia and nitric acid (forming nitrogen and water) is -86.98 kcal mole^{-1} (NH_3) what is the standard heat of formation of nitric acid?

Compound	ΔH°_f (kcal mole^{-1})
NH_3	-11.0
H_2O	-68.3

*A.10. In cases where the oxygen content of ocean water is completely exhausted, the formation of toxic hydrogen sulfide will occur by a sulfate reduction process illustrated by the equation:

$$CH_3OH_{(\ell)} + H_2SO_{4(aq)} \rightarrow CO_{2(g)} + H_2S_{(g)} + H_2O_{(\ell)}$$

Balance the equation and calculate the standard enthalpy change for oxidation of 160 g of methyl alcohol. [For $H_2SO_{4(aq)}$, $\Delta H^\circ_f = -216.9$ kcal mole^{-1}. See Appendix F for other data.]

ANSWERS:

(A.1.) -82.3 kcal mole^{-1}, (A.2.) -73.4 kcal mole^{-1}, (A.3.) +36.4 kcal mole^{-1}, (A.4.) -263.9 kcal mole^{-1}, (A.5.) -20.5 kcal mole^{-1}, (A.6) +22.5 kcal, (A.7.) -284 kcal, (A.8.) -320 kcal, (A.9.) -41.6 kcal mole^{-1}, (A.10.) -72.5 kcal

---

RELEVANT
PROBLEMS

Unit 7: Enthalpy Problems

Set B: Biological & Medical Sciences

B.1. The most "successful" animals on earth are insects. There are more than 800,000 species of insects known today. By comparison, all the species of vertebrate animals (fish, amphibians, reptiles, bird and mammals) total less than 36,000. The total number of beetle species alone is greater than 240,000. One characteristic of insects that contributes to their unparalleled success is their diversified defense mechanisms. Many insects possess chemical armaments as their first line of defense. These include secretions that produce evil smell and taste, poisons such as formic acid, and liquids that can raise blisters. One of the most elegant chemical defense mechanisms belongs to the bombardier beetle. The last segment of the beetle's abdomen can telescope and be aimed in any direction. When a predator attacks, the beetle instantly aims its "turret" and discharges a vapor which repels the enemy. The chemical "ammunition", hydroquinone, and hydrogen peroxide are secreted by glands and stored in separate reservoirs. To fire, the beetle allows the chemicals to flow into a strong-walled compartment where, with appropriate catalysis, the mixture reacts explosively and is forcibly expelled. On occasion, bombardiers have fired as many as 30 times in five minutes. The hot repellant spray of the bombardier beetle is generated by the reaction:

$$HO-\bighexagon-OH + H_2O_2 \rightarrow O=\bighexagon=O + 2H_2O + heat$$

The irritating mist of quinone ($C_6H_4O_2$) is fired at the enemy by pressure resulting from gaseous oxygen, produced by decomposition of excess hydrogen peroxide:

$$H_2O_{2(\ell)} \rightarrow H_2O_{(\ell)} + O_{2(g)}$$

Balance the latter equation, then calculate the standard enthalpy change in kcal mole^{-1} (O_2). [ΔH_f° for H_2O_2 is -44.8 kcal mole^{-1}, for other ΔH_f° data see Appendix F.]

B.2. Gas gangrene occurs through the infection of tissues around a wound by certain anaerobic bacteria which grow best in tissues away from air. The factors that predispose to the invasion of tissues by the bacilli are impaired local circulation, owing to pressure from foreign bodies (casts or tourniquets), presence of metallic bodies or rust, clothing or dirt in the wound, or occurrence of necrotic tissue or hemorrhage. The infection is necrotic and rapidly spreading. It is accompanied by edema and gaseous infiltration of the tissues. The most common organism found in gas gangrene infections is Bacillus Welchii. These organisms are normal inhabitants of the human and animal intestinal tracts and are present in the soil. Treatment of this infection must be prompt and vigorous. Most important is thorough excision of all devitalized tissue. Involved extremities often require amputation. The organisms which cause gas gangrene liberate a toxin which destroys tissue, particularly muscle, and they produce gas by fermenting "muscle sugar". This fermentation process may be described by the following equation:

$$C_6H_{12}O_{6(s)} \xrightarrow{\text{(Bacillus Welchii)}} C_2H_5OH_{(\ell)} + CO_{2(g)}$$

Calculate the standard enthalpy change for this fermentation, after supplying the coefficients needed to balance the equation. Report the answer in kcal mole^{-1} (CO_2).

Compound	ΔH°_f (kcal mole^{-1})
$C_6H_{12}O_6$	-300
C_2H_5OH	- 66.4
CO_2	- 94.0

B.3. The energy released during animal respiration results primarily from oxidation of glucose. Most of the oxidation in a cell is brought about by the removal of hydrogen from a molecule. In aerobic respiration, where atmospheric oxygen is the hydrogen acceptor, molecules of glucose ($C_2H_{12}O_6$) are first split into two molecules of a 3-carbon compound ($C_3H_6O_3$) by enzyme action. Other respiratory enzymes are utilized in the oxidation of these molecules to pyruvic acid ($C_3H_4O_3$) and water. Further energy is released when the pyruvic acid is oxidized to carbon dioxide and water. The overall process, in greatly simplified form, occurring in the human body is the oxidation of glucose:

$$C_6H_{12}O_6(s) + O_2(g) \rightarrow CO_2(g) + H_2O(\ell)$$

Balance the equation, then calculate the standard enthalpy change in kcal mole^{-1} ($C_6H_{12}O_6$).

Compound	ΔH°_f (kcal mole^{-1})
$C_6H_{12}O_6$	-300.0
CO_2	- 94.0
H_2O	- 68.3

B.4. Most chemical reactions in organisms take place in a narrow range of temperatures. These temperatures are not sufficiently high to provide the activation energy to "start" most reactions. Therefore, organisms require catalysts (substances to reduce the activation energy) in almost all their chemical reactions. Several simple inorganic compounds can act as catalysts, however, many are complex organic molecules called enzymes. Enzymes are very specific in the reactions which they catalyze. For example, tissue cells are the main sites for reactions with oxygen. Here the body can oxidize a wide variety of chemicals that are stable in air. These reactions are quite productive of energy, and any enzyme participating in them is classified as a respiratory enzyme. Hydrogen peroxide is a temporary product in certain metabolic pathways, and hydroperoxidase is the enzyme that will catalyze its further breakdown as summarized in the equation:

$$H_2O_{2(\ell)} \rightarrow H_2O_{(\ell)} + O_{2(g)}$$

Supply the coefficients needed to balance the equation, then calculate the standard enthalpy change in kcal mole^{-1} (H_2O_2). [Compare your answer with that for problem B.1.]

Compound	ΔH°_f (kcal mole^{-1})
$H_2O_{2(\ell)}$	-44.9
$H_2O_{(\ell)}$	-68.3

B.5. Arsenic is a strong protoplasmic poison capable of causing death within 24 hours when ingested in large amounts. The progressive accumulation of small doses may give rise to chronic poisoning. Acute arsenic poisoning may occur accidentally, since arsenic is commonly employed in some rat poisons and insecticides that might be innocently ingested by children. This element is also used in paint pigments, and wall paper pastes, especially in older homes. Such applications permit the absorption over long periods of time of minute cumulative doses leading to chronic poisoning. Arsenic poisons such as arsenic acid (H_3AsO_4) or sodium arsenate Na_3AsO_4) function by blocking enzymes that contain phosphate units. Arsenate ions (AsO_4^{3-}) closely resemble phosphate ions (PO_4^{3-}) and can sometimes take their place. When that happens certain enzymes do not function properly. Arsenic acid is formed when arsenic(III) oxide is oxidized by concentrated nitric acid according to the equation:

$$As_4O_{6(s)} + HNO_{3(\ell)} + H_2O_{(\ell)} \rightarrow H_3AsO_{4(s)} + NO_{2(g)}$$

Calculate the standard enthalpy change for the preparation of arsenic acid, after supplying the coefficients needed to balance the equation. Report the answer in kcal mole^{-1} (H_3AsO_4).

Compound	ΔH°_f (kcal mole^{-1})
As_4O_6	-314.0
H_3AsO_4	-216.6
NO_2	7.9
HNO_3	- 41.6
H_2O	- 68.3

*B.6. Anesthetics are substances used to reduce or eliminate the sense of pain as well as to relax muscles and produce the loss of consciousness. A general anesthetic is

*Proficiency Level

one which produces in the patient a complete loss of both consciousness and sensation of pain. Commonly used general anesthetic drugs are diethyl ether, chloroform, nitrous oxide (laughing gas) and cyclopropane. Anesthesia produced by these substances is effected by inhalation of the gases or vapors. One of the problems with certain anesthetics, particularly cyclopropane and diethyl ether, is their flammability. Mixtures of the gases or vapors with air may ignite explosively with a tiny spark. Careful precautions must be followed to avoid such explosions in operating rooms, including special care to eliminate possible electrostatic sparking. Supply the coefficients to balance the following equation, then estimate the amount of heat produced by the accidental combustion of 400 g of diethyl ether.

$$C_2H_5OC_2H_{5(g)} + O_{2(g)} \rightarrow CO_{2(g)} + H_2O_{(g)}$$

Compound	ΔH°_f (kcal mole^{-1})
diethyl ether	-88.6
carbon dioxide	-94.0
water	-57.8

*B.7. Methyl alcohol (CH_3OH, "wood alcohol") may be absorbed by the accidental or suicidal ingestion of the liquid alcohol or by the inhalation of its vapors. It has been estimated that it requires in the range of 100 to 150 ml of this compound to cause death, although toxic effects have been observed from the rapid absorption of as little as 10 ml. Local injuries are produced at the site of absorption. When ingested, methyl alcohol causes edema and hemorrhages in the stomach. On inhalation edema and hemorrhages occur in the lung tissues. However, methyl alcohol exerts its main toxic effect after absorption by its oxidation to formaldehyde and formic acid, both of which are more toxic than the parent substance. These derivatives cause degeneration of the receptor cells of the retina, with associated degeneration of the optic disc and nerve. Swelling of the brain and brain stem may also occur, accompanied by marked congestion of the cerebral vessels.

The ingestion of methyl alcohol may result in the formation of the very toxic formic acid, as indicated in simplified terms by the following equation:

$$CH_3OH_{(\ell)} + O_{2(g)} \xrightarrow{\text{(enzymes)}} HCO_2H_{(\ell)} + H_2O_{(\ell)}$$

Calculate the standard enthalpy change that occurs when a 10 ml amount of methyl alcohol is oxidized. (Assume 100% efficiency.) Methyl alcohol has a density of 0.792 g ml^{-1}.

Compound	ΔH°_f (kcal mole^{-1})
CH_3OH	-57.0

HCO$_2$H $\qquad\qquad\qquad$ -96.2

H$_2$O $\qquad\qquad\qquad$ -68.3

*B.8. Mercury(II) salts are violent poisons, having a protein-precipitating action which results in the coagulative necrosis of tissues exposed to toxic levels. It is estimated that as little as 0.1 gram may cause severe tissue damage and one gram, death. Absorbed through the stomach, the mercury(II) ion is transiently stored in the liver and thence removed from the blood stream through the kidneys. Mercury poisoning is almost invariably due to the absorption of the highly soluble toxic mercury(II) ion. Metallic mercury is highly insoluble and relatively unreactive, and therefore has little toxic effect when swallowed. Quite rarely, mercury(I) salts may undergo oxidation in the body to produce the toxic mercury(II) ion. When incorporated into therapeutic agents such as laxatives (calomel), the mercury(I) ion slowly accumulates and may occasionally give rise to tissue toxicity by its oxidation.

Write the balanced equation for the formation of mercury(II) chloride (used in dilute solutions as a topical antiseptic) from sodium chloride and mercury(II) sulfate and calculate the standard enthalpy change involving the production of 952 g of mercury(II) chloride.

Compound	ΔH°_f (kcal mole^{-1})
NaCl	-98.2
Na$_2$SO$_4$	-330.9
HgSO$_4$	-169.1
HgCl$_2$	-53.6

*B.9. Sulfur is a plant nutrient because some of the amino acids which are necessary for proteins contain sulfur. The green plants absorb minerals from the soil and use them in synthesizing organic compounds. Animals consume the plants and rearrange these compounds to meet their specific needs. When a plant or animal dies, the chemical substances composing it are left in a complex form - products of life, yet not in a chemical form to supply the needs of another generation of plants and animals. Of course the constituents of these complex molecules are recycled through bacterial decay and putrefaction. Through these natural biological decay processes a large proportion of the sulfur from the proteins is converted to hydrogen sulfide which reaches the atmosphere and is subsequently oxidized to water and sulfur dioxide. (a) Write the balanced equation for the oxidation of hydrogen sulfide to water and sulfur dioxide and calculate the standard enthalpy change for the oxidation of 85.2 g of hydrogen sulfide. (b) Calculate for contrast the heat of reaction of 85.2 g of hydrogen sulfide for the reaction represented by:

$$2H_2S_{(g)} + 4 O_{2(g)} \rightarrow 2H_2O_{(\ell)} + 2SO_{3(g)}$$

[Use Appendix F for necessary data.]

*B.10. The greatest number of bacteria are facultative anaerobes, which grow best as aerobes but may grow, at least to some extent, as anaerobes. Among these organisms are <u>Escherichia coli</u> (common bacillus of the human intestine), thyphoid, diphtheria, and scarlet fever bacteria. Other bacteria, obligate aerobes, require atmospheric oxygen for respiration just as most plants and animals. These include tuberculosis bacilli and cholera bacteria. At the other extreme are organisms classed as obligate anaerobes. These bacteria cannot grow in the presence of atmospheric oxygen and include the tetanus and botulism bacteria. One of the best known anaerobic reactions occurs in the bacterially mediated formation of methane and water from hydrogen and carbon dioxide. If the heat of reaction between hydrogen and carbon dioxide (forming methane and $H_2O_{(\ell)}$) is -60.5 kcal mole^{-1} (CH_4), what is the standard heat of formation of carbon dioxide? [ΔH_f° for CH_4 is -17.9 kcal mole^{-1} and $H_2O_{(\ell)}$ is -68.3 kcal mole^{-1}.]

ANSWERS:

(B.1.) -47.0 kcal mole^{-1}, (B.2.) -10.4 kcal mole^{-1}, (B.3.) -674 kcal mole^{-1}, (B.4.) -23.4 kcal mole^{-1}, (B.5.) -5.0 kcal mole^{-1}, (B.6.) -3.1×10^3 kcal, (B.7.) -27 kcal, (B.8.) -66.6 kcal, (B.9.)(a) -336 kcal, (b) -395 kcal, (B.10.) -94.0 kcal mole^{-1}

RELEVANT
PROBLEMS

Unit 7: Enthalpy Problems

Set I: Industrial Chemistry

I.1. The fertility of soil (its ability to bear crops) has always been an important consideration in the history of man. As soon as early farmers learned that manure, bones, and ashes increased the fertility of their land, they began saving such materials and using them to increase crop production. Then, as specialization became a part of community economy, the practice of selling fertilizer materials began. The development of the fertilizer industry has been characterized by the shift from one product to another in a continuing effort to increase nutrient content. Currently, the major fertilizer complexes normally involve ammonium phosphate and urea as the principal constituents. The United States Department of Agriculture reported that 5.96×10^5 tons of urea were consumed in fertilizers during 1973. The agricultural advantages of urea include its high available nitrogen content, good plant response, and simplicity of application and storage. Some plants are capable of producing 1500 tons of urea per day. The compound is synthesized by the reaction of ammonia and carbon dioxide according to the following reaction.

$$2\ NH_{3(g)} + CO_{2(g)} \rightarrow \underset{}{H_2N\text{-}\overset{\overset{O}{\|}}{C}\text{-}NH_{2(s)}} + H_2O_{(\ell)}$$

Calculate the standard enthalpy change, in kilocalories per mole of urea, for this synthesis. [ΔH°_f (urea) = -35.8 kcal mole^{-1}. See Appendix F for other data.]

I.2. Silicon carbide, SiC, is better known as carborundum. It is a crystalline substance varying in color from nearly clear, through pale yellow or green, to black, depending upon the nature and amount of impurities. Hardness is probably the best known characteristic of silicon carbide. It ranks just below diamond in hardness. Silicon carbide first became important as an abrasive. It is used loose, mixed with a vehicle to form abrasive pastes, mixed with organic or inorganic binders, shaped and cured to form abrasive wheels or rubs, or adhered to paper or cloth backings to form abrasive sheets, discs or belts. The abrasive and refractory industries are the largest users of carborundum. It is also used for heating elements in electric furnaces, in electronic devices, and in applications where its resistance to damage by nuclear radiation is desirable. Silicon carbide is produced commercially by the reaction in an electric furnace of a charge of high grade silica sand and slightly more than the stoichiometric quantity of carbon in the form of coke or anthracite coal. For the reaction described by the equation

$$SiO_{2(s)} + 3\ C_{(s)} \rightarrow SiC_{(g)} + 2\ CO_{(g)}$$

Calculate the standard enthalpy change, in kilocalories per mole of SiC, for the synthesis of carborundum.

Compound	ΔH°_f (kcal mole^{-1})
SiC	- 26.7
CO	- 26.4
SiO$_2$	-205.1

I.3. One promising method of converting solar energy to usable energy for man is the Solchem process. Sulfur trioxide gas is pumped into a chamber at the focal point of a parabolic mirror. The temperature in this chamber is raised by the sun's rays to 800°C. This temperature is enough to cause the dissociation reaction to occur:

$$2 \ SO_3 \rightarrow 2 \ SO_2 + O_2$$

The product gases are then pumped to a heat exchanger where the following catalyzed reaction occurs:

$$2 \ SO_2 + O_2 \rightarrow 2 \ SO_3 + heat$$

The heat is used to melt a solid salt for storage or to superheat steam to produce electricity. What is the standard heat of reaction for:

$$2 \ SO_2 + O_2 \rightarrow 2 \ SO_3$$

[See Appendix F for data.]

I.4. "Water gas" is an impure mixture of hydrogen and carbon monoxide. It is a good heat source but can not be used in homes because of the danger of carbon monoxide poisoning. "Water gas" can be made by the action of steam on coal or coke:

$$H_2O_{(g)} + C \rightarrow CO + H_2$$

This reaction is endothermic. The conversion of carbon to carbon dioxide is exothemic:

$$C + O_2 \rightarrow CO_2$$

A process has been worked out so that by combining the correct ratios of the two reactions, the necessary heat for the first reaction is supplied by the second reaction. What is the standard heat of reaction of each process and what is the correct reaction ratio? [See Appendix F for data.]

I.5. Nitrous oxide, "laughing gas", is used as a dental anesthetic, usually mixed with oxygen, and as a propellant in some aerosol spray cans. The N_2O is prepared commercially by heating ammonium nitrate very cautiously:

$$NH_4NO_{3(s)} \rightarrow N_2O_{(g)} + 2 \ H_2O_{(g)}$$

What is the standard heat of reaction for the production of nitrous oxide?

*I.6. Ammonium nitrate if improperly heated, will decompose explosively to nitrogen gas, oxygen gas and water vapor. This presents a very real safety problem in the industrial production of N_2O (problem I.5). What amount of heat is produced by explosive decomposition of 138 kg of ammonium nitrate? [See Appendix F for data.]

*I.7. Many years ago "carbide lanterns" were used by miners because of the high-intensity white light provided and the low fuel weight required (so that the lamps could be

*Proficiency Level

170

mounted on headgear). The fuel itself was acetylene, released at a rate regulated by the addition of water to calcium carbide:

$$CaC_{2(s)} + 2 H_2O_{(\ell)} \rightarrow C_2H_{2(g)} + Ca(OH)_{2(s)}$$

Open flame mining lights have long been outlawed as safety hazards, but carbide camping lanterns are still occasionally seen. Calcium carbide is now used industrially to prepare calcium cyanamide, $CaCN_2$, an important chemical for the fertilizer industry. The carbide is also a valuable source of acetylene for welding and metal-cutting operations. For the reaction of calcium carbide with water to form acetylene and calcium hydroxide, the standard enthalpy change is -30.0 kcal mole^{-1}. What is the standard heat of formation for acetylene?

Compound	ΔH°_f (kcal mole^{-1})
CaC_2	- 15.0
H_2O	- 68.3
$Ca(OH)_2$	-235.8

*I.8. The complete combustion of acetylene is formulated by:

$$2 H-C\equiv C-H_{(g)} + 5 O_{2(g)} \rightarrow 4 CO_{2(g)} + 2 H_2O_{(g)}$$

Use the information from problem I.7 to calculate the standard heat of combustion for acetylene, in kcal mole^{-1} (C_2H_2).

*I.9. The industrial synthesis of sulfuric acid typically involves a stage at which sulfur trioxide gas is absorbed in slightly diluted H_2SO_4 to produce the concentrated commercial acid (~18 M). The reaction of SO_3 with water is sufficiently exothermic to require, in most commercial operations, a significant investment in cooling equipment. How much heat would be released by the reaction of 2.5 metric tons of SO_3 with an equimolar amount of water?

Compound	ΔH°_f (kcal mole^{-1})
SO_3	- 94.6
H_2O	- 68.3
H_2SO_4	-194.6

I.10. Aniline is an important intermediate in the production of dyes, but it has a larger commercial market in the rubber industries. Aniline can be made industrially by iron reduction of nitrobenzene, by ammonolysis of chlorobenzene, or by vapor-phase hydrogenation of nitrobenzene. The equation for the first process is:

$$4 \; \underset{}{\overset{NO_2}{\bigcirc}} + 9 \; Fe + 4 \; H_2O_{(\ell)} \rightarrow 4 \; \underset{}{\overset{NH_2}{\bigcirc}} + 3 \; Fe_3O_4$$

If the standard heat of reaction for this process is -130 kcal mole^{-1} ($C_6H_5NH_2$), what is the standard heat of formation of nitrobenzene, in kcal mole^{-1} (nitrobenzene)?

Compound	ΔH°_f (kcal mole^{-1})
Fe_3O_4	-267.3
$C_6H_5NH_2$	$+ \; 7.3$

ANSWERS:

(I.1.) 11.9 kcal/mole, (I.2.) 125.6 kcal/mole, (I.3.) -47.4 kcal, (I.4.)(a) 31.4 kcal, (b) -94.0 kcal, 1 CO_2 production for each 3 CO formations, (I.5.) -8.6 kcal, (I.6.) 9.9, 4.86×10^4 kcal,   (I.7.) 54.2 kcal/mole, (I.8.) - 300 kcal/mole, (I.9.) 8.1×10^5 kcal, (I.10.) 5.1 kcal/mole^{-1}

RELEVANT
PROBLEMS

Unit 7: *Enthalpy Problems*

Set E: *Environmental Sciences*

E.1. Certain substances undergo slow oxidation at ordinary temperature with the liberation of heat. If the material being oxidized is confined in an enclosed space, the heat released by oxidation cannot be dissipated and gradually causes the temperature to rise. When the temperature reaches the kindling point, combustion occurs. Spontaneous combustion has been responsible for fires and explosions in flour mills, coal mines, and places where oily materials are stored without having proper ventilation. When any combustible substance is finely subdivided into a powder, the rate of oxidation increases tremendously due to the increase in surface area where oxidation may occur. Thus a chunk of coal may take several hours to burn, whereas the same amount of coal ground to a fine dust may burn very rapidly. The slow oxidation of pyrite, FeS_2, associated with coal dust in a mine, may generate enough heat to raise the temperature to the kindling point and set off a violent explosion. Estimate the heat released by the slow oxidation of 1.2 kg of iron pyrite, FeS_2. The equation for the reaction is given by:

$$4 \; FeS_{2(s)} + 11 \; O_{2(g)} \rightarrow 2 \; Fe_2O_{3(s)} + 8 \; SO_{2(g)}$$

Compound	ΔH°_f (kcal mole^{-1})
$FeS_{2(s)}$	- 42.5
$Fe_2O_{3(s)}$	-196.5

E.2. Stationary sources such as power plants form a major source of nitrogen oxide pollutants. The burning of fuels in air produces a quantity of nitrogen oxides from the
combustion of some of the nitrogen in the air. The quantity of nitrogen oxides
formed during the combustion process is primarily a function of the temperature developed during the combustion of the fuel. Recently, it has been determined that
irritating atmospheric contaminants observed during smog conditions are primarily
due to a photochemical reaction with organic materials, mostly hydrocarbons, in the
presence of oxides of nitrogen. Thus, one desirable means for reducing atmospheric
contamination, such as smog, is to reduce the amount of nitrogen oxides discharged
into the atmosphere. A reduction of the amount of nitrogen oxide produced can be
effected by controlling the maximum temperature of the burner flame in the power
plant. In addition, the oxides of nitrogen that are formed may be removed by sub-
jecting stack gases containing nitrogen oxides to contact at elevated temperatures
and in the presence of a reducing gas, such as methane, with a catalyst for reduction.
The stoichiometric reducing gas requirements are based on the equations:

$$CH_{4(g)} + 2 \; NO_{2(g)} \rightarrow CO_{2(g)} + N_{2(g)} + 2 \; H_2O_{(g)}$$

$$CH_{4(g)} + 4 \; NO_{(g)} \rightarrow CO_{2(g)} + 2 \; N_{2(g)} + 2 \; H_2O_{(g)}$$

What are the standard enthalpy changes for each of the reactions, in kcal mole^{-1} of
methane consumed? [ΔH°_f ($NO_{2(g)}$) = 9.1 kcal mole^{-1}. For other data, see Appendix F.]

E.3. Numerous organic pollutants such as acetate salts, detergents, pesticides, and hydrocarbons are found in all the major rivers and lakes of the continental United States.
Many of these produce objectionable characteristics ranging from foaming to potential
toxicity. Water treatment plants normally remove organic contaminants by oxidation,
either by atmospheric oxygen in aeration treatments or by addition of chemical oxidizing agents. Such oxidations release heat. Even the slower oxidations occurring
within natural water systems may produce sufficient heat to damage sensitive water
ecosystems, perhaps already jeopardized by the oxygen demanded for pollutant oxidation. Calculate the standard enthalpy change for oxidation of 180 g of acetate ion:

$$CH_3CO_2^-{}_{(aq)} + 2 \; O_{2(g)} \rightarrow HCO_3^-{}_{(aq)} + H_2O_{(\ell)} + CO_{2(g)}$$

Compound	ΔH°_f (kcal mole^{-1})
$CH_3CO_2^-(aq)$	-116.8
$HCO_3^-(aq)$	-165.2

E.4. Sulfur dioxide from certain smelters or from plants burning "high sulfur" coal can be a major air pollutant. In addition, the atmospheric oxidation of SO_2 to SO_3 provides a source of "acid rains" from the reaction of SO_3 with water vapor. Calculate the standard heat of reaction for the process, in kcal mole^{-1} (H_2SO_4).

$$SO_{3(g)} + H_2O_{(g)} \rightarrow H_2SO_{4(aq)}$$

Compound	ΔH°_f (kcal mole^{-1})
$SO_{3(g)}$	- 94.6
$H_2O_{(g)}$	- 57.8
$H_2SO_{4(aq)}$	-207.5

E.5. The "nitrogen cycles" of nature provide nitrogen for plant nutrition in many ways. One such way is in the form of very dilute nitric acid in rainfall following a "thunderstorm" in which some atmospheric N_2 is oxidized during lightning discharge and converted to HNO_3. Nitrogen oxide pollutants may add appreciably to this natural source of nitrate. Excess nitrates may be reconverted to atmospheric nitrogen by the reducing action of organic matter in soils, in a process involving certain micro-organisms. Calculate the standard enthalpy change associated with the reduction of nitrate by glucose, in kcal mole^{-1} (N_2):

$$5 C_6H_{12}O_6 + 24 HNO_3 \rightarrow 30 CO_2 + 42 H_2O + 12 N_2$$

Compound	ΔH°_f (kcal mole^{-1})
$C_6H_{12}O_6$	-300.0
CO_2	- 94.0
HNO_3	- 49.4
H_2O	- 68.3

*E.6. Under conditions of large nitrate excess, sufficient organic matter may not be available in soils for the complete reduction of nitrate to N_2 (problem E.7). In such cases nitrogen oxides may be formed, for example:

$$C_6H_{12}O_6 + 6 HNO_3 \rightarrow 6 CO_2 + 9 H_2O + 3N_2O$$

*Proficiency Level

If the standard enthalpy change for this process is -570.3 kcal mole^{-1} (glucose), what is the standard heat of formation of nitrous oxide?

*E.7. Methanol ("wood alcohol") is widely recognized as a poison, if swallowed. It is equally true, but less generally known, that methanol vapors are hazardous. Industrial operations producing or using significant amounts of methanol must use care to avoid the release of methanol vapor into the atmosphere or into laboratory or plant working areas. Inhaled methanol may be oxidized in the body to formic acid, a chemical particularly destructive to retinal cells and to the central nervous system. What standard enthalpy change is associated with the oxidation of 5.0 g of methanol by the process represented as:

$$CH_3OH_{(\ell)} + O_{2(g)} \rightarrow HCO_2H_{(aq)} + H_2O_{(\ell)}$$

Compound	ΔH°_f (kcal mole^{-1})
CH_3OH	-57.0
HCO_2H	-98.0
H_2O	-68.3

*E.8. The "acid rains" resulting in certain areas heavily polluted by sulfur dioxide (problem E.4) are also involved in exothermic atmospheric processes:

$$2\ SO_{2(g)} + O_{2(g)} \rightarrow 2\ SO_{3(g)}$$

$$SO_{3(g)} + H_2O_{(\ell)} \rightarrow H_2SO_{4(aq)}$$

What standard enthalpy change is associated with the complete conversion of 12 metric tons of SO_2 to H_2SO_4? [For data, see Appendix F and problem E.4.]

*E.9. Thermal pollution is the impairment of the quality of the environmental air or water by raising its temperature. It is difficult to evaluate the intensity of thermal pollution with a thermometer because what is pleasantly warm for man or some animals can be instant death for other members of the ecosystem. All life processes involve chemical reactions, and the rates of these reactions are very sensitive to changes in temperature. As a rough approximation, the rate of a reaction doubles for every rise in temperature of 10°C. Most warm-blooded animals are able to maintain a constant body temperature over a wide range of environmental temperatures, but the entire aquatic ecosystem is very sensitive to temperature changes. All of these organisms respond to temperature increase by speeding up their metabolic processes. Consequently, their need for oxygen and rate of respiration are increased, and as temperatures increase, the solubility of gases in their aquatic environment decreases.

As the thermal pollution load in a body of water increases, so will the potential for increased loss of the various members of the aquatic ecosystem. An industrial electroplating firm dumps 4000 liters per week of 2.0% (20 g liter^{-1}) waste sulfuric acid into a holding pond where sufficient lime slurry, $Ca(OH)_2$, is added to neutralize the acid. After the resulting $CaSO_4$ has settled, the pond water is drained into a nearby lake. The draining is done at two-week intervals. Assuming that the waste water drainage adds heat equivalent to 25% of the standard heat of reaction $[H_2SO_{4(aq)} + Ca(OH)_2]$ for the sulfuric acid waste, how much heat is added to the lake each two weeks? (Hint: Write the complete balanced equation.)

Compound	ΔH_f° (kcal mole^{-1})
$H_2SO_{4(aq)}$	-207.5
$Ca(OH)_2$	-235.8
$CaSO_4$	-342.4
$H_2O_{(\ell)}$	- 68.3

*E.10. The burning of gasoline by automobile engines is a major source of carbon monoxide pollution. Of comparable importance is the fact that the incomplete combustion associated with this process is an inefficient use of energy resources. For comparison, calculate: (a) the standard enthalpy change for the complete combustion of 3.5 kg of octane (to $CO_{2(g)}$ and $H_2O_{(g)}$) and (b) the standard enthalpy change for the incomplete combustion of the same amount of octane, as represented by:

$$C_8H_{18(\ell)} + 11\ O_{2(g)} \rightarrow 5\ CO_{2(g)} + 3\ CO_{(g)} + 9\ H_2O_{(g)}$$

[For octane, ΔH_f° = -59.7 kcal mole^{-1}. See Appendix F for other data.]

ANSWERS:
(E.1.) -2000 kcal, (E.2.) -210 kcal mole^{-1}, -278 kcal mole^{-1}, (E.3.) -642 kcal, (E.4.) -55.1 kcal mole^{-1}, (E.5.) -250 kcal mole^{-1}, (E.6.) +19.6 kcal mole^{-1}, (E.7.) -17 kcal, (E.8.) -1.3 x 10^7 kcal, (E.9.) 1.5 x 10^4 kcal, (E.10.)(a) 3.7 x 10^4 kcal, (b) 3.1 x 10^4 kcal

STOICHIOMETRY

UNIT 8: ELECTROCHEMISTRY

Electrochemical processes involve oxidation-reduction reactions, either resulting from the application of an external electric current (in "electrolytic" cells, such as those used in aluminum plating) or proceeding spontaneously to <u>produce</u> an electric current (in "Galvanic" cells, such as the familiar flashlight cell). All such processes utilize the indirect transfer of electrons through an external circuit, with oxidation and reduction half-reactions occurring at separate electrodes.

As a result, it is quite convenient to consider electrochemical reactions as two distinct half-reactions, each of which may be represented by an equation which includes electron symbols (e^-). Consider, for example, the familiar automobile ("lead storage") battery. During discharge, the net equation for the chemical reaction producing an electric current is given by:

$$Pb_{(s)} + PbO_{2(s)} + 2\ H^+_{(aq)} + 2\ HSO_4^-{}_{(aq)} \rightarrow 2\ PbSO_{4(s)} + 2\ H_2O_{(\ell)}$$

To see the role of this reaction in trasferring electrons (i.e., producing "current"), it is helpful to consider formulations for the separate electrode processes:

> <u>at the lead electrode</u>
>
> $$Pb_{(s)} + HSO_4^-{}_{(aq)} \rightarrow PbSO_{4(s)} + H^+_{(aq)} + 2e^-$$
>
> <u>at the lead dioxide electrode</u>
>
> $$PbO_{2(s)} + HSO_4^-{}_{(aq)} + 3H^+_{(aq)} + 2e^- \rightarrow PbSO_{4(s)} + 2H_2O_{(\ell)}$$

These equations show that current results from electrons leaving the cell at the external terminal of the lead electrode and passing through a conductor (the circuitry of the automobile, for example) to reenter the cell at the terminal of the lead dioxide electrode.

We can use a balanced equation for a half-reaction to establish a mathematical relationship between the quantity of electric charge transferred and quantity of any chemical produced or consumed as a result of electron-transfer.

Since we found mole/mass unity factors useful in dealing with mass/mass problems and mole/volume unity factors useful with gas volumes, it would seem desirable to adopt such an approach for electrochemical stoichiometry. What we need is a mole/charge relationship. This can easily be derived from the experimentally-determined value of the charge on a single electron, 1.6021×10^{-19} coulomb (C), and the number of unit particles in a MOLE, 6.0225×10^{23}.

$$\frac{1.6021 \times 10^{-19}C}{1\ electron} \times \frac{6.0225 \times 10^{23}\ electrons}{1\ mole(e^-)} = 9.6486 \times 10^4\ C\ mole^{-1}(e^-)$$

We shall find this value (sometimes called the "Faraday", in honor of Michael Faraday) of

177

considerable utility in our work with electrochemistry. For convenience, we will use three-place accuracy:

$$1 \text{ mole}(e^-) = 9.65 \times 10^4 \text{ coulombs}$$

From a practical point of view, we are often less concerned with the <u>quantity</u> of charge transferred than with the <u>rate</u> of charge transfer, i.e., "current". A convenient unit of electrical current is the ampere (amp), defined as a charge transfer rate of one coulomb per second. We will use this relationship primarily in the form:

(no. amps) x (no. seconds) = no. coulombs

or as one of the unity factors:

$$\frac{1 \text{ amp sec}}{1 \text{ C}} \qquad \frac{1 \text{ C}}{1 \text{ amp sec}}$$

The relationships between coulombs and chemical quantity, as indicated by the balanced half-reaction equation, and between coulombs, current, and time provide us the essential keys to electrochemical stoichiometry.

- References -

1. O'Connor, Rod. 1977. <u>Fundamentals of Chemistry</u>, second edition. New York: Harper & Row (Units 22, 23)

2. Brown, Theodore, and H. Eugene LeMay, Jr. 1977. <u>Chemistry: The Central Science</u>. Englewood Cliffs, N.J.: Prentice-Hall (Chapter 19)

3. Masterton, William, and Emil Slowinski. 1977. <u>Chemical Principles</u>, fourth edition. Philadelphia: W.B. Saunders (Chapter 22)

4. Nebergall, William, F.C. Schmidt, and H.F. Holtzclaw, Jr. 1976. <u>College Chemistry</u>, fifth edition. Lexington, MA: D.C. Heath (Chapters 16, 22)

5. Lyons, Ernest H., Jr. 1967. <u>Introduction to Electrochemistry</u>. Lexington, MA: Heath/Raytheon.

OBJECTIVES:

(1) *Given the description of an electrochemical process (with access to balanced equations for electrode reactions, Appendix E), including the time and average rate (amps) of current flow, be able to calculate the amount of a chemical substance consumed or formed, assuming 100% efficiency.*

(2) *Given the description of an electrochemical process (with access to balanced equations for electrode reactions, Appendix E), including the quantity of a chemical substance consumed or formed and either the time or rate (amps) of*

current flow, be able to calculate the current or time, respectively, assuming 100% efficiency.

*(3) Given a detailed description of an electrochemical process, including the percentage efficiency, be able to calculate a specified quantity such as rate of chemical change or current.

PRE-TEST:

Necessary Atomic Weights and Electrode Equations:	
Cr (52.00)	$Cr^{3+} + 3\ e^- \rightarrow Cr$
Cl (35.45)	$2\ Cl^- \rightarrow Cl_2 + 2\ e^-$
Pb (207.2)	$Pb + HSO_4^- \rightarrow PbSO_4 + H^+ + 2\ e^-$

(1) Chrome-plating for the protection of more reactive metals or for decorative purposes, as in the chromium strips on some automobiles, is a relatively expensive process. Not only is chromium a fairly costly metal, but its reduction from the +3 ion requires a high electrical energy consumption per kilogram of chromium formed. How many grams of chromium could be produced from a chromium(III) salt by use of a constant current of 12.0 amps flowing for 1.00 hour (3600 sec)? _____

(2) Chlorine for use in water purification systems can be prepared on a commercial scale by the electrolysis of brine (concentrated seawater, primarily aqueous NaCl). How long (in hours) would an electrolysis require to produce 50.0 lb (22.7 kg) of chlorine at an average current of 30.0 amps, assuming 100% efficiency? _____

*(3) The automobile ("lead storage") battery is not designed for a prolonged discharge, especially under a high current load. The particular utility of this battery lies in its design for sequential discharge/recharge operation. The lead sulfate formed during discharge adheres to the surfaces of both electrodes. During recharge by the automobile generator or alternator, the electrode reactions are reversed and the lead sulfate coatings revert to lead and lead dioxide at the respective electrodes, also regenerating the sulfuric acid of the electrolyte solution. If the discharge is maintained too long under conditions of high current load, the lead sulfate coatings may form in irregular deposits, making it difficult to recharge the battery to its original condition. What is the maximum current that should be drawn from a battery whose specifications indicate that the lead electrode should not be oxidized faster than 85 mg (Pb) per minute? Under these conditions current measurement records only 94% of the true charge transfer rate. _____

Answers and Directions:

(1) 7.76 g   (2)   572 hours

*If both are correct, go on to question *3. If you missed either, study METHODS, sections 8.1 and 8.2.*

*(3) 1.3 amps

If your answer was correct, go on to RELEVANT PROBLEMS, Unit 8. If you missed it, study METHODS, section 8.3.

METHODS

8.1 Finding Chemical Change from Current and Time

The same "proportion method" we found useful for mass/mass (Unit 4) and gas volume (Unit 5) problems can be employed in electrochemical stoichiometry. This time, however, we are involving a measurement of <u>electrons</u> and neither mass nor volume units would be appropriate. The key is to note that a measurement of electrical <u>charge</u> can be directly related to "counting electrons". If we remember that each mole of electrons corresponds to 9.65×10^4 coulomb (C) of charge and the product of current (in amps) and time (in seconds) is expressed in coulombs, we can see how to form a ratio useful in a stoichiometric calculation.

The method is similar in many respects to that used in gas volume problems. Below the formula of the chemical species of interest we write the formula weight <u>in grams</u>, multiplied by the coefficient from the balanced equation. The numerator for our mass ratio must also be expressed in grams. The other ratio is obtained by writing below the e^- symbol a product of e^- coefficient and (9.65×10^4 coulombs) and above the e^- symbol the product of current (in amps) and time (in seconds). The mass ratio and charge ratio thus defined are equated and the "unknown" mass is calculated.

EXAMPLE 1

Peroxyborate bleaches, such as BORATEEM, have found extensive markets in replacing the older "chlorine" (hypochlorite) bleaching agents, both for home and industrial uses. Sodium peroxyborate, the active ingredient of most such bleaching agents, may be prepared economically by the electrolytic oxidation of borax solutions, according to the electrode equation:

$$Na_2B_4O_7(aq) + 10NaOH(aq) \rightarrow 4NaBO_3(aq) + 5H_2O(\ell) + 8Na^+(aq) + 8e^-$$

How many grams of sodium peroxyborate could be prepared by this process using a constant current of 25.0 amps for a 24 hour period, assuming 100% efficiency?

SOLUTION:

(Using a method similar to that employed in Units 4 and 5.)

$$\dots \rightarrow \boxed{4\ NaBO_3} + \dots + \boxed{8\ e^-}$$

$$(4\ moles \times 81.8\ g\ mole^{-1}) \qquad (8\ moles \times 9.65 \times 10^4\ C\ mole^{-1})$$

with the figure $(25.0 \times 24 \times 3600)\ C[1]$ above the $8\ e^-$ box.

$$\frac{w}{(4 \times 81.8)\ g} = \frac{(25.0 \times 24 \times 3600)\ C}{(8 \times 9.65 \times 10^4)\ C}$$

$$w = \frac{(4 \times 81.8)\ g \times (25.0 \times 24 \times 3600)\ C}{(8 \times 9.65 \times 10^4)\ C} = \underline{915\ g}$$

EXERCISE 1

In the common "dry cell" of the type used for relatively inexpensive flashlight "batteries", one electrode is the zinc can itself. The other (central) electrode is a graphite rod packed in manganese dioxide. The remaining space within the zinc can is filled with a moist paste of zinc chloride, ammonium chloride, and an inert filler. A cell of this type, when new, produces about 1.5 volts. Such cells are designed for low current flow and under typical use the reaction at the inert graphite electrode is formulated as:

$$MnO_2 + NH_4^+ + e^- \rightarrow MnO(OH) + NH_3$$

The electrons for this reaction are furnished by the conversion of zinc from the can to Zn^{2+} ions (see Appendix E). How many grams each of zinc and manganese dioxide are consumed during discharge of a dry cell at a current of 0.30 amp for 6.0 hours?

(answer, page 186)

8.2 Finding Time or Current from Chemical Change

If the quantity of chemical change is known for an electrochemical process and either the time or current is specified, then the remaining variable can be calculated by our familiar "proportion method". We must remember, of course, that the product of current and time will be expressed in coulombs only if current is in amps and time is in units of seconds. All we have to do is to modify the numerator of our charge ratio term by using some symbol for the "unknown" (e.g., t for time, in seconds, or a for amperes). For units in this ratio, we will use (amp sec) in place of coulombs. The conversion to some other indicated units, such as milliamperes or hours, can be done after solution of the equation made by equating the mass and charge ratios.

---

[1]amps x hours x $(sec\ hr^{-1})$ = C

EXAMPLE 2

How long, in hours, would be required for production of 1.00 lb (454 g) of sodium peroxyborate (EXAMPLE 1) by the electrolysis of a borax solution with a current of 15.0 amps?

SOLUTION:

(Based on the electrode equation given in EXAMPLE 1 and using t to represent time, in seconds:)

$$454 \text{ g} \qquad\qquad (15.0 \text{ amp} \times t)$$

$$.... \rightarrow \boxed{4 \text{ NaBO}_3} + ... + \boxed{8 e^-}$$

$$(4 \text{ moles} \times 81.8 \text{ g mole}^{-1}) \qquad (8 \text{ moles} \times 9.65 \times 10^4 \text{ amp sec mole}^{-1})$$

$$\frac{454 \text{ g}}{(4 \times 81.8) \text{ g}} = \frac{(15.0 \text{ amp} \times t)}{(8 \times 9.65 \times 10^4) \text{ amp sec}}$$

$$t = \frac{454 \text{ g} \times (8 \times 9.65 \times 10^4) \text{ amp sec}}{(4 \times 81.8) \text{ g} \times 15.0 \text{ amp}} = 7.14 \times 10^4 \text{ sec}$$

Then, to convert to hours:

$$\frac{7.14 \times 10^4 \text{ sec}}{1} \times \frac{1 \text{ hr}}{3600 \text{ sec}} = \underline{19.8 \text{ hours}}$$

EXERCISE 2

With soaring costs of electrical power, electroplating industries must inevitably raise product prices to compensate for increasing process costs. The effect of electrical price changes is not felt in the same way by all industries. The charge of the ion being electrolytically reduced and the atomic weight of the metal are both major considerations. For the same amount of electrical energy, for example, a silver plating industry could produce twelve metric tons of silver for each metric ton of aluminum that an aluminum plating company could produce. (Compare $Ag^+ + e^- \rightarrow Ag$, atomic weight 107.9, with $Al^{3+} + 3 e^- \rightarrow Al$, atomic weight 26.98.) What average current, in amps, must be maintained to electroplate 25.0 lb (11,340 g) of copper from a solution of a copper(II) salt during an 8.0 hour plant run?

(answer, page 186)

[See Appendix E for the equation.]

==

Extra Practice

EXERCISE 3

How many grams of lead sulfate are formed on the lead electrode of an automobile battery during a 10.0 hour discharge at an average current of 6.0 amps resulting from the failure of the automobile's owner to switch off his car lights one night? [$Pb + HSO_4^- \rightarrow PbSO_4 + H^+ + 2e^-$]

(answer, page 187)

EXERCISE 4

How many hours would the battery in EXERCISE 3 have to be connected to a 0.50 amp battery charger to restore the lead electrode to its condition prior to the accidental discharge, assuming 100% recharge efficiency? [$PbSO_4 + H^+ + 2e^- \rightarrow Pb + HSO_4^-$]

(answer, page 187)

==

At this point you should try the competency level Self-Test questions on page 185.

==

*8.3 Inefficiency of Electrochemical Processes

Few real chemical processes are as simple as a single balanced equation seems to suggest. Electrochemical processes, particularly those involving aqueous solutions or other mixtures, are susceptible to a number of factors which reduce their efficiency below the theoretical limits. The "resistance" of an electrical circuit, for example, reduces the amount of work that can be obtained from a Galvanic cell. Electrolytic cells using aqueous solutions, as in the recharging of an automobile battery, often utilize some of the current for the electrolysis of water, rather than for the intended electrochemical reaction. These and other problems suggest that calculations in electrochemical stoichiometry may represent idealized situations. Actual efficiencies, usually indicated in terms of percentage of some "theoretical" quantity, must be determined for each real system by experiment.

An efficiency of less than 100% indicates, as in the case of percentage yields for mass/mass problems (Unit 4), that either we will obtain less than the "expected" amount of product or current "output" or we will require more than the theoretically calculated "input". Efficiency "corrections" may be applied to stoichiometric calculations for electrochemical reactions in the same way as in simpler problems (Unit 4). That is, they may be included as part of the "set-up" in a unity factor calculation.

*Proficiency Level

*EXAMPLE 3

Aluminum was a relatively rare and expensive metal until the development of the Hall process in 1886 for electrolytic aluminum production. Two aluminum ores are required in this method, cryolite (Na_3AlF_6) and bauxite (hydrated Al_2O_3). The process begins by melting a charge of powdered cryolite with heat generated by carbon arcs. Purified Al_2O_3 is added as electrolysis begins and the molten mixture is maintained around $1000°C$ during the electrolysis, with the liquid aluminum formed (aluminum melts around $660°C$) being drained from the reaction vessel. Thus electrical energy is needed both for the electrolysis itself and for the heating of the reaction mixture. What average current, in amps, must be maintained for an aluminum production by the Hall process at a rate of 2.50 lb (1134 g) of aluminum per hour, if only 68% of the current is actually effective in producing aluminum metal?

SOLUTION:

To establish proper unity factors, we must first note that metallic aluminum is produced from aluminum in the +3 oxidation state. Thus, 3 moles of electrons are required per mole of aluminum ($Al^{3+} + 3e^- \rightarrow Al$). Then, remembering the basic electrochemical "conversion factors", we can solve the problem by stepwise application of appropriate unity factors:

$$\frac{1134 \text{ g(Al)}}{1 \text{ hour}} \times \frac{1 \text{ mole(Al)}}{26.98 \text{ g(Al)}} \times \frac{3 \text{ moles(e}^-)}{1 \text{ mole(Al)}} \times \frac{9.65 \times 10^4 \text{ C}}{1 \text{ mole(e}^-)} \times \frac{1 \text{ amp sec}}{1 \text{ coul}} \times \frac{1 \text{ hour}}{3600 \text{ sec}} \times \frac{100}{68}$$

[g(Al)/hr → moles(Al)/hr → moles(e⁻)/hr → coulombs/hr → amp sec/hr → amp]   ["efficiency correction"]

answer: 4.97×10^3 amps

---

*EXERCISE 5

During the periodic recharging of an automobile battery by the generator or alternator, some of the current produces hydrogen and oxygen by electrolysis of the "battery acid" (aq. H_2SO_4). This has two results. It suggests that an automobile battery might be the source of an explosive gas mixture, so that checking the electrolyte level in a battery at night by matchlight would be inadvisable. (Explosions from such attempts have been reported.) In addition, the "waste current" used in the decomposition of the electrolyte water reduces the efficiency of the recharging process. How many hours would be required for restoration of 50.0 g of PbO_2 to the lead dioxide electrode by a recharging current of 1.25 amp, operating at a recharging efficiency of 87%? [For the electrode equation, see Appendix E.]

(answer, page 187)

184

==

Extra Practice

*EXERCISE 6

Potassium permanganate is a valuable commercial chemical used as an oxidizing agent in a number of industrial processes, as an analytical reagent in several "quality-control" procedures, and in the preparation of a disinfectant solution (Condy's Liquid). The industrial production of $KMnO_4$ is a multi-step process beginning with the alkali fusion of MnO_2 and concluding with the electrolytic oxidation of potassium manganate (K_2MnO_4). What average current, at an electrode efficiency of 93%, should be maintained to produce 5.00 lb (2268 g) per hour of $KMnO_4$ from the manganate salt?

(answer, page 187)

==

SELF-TEST (UNIT 8)   [answers, page 187]

8.1 The older "dry cells used in flashlights could not be recharged efficiently because the products of the discharge reaction, unlike the $PbSO_4$ in the automobile battery, did not adhere well to the electrode surfaces. In recent years rechargeable cells, such as the "nickel-cadmium" cell, have found wide markets in products ranging from rechargeable flashlights to electric toothbrushes. Like other electrochemical cells, the "nickel-cadmium" system is designed for relatively low current production. How many grams of cadmium would be consumed during the discharge of a set of "nickel-cadmium" cells used in a flashlight operating at 0.15 amp for 4.5 hours?

Electrode Equations:

$$Cd_{(s)} + 2OH^-_{(aq)} \rightarrow Cd(OH)_{2(s)} + 2e^-$$

$$NiO_{2(s)} + 2H_2O_{(\ell)} + 2e^- \rightarrow Ni(OH)_{2(s)} + 2OH^-_{(aq)}$$

8.2 How long, in minutes, could the "nickel-cadmium" cells described in question 8.1 be used in a child's mechanical toy, requiring a current of 1.8 amps, before 2.5 g of the NiO_2 would be consumed?

- -

If you completed Self-Test questions 8.1 and 8.2 correctly, you may go on to the proficiency level, try the RELEVANT PROBLEMS (Unit 8), or stop here. If not, you should consult your instructor for suggestions of further study aids.

- -

*8.3 One of the most efficient ways of using methane as a fuel, in a situation re-
quiring only a low <u>rate</u> of energy production, is in a methane-oxygen fuel cell.
The electrode reactions in this cell, during discharge, may be formulated as:

$$CH_{4(g)} + 10\ OH^-_{(aq)} \rightarrow CO_{3(aq)}^{2-} + 7\ H_2O_{(\ell)} + 8\ e^-$$

$$O_{2(g)} + 2\ H_2O_{(\ell)} + 4\ e^- \rightarrow 4\ OH^-_{(aq)}$$

What current could be generated by a cell of this type using a fuel consumption
rate of 125 ml (<u>STP</u>) of methane per hour, if the conversion efficiency is 78%?

---

*If you answered this question correctly, you may go on to the RELEVANT PROBLEMS for
Unit 8. If not, you should consult your instructor for suggestions of further study aids.*

ANSWERS to EXERCISES, Unit 8

1. [2.2 g(Zn), 5.8 g (MnO$_2$)] <u>Solution:</u>

for zinc

(Using the "reverse form" of an equation from Appendix E.)

$$\boxed{\underset{(1\ mole\ \times\ 65.4\ g\ mole^{-1})}{\overset{w}{Zn}}} \rightarrow Zn^{2+} + \boxed{\underset{(2\ moles\ \times\ 9.65\ \times\ 10^4 C\ mole^{-1})}{\overset{(0.30\ \times\ 6.0\ \times\ 3600)C}{2\ e^-}}}$$

$$w = \frac{65.4\ g\ \times\ (0.30\ \times\ 6.0\ \times\ 3600)C}{(2\ \times\ 9.65\ \times\ 10^4)C}$$

for manganese dioxide

$$\boxed{\underset{(2\ moles\ \times\ 86.94\ g\ mole^{-1})}{\overset{w}{2\ MnO_2}}} + ... + \boxed{\underset{(2\ moles\ \times\ 9.65\ \times\ 10^4 C\ mole^{-1})}{\overset{(0.30\ \times\ 6.0\ \times\ 3600)C}{2\ e^-}}} \rightarrow .....$$

$$w = \frac{(2\ \times\ 86.94)\ g\ \times\ (0.30\ \times\ 6.0\ \times\ 3600)C}{(2\ \times\ 9.65\ \times\ 10^4)C}$$

2. (1.2 × 10^3 amps) <u>Solution:</u>

$$\frac{8.0\ hours}{1} \times \frac{3600\ sec}{1\ hr} = 2.9\ \times\ 10^4\ sec$$

$$Cu^{2+} + \boxed{\underset{(2\ moles\ \times\ 9.65\ \times\ 10^4\ amp\ sec\ mole^{-1})}{\overset{(a\ \times\ 2.9\ \times\ 10^4\ sec)}{2e^-}}} \rightarrow \boxed{\underset{(1\ mole\ \times\ 63.54\ g\ mole^{-1})}{\overset{11,340\ g}{Cu}}}$$

$$\frac{a \times (2.9 \times 10^4) \text{ sec}}{(2 \times 9.65 \times 10^4) \text{ amp sec}} = \frac{11,340 \text{ g}}{63.54 \text{ g}}$$

$$a = \frac{11,340 \text{ g} \times (2 \times 9.65 \times 10^4) \text{ amp sec}}{63.54 \text{ g} \times (2.9 \times 10^4) \text{ sec}}$$

3. (340 g) Solution:

$$..... \rightarrow \boxed{\overset{w}{PbSO_4}} + ... + \overset{(6.0 \times 10.0 \times 3600)C}{\boxed{2e^-}}$$

(1 mole × 303.3 g mole^{-1}) (2 mole × 9.65 × 10^4 C mole^{-1})

$$w = \frac{303.3 \text{ g} \times (6.0 \times 10.0 \times 3600)C}{(2 \times 9.65 \times 10^4)C}$$

4. (120 hours) Solution:

$$\overset{340 \text{ g}}{\boxed{PbSO_4}} + ... + \overset{(0.50 \text{ amp} \times t)}{\boxed{2e^-}} \rightarrow Pb + ...$$

(1 mole × 303.3 g mole^{-1}) (2 moles × 9.65 × 10^4 amp sec mole^{-1})

$$t = \frac{340 \text{ g} \times (2 \times 9.65 \times 10^4) \text{ amp sec}}{303.3 \text{ g} \times 0.50 \text{ amp}} = 4.3 \times 10^5 \text{ sec}$$

$$\frac{4.3 \times 10^5 \text{ sec}}{1} \times \frac{1 \text{ hour}}{3600 \text{ sec}}$$

- -

*5. (10.3 hours) Solution:

[From Appendix E, 2 moles (e$^-$) per mole (PbO$_2$).]

$$\frac{50.0 \text{ g(PbO}_2)}{1} \times \frac{1 \text{ mole(PbO}_2)}{239.2 \text{ g(PbO}_2)} \times \frac{2 \text{ moles(e}^-)}{1 \text{ mole(PbO}_2)} \times \frac{(9.65 \times 10^4) \text{ amp sec}}{1 \text{ mole(e}^-)} \times \frac{1}{1.25 \text{ amp}} \times \frac{1 \text{ hr}}{3600 \text{ sec}} \times \frac{100}{87}$$

[g(PbO$_2$) → moles(PbO$_2$) → moles(e$^-$) → amp sec → sec → hr]

["efficiency correction"]

*6. (414 amps) Solution:

$$\frac{2268 \text{ g(KMnO}_4)}{1 \text{ hour}} \times \frac{1 \text{ mole(KMnO}_4)}{158 \text{ g(KMnO}_4)} \times \frac{1 \text{ mole(e}^-)}{1 \text{ mole(KMnO}_4)} \times \frac{(9.65 \times 10^4) \text{ amp sec}}{1 \text{ mole(e}^-)} \times \frac{1 \text{ hour}}{3600 \text{ sec}} \times \frac{100}{93}$$

[g hr^{-1}(KMnO$_4$) → mole hr^{-1}(KMnO$_4$) → mole hr^{-1}(e$^-$) → amp sec hr^{-1} → amps]

["efficiency correction"]

- -

ANSWERS to SELF-TEST, Unit 8

8.1. 1.4 g
8.2. 49 minutes
*8.3. 0.94 amp

A.1. During the bacterial decomposition of organic sulfur compounds, sulfides are formed
along with a variety of other incompletely oxidized species including elemental sulfur, sulfites, and thiosulfates. These reduced substances are subject to further
oxidation, in some cases by ordinary chemical reactions. However, most of the sulfur
oxidation occurring in soils and marine sediments is believed to be mediated by
bacteria. Five species of the genus <u>Thiobacillus</u> that mediate this process have been
characterized. The oxidation of hydrogen sulfide is a soil acidifying process. For
each sulfide ion oxidized two hydrogen ions are formed as:

$$H_2S \rightarrow S + 2 H^+ + 2e^-$$

How many grams each of elemental sulfur and hydrogen ion are formed at a "biological
current" of 0.25 amp in an interval of 10.0 hours?

A.2. The acidifying effect of sulfur oxidation is even more pronounced when elemental sulfur is converted to sulfate according to the equation:

$$S + 4 H_2O \rightarrow SO_4^{2-} + 8 H^+ + 6e^-$$

What average "biological current", in amps, must be maintained to form 25.0 g of
sulfate during a 5.0 hour period of biooxidation?

A.3. The inorganic end product of oxidation of organic nitrogen compounds in the sea is the
nitrate ion, with ammonia and nitrite as successive intermediates. Most inorganic
forms of nitrogen are consumed by phytoplankton, frequently to near zero concentration in the euphotic zone. Bacterial action is responsible for the major part of
nitrogen regeneration. For some simple substances, such as urea, chemical hydrolysis
to ammonia is possible, although bacteria can also liberate ammonia from urea.
Ammonia and nitrite generally appear where plankton decompose in quantity. In shallow coastal waters of temperate regions, ammonia is present in very small amounts at
the end of winter, but it increases in the spring and sometimes becomes the major
form in which inorganic nitrogen is available. The bacterial oxidation of ammonium
ion to form nitrite occurs according to the equation:

$$NH_4^+ + 2 H_2O \rightarrow NO_2^- + 8 H^+ + 6e^-$$

How many grams of nitrite are formed during a 48.0 hour oxidation period at an average "biological current" of 0.50 amp?

A.4. In the sediments of many coastal areas and marine lagoons, there are numerous distinguishable layers as a result of sedimentation processes. Organic matter layered

188

with the sediment provides an energy source for microorganisms. As a result of the microbial activity, oxygen is consumed and anaerobic conditions are eventually established, accompanied by the formation of hydrogen sulfide. The sulfate-reducing bacteria, strict anaerobes, produce the sulfide which, in the presence of iron(II) ion, reacts to form a black precipitate, iron(II) sulfide. If a particular bacterial community produced 400 g of bisulfide over a 24 hour period as a result of anaeorbic oxidative metabolism, what average "biological current" (in amps) must have been maintained, if we assume the bisulfide formation is represented entirely by:

$$SO_4^{2-} + 9 \ H^+ + 8e^- \rightarrow HS^- + 4 \ H_2O$$

A.5. If we assumed the electrons used in the reduction of sulfate (problem A.4) were generated by the anaerobic oxidation of glucose, how many kilograms of glucose must have been oxidized over the 24 hour period?

$$C_6H_{12}O_6 \rightarrow 2 \ CH_3\overset{\overset{\displaystyle O}{\|}}{C}CO_2H + 4 \ H^+ + 4e^-$$

      (glucose)   (pyruvic acid)

*A.6. Iron is present in sea water in the oxidized state in several forms. Relatively little is present in true solution because of the instability of the Fe^{2+} ion and the relative insolubility of $Fe(OH)_3$ and its various hydrates. However, stable colloids of hydrated iron(III) oxide occur in all parts of the ocean. The active surfaces of these colloidal particles probably play an important role in geochemical and sedimentary processes of other minor elements in the ocean. The existence of autotrophic bacteria which mediate Fe^{2+} oxidation in acid solution (e.g., Ferro-bacillus ferrooxidans) play an important role in depositing iron(III) oxides from solutions of iron(II) ion. What average "biological current", at an efficiency of 65%, should be maintained to produce 100 g per hour of Fe^{3+} from Fe^{2+}?

*A.7. Nitrate in farm waters such as stock ponds or wells is a common, and dangerous, manifestation of pollution from chemical fertilizers and feedlots because of the particular susceptibility of ruminant animals to nitrate poisoning. The stomach contents of animals such as cattle and sheep constitute a reducing medium and contain bacteria capable of reducing nitrate ion to toxic nitrite ion. Adults have a high tolerance for nitrate, however, in the stomachs of infants nitrate is reduced to nitrite. The nitrite ion inactivates hemoglobin, producing a condition described as methemoglobinemia (blue babies). Several cases of this disease, some resulting in death, have been traced to nitrate in farm well water. How many hours would be re-

*Proficiency Level

quired to form 75.0 g of NO_2^- from NO_3^- by a "bacterial reducing current" of 1.50 amps, operating at an efficiency of 89%?

*A.8. Agricultural interest in selenium was initiated by the discovery of certain plants that absorb large quantities of the element from seleniferous soils. A wider interest in the role of selenium in agriculture was stimulated by its association with the poisoning of a variety of farm animals. Ranchers in certain areas experienced losses of livestock resulting from a malady called "alkali disease". The affected animals lost vitality, became emaciated, and eventually died with characteristic pathological conditions which included erosion of bones, loss of hair, anemia, and loss of hoofs. Interest has intensified recently with the discovery that selenium in trace amounts is an essential dietary element. Selenium commonly occurs in soils as a selenide (Se^{2-}) mineral. However, its availability to plants requires the formation of soluble selenites (SeO_3^{2-}). How many grams of selenous acid (H_2SeO_3) would be formed by the bacterial oxidation of selenide (Se^{2-}) using a "biological current" of 1.20 amps for 24 hours, if only 87% of the current is actually effective in producing H_2SeO_3?

*A.9. One of the most striking observations about the vast array of selenium intoxication studies is the wide variance in pathological changes between species. For example, almost all selenium poisoned sheep have severe lesions of the myocardium, a condition not found in cattle. On the other hand, cattle have lesions of the central nervous system, but these lesions are absent in sheep. In spite of these marked differences, some conditions develop which seem to be common to all animals that experience selenium intoxication. For example, the organ most affected is the liver, the spleen becomes enlarged, and the gastrointestinal tract shows hemorrhages. The very toxic hydrogen selenide (15 times more dangerous than hydrogen sulfide) is a metabolite of several plants species, however, few farm animals have died by hydrogen selenide intoxication. This is probably bacause the compound is readily oxidized to the non-toxic elemental "red selenium". What average "biological current", at an efficiency of 95%, should be maintained to produce 2.0 g per hour of "red selenium" from hydrogen selenide?

*A.10. The natural abundance of selenium in sea water is approximately 5.0×10^{-6} g liter^{-1}. The native element and a number of selenide minerals occur naturally and appear to have been formed under conditions of low redox potential. Selenium(IV) is the thermodynamically most probable state of the element under normal aerobic conditions and a number of selenide minerals are known to occur. It is probable that selenium(IV) is formed during the weathering of rocks and that this is the form in which the element enters the sea and participates in the marine sedimentary cycle. The only selenium(VI) speices found naturally are in Chilean nitrate beds. These beds appear to have been formed under conditions of very high redox potential and contain other characteristic ions such as iodate (IO_3^-) and perchlorate (ClO_4^-). How many hours would be required for the production of 100 g of SeO_4^{2-} from SeO_3^{2-} (in a Chilean nitrate bed) by a current of 1.4 amps, operating at an efficiency of 86%?

ANSWERS:

(A.1.) 1.5 g [S], 9.4×10^{-2} g [H^+], (A.2.) 8.4 amps, (A.3.) 6.9 g, (A.4.) 108 amps, (A.5.) 4.4 kg, (A.6.) 74 amps, (A.7.) 65 hr, (A.8.) 20 g, (A.9.) 1.4 amps, (A.10.) 31 hr

---

RELEVANT
PROBLEMS

Unit 8: *Electrochemistry*

Set B: *Biological & Medical Sciences*

B.1. Electric current is basically the movement of electric charge (as electrons or ions) from one location to another. In many biological systems, charge transfer processes occur which may be thought of, in simple terms, as "biological current". There are, of course, no wires in these systems and charge transfer takes place over very short distances. However, the relationship between charge transfer and amount of chemical change follows Faraday's laws in exactly the same way as in simple electrochemical cells. Oxidative metabolism of foods occurs in two phases, each involving a series of complex interdependant reactions. In glucose metabolism, for example, an anaerobic phase (without oxygen) converts glucose to pyruvic acid. The subsequent aerobic phase utilizes oxygen to complete the conversion to carbon dioxide and water. Special chemical systems participate in various steps of each phase as "electron-carriers" and "hydrogen-carriers". Simple steps in metabolism processes may be represented as oxidation or reduction half-reactions and the equations for such steps establish the relationship between chemical change and charge transfer ("biological current"). If a particular individual produced 393 grams of water over a 24-hour period as a result of oxidative metabolism, what average biological current (in amps) must have been maintained if we assume formation of the water to be represented entirely by:

$$O_2 + 4\ H^+ + 4e^- \rightarrow 2H_2O$$

B.2. If we assumed the electrons used in the reduction of oxygen (problem B.1) were generated by the anaerobic oxidation of glucose, at what average rate (in grams of glucose per hour) must the oxidation occur?

$$C_6H_{12}O_6 \rightarrow 2\ CH_3\text{-}\overset{\overset{\textstyle O}{\textstyle \|}}{C}\text{-}CO_2H + 4\ H^+ + 4e^-$$

   (glucose)   (pyruvic acid)

B.3. Light interacts with living organisms in such diverse processes as vision, bioluminescence, and photosynthesis. In photosynthesis, chlorophyll molecules are activated by light energy from the sun. These activated molecules transfer energy through chemical processes involving the oxidation of water and the reduction of carbon dioxide, with the eventual formation of carbohydrates. These may be represented for simple purposes by the empirical formula $[CH_2O]$, so that the net photosynthetic process can be formulated by:

$$H_2O + CO_2 \rightarrow [CH_2O] + O_2$$

Photosynthesis is, of course, far from a simple process, but we can describe three general phases. In the first, water is oxidized to oxygen, with a release of electrons. The second stage may be thought of as "pumping" electrons against a potential barrier by use of solar energy, and the final phase as the reduction of carbon dioxide. In simplest terms:

(1) $2\ H_2O \rightarrow O_2 + 4\ H^+ + 4e^-$

(2)   "electron pumping"

(3) $4e^- + 4\ H^+ + CO_2 \rightarrow [CH_2O] + H_2O$

A large tree maintained an average photosynthetic rate equivalent to an electron transfer rate of 23.4 amps. How many kilograms of oxygen were formed by this tree in one week?

B.4. What volume of carbon dioxide (at STP) would have been reduced by the tree (problem B.3) each 24 hours?

B.5. What mass of carbohydrate would have been formed by the tree (problem B.3) over a 30 day period?

*B.6. When the body's glucose supply is depleted by starvation or in diabetes, fatty acids are used as an energy source. The fatty acids are catabolized in the liver at a very rapid rate and, in the process, excess acetyl coenzyme A is produced. The ex-

*Proficiency Level

cess coenzyme is converted into "ketone bodies" (acetoacetic acid, β-hydroxybutyric acid, and acetone). Ketosis results. If the condition is allowed to continue long enough, acidosis ensues and leads to dehydration, disturbances of the central nervous system, and eventually coma and death. The production of β-hydroxybutyric acid by reduction of acetoacetic acid may be formulated as:

$$CH_3-\overset{\overset{\displaystyle O}{\|}}{C}-CH_2\overset{\overset{\displaystyle O}{\|}}{C}-OH + 2\ H^+ + 2e^- \rightarrow CH_3\overset{\overset{\displaystyle OH}{|}}{C}HCH_2\overset{\overset{\displaystyle O}{\|}}{C}-OH$$

How many hours would be required for reduction of 10.0 g of acetoacetic acid to β-hydroxybutyric acid by a "biological current" of 1.35 amp, operating at a reduction efficiency of 78%?

*B.7. Symbiosis is a mode of living, characterized by intimate and constant association or close union of two dissimilar organisms. The results of this association may be beneficial, indifferent, or harmful to either partner. Frequently one organism can nutritionally utilize a substance from the environment which is normally inhibitory to another. This is termed passive stimulation. For example, the so-called "S organism", is a motile rod that produces hydrogen from ethanol as shown by the equation, and the hydrogen is autotoxic to the cell:

$$CH_3CH_2OH + H_2O \xrightarrow{\text{(S organism)}} CH_3CO_2H + 2\ H_2$$

Another bacterium, Methanobacterium, a nonmotile rod, uses the hydrogen produced by the S organism to reduce carbon dioxide to methane, as shown by the equation:

$$4\ H_2 + CO_2 \rightarrow CH_4 + 2\ H_2O$$

thereby allowing the S organism to proliferate. What average "biological current", in amps, must be maintained by an S organism community for production of 112.0 liters (at STP) of hydrogen during a 10 day period if only 82% of the current is actually effective in producing hydrogen?

*B.8. What volume of carbon dioxide (at STP) would have been reduced by the Methanobacterium members of the community (problem B.7) each 24 hours at 1.2 amps (90% efficient)?

*B.9. What mass of methane would have been formed by the bacteria (problem B.7) over a 30 day period, at a "biological current" of 1.8 amps at 90% efficiency?

*B.10. The cycling of nutrient elements is the essential part of all ecological considerations because of their limited availability in the biosphere. The continued utilization of one element as a nutrient by a given category of living organism would eventually tie up all the available supply and no further growth could occur once the supply was depleted. For example, if all the carbon dioxide available to photosynthetic organisms was "fixed" as carbohydrate, no further photosynthesis could take place. If it were not for the continued renewal of the supply of carbon dioxide by the respiratory activities of other types of organisms, all life would stop soon after photosynthesis stopped and all of the available carbon would be "fixed" in the form of dead carcasses, leaves, and wood to be preserved forever. The autotrophic bacterium Methanomonas oxidizes methane according to the half-reaction equations:

$$CH_4 + 2 H_2O \rightarrow CO_2 + 8 H^+ + 8e^-$$

$$O_2 + 4 H^+ + 4e^- \rightarrow 2 H_2O$$

What average "biological current", at an oxidation efficiency of 91%, should be maintained to produce 22.4 g per hour of CO_2 from methane?

ANSWERS:

(B.1.) 49 amps, (B.2.) 82 g hr^{-1}, (B.3.) 1.17 kg, (B.4.) ~120 liters, (B.5.) 4.7 kg, (B.6.) 5.0 hr, (B.7.) 1.4 amps, (B.8.) 2.7 liters, (B.9.) 92 g, (B.10.) 120 amps

---

RELEVANT PROBLEMS

Unit 8: Electrochemistry

Set I: Industrial Chemistry

I.1. In the electrolytic refining of crude "blister copper", the impure metal is used as the anode with a pure copper cathode and an electrolyte solution of copper(II) sulfate. Copper is transferred from anode to cathode via the intermediate Cu^{2+}. For a large copper refining plant using an average net current of 375 amp, what mass of pure copper is produced during each 8.0 hour work shift?

I.2. Any of a number of "fuels" can serve as more efficient energy sources in electrochemical processes than in simple combustion systems. In an acetylene fuel cell, for example, much more energy is available for useful work than could have been obtained by burning the acetylene. The anode process for one type of acetylene fuel cell is represented by:

$$C_2H_{2(g)} + 4 H_2O_{(\ell)} \rightarrow 2 CO_{2(g)} + 10 H^+_{(aq)} + 10e^-$$

What volume of carbon dioxide (measured at STP) would be formed by this process, per hour, at a net average current of 0.45 amp?

I.3. The mining and refining of "fuel element" materials is a significant part of the nuclear industry. Although the cost of uranium, like most prices, has increased dramatically within the last few years, nuclear power plants are still relatively economical energy sources. With improved methods for uranium purification, nuclear power could become an even better dollar bargain. One method tested for producing pure uranium is an electrolytic process for which the cathode reaction may be represented as:

$$UO_2 + 4e^- \rightarrow U + O^{2-}$$

What mass of uranium could be produced in 48.0 hours by a plant operation using a net average current of 138 amps?

I.4. The U.S. production of zinc is estimated as more than 800,000 metric tons annually. One of its principal commercial uses is in the preparation of "galvanized" steel by deposition of a thin coating of metallic zinc on the surface of the steel. Since zinc is much more readily oxidized than iron, the zinc layer "protects" the iron against corrosion until the zinc has all been used up. "Galvanized" steel may be made by the electrolysis of aqueous zinc sulfate, using the steel item as the cathode. This is a relatively profitable process, since the zinc sulfate required can be obtained easily by treating relatively inexpensive <u>smithsonite</u> ore ($ZnCO_3$) with sulfuric acid. How long will be required for the electrodeposition of 23 metric tons of zinc from zinc sulfate solution by an average net operations current of 1.32×10^3 amps?

I.5. Strong, lightweight alloys are especially valued in the aerospace industries. For many purposes, magnesium alloys are replacing aluminum alloys as structual metals since, for equal strengths, the magnesium alloy is typically at least 30% lighter in weight. One commercial method for magnesium production utilizes electrolysis of molten magnesium chloride, with chlorine gas as a valuable byproduct. How many kilograms per hour of magnesium could be produced by this method in a plant operation using a net average current consumption rate of 9.1×10^4 amps?

*I.6. Nearly 3% of the total U.S. production of electrical energy is used by the aluminum industry. This exceptionally high demand makes aluminum particularly susceptible to the problems of the "energy crisis", and prices of the metal can be expected to reflect increasing energy costs. Aluminum is the most abundant metal in the earth's crust but, unlike many other needed metals (e.g., gold, silver, or copper), it is available, for all practical purposes, only as the oxide. Its principal ore is bauxite, which contains Al_2O_3 along with appreciable quantities of iron oxides and

*Proficiency Level

other impurities. The first step in the production of metallic aluminum is the purification of the alumina (Al_2O_3) from the crude ore. The metal is then prepared by electrolytic reduction. Since alumina is insoluble in water, it must be reduced in the molten stage, using electrical energy to heat the furnace. Pure alumina melts around 2000°C, but addition of a mixture of NaF, CaF_2, and AlF_3 reduces the melting range to below 1000°C so that energy demands for heating are minimized. It is fortunate that aluminum ores are relatively cheap, since the electrolytic reduction of aluminum is expensive for two reasons. Aluminum occurs in nature in the +III oxidation state so that charge transfer requirements are high, and aluminum is a light metal so that the cost of electricity per <u>metric ton</u> is still higher. Calculate and compare the current requirements for production of 1.00 kg per hour of each of these metals:

$$Al^{3+} + 3e^- \rightarrow Al$$
$$Cu^{2+} + 2e^- \rightarrow Cu$$
$$Ag^+ + e^- \rightarrow Ag$$

(assume 100% conversion efficiency)

*I.7. An aluminum plant produces 125 kg of aluminum per hour from Al_2O_3. What average current must be maintained if the electrode efficiency is 85%?

*I.8. A byproduct of the aluminum industry is oxygen, a valuable commercial product itself. What volume of oxygen (at STP) is generated per week by the plant if operation is uniformly maintained 24 hours per day at 57,000 amps and 78% efficiency?

$$[2\ Al_2O_3 \xrightarrow{\text{elec.}} 4\ Al + 3\ O_2]$$

*I.9. Recent plant operations to produce very pure sodium hydroxide utilize a special "mercury cathode" cell. In this cell sodium is formed at the cathode and protected from reaction with water by formation of a mercury amalgam:

$$2\ NaCl + 2n\ Hg \xrightarrow{\text{elec.}} 2\ Na(Hg)_n + Cl_2$$

The amalgam is transferred to a separate compartment where it is decomposed by hot water to liberate hydrogen, regenerate mercury, and form very pure caustic soda:

$$2\ Na(Hg)_n + 2\ H_2O \rightarrow 2\ NaOH + H_2 + 2n\ Hg$$

A new mercury cell has an electrode efficiency of 97% and operates at an average current of 3.0×10^4 amps. How much sodium hydroxide (in kg) is produced each 24 hours, if the amalgam decompostion stage is 92% efficient, assuming continuous operation of the complete process?

*I.10. Sodium chlorate is used for bleaching pulp, for preparing certain herbicides and defoliants, and for producing ammonium perchlorate for "solid fuel" rockets. In the electrolytic production of sodium perchlorate (the Foerster method), hydrogen is a valuable byproduct, as shown by the equation for the overall process:

$$NaCl_{(aq)} + 3 H_2O_{(\ell)} \xrightarrow{\text{(current)}} NaClO_{3(aq)} + 3 H_{2(g)}$$

This process is typically 75% efficient in terms of current utilization. What is the rate of sodium chlorate production in a typical process with an average current demand of 2500 amps?

ANSWERS:

(I.1.) 3.6 kg, (I.2.) 75 ml, (I.3.) 14.7 kg, (I.4.) 1.4×10^4 hr, (I.5.) 41 kg hr^{-1}, (I.6.) 2980 amps, 843 amps, 248 amps, (I.7.) 4.4×10^5 amps, (I.8.) 1.6×10^6 liters, (I.9.) 960 kg, (I.10.) 1.2 kg hr^{-1}

RELEVANT PROBLEMS

Unit 8: Electrochemistry

Set E: Environmental Sciences

E.1. It is estimated that the corrosion of iron and steel products in the United States alone costs in excess of $5 billion annually. Nearly a fourth of the iron produced each year is needed for replacement of iron lost through corrosion and this represents an enormous consumption of fuel and a large production of waste byproducts. Improved methods for corrosion prevention will, therefore, have a number of benefits. Electrolytic corrosion is much more rapid than the formation of rust by direct reaction of iron with moist air. All that is needed for electrolytic corrosion is some aqueous electrolyte and two different regions of the iron, one exposed to the air and one protected from the air. Corrosion is enhanced near bimetallic junctions, as in a connection between a copper pipe and an iron pipe in moist soil. Typical examples of common sources of electrolytic corrosion are encountered with garden tools and farm implements stored partially covered with mud. An iron drainage pipe was laid through an irregular area so that some segments of the pipe were exposed to moist air and others were covered by soil. It was found that electrolytic corrosion was occurring at a rate equivalent to an average current of 0.12 amp. What mass of iron would be lost by conversion to Fe^{2+} over a thirty day period?

E.2. A steel beam on a saltwater pier was accidentally installed without corrosion protection so that electrolytic corrosion set in at a rate equivalent to 0.28 amp. How much iron would be lost as Fe^{2+} in one year, assuming average corrosion at this rate for 24 hours per day?

E.3. The corrosion of iron is a complex multi-step process, resulting in the formation of "rust" (hydrated Fe_2O_3). We can represent three phases of the process by simple equations:

(I) simultaneous $\begin{cases} O_{2(g)} + 2\ H_2O_{(\ell)} + 4e^- \rightarrow 4\ OH^-_{(aq)} \\ Fe_{(s)} \rightarrow Fe^{2+}_{(aq)} + 2e^- \end{cases}$

(II) $\qquad\qquad Fe^{2+}_{(aq)} + 2\ OH^-_{(aq)} \rightarrow Fe(OH)_{2(s)}$

(III) $\qquad\qquad 4\ Fe(OH)_{2(s)} + O_{2(g)} \rightarrow 2\ Fe_2O_3(H_2O)_{2(s)}$

$\qquad\qquad\qquad\qquad\qquad\qquad\qquad\qquad\qquad$ ("rust")

The initial steps in the corrosion process may, and often do, occur at two different locations on a piece of iron or steel. In electrolytic corrosion, an area of the metal exposed to air and moisture acts as a cathode for reduction of atmospheric oxygen. Some other area protected from the atmosphere can serve as the anode for oxidation of iron to iron(II) ion. Electrons are transferred through the metal and Fe^{2+} and OH^- ions migrate through any surrounding electrolyte solution such as sea water or moist soil. The iron becomes pitted in the anode region and rust begins to form near the cathode region. How much iron is changed to Fe^{2+} in 24 hours by an average current of 0.19 amp?

E.4. What mass of water was consumed during the corrosion of iron described in problem E.3?

E.5. One method of reducing corrosion of iron and steel objects buried in the ground is "cathodic protection". A more reactive metal, such as magnesium or zinc, is connected to the steel object. The reactive metal acts as the anode and thus is preferentially oxidized. The steel is the cathode and is thus protected from oxidation. Pipelines, storage tanks, and bridge supports are protected in this way. If magnesium is used as the reactive metal, and the "corrosion current" is .083 amp, how long (in years) will a 1.2 kilogram block of magnesium protect a steel object?

*E.6. Another method of protecting metal objects from corrosion is by electroplating. Zinc, cadmium, tin, chromium, nickel and copper are among metals used for this type of corrosion protection. If 43,000 objects are to be protected from corrosion, each needing 0.78 g of nickel plating, how much time (in days) will it take to electroplate the objects from a nickel(II) solution at a 1.3 amps average current?

*Proficiency Level

*E.7. One of the environmental concerns associated with the metals industries is the increasing demand for farm and ranch lands or wilderness areas for mining operations. This is a direct result of the depletion of mineral resouces by the massive metal requirements of a modern industrial society. Magnesium is one important metal that, at least in theory, should require no mining lands. It is estimated that a reasonably efficient method of recovering magnesium from sea water could supply the world demands for this metal for the next million years without consuming more than a tiny percentage of the magnesium in the oceans. If an electrolytic process were developed for producing 25 kg hr^{-1} of magnesium from Mg^{2+} at an overall current use efficiency of 80%, what average current would be required?

*E.8. Electroplating industries are among the nation's largest users of electrical power and, as such, they are intimately connected with the depletion of fuel reserves and the formation of air contaminants from fossil fuel combustion. Such industries are, of course, essential to our metal-dependent society and electrochemical processes themselves are among the most efficient and clean of all industrial chemical methods. Energy requirements for electroplating are directly related to the oxidation state of the metal ion being reduced. Thus, aluminum production (from Al^{3+}) has a much greater electrical energy demand than does silver production (from Ag^{+}). Other steps in metal production, such as the smelting of ores, may offset these differences to some extent, but the aluminum-plating industries still have a very high electrical energy bill. Calculate the number of kilowatt-hours of electrical energy required per day for an aluminum manufacturing plant using 220 volt lines for production of 9.0 kg hr^{-1} of aluminum from a molten aluminum salt, at an electrode efficiency of 89%. Assume steady operation for a 24 hour day. [watts = amps x volts]

*E.9. How many kilowatt-hours per day would the industry require if it were to switch to production of silver from a silver(I) complex at the rate of 9.0 kg hr^{-1}, assuming the same line voltage and electrode efficiency (problem E.8)?

*E.10. The impact of the "energy crunch" on electroplating industries is particularly severe for processes that are relatively inefficient. Typical lead plating operations, for example, can only produce 5.00 kg of lead per kilowatt-hour. If the process were 100% efficient at a plating potential of 0.60 volt for conversion of Pb^{2+} to lead, how many kilowatt-hours would be required to produce 1.00 kg of lead? What is the percentage efficiency of the "typical" operation? [watts = amps x volts]

ANSWERS:

(E.1.) 90 g, (E.2.) 2.6×10^3 g, (E.3.) 4.8 g, (E.4.) 0.77 g, (E.5.) 3.6 years, (E.6.) 980 days, (E.7.) 6.9×10^4 amps, (E.8.) 1.6×10^5 kilowatt-hours, (E.9.) 1.3×10^4 kilowatt-hours, (E.10.) 0.16 kilowatt-hour, 78%

STOICHIOMETRY

UNIT 9: ORGANIC CHEMISTRY

The stoichiometry of organic processes is no different from that of the "simpler" inorganic reactions. It is only that the chemical species appear more complex in their formulations. If we learn to recognize the similarity between organic and inorganic stoichiometric calculations, it will be obvious that the methods we have studied can all be applied to what, at first glance, may seem to be very complicated chemical equations. Consider for example a reaction between "trimethyl amine" and "methyl salicylate":

$$(CH_3)_3N + \underset{OH}{\bigcirc}CO_2CH_3 \rightarrow (CH_3)_3NH^+ + \underset{O^-}{\bigcirc}CO_2CH_3$$

This is just an ordinary acid/base reaction, no different in type from that of ammonia with hydrofluoric acid:

$$NH_3 + HF \rightarrow NH_4^+ + F^-$$

The same mole ratios, formula weight calculations, percentage yield corrections, etc. that we have used with simple compounds can be applied to the reactions of organic species.

As with most situations, there are some "tricks" to handling the stoichiometry of organic reactions that can simplify the operations appreciably. First of all, we need to recognize some of the chemical "shorthand" of the organic chemist.

"Ring" structures are often encountered and we need to know that a "corner" on a line-structure ring formula represents a carbon atom. Carbon atoms in organic molecules have four covalent bonds. Some of these may be shown as lines or, in the case of so-called "aromatic" rings, as a circle. Hydrogen atoms bonded to ring carbons are not shown in the abbreviated formulas, so we count them whenever the other bonds on carbon add up to less than four. Some examples will illustrate the common uses of ring notations:

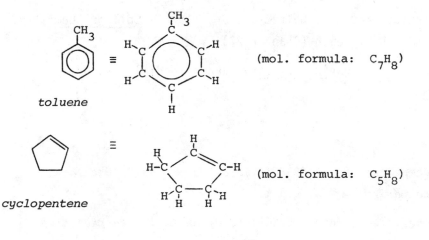

toluene   (mol. formula: C_7H_8)

cyclopentene   (mol. formula: C_5H_8)

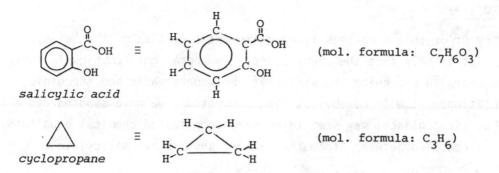

The second "trick" is to reduce a structural formula equation to one using only the molecular formulas. It is simpler and faster to calculate formula weights or balance an equation with the molecular formulas than it is with the more "cumbersome" structural formulas.

With these two "tricks", the various stoichiometric calculations dealing with organic reactions should pose no difficulties. This Unit, then, serves mainly as a review of some of the problem solving methods introduced in the preceding eight Units.

- References -

1. O'Connor, Rod. 1977. Fundamentals of Chemistry, second edition. New York: Harper & Row (Units 7, 8, 28, 29 and Excursion 5)

2. Brown, Theodore, and H. Eugene LeMay, Jr. 1977. Chemistry: The Central Science. Englewood Cliffs, NJ: Prentice-Hall (Chapters 8, 24)

3. Masterton, William, and Emil Slowinski. 1977. Chemical Principles, fourth edition. Philadelphia: W.B. Saunders (Chapters 8, 10)

4. Nebergall, William, F.C. Schmidt, and H.F. Holtzclaw, Jr. 1976. College Chemistry, fifth edition. Lexington, MA: D.C. Heath (Chapter 27)

5. Stille, John K. 1968. Industrial Organic Chemistry. Englewood Cliffs, NJ: Prentice-Hall.

OBJECTIVES:

(1) Be able to apply the competency level OBJECTIVES of Units 1-8 to processes involving organic compounds.

*(2) Be able to apply the proficiency level OBJECTIVES of Units 1-8 to processes involving organic compounds.

There are no PRE-TESTS for this Unit. Rather, we will trace through a few examples and exercises to review earlier methods as applied to organic reactions. The final Self-Test will let you determine how well you have accomplished your goals of "competency" or "proficiency" in the area of chemical STOICHIOMETRY.

EXAMPLE 1

One of the more interesting groups of natural products are the <u>terpenes</u>. The name was originated to describe a "family" of compounds isolated by steam distillation of turpentine, but the class now includes a host of substances ranging from relatively simple ring compounds to enormous polymeric species such as the compounds of natural rubber. Some of the more familiar terpenes are:

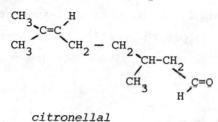

citronellal

(from eucaluptus leaves)

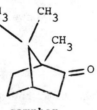

limonene

(from lemon peel)

camphor

An active principal of that feline favorite, <u>catnip</u>, is a terpene containing only carbon, hydrogen, and oxygen. Analysis of the compound, "nepatolactone", shows it to consist of 73.30%C and 8.95%H. What is the empirical formula (same as the molecular formula) of "nepatolactone"?

SOLUTION: *(following the method introduced in Unit 1)*

%O = 100% - (73.30% + 8.95%) = 17.75%

We'll find the carbon and hydrogen ratios with respect to oxygen:

$$\frac{73.30 \text{ amu(C)}}{100 \text{ amu(cpd)}} \times \frac{100 \text{ amu(cpd)}}{17.75 \text{ amu(O)}} \times \frac{1 \text{ atom(C)}}{12.01 \text{ amu(C)}} \times \frac{16.00 \text{ amu(O)}}{1 \text{ atom(O)}} = 5.50 \frac{\text{atom(C)}}{\text{atom(O)}}$$

$$\frac{8.95 \text{ amu(H)}}{100 \text{ amu(cpd)}} \times \frac{100 \text{ amu(cpd)}}{17.75 \text{ amu(O)}} \times \frac{1 \text{ atom(H)}}{1.008 \text{ amu(H)}} \times \frac{16.00 \text{ amu(O)}}{1 \text{ atom(O)}} = 8.00 \text{ atom(H)/atom(O)}$$

$$C_{(5\frac{1}{2} \times 2)} H_{(8 \times 2)} O_{(1 \times 2)} = C_{11}H_{16}O_2$$

The structural formula of nepatolactone is:

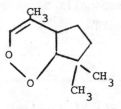

EXERCISE 1

Acetylsalicyclic acid ("aspirin") does not dissolve very rapidly in the aqueous "juices" of the stomach, so most preparations use a more soluble salt of the acid. Although the sodium salt is marketed by most aspirin manufacturers, one well-known firm sells the calcium salt, claimed to dissolve more rapidly. Supply the coefficients necessary to balance the equation for formation of calcium acetylsalicylate, given that one mole of the acid reacts with only one mole of hydroxide ion.

$$ + \ Ca(OH)_2 \ \rightarrow \ H_2O + Ca(C_9H_7O_4)_2 $$

(answer, page 216)

EXAMPLE 2

Monosodium glutamate (MSG) was at one time the most widely sold food additive, used to enhance the "meat" flavor of protein foods. Evidence of possible health hazards associated with excessive use of MSG have now reduced its economic value appreciably. Monosodium glutamate is formulated as:

$$ Na^+ \quad {}^-O_2CCH_2CH_2\overset{\overset{H}{|}}{C}CO_2{}^- $$
$$ \underset{+NH_3}{} $$

What fraction of a mole of this salt is contained in a 4.0 oz (113 g) package of MSG?

SOLUTION: *(following the method introduced in Unit 3)*

It is most convenient to change the structural formula to a "molecular" formula for use in calculating the formula weight. Counting all atoms carefully gives:

$$ C_5H_8NO_4Na $$

for which:

$$C_5: \quad 5 \times 12.01 = 60.05$$
$$H_8: \quad 8 \times 1.008 = 8.064$$
$$N: \quad 1 \times 14.01 = 14.01$$
$$O_4: \quad 4 \times 16.00 = 64.00$$
$$Na: \quad 1 \times 22.99 = \underline{22.99}$$
$$form.\ wt.:169.1 \qquad (to\ 4\ digits)$$

then, using a mole/mass unity factor:

$$\frac{113\ g\ (MSG)}{1} \times \frac{1\ mole\ (MSG)}{169.1\ g\,(MSG)} = \underline{0.668\ mole}$$

EXERCISE 2

Estradiol is a female hormone which acts to control the ovulation cycle. It may be converted by a sequence of reactions to mestranol, one of the minor components of several common oral contraceptive preparations. If a particular reaction sequence started with 250 g of estradiol and 192 g of mestranol was eventually isolated, what was the percentage yield for the overall sequence? (The two compounds are related on a 1:1 mole basis as indicated by the following formulation.)

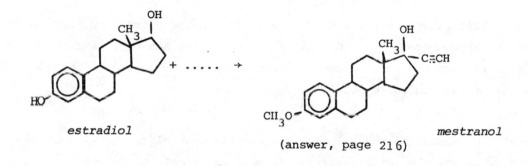

estradiol   mestranol

(answer, page 216)

EXAMPLE 3

When animal flesh decays, some of the amino acids from decomposing proteins are converted enzymatically into odoriferous amines by loss of carbon dioxide. "Putrescine" and "cadaverine" are two such amines, whose names provide a rough idea of their odors. How many grams of cadaverine were formed during the decomposition of a lysine sample if the CO_2 formed was measured to be 7.82 liters at 37°C and 700 torr? The balanced equation for this reaction is:

$$\overset{+}{H_3}NCH_2CH_2CH_2\overset{\overset{H}{|}}{\underset{\underset{NH_2}{|}}{C}}CO_2^- \rightarrow H_2N(CH_2)_5NH_2 + CO_2$$

lysine   cadaverine

205

SOLUTION: (_following the method introduced in Unit 5_)

In order to work with a volume/molar volume ratio using 22.4 liters mole^{-1}, we must first "correct" the measured CO_2 volume to STP:

(37°C = 310°K)

$$V_{STP} = 7.82 \text{ liters} \times \frac{273°}{310°} \times \frac{700 \text{ torr}}{760 \text{ torr}} = 6.34 \text{ liters}$$

Then, using a simplified molecular formula for cadaverine:

$$\begin{array}{ccc} & w & 6.34 \text{ liters} \\ \cdots & \boxed{C_5H_{14}N_2} + & \boxed{CO_2} \\ & (1 \text{ mole} \times 102.2 \text{ g mole}^{-1}) & (1 \text{ mole} \times 22.4 \text{ liters mole}^{-1}) \end{array}$$

$$w = \frac{102.2 \text{ g} \times 6.34 \text{ liters}}{22.4 \text{ liters}} = \underline{28.9 \text{ g}}$$

- -

EXERCISE 3

All complex animal organisms take in more nitrogen in their foods than they can utilize biochemically, so some nitrogen is invariably excreted as waste. Most healthy mammals eliminate waste nitrogen primarily as urea, H_2NCNH_2. Insects, on the other hand, excrete uric acid (a complex cyclic compound, $C_5H_4N_4O_3$), while fish excrete ammonia. A sample of aquarium water was analyzed for its ammonia content, after first removing ionic bases by ion exchange, by titration with sulfuric acid:

$$2 \text{ NH}_3 + \text{H}_2\text{SO}_4 \rightarrow (\text{NH}_4)_2\text{SO}_4$$

If the ammonia of a 25.0 ml sample of the solution was exactly neutralized by 21.6 ml of 0.178 M sulfuric acid, what was the molarity of the ammonia solution?

(answer, page 217)

- -

EXAMPLE 4

In 1953, Stanley Miller, investigating electrical discharge through a mixture of methane, ammonia, hydrogen, and water vapor (believed to simulate the "primitive earth" atmosphere), detected some of the natural-occuring amino acids among his reaction products. These classic experiments have often been cited by proponents of chemical evolution as "evidence" for the theory of accidental origin of living systems. Those of us who believe in God as the creator of life are not overly disturbed by such "evidence", although we do find it disturbing to hear some "scientists" present an interesting mechanistic theory as though it were "scientific fact". Apart from their philosophical interpretations, Miller's experiments provide some very exciting chemistry, providing as they do some glimpses of the formation of relatively complex molecules from simple chemical species. Using the following simplified equation for the formation of an amino acid, glycine, calculate the standard heat of reaction in kcal $mole^{-1}(CH_4)$. [For glycine, $\Delta H_f^\circ = -126.3$ kcal $mole^{-1}$. Use Appendix F for other necessary data.]

$$2\ CH_{4(g)} + NH_{3(g)} + 2\ H_2O_{(g)} \rightarrow H_3\overset{+}{N}CH_2CO_{2(s)}^- + 5\ H_{2(g)}$$

SOLUTION: *(following the method introduced in Unit 7)*

$$\Delta H_{reaction}^\circ = [\Delta H_f^\circ(glycine) + 5\Delta H_f^\circ(H_2)] - [2\Delta H_f^\circ(CH_4) + \Delta H_f^\circ(NH_3) + 2\Delta H_f^\circ(H_2O)]$$

$$= [(-126.3) + 5(zero)] - [2(-17.9) + (-11.0) + 2(-57.8)]$$

$$= +36.1\ kcal$$

to convert to kcal $mole^{-1}(CH_4)$:

$$\frac{+36.1\ kcal}{2\ moles\,(CH_4)} = +18.05\ kcal\ mole^{-1}(CH_4)$$

- -

EXERCISE 4

An experimental program seeking better utilization of fossil fuels has devised a tractor operated by an isooctane-oxygen fuel cell, similar in principle to the simpler methane-oxygen fuel cell, but using a fuel more like that of an ordinary gasoline engine. The electrode reaction for the oxidation of isooctane may be formulated as:

$$CH_3\underset{\underset{CH_3}{|}}{CH}CH_2CH_2CH_2CH_3 + 66\ OH^- \rightarrow 8\ CO_3^{2-} + 42\ H_2O + 50\ e^-$$

How many grams of octane would be consumed by this cell, assuming 100% efficiency, to produce a steady current of 2.5 amps for 3.0 hours?

(answer, page 217)

==

*EXAMPLE 5

Several innovative schemes have been proposed to reduce man's dependence on chlorinated insecticides, which may have serious long term ecological effects. Among these attempts have been insect sterilization by radiation, use of juvenile sex hormones, and the "baiting" of insect traps with chemical "sex attractants". Sex attractants have been used effectively in control of the gypsy moth and show considerable promise with a number

*Proficiency Level

of other insect pests. A sex attractant isolated from a common insect was identified as 2,2-dimethyl-3-isopropylidenecyclopropyl butanoate. Analysis shows this compound to consist of 73.42%C, 10.27%H, and 16.30%O. Its vapor density (corrected to STP) is 8.8 g liter^{-1}. What is its molecular formula?

SOLUTION: *(following the method introduced in Unit 1)*

mass ratios:

73.42 amu(C):   10.27 amu(H):   16.30 amu(O)

atom ratios:

$$\frac{73.42\ amu(C)}{12.01\ amu\ atom^{-1}} : \frac{10.27\ amu(H)}{1.008\ amu\ atom^{-1}} : \frac{16.30\ amu(O)}{16.00\ amu\ atom^{-1}}$$

$C_{6.113}H_{10.19}O_{1.019}$

whole-number ratios:

$$\frac{6.113\ atom(C)}{1.019} : \frac{10.19\ atom(H)}{1.019} : \frac{1.019\ atom(O)}{1.019}$$

$C_6H_{10}O$ [empirical formula]

approximate mol. wt.:

[@STP] $\dfrac{8.8\ g}{1\ liter} \times \dfrac{22.4\ liters}{1\ mole}$ = $\underline{197\ g\ mole^{-1}}$

$$\frac{formula\ wt.}{emp.\ formula\ wt.} = \frac{197}{98} \approx 2$$

mol. formula:

$$C_{(6\ \times\ 2)}H_{(10\ \times\ 2)}O_{(1\ \times\ 2)} = \underline{C_{12}H_{20}O_2}$$

The proposed structure was:

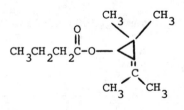

---

*EXERCISE 5

Polymers are very large molecules ("macromolecules") produced by repetitive reactions among smaller molecules. Both DACRON and MYLAR are trade names for the same basic polymer, the former referring to the polymeric material when extruded into a fibre and the latter to the rolled sheets of polymer film. The polymerization reaction from the "monomers" used as starting materials, terephthalic acid and ethylene glycol, may be represented as:

$$(n)\ \text{HO}_2\text{C-}\bigcirc\text{-CO}_2\text{H} + (n)\text{HOCH}_2\text{CH}_2\text{OH} \xrightarrow[\text{reaction}]{\text{repetitive}} \left[\begin{matrix}\overset{O}{\overset{\|}{-}}\text{C-}\bigcirc\overset{O}{\overset{\|}{-}}\text{C-OCH}_2\text{CH}_2\text{O-}\end{matrix}\right]_n + n\text{H}_2\text{O}$$

Terephthalic acid for the production of DACRON and MYLAR may be prepared by the oxidation of p-xylene (from coal tar) with hot alkaline potassium permanganate, followed by acidification and extraction of the acid. Complete and balance the equation for the oxidation process:

$$\text{CH}_3\text{-}\bigcirc\text{-CH}_3 + \text{MnO}_4^- \longrightarrow \left[\text{O}_2\text{C-}\bigcirc\text{-CO}_2\right]^{2-} + \text{MnO}_2$$

(answer, page 217)

----

*EXAMPLE 6

One of the more common synthetic hormones used in oral contraceptive preparations is norethinodrone:

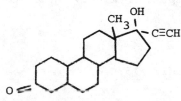

A typical "birth-control" tablet containing this hormone consists of 2.0 mg of nor-ethinodrone and 0.10 mg of mestranol (page 205). A normal prescription calls for one tablet per day for twenty days each month. How many mmols of norethinodrone would be required for a typical year's supply?

SOLUTION: (following the method introduced in Unit 3)

To find: mmols of norethinodrone per year

Information available: tablet contents, dosage and directions, formula of nor-ethinodrone

Information needed, not given: Months per year (12), formula wt.

Procedure: first "count atoms" (page 204) to find the formula wt., then use unity factors

formula from "counting atoms"

$C_{20}H_{26}O_2$ (form. wt. 298.4)

set-up

$$\frac{2.0 \text{ mg}}{1 \text{ tablet}} \times \frac{20 \text{ tablets}}{1 \text{ month}} \times \frac{12 \text{ months}}{1 \text{ year}} \times \frac{1 \text{ mmol}}{298.4 \text{ mg}} = \underline{1.6 \text{ mmol}}$$

----

*EXERCISE 6

Acetone is a simple compound of immense commercial importance. It is used as a solvent for a wide variety of purposes, particularly since it is miscible with water as well as with most common organic liquids. In addition, acetone is used in the production of a number of synthetic chemicals. Prior to World War I, most acetone was prepared by the thermal decomposition of calcium acetate, a relatively expensive procedure. Since acetone is used in making an explosive gel (cordite) from a mixture of nitrocellulose, nitroglycerine, and petroleum jelly, an inexpensive preparation of acetone became of considerable importance with the advent of the "Great War". Such a preparation was developed, using a special bacterial fermentation of molasses, by a British team headed by Chaim Weizmann[1]. As a result of this, and later synthetic developments, acetone is now a relatively inexpensive chemical. Calculate the number of grams of iodine which would be required for the production of 1.00 lb (454 g) of iodoform, an antiseptic chemical, by a reaction which is 82% efficient, as indicated by:

$$CH_3\overset{\overset{\textstyle O}{\textstyle \|}}{C}CH_3 + 3\ I_2 + 4\ OH^- \rightarrow CH_3CO_2^- + 3\ I^- + CHI_3 + 3\ H_2O$$

(answer, page 218)

----

*EXAMPLE 7

Fats and oils are primarily esters of glycerol with various long chain carboxylic acids. The carbon chains of most of these acids contain one or more carbon-carbon double bonds ("unsaturation"). There is a general trend in natural fats and oils for the melting point to increase as the number of double bonds is decreased, hence the conversion of liquid vegetable oils by addition of hydrogen to "solid" fats for "shortenings" such as CRISCO, as shown in the following equation:

----

[1] Later to become the first President of Israel.

$$H-C-O-C(CH_2)_7CH=CH(CH_2)_7CH_3$$

(structural diagram, left)

$$+ 3H_2 \xrightarrow{\text{Ni(catalyst)}}$$

(structural diagram, right)

triolein (a liquid "oil")   tristearin (a solid "fat")

How many liters of hydrogen gas, measured under the experimental conditions of 200°C and 2000 torr, would be required for the preparation of 1.00 lb (454 g) of tristearin (a "hydrogenated vegetable oil" product), if the reaction is 91% efficient?

SOLUTION: (following the method introduced in Unit 5)

To find: liters of H_2 at 200°C(473°K), 2000 torr

Information available: mole ratio, tristearin formula and mass, % yield

Information needed, not given: formula wt. of tristearin (from "counting atoms", 891)

Set-up: (using unity factors and gas "corrections")

$$\frac{454 \text{ g(tristearin)}}{1} \times \frac{1 \text{ mole(tristearin)}}{891 \text{ g(tristearin)}} \times \frac{3 \text{ mole}(H_2)}{1 \text{ mole(tristearin)}} \times \frac{22.4 \text{ liters}}{1 \text{ mole } (H_2)} \times \frac{473°}{273°} \times \frac{760 \text{ torr}}{2000 \text{torr}}$$

[g(tristearin) → mole(tristearin) → moles (H_2) → STP vol.(H_2)→"corrected"vol.]

$$\times \frac{100}{91} = 24.8 \text{ liters}$$

["efficiency correction"]

- -

*EXERCISE 7

Acetylcholine is a chemical involved in the intricate processes of transmission of nerve impulses. A somewhat similar chemical, succinylcholine, has found use in the tranquilizing darts used in many research studies of wild animals. The drug acts rapidly to immobilize the animal by interfering with normal acetylcholine function, but causes no permanent damage. A solution of succinylcholine, as its chloride salt, was checked for proper concentration by titration with standardized sodium hydroxide, according to the equation:

$$[HO_2C(CH_2)_2\overset{O}{\overset{\|}{C}}OCH_2CH_2N(CH_3)_3^+ + Cl^-] + OH^- \rightarrow [^-O_2C(CH_2)_2\overset{O}{\overset{\|}{C}}O(CH_2)_2N(CH_3)_3^+] + H_2O + Cl^-$$

If a 25.0 ml aliquot of the solution required 22.6 ml of 0.118 $\underline{M}$ sodium hydroxide, how many grams of succinylcholine chloride were contained in 1.00 pt (473 ml) of the solution?

(answer, page 218)

----

*EXAMPLE 8

Glucose is one of the major "fuels" for producing energy needed for various functions of the human body. Its biochemical "combustion" involves a number of complex steps, but the overall process of glucose oxidation may be represented by the simplified equation:

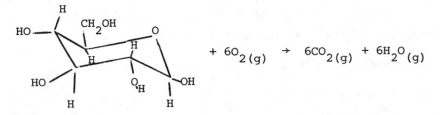 $+ 6O_{2(g)} \rightarrow 6CO_{2(g)} + 6H_2O_{(g)}$

What would be the standard enthalpy change for the complete combustion of 4.0 oz (113 g) of glucose? $[\Delta H_f^\circ$(glucose) = -300.8 kcal mole^{-1}. For other necessary data, see Appendix F.]

SOLUTION: (following the method introduced in Unit 7)

on a molar scale:

$$\Delta H^\circ_{comb} = [6\Delta H_f^\circ(CO_2) + 6\Delta H_f^\circ(H_2O)] - [\Delta H_f^\circ(glucose) + 6\Delta H_f^\circ(O_2)]$$

$$= [6(-94.0) + 6(-57.8)] - [(-300.8) + 6(zero)]$$

$$= -610.0 \text{ kcal}$$

per mole (glucose):

$$\frac{-610.0 \text{ kcal}}{1 \text{ mole(glucose)}} = -610.0 \text{ kcal mole}^{-1}\text{(glucose)}$$

from "counting atoms":

glucose = $C_6H_{12}O_6$, form. wt. = 180.2

using unity factors:

$$\frac{113 \text{ g(glucose)}}{1} \times \frac{1 \text{ mole(glucose)}}{180.2 \text{ g(glucose)}} \times \frac{(-610.0)\text{kcal}}{1 \text{ mole(glucose)}} = \underline{-383 \text{ kcal}}$$

----

*EXERCISE 8

The actual stages of the biochemical oxidation of glucose involve a number of electron-transfer steps. Since electrons are actually moving from one atom to another, although obviously not via a typical "electrical circuit", we could express the rate of glucose oxidation in terms of a rate of electron migration, a sort of "biochemical current". For each molecule of glucose converted to CO_2 and H_2O, there is a transfer of 24 electrons. What "biochemical current", in amps, would correspond to the complete combustion of 5.0 g of glucose per hour?

(answer, page 218)

===

SELF-TEST (UNIT 9)   [answers, page 218]

The "sulfa drugs" developed during the 1930's are credited with saving thousands of lives during World War II that would otherwise have been lost from bacterial infections of wounds. The "sulfa drugs" are believed to function by competing with p-aminobenzoic acid, an essential biochemical in a number of pathogenic microorganisms. The "sulfa drug" apparently interferes with an enzyme-catalyzed process for normal bacterial utilization of the p-aminobenzoic acid.

The compentency level Self-Test questions will deal with this subject in testing your achievement of the OBJECTIVES for Units 1-8, as applied to the stoichiometry of organic compounds.

9.1.   The bacterial metabolite p-aminobenzoic acid consists of 61.30%C, 5.14%H, 10.22%N, and 23.33%O. Calculate the empirical formula of this compound. (This is also its molecular formula. The structural formula is shown with the answer to this question, page 218.)

9.2.   The simplest of the "sulfa drugs", sulfanilamide, was the first such compound found to be an effective antibiotic. Once the value of the drug was established, a number of possible syntheses were investigated to find the most efficient route to a high-purity product. Supply the necessary coefficients to balance the equation for one such synthesis:

$$O_2N-\langle\bigcirc\rangle-SO_2NH_2 + H_2 \xrightarrow{\text{Pt (catalyst)}} H_2N-\langle\bigcirc\rangle-SO_2NH_2 + H_2O$$

(sulfanilamide)

9.3.   How many moles of sulfanilamide (question 9.2.) are contained in an 8.0 oz (227 g) packet of sulfanilamide powder?

9.4.   In an experiment testing the synthesis of sulfanilamide as described in question 9.2, 25.0 g of the nitro compound was treated with excess hydrogen and 13.2 g of sulfanilamide was recovered. What was the percentage yield in this experiment?

9.5. What volume of hydrogen gas, measured at 27°C and 3040 torr, would have been needed for the preparation of 13.2 g of sulfanilamide by the method described in question 9.2 if the reaction had been 100% efficient?

9.6. Sulfanilamide is basic and a solution of the drug can be titrated with an aqueous acid. Calculate the molarity of a sulfanilamide solution from which a 25.0 ml aliquot consumes 20.9 ml of 0.308 $\underline{M}$ sulfuric acid, according to the equation:

$$2\ H_2N-\bigcirc-SO_2NH_2 + H_2SO_4 \rightarrow (H_3N-\bigcirc-SO_2NH_2)_2SO_4$$

9.7. Some drugs are eventually metabolized in the body to simple chemical waste products. Calculate the standard heat of reaction for the theoretical decomposition of sulfanilamide, as described by:

$$H_2N-\bigcirc-SO_2NH_2 + 7\ O_2 \rightarrow 6CO_{2(g)} + 2NH_{3(g)} + H_2O_{(\ell)} + SO_{3(g)}$$

[ΔH_f° for sulfanilamide is -1772 kcal mole^{-1}. For other necessary data, see Appendix F.]

9.8. Since the early studies of sulfanilamide, a number of other "sulfa drugs" have been prepared. Several of these have properties superior, for some purposes, to those of sulfanilamide itself. Among these is sulfathiazole. One of the methods tested for manufacturing sulfathiazole used an electrolytic reduction process, for which the electrode equation may be written as:

$$O_2N-\bigcirc-SO_2N-\langle\ \rangle + 6H^+ + 6e^- \rightarrow H_2N-\bigcirc-SO_2N-\langle\ \rangle + 2H_2O$$

(sulfathiazole)

How many hours would a current of 1.50 amps have to be maintained to product 25.0 g of sulfathiazole by this method, assuming 100% efficiency?

==

*The opium poppy produces a number of nitrogen-containing organic bases (<u>alkaloids</u>). Among these, the most familiar is morphine, one of the most effective analgesic agents ("pain-killers") ever discovered. Morphine also produces a distinct euphoria and, unfortunately, is a very addictive drug. Chemical modifications have been attempted to alter the molecular structure of morphine in such a way as to reduce its addictive properties without loss of analgesic activity. Few of these attempts have shown much promise and morphine is still used extensively.

The _proficiency_ level Self-Test questions will deal with morphine and some of its modifications in testing your achievement of the OBJECTIVES for Units 1-8, as applied to the stoichiometry of organic compounds.

---

*Proficiency Level

Step 6 - $C_8H_{10} + 4H_2O \rightarrow C_8H_4O_4^{2-} + 14H^+ + \underline{12e^-}$

$\underline{4}MnO_4^- + \underline{16}H^+ + \underline{12e^-} \rightarrow \underline{4}MnO_2 + \underline{8}H_2O$

Step 7 - (eliminating duplication of $4H_2O$, $14H^+$, $12e^-$)

$C_8H_{10} + 4MnO_4^- + 2H^+ \rightarrow C_8H_4O_4^{2-} + 4MnO_2 + 4H_2O$

Step 8 - CHECK $\quad\boxed{(+2OH^- \rightarrow 2H_2O)}\qquad\boxed{(+2OH^-)}$

Step 9 -

$C_8H_{10} + 4MnO_4^- + \underline{2H_2O} \rightarrow C_8H_4O_4^{2-} + 4MnO_2 + 4H_2O + \underline{2OH^-}$

Step 10 - (Eliminating $2H_2O$ gives the answer, which is then reconverted to the original formulation.)

*6. (1070 g) Solution: (following the method introduced in Unit 4)

$$\frac{454 \text{ g}(CHI_3)}{1} \times \frac{100}{82} \times \frac{1 \text{ mole}(CHI_3)}{393.7 \text{ g}(CHI_3)} \times \frac{3 \text{ moles}(I_2)}{1 \text{ mole}(CHI_3)} \times \frac{253.8 \text{ g}(I_2)}{1 \text{ mole}(I_2)}$$

[actual g(CHI_3)→theor. g(CHI_3)→moles (CHI_3)→moles (I_2)→g(I_2)]

*7. (10.5 g) Solution: (following the method introduced in Unit 6)

[From "counting atoms", succinylcholine chloride (SCC) has the molecular formula $C_9H_{18}NO_2Cl$, form. wt. 207.7.]

$$\frac{0.118 \text{ mmol}(OH^-)}{1 \text{ ml}(OH^-)} \times \frac{22.6 \text{ ml}(OH^-)}{25.0 \text{ ml}(SCC)} \times \frac{1 \text{ mmol}(SCC)}{1 \text{ mmol}(OH^-)} \times \frac{473 \text{ ml}(SCC)}{1} \times \frac{207.7 \text{ mg}(SCC)}{1 \text{ mmol}(SCC)} \times \frac{1g}{10^3 \text{ mg}}$$

*8. (18 amps) Solution: (following the method introduced in Unit 8)

[Note: form. wt. of glucose is 180.2, page .]

$$\frac{5.0 \text{ g}(glucose)}{1 \text{ hour}} \times \frac{1 \text{ mole}(glucose)}{180.2 \text{ g}(glucose)} \times \frac{24 \text{ moles}(e^-)}{1 \text{ mole}(glucose)} \times \frac{1 \text{ hour}}{3600 \text{ sec}} \times \frac{(9.65 \times 10^4) \text{ amp sec}}{1 \text{ mole }(e-)}$$

= =

ANSWERS *to* SELF-TEST, Unit 9

9.1. $C_7H_7NO_2$ (The structural formula is H_2N-⬡-CO_2H.)

9.2. O_2N-⬡-$SO_2NH_2 + \underline{3}H_2 \xrightarrow{\text{Pt (catalyst)}} H_2N$-⬡-$SO_2NH_2 + \underline{2}H_2O$

9.3. 1.32 moles

9.4. 62%

9.5. 1.41 liters

---

*9.9. A morphine derivative called codeine, less addictive and a milder analgesic than morphine itself, is sometimes prescribed in tablets for severe headaches or as a component of a "cough syrup". Codeine is found to contain 72.21%C, 7.07%H, 4.68%N, and 16.03%O. Although the drug is not very volatile (boiling at 250°C at a pressure of 12 torr), careful measurements show it to have an approximate vapor density (corrected to STP) of 13 g liter^{-1}. What is the molecular formula of codeine? [The structural formula is given with the answer to this question.]

*9.10. Codeine has been oxidized in an attempt to synthesize a drug of improved properties. Using the formula found for question 9.9, water, and hydrogen ion as necessary, complete and balance the following equation for the oxidation of codeine in acidic solution.

codeine + $Cr_2O_7^{2-} \rightarrow Cr^{3+}$ +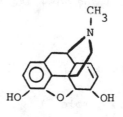

*9.11. A particular prescription "cough syrup" contains 3.50 g of codeine per liter. How many moles of codeine would be required by a pharmaceutical supply house to prepare 100 liters of a stock supply of this "cough syrup"?

*9.12. How many grams of potassium dichromate ($K_2Cr_2O_7$) would be required for the preparation of 50.0 g of the oxidation product ($C_{18}H_{19}NO_3$) for the reaction described in question 9.10 if the product yield is 73%?

*9.13. Before the structure of morphine had been determined, numerous tests had been performed to identify various structural features of the molecule. One analysis which can be used to detect O-H and N-H groups (having so-called "active hydrogens") uses dimethyl magnesium, $Mg(CH_3)_2$. Each O-H or N-H group in a molecule reacts with this reagent to liberate one molecule of methane. How many milliliters of methane (at 27°C and 690 torr) would have been formed by reaction of excess dimethyl magnisium with 125 mg of morphine?

morphine

*9.14. Morphine is only slightly soluble in water, so aqueous solutions for injection of the drug are prepared from one of its water-soluble salts, usually the sulfate or chloride. Since morphine contains only a single basic group (the nitrogen function), two moles of morphine react with one of sulfuric acid. What volume of

0.356 $\underline{M}$ H_2SO_4 would be required to convert 1.35 g of morphine to its sulfate salt?

*9.15. Illegal drugs confiscated by narcotics agents are usually destroyed by burning. What would be the standard enthalpy change for the complete combustion of 1.00 kg of heroin? [ΔH_f^o for heroin is -73.7 kcal mole^{-1}. For other necessary data, see Appendix F.]

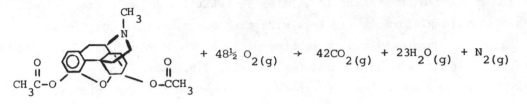

(heroin)

$+ 48\frac{1}{2} O_2(g) \rightarrow 42CO_2(g) + 23H_2O(g) + N_2(g)$

*9.16. One of the reactions proposed in an attempt to modify the structure of morphine for improved physiological properties was the electrolytic reduction of the carbon-carbon double bond. Two electrons would be required per molecule of morphine reduced. How many grams of morphine could be reduced per hour by a current of 1.50 amps if the process were 63% efficient? [The structural formula of morphine is given in question 9.13.]

===

ANSWERS to EXERCISES, Unit 9

1. [$\underline{2}$ $C_9H_8O_4$ + Ca(OH)$_2$ $\rightarrow$ $\underline{2}$ H_2O + Ca(C$_9$H$_7$O$_4$)$_2$]

    Solution: (following the method introduced in Unit 2)

    Since each molecule of "aspirin" reacts with only one hydroxide ion and calcium hydroxide furnishes two hydroxide ions, two "aspirins" are needed per Ca(OH)$_2$. Atom (symbol) balance then requires a coefficient of 2 for H_2O.

2. (67.4%) Solution: (following the method introduced in Unit 4)

    To simplify the process, the structures (page 205) are represented by molecular formulas derived as described on page 201.

    250 g   t

    $\boxed{C_{18}H_{24}O_2}$ + ... $\rightarrow$ $\boxed{C_{21}H_{26}O_2}$ + ...

    (1 mole x 272.4 g mole^{-1})   (1 mole x 310.4 g mole^{-1})

    $t = \dfrac{250 \text{ g} \times 310.4}{272.4} = 285$ g

    % yield $= \dfrac{192 \text{ g}}{285 \text{ g}} \times 100\%$

---

3. (0.308 $\underline{M}$) Solution: (following the method introduced in Unit 6)

    25.0 ml x C   21.6 ml x 0.178 $\underline{M}$

    $\boxed{2 \text{ NH}_3}$ + $\boxed{H_2SO_4}$ $\rightarrow$ ....

    2 mmols   1 mmol

    $C = \dfrac{2 \times 21.6 \text{ ml} \times 0.178 \ \underline{M}}{25.0 \text{ ml}}$

4. (0.64 g) Solution: (following the method introduced in Unit 8)

    To simplify calculations, isooctane is represented by the molecular formula.

    w   (2.5 x 3.0 x 3600)C

    $\boxed{C_8H_{18}}$ + ... $\rightarrow$ .... + $\boxed{50e^-}$

    (1 mole x 114.2 g mole^{-1})   (50 moles x 9.65 x 10^4 C mole^{-1})

    $w = \dfrac{114.2 \text{ g} \times (2.5 \times 3.0 \times 3600)C}{(50 \times 9.65 \times 10^4)C}$

*5. [CH$_3$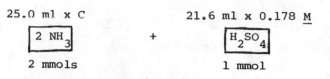-CH$_3$ + 4MnO$_4^-$ $\rightarrow$ [O$_2$C-⬡-CO$_2$]$^{2-}$ + 4MnO$_2$ + 2H$_2$O + 2OH$^-$]

    Solution: (following the method introduced in Unit 2)

    The procedure is simplified considerably by using molecular formulas, rather than structural formulas. Reconversion to the origianl formulation may be made after establishing the proper coefficients.

    Step 1 - C$_8$H$_{10}$ $\rightarrow$ C$_8$H$_4$O$_4$$^{2-}$

               MnO$_4^-$ $\rightarrow$ MnO$_2$

    Step 2 - (same as Step 1, since C and Mn are already balanced.)

    Step 3 - C$_8$H$_{10}$ + $\underline{4}$H$_2$$\underline{O}$ $\rightarrow$ C$_8$H$_4$O$_{\underline{4}}$$^{2-}$

               MnO$_{\underline{4}}^-$ $\rightarrow$ MnO$_{\underline{2}}$ $\underline{2H_2O}$

    Step 4 - C$_8$H$_{\underline{10}}$ + $\underline{4}$H$_2$O $\rightarrow$ C$_{\underline{8}}$H$_{\underline{4}}$O$_4$$^{2-}$ + $\underline{14H^+}$

               MnO$_4^-$ + $\underline{4H^+}$ $\rightarrow$ MnO$_2$ + $\underline{2H_2O}$

    Step 5 - C$_8$H$_{10}$ + 4H$_2$O $\rightarrow$ C$_8$H$_4$O$_4$$^{2-}$ + 14H$^+$ + $\underline{12e^-}$

               [(zero) = (2-) + (14+) + (12-)]

               MnO$_4^-$ + 4H$^+$ + $\underline{3e^-}$ $\rightarrow$ MnO$_2$ + 2H$_2$O

               [(1-) + (4+) + (3-) = (zero)]

9.6. 0.515 $\underline{M}$

9.7. 1023 kcal mole^{-1}

9.8. 10.5 hours

If you completed Self-Test questions 9.1-9.8 correctly, you have demonstrated a reasonable competency in chemical stoichiometry. You may wish to try some of the RELEVANT PROBLEMS for Units 1-9 or to extend your skills through studying the proficiency level sections of these Units. If you missed any, you should review the corresponding Units. Part II of this book will explore some chemical applications of equilibrium systems.

----

*9.9. $C_{18}H_{21}NO_3$

structural formula:

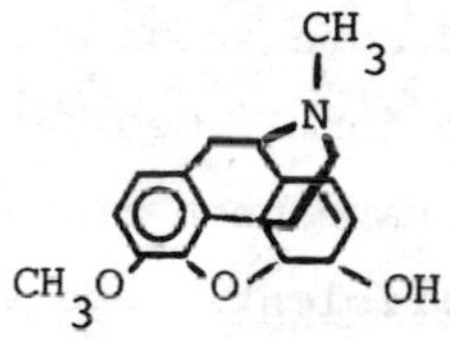

*9.10. $3C_{18}H_{21}NO_3 + Cr_2O_7^{2-} + 8H^+ \rightarrow 3C_{18}H_{19}NO_3 + 2Cr^{3+} + 7H_2O$

*9.11. 1.17 moles (codeine)

*9.12. 2.26 g ($K_2Cr_2O_7$)

*9.13. 23.7 ml (CH_4)

*9.14. 6.64 ml

*9.15. 6944 kcal

*9.16. 5.03 g (morphine) hour^{-1}

If you completed Self-Test questions 9.9-9.16 correctly, you have demonstrated real proficiency in chemical stoichiometry. You may wish to try the RELEVANT PROBLEMS (UNIT 9) for additional practice. If you missed any questions, you should review the corresponding Unit. Part II of this book will introduce some aspects of chemical equilibrium.

A.1. Dissolved organic compounds in the sea originate from several sources, such as plant
and animal excretions, bacterial decomposition, and autolysis of dead carcasses.
Additional increments are introduced by river drainage, land run-off, and biogeo-
chemical cycles. Most of these sources are expected to increase the concentration
of organic compounds more in surface layers than in deep water, and they are expected to influence given areas of the ocean far more than others. In spite of this,
the absolute amount of carbon in both surface and deep water is very similar in all
areas of the sea, indicating the participation of carbon in elemental cycling processes. This is accomplished in terms of balance among the oxidative processes of
respiration and combustion and the reductive processes which "fix" carbon dioxide
into an organic form. Geological records suggest that large quantities of carbon
from carbonate have been transferred to "organic" carbon in marine reservoirs.
Supply the coefficients necessary to balance the overall equation used to describe
this complex process, in which $[CH_2O]$ represents "organic" carbon:

$$FeS_2 + CaCO_3 + MgCO_3 + SiO_3 + H_2O \rightarrow CaSO_4 \cdot 2H_2O + Fe_2O_3 + MgSiO_3 + [CH_2O]$$

A.2. Indoleacetic acid, a well known plant growth hormone, is closely related to the
amino acid tryptophan. This substance, and a number of closely related structural
analogs, is found in very high concentration in rapidly growing tissues such as root
tips and the tips of growing shoots. It exerts its growth-promoting effects through
control of cellular growth and elongation, rather than through changes in cell division rates. This hormone is involved in simple plant growth, phototropism in
plants, prevention of premature fruit drop, and in the initiation of flowering.
Indolebutyric acid, another plant growth hormone, can be produced from indoleacetic
acid by a series of reactions in which the two compounds are related on a 1:1 mole
basis as indicated by the following formulation:

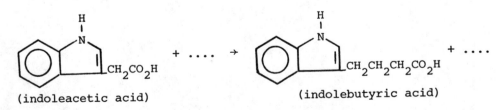

(indoleacetic acid)   (indolebutyric acid)

In a particular reaction sequence started with 350.0 g of indoleacetic acid, 312.0 g
of indolebutyric acid was eventually isolated. What was the percentage yield for
the overall process?

A.3. Beneath the marine water column anoxic conditions are frequently established at the water-sediment interface as a consequence of the accumulation of organic matter and the restrictions imposed on the rate of oxygenation. In this environment the species of bacteria involved in sulfate reduction assumes the dominate role in degradation of organic matter within the sediments. Some of the metabolic products of these bacteria (CO_2, H_2S, NH_3, and $H_2PO_4^-$) are chemically reactive and will therefore influence subsequent diagenetic processes. Metal sulfide precipitation, carbonate precipitation, pH modification, and control of methane generation are associated with bacterial activity. The degradation products of carbohydrates and proteins include ethanol, acetic acid, lactic acid, ammonia, and carbon dioxide. An acetic acid solution isolated by steam distillation of a marine sediment sample was titrated with sodium hydroxide. If the acetic acid of a 25.0 ml aliquot of the distillate was exactly neutralized by 29.2 ml of 0.184 $\underline{M}$ sodium hydroxide, what was the molarity of the acetic acid solution?

A.4. Nitrogen-fixing microorganisms supply about two-thirds of the world's "fixed" nitrogen used for growing crops. One of the most important nitrogen-fixing bacteria is Rhizobium japonicum, which converts atmospheric nitrogen into ammonia in the root nodules of soybeans. The soybean plant is able to incorporate the ammonia into nucleic acids, proteins, and chlorophyll. The plant proteins, consumed by animals, are converted to animal proteins by rearranging the amino acid building blocks. These proteins are used by the animal to build new tissues or as an energy source. Using the following simplified equation for the oxidative catabolism of the amino acid glycine calculate the standard heat of reaction in kcal mole^{-1} (glycine). [For glycine, ΔH_f° = -126.3 kcal mole^{-1}. Use Appendix F for other necessary data.]

$$2\ H_3\overset{+}{N}CH_2CO_2^-{}_{(s)} + 3\ O_{2(g)} \rightarrow 2NH_{3(g)} + 4CO_{2(g)} + 2H_2O_{(\ell)}$$

A.5. The resources required to produce food for human beings fall into two broad categories: natural (including physical and biological resources) and social. The physical resources, although they are very large, are ultimately limited. They include the four basic elements of the Greeks: earth, air, fire (energy), and water. The biological resources, on the other hand, are quite variable. They include the plants and animals farmers grow and the microbes and other organisms that play diverse roles in the food system. The social resources are also variable. They include the capital for agricultural investment, the social institutions that help the farmers to do their job, human labor and skills, and the growing store of scientific and practical knowledge that has transformed agriculture into a science. One of the biological resources, acetic acid bacteria (Acetobacter), can convert ethanol to

acetic acid. Vinegar, which contains about 4.0% acetic acid, is used in food preservation processes. The electrode reaction for the biological oxidation of ethanol
may be formulated as:

$$CH_3CH_2OH + H_2O \xrightarrow{\text{(Acetobacter)}} CH_3CO_2H + 4H^+ + 4e^-$$

How many grams of ethanol would be consumed by this cell, assuming 100% efficiency,
to produce an average "biological current" of 2.8 amps for 5.0 hours?

*A.6. Chlorophyll, the principal agent of photosynthesis, is produced in plants exposed to
sunlight. These molecules are embedded in protein disks held together by layers of
fatlike molecules. Frequently, four different types of chlorophyll are found in the
plant cells, The most abundant form, as well as the most important in photosynthesis,
is chlorophyll _a_. This bright bluish-green pigment has molecules containing carbon,
hydrogen, oxygen, nitrogen, and a single atom of magnesium (in the center of the
molecule). The molecular formula for chlorophyll _a_ is $C_{55}H_{72}O_5N_4Mg$. A second yel-
lowish-green pigment is called chlorophyll _b_. Its molecular formula is $C_{55}H_{70}O_6N_4Mg$.
A "standard tomato plant" leaf contains 1.4 mg of chlorophyll _a_. A typical tomato
plant has 750 leaves. How many __mmols__ of chlorophyll _a_ would be produced by a typical
tomato plant?

*A.7. Only a minority of the world's insect population is detrimental to man, yet the
"obnoxious" forms are so prominent and well known that popular opinion is apt to
condemn all insects. Their harmful activities include destruction of grain, vegetables, and fruit, injury to shade trees, transmission of disease between man and
animals, participation in the spread of plant diseases, destruction of buildings,
injury by bites and stings, destruction of food and clothing, and parasitic action
on domestic animals and man. The war against insects is fought with quarantine,
biological control, and chemical control. Calculate the number of grams of sulfur
which would be required for the production of 5.00 lbs (2270 g) of __parathion__, an
organophosphorus insecticide, by a reaction which is 73.0% efficient, as indicated by:

$$PCl_3 + S + 2NaOC_2H_5 + NaO-\bigcirc-NO_2 \rightarrow \underset{C_2H_5-O}{\overset{C_2H_5-O}{}}\overset{S}{\underset{}{P}}-O-\bigcirc-NO_2 + 3NaCl$$

*A.8. Cattlemen have attempted to keep the price of beef down by adopting technical in-
novations, such as implants of the hormone mimic diethylstilbestrol (DES), to improve the efficiency with which cattle convert their feed to added weight. However,
DES is generally considered to be a carcinogen, and recent evidence indicates that

*Proficiency Level

222

traces of DES remain as metabolites in the livers of cattle that have been given this drug. To replace DES, cattlemen are seeking other methods to improve the efficiency with which cattle gain weight. Perhaps the most effective of the new products is Rumensin, the trade name for monensin, a fermentation product from Streptomyces cinnamonensis. It is thus formally classed as an antibiotic, but it actually has only a very limited antibiotic activity. It works by altering the metabolic products of microbes in the rumens of cattle. These microbes convert the ruminants' feed, whether grass, hay, silage, or grain, into a form that is more efficiently digested. Plant fibers and starches are first broken down into sugars. These sugars are then converted into small fatty acids, such as acetic acid, which are the ruminant's principal source of energy. A typical sugar, $C_6H_{12}O_6$ is converted to the acetic acid according to the equation:

$$C_6H_{12}O_6(aq) \xrightarrow{\text{(Rumensin)}} 2CH_3COOH_{(aq)} + CH_{4(g)} + CO_{2(g)}$$

What total volume of gas, measured under the experimental conditions of 37°C and 780 torr, would be formed during the conversion of 901 g of glucose to acetic acid if the reaction is 60.0% efficient?

*A.9. The administration of monensin (Problem A.8) does not appear to alter the microbial species in the cattles' rumen, nor does it leave detectable residues or metabolites in the meat obtained from animals that have consumed it. Monensin and its metabolites degrade rapidly when they are released to the soil in manure, and they do not affect the growth of crops fertilized with the manure. Finally, the additive does not affect the quality of beef carcasses, their composition, or the ease with which they can be cut. A rumen solution was titrated with standardized sodium hydroxide to determine the acetic acid content. If a 25.0 ml aliquot of the rumen solution required 27.1 ml of 0.125 M sodium hydroxide, how many grams of acetic acid were contained in 1.00 pint (473 ml) of the solution?

*A.10. Feedlot cattle normally eat only enough to satisfy their energy needs, so monensin (Problem A.8) does not increase their average daily weight gain. But since they obtain more energy per unit of feed, it does decrease their food consumption by more than 10 percent, or about 1 kilogram of feed for every kilogram of weight gained. The biochemical oxidation of acetic acid serves as an energy source in feedlot cattle. For each molecule of acetic acid converted to CO_2 and H_2O, there is a transfer of 8 electrons. What "biochemical current", in amps, would correspond to the complete oxidation of 20.0 g of acetic acid per hour?

ANSWERS:

(A.1.) $6FeS_2 + 12CaCO_3 + 7MgCO_3 + 7SiO_3 + 43H_2O \rightarrow 3Fe_2O_3 + 12CaSO_4 \cdot 2H_2O + 7MgSiO_3 + 19 [CH_2O]$

(A.2.) 76.8%, (A.3.) 0.215 $\underline{M}$, (A.4.) -141 kcal mole^{-1}, (A.5.) 6.0 g, (A.6.) 1.2 mmol,

(A.7.) 342 g, (A.8.) 149 liters, (A.9.) 3.85 g, (A.10.) 71.5 amps

RELEVANT PROBLEMS

Unit 9: Organic Chemistry

Set B: Biological & Medical Sciences

B.1. Rancidity in fats is of two different kinds, oxidative and hydrolytic. The latter is the result of the hydrolysis of the glycerides with the liberation of free fatty acids. This type of rancidity is usually produced by the action of lipase enzymes of certain microorganisms. Since the fatty acids of high molecular weight are tasteless, this kind of rancidity is noticeable only in fats, such as butter and coconut oil, containing the low molecular weight fatty acids. In butter, 10 to 14 percent of the fatty acids are of the "volatile" type, such as butyric acid, $C_3H_7CO_2H$. An assay revealed that a pound (454 g) of butter was composed of 12.5%, by weight, glyceryl tributyrate. If the pound of butter is allowed to rancidify according to the equation,

$$CH_2-O-\overset{\overset{O}{||}}{C}-C_3H_7 \atop CH-O-\overset{\overset{O}{||}}{C}-C_3H_7 \atop CH_2-O-\overset{\overset{O}{||}}{C}-C_3H_7 \quad + \ 3 \ H_2O \ \rightarrow \quad {CH_2OH \atop CH-OH \atop CH_2OH} \quad + \quad 3 \ C_3H_7COOH$$

how much butyric acid would be produced?

B.2. If the butyric acid (problem B.1) of a 25.0 ml test sample was exactly neutralized by 31.2 ml of 0.516 $\underline{M}$ sodium hydroxide, what was the molarity of the solution?

$$C_3H_7COOH + OH^- \rightarrow C_3H_7COO^- + H_2O$$

B.3. Aldehydes occur widely in nature and are important in many biological processes. With the exception of formaldehyde, which is a gas, the short-chain aldehydes are colorless liquids at room temperature and have unpleasant, penetrating odors. The longer-chain aldehydes tend to have more fragrant odors, particularly those that contain an aromatic ring. Consequently, many of the aromatic aldehydes are used for flavoring and perfumes. Benzaldehyde, for example, has the odor and flavor of almonds and is found in the seeds of members of the prune family. Vanillin, the principle odorous component of the vanilla bean, has the structure

Vanillin is one of the most widely used flavoring agents in foods and confectionery. A reaction peculiar to glycerol is its facile dehydration by heating to form acrolein, $CH_2=CHCHO$, especially in the presence of metals. This compound is an aldehyde with a pungent, disagreeable odor and is responsible for the acrid odor of overheated bacon and other fat-containing substances. This dehydration is shown by the following equation:

$$\begin{array}{ccc} CH_2OH \\ | \\ CHOH & \xrightarrow{(Fe)} & \\ | \\ CH_2OH \end{array} \quad + \ 2\ H_2O$$

A certain well-known brand of bacon contains 17.5% glycerol, by weight. How much acrolein could be produced from one kilogram of this bacon, assuming 75% efficiency of conversion?

B.4.   One of the most widely distributed organic acids in the plant kingdom is a monohydroxytricarboxylic acid, citric acid. Found in highest concentration in citrus fruits, it is present also in many other fruits, such as raspberries, cranberries, strawberries, and pineapple. The recovery of citric acid from lemon juice or pineapple waste is a complicated process and to obtain satisfactory yields and economic operation the industrial plants must be large. Consequently, more than 95 percent of the citric acid produced in the United States is obtained from beet sugar solutions by microbiological methods. Citric acid may be represented by the structure:

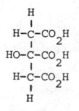

How many kilograms of sucrose ($C_{12}H_{22}O_{11}$) are required for conversion to 50.0 kg of citric acid, assuming production of two moles of citric acid per mole of sucrose?

B.5. Citric acid is a solid, very soluble in alcohol and water. It is used for the preparation of synthetic fruit beverages and is the organic acid most widely used in the manufacture of "soft drinks". Magnesium citrate, mixed with sodium bicarbonate and sugar, is sold as a mild laxative. Sodium citrate is used to prevent the clotting of blood stored in blood banks. A commercial lemon-growing operation sells its poorer quality lemons to a processing firm which produces 35.0 kg of citric acid per metric ton of lemons. If this is to be converted to sodium citrate ($Na_3C_6H_5O_7$) for use in blood preservation, how many liters of blood could be protected from clotting by the sodium citrate produced from the original metric ton of lemons? (4.4 g of the salt must be used per liter of blood.)

*B.6. Deamination, the removal of the amine group from the carbon chain, occurs very early in the catabolism of amino acids. The manner in which the body disposes of the nitrogen, once it is separated from the carbon chain, is generally the same regardless of the particular amino acid from which it is derived. Removal of the amine group from the carbon chain may be effected by <u>oxidative deamination</u>, so called because the amine group is replaced by oxygen. The oxidative deamination of alanine is catalyzed by the enzyme <u>L-amino acid oxidase</u> according to the equation:

$$\overset{\overset{+}{N}H_3}{\underset{|}{CH_3-CH-CO_2^-}} + O_2 + H_2O \rightarrow \overset{\overset{O}{||}}{CH_3-C-COOH} + NH_3 + H_2O_2$$

The hydrogen peroxide produced by this reaction is decomposed by the action of <u>catalase</u>. How many liters of ammonia, measured as a gas under the experimental conditions of 37°C and 650 torr, would be produced by the oxidative deamination of 150 g of alanine, if the reaction is 93% efficient?

*B.7. Excess heat absorbed through the skin causes tissue changes, burns, which vary in extent and depth of damage, depending upon the level of the temperature and the duration of exposure. The nature of the heat modifies the character of the injury. Dry heat causes desiccation and charring of tissues, while wet heat essentially "boils" tissues. In the usual pathway of absorption, the surface tissue suffers the most intense effects. Sunburns are generally first or second degree burns. The first degree burn involves only the epidermis, resulting in a localized area of redness caused by vascular dilatation and hyperemia of the vessels in the skin. The second degree burn includes first degree changes plus necrosis of the epidermal cells, accompanied by the exudation of blood serum to form vesicles or blisters. The compound p-aminobenzoic acid (PABA), the active ingredient in several suntan preparations, is used to prevent the sun's ultraviolet radiation from reaching the skin. A solution of p-aminobenzoic acid extracted for analysis of a suntan preparation was titrated with standardized sodium hydroxide, according to the equation:

$$H_2N-\langle \bigcirc \rangle-COOH + OH^- \rightarrow H_2N-\langle \bigcirc \rangle-COO^- + H_2O$$

If a 25.0 ml aliquot of the solution required 19.2 ml of 0.052 $\underline{M}$ sodium hydroxide, how many grams of p-aminobenzoic acid were contained in 1.00 pt (473 ml) of the solution?

*B.8. Two classes of foods, the carbohydrates and lipids, provide man with practically all his energy for living. Each class of food follows its own unique pathway for several steps, and then the pathways converge into two of the most important energy producing metabolic sequences in the body, the citric acid cycle and the respiratory chain. Lipids, our major energy storage form, contain as a part of their structure one or more fatty acids. In a series of reactions, the fatty acid molecules have their chains shortened two carbons at a time. The two-carbon fragments, as acetyl units, are involved in the citric acid cycle for the production of energy through oxidation of the acetyl units to carbon dioxide. The shortened fatty acids continue through the fatty acid cycle repeatedly, until they are entirely converted to two-carbon units. The complete oxidation of a typical fatty acid, such as, palmitic acid, and the simultaneous production of "high energy phosphate bonds" is a complex process involving electron transfer, however, the overall process for the oxidation of palmitic acid in the body may be represented simply as:

$$CH_3(CH_2)_{14}CO_2H + 23\ O_2 \rightarrow 16\ CO_2 + 16\ H_2O$$

For each molecule of palmitic acid converted to CO_2 and H_2O, there is a transfer of 92 electrons. What "biological current", in amps, would correspond to the complete biochemical oxidation of 6.0 g of palmitic acid per hour?

*B.9. Ammonia is produced in relatively large amounts by the oxidative deamination of amino acids. The tissue toxicity of ammonia makes its accumulation in cells or extracellular fluids in appreciable concentrations incompatible with life. Organisms living in aqueous environments such as a lake or the sea can without difficulty excrete ammonia directly. Terrestrial organisms, however, are prevented from the direct excretion of ammonia by the lack of sufficient water to dilute the ammonia to nontoxic levels. For most land animals the mechanism for ammonia removal lies in the formation of urea, $H_2N-\overset{\overset{\displaystyle O}{\|}}{C}-NH_2$, a compound which can be tolerated at much higher levels than ammonia. The metabolic synthesis of urea can be summarized as:

$$2 NH_3 + CO_2 \rightarrow H_2N-\overset{\overset{O}{\|}}{C}-NH_2 + H_2O$$

What would be the standard enthalpy change for the formation of 28 g of urea (the average amount excreted by an adult human per day)? [ΔH_f° (urea) = -35.8 kcal mole^{-1}. For other necessary data, see Appendix F.]

*B.10. Arteriosclerosis literally means "hardening of the arteries", but more accurately it refers to a group of processes which have in common thickening and loss of elasticity of arterial walls. Three distinctive morphologic variants are included within the term arteriosclerosis: atherosclerosis, characterized by the formation of lipid deposits; calcific sclerosis, characterized by calcification of the muscular arteries; and arteriolosclerosis, marked by fibromuscular thickening of the walls of small arteries and arterioles. Approximately 60 percent of all deaths in the United States result from cardiovascular disease. Almost 85 percent of these are linked to atherosclerosis. Hypertension (high blood pressure) develops when the blood must be pumped harder to make it circulate through the "clogged" arteries. When hypertension is detected, usually by monitoring the blood pressure, it is frequently necessary for the physician to prescribe a drug that will function as a coronary vasodilator to reduce the blood pressure. One of the common coronary vasodilator drugs in use is Persantin:

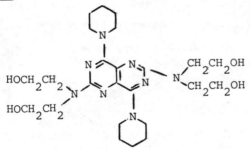

A typical "high blood pressure" tablet contains 25 mg of this drug. A particular prescription called for one tablet four times a day for thirty days. How many mmols of Persantin would be required for a thirty day supply?

ANSWERS:

(B.1.) 49.7 g, (B.2.) 0.644 M, (B.3.) 0.08 kg, (B.4.) 45 kg, (B.5.) 11 thousand liters, (B.6.) 47 liters, (B.7.) 2.6 g, (B.8.) 58 amps, (B.9.) 5.5 kcal, (B.10.) 5.9 mmol

RELEVANT

PROBLEMS

I.1. Phthalate esters are the backbone of the plasticizer industry. They contribute low temperature flexibility, resilience, high impact strength, and good electrical properties to vinyl resins. Phthalic anhydride is used for the production of most phthalate esters. Balance the equation for the industrial production of phthalic anhydride from naphthalene.

I.2. How much carbon dioxide, in liters at STP, results from consumption of 3.4×10^6 liters of oxygen, measured at 700 torr and 50°C, during the production of phthalic anhydride (problem I.1)?

I.3. The specific ethanolamine of greatest demand varies from year to year. The three ethanolamines, mono-, di-, and triethanolamine, are products of those industries using ethylene as a raw material. Chemically, the synthetic process is simple, but serious corrosion problems greatly reduce the life of the equipment used in production. Even though they are organic compounds the ethanolamines are strongly alkaline. The principal uses for the ethanolamines include detergents, gas scrubbing, textile chemicals, and cosmetics. Ethanolamines are produced by the exothermic reaction of ammonia with ethylene oxide. The ratio of the three products depends upon the reaction conditions. Under the conditions of 32°C and 15 psi a 10:1 ammonia/ethylene oxide mole ratio yields 75% monoethanolamine, 21% diethanolamine and 4% triethanolamine. A 30:1 ammonia/ethylene oxide mole ratio yields almost exclusively monoethanolamine whereas an equimolar mixture of reactants produces a 12:23:65 mixture of (mono:di:tri) products. It is relatively simple for industry to vary these conditions to meet changing consumer demands for any of the products. The industrial process may be represented as:

$$n \; H_2C\overset{O}{\overset{\triangle}{C}}H_2 + m \; NH_3 \rightarrow x \; H_2NCH_2CH_2OH + y \; HN(CH_2CH_2OH)_2 + z \; N(CH_2CH_2OH)_3$$

Because of changed consumer demands, a plant switched to reaction conditions using 1600 kg of ethylene oxide and 1560 kg of ammonia. How many moles of each reactant were employed?

I.4. Under the conditions described for the 30:1 ammonia/ethylene oxide mole ratio (problem I.3), how much of each product would be expected for a reaction starting with 1600 kg of the ethylene oxide?

I.5. Ethylene oxide (problem I.3) is produced industrially from pure ethylene and oxygen under catalytic conditions. Ethylene oxide derivatives include ethylene glycol ethers and esters which are used as solvents and polyethylene oxide chains which are used, for example, to stabilize synthetic whipped cream. What is the standard heat of reaction for ethylene oxide production, in kcal mole^{-1} ($H_2C = CH_2$)

Compound	ΔH_f° (kcal mole^{-1})
ethylene	12.5
ethylene oxide	-12.2

*I.6. Mixtures of air and ethylene oxide can be explosive if accidentally ignited. This presents a special safety problem for industries engaged in the production or use of ethylene oxide. What product gas volume, at 1300°C and 860 torr, would result from the explosive complete combustion of 23 kg of ethylene oxide?

*I.7. The oxo process is a method of oxygenating an olefin (alkene) and increasing its chain length by one carbon atom. Thus, ethylene yields propionaldehyde, propylene yields butyaldehyde and 1-nonene yields decyl alcohols upon reduction of the intermediate aldehydes. Most of the oxo aldehydes are converted directly to primary alcohols. However, some of the aldehydes are oxidized to the corresponding carboxylic acids. The principal use of the oxo process has been in the production of alcohols containing 8 to 16 carbon atoms, which are very useful in the manufacture of detergents such as the alkyl sulfonates. The major oxo process users are petroleum refiners and chemical companies with petrochemical experience and facilities. The raw materials, olefins and sythesis gas, which is made by reforming refinery gases, are readily available in petroleum refineries or related industries. Also, the oxo process is adapted to large-scale processing. The oxo plants range in size from 10 million to 100 million kilograms of aldehydes and alcohols annually. The preparation of butyraldehyde by the oxo process is illustrated by the following equation:

$$CH_3-CH=CH_2 + CO + H_2 \xrightarrow{\text{(catalyst, heat, pressure)}} CH_3-CH_2-CH_2-CHO$$

What is the amount of olefin that would be needed to obtain 2500 grams butyraldehyde if the reaction has an efficiency of 70%?

*I.8. Melamine is used industrially with formaldehyde to produce a thermosetting resin. The colorless resins are stable to heat, moisture, and light. The largest use is in tableware, but applications also include decorative laminates, adhesives in speciality chipboard, and certain automobile surface coatings. The products of complete combustion of a 1.785 g of <u>melamine</u> were 1.870 g of CO_2 and 0.771 g of H_2O. It is known that <u>melamine</u> also contains nitrogen and has a molecular weight of 126. What are the empirical and the molecular formulas of <u>melamine</u>?

I.9. Anthraquinone vat dyes are remarkable in their properties of fastness and permanency and are therefore a large portion of the total dye production. One of the steps in production of these dyes is the halogenation of β-anthraquinone-sulfonate. In this process four moles of NaCl are produced for every three moles of β-anthraquinone-sulfonate reacted. How many liters of 0.56 M NaCl solution will be made from a batch charge of 23 liters of 1.38 M β-anthraquinone-sulfonate, if the process is 87% efficient?

I.10. Benzaldehyde is used as a flavoring agent, as an ingredient in certain pharmaceuticals, and as an intermediate in many chemical syntheses. The refined, chlorine-free grade of benzaldehyde is produced by the direct vapor-phase oxidation of toluene.

$$C_6H_5CH_{3(g)} + O_{2(g)} \rightarrow C_6H_5CHO_{(g)} + H_2O_{(g)}$$

The standard heat of reaction is -79.4 kcal $mole^{-1}$ (benzaldehyde). The standard heat of formation of toluene is 12.0 kcal $mole^{-1}$. What is the standard heat of formation of gaseous benzaldehyde? (For other data, see Appendix F.)

ANSWERS:

(I.1.) (I.2.) 1.2×10^6 liters, (I.3.)

3.6×10^4 moles, 9.2×10^4 moles, (I.4.) 2200 kg of monoethanolamine, with only trace amounts of other products, (I.5.) -24.7 kcal/mole, (I.6.) 2.4×10^5 liters, (I.7.) 2.1×10^3 g, (I.8.) CH_2N_2, $C_3H_6N_6$, (I.9.) 66 liters, (I.10.) -9.6 kcal $mole^{-1}$

RELEVANT
PROBLEMS

Unit 9: Organic Chemistry

Set E: Environmental Sciences

E.1. Dioxane is a solvent that has been widely used in the chemical industry, especially in the manufacture of cosmetics, glues and deodorants. Dioxane has recently been identified as a carcinogen, so its industrial use has been sharply curtailed. Dioxane is 9.15% H, 54.53% C and 36.32% O. The formula weight is 88. What are the empirical and the molecular formulas?

E.2. How many moles of dioxane (problem E.1) correspond to 2300 kg?

E.3. Vehicle exhaust, particularly from automobiles, appears to be the major source of atmospheric aldehydes. However, significant amounts may also be produced from other combustion sources such as open burning and incineration of solid waste materials. Industrial plants that manufacture formaldehyde may be local sources of aldehyde pollution. However, the major amount of aldehyde pollution in most areas of the United States is the photochemical reaction between nitrogen oxides and olefinic hydrocarbons. Among the various aldehydes found in photochemical smog, one of the most irritating is acrolein ($H_2C=C-C=O$). One mechanism proposed for its formation is:
$$\begin{array}{cc} & H\ H \end{array}$$

$$NO_2 \xrightarrow{\text{(light)}} NO + O$$

$$O + O_2 \rightarrow O_3$$

$$2\ H_2C=CHCH_3 + O_3 \rightarrow 2\ H_2C=C-C=O + 2\ H_2O$$
$$\ H\ H$$

What mass of acrolein could be formed by the photochemical conversion of 150 g of propylene ($H_2C=CHCH_3$)?

E.4. Ozone itself is a significant air pollutant. What volume of ozone, at 47°C and a partial pressure of 38 torr, would be consumed in the photochemical conversion of 150 g of propylene (problem E.3)?

E.5. Aromatic hydroxyl compounds, phenols, are serious water pollutants. At very low concentrations they are toxic to a broad spectrum of bacteria, including many which are essential components of natural biodegradation systems. At high concentrations they are toxic to most living organisms. Many phenols impart undesirable odor or taste to water, even in very dilute solution. The peculiar flavor of p-chlorophenol, for example, is detectable at a level of 1 part per ten billion. A number of industrial concerns dump phenolic residues into their aqueous wastes. Insecticide, resin, and "cattle dip" manufacturers, along with the petrochemical, explosives, and photographic industries, typically produce large amounts of phenolic wastes. Removal of phenolic products in waste water treatment plants usually requires their oxidation by air or by chlorine. Bacterial oxidations are often ineffective because of the bacteriocidal properities of many phenols. What is the molarity of a solution that is 530 mg liter^{-1} in phenol (⬡-OH)?

*E.6. To remove certain phenols from waste water (problem E.5), such as some of the chlorophenols, a chemical oxidizing agent must be added.

$$[Cl-\bigcirc-OH + Cr_2O_7^{2-} + H^+ \rightarrow Cr^{3+} + CO_2 + H_2O + Cl_2]$$

Balance the equation for the oxidation of p-chlorophenol by dichromate.

*E.7. What molarity of dichromate solution will be required to oxidize 7520 liters of 1.37 M p-chlorophenol (problem E.6) if 6380 liters of dichromate solution are to be used?

*E.8. Aldehydes themselves may undergo photochemical reactions. They may produce, at low partial pressure in the presence of nitrogen oxides, other products such as carbon monoxide, nitrates, ozone, and alkyl hydroperoxides. Formaldehyde ($H_2C=O$) is a common eye-irritant in photochemical smogs. It is also associated with the production of formic acid, an even more irritating compound (HCO_2H). What concentration of formic acid (in ppm) would result from the conversion of the formaldehyde in a smog layer containing 38 mg H_2CO per kg of smog if the process is 77% efficient.

*E.9. The environmental problem of excess heat is always important. The rise of temperature by a few degrees can destroy a carefully balanced ecological cycle. Industrial "heat exchangers" often transfer deleterious amounts of heat to the environment. Process control by heat exchanger is especially important with highly exothermic reactions. During the manufacture of ethylene oxide from ethylene and oxygen, a side reaction forming CO_2 and water vapor also occurs and is highly exothermic. The standard heat of this reaction is -316.1 kcal mole^{-1} (C_2H_2). What is the standard heat of formation of ethylene? How much heat would be generated by 1200 kg of ethylene involved in the side reaction?

*E.10. Benzene is a raw material of major importance. It is used in the manufacture of many dyes, polyester fibers, recording tapes, tire cords, and - in fact - almost all chemicals that include a benzene ring. The drug industry is very dependent on benzene as it is a raw material for the production of acetylsalicylic acid (aspirin), sulfanilamide, and many of the other sulfa drugs. Reports in 1977 link benzene vapor inhalation with leukemia. Persons who work regularly in areas of moderate benzene vapor concentrations were found to have higher incidence of leukemia than the normal population. An air level of benzene contamination based on these recent reports will be significantly less than the pre-1977 tolerance limit of 25 ppm. What is the concentration, in ppm, of benzene in a sealed industrial work area containing 45 mmol of benzene in a total air volume of 1.2×10^5 liters, if the air density is 0.61 g liter^{-1}?

ANSWERS:

(E.1.) C_2H_4O, $C_4H_8O_2$, (E.2.) 2.6×10^{-4} moles, (E.3.) 200 g, (E.4.) 937 liters, (E.5.) 5.6×10^{-3} $\underline{M}$, (E.6.) $2\ Cl\text{-}\langle\bigcirc\rangle\text{-}OH + 9\ Cr_2O_7^{2-} + 72\ H^+ \rightarrow 12\ CO_2 + Cl_2 + 18\ Cr^{3+} + 41\ H_2O$,

(E.7.) 7.27 $\underline{M}$, (E.8.) 45 ppm, (E.9.) 12.5 kcal mole^{-1}, 1.4×10^7 kcal, (E.10.) 48 ppm

EQUILIBRIUM

UNIT 10: HETEROGENEOUS EQUILIBRIA

When we were dealing with stoichiometric calculations, a chemical process could be described quite adequately by a simple chemical equation showing the conversion of original reactants to final products. The addition of a little strontium carbonate to water, for example, could be formulated by a "net" equation (Unit 2), showing the principal species present, as determined by careful chemical analysis:

$$2 \; SrCO_{3(s)} + H_2O_{(\ell)} \rightarrow 2 \; Sr^{2+}_{(aq)} + CO_3^{2-}{}_{(aq)} + HCO_3^-{}_{(aq)} + OH^-_{(aq)}$$

We know, of course, that many reactions yield less than 100% of the theoretically-calculated products, so we would not be too surprised to find that a solution prepared by adding 0.10 mole of $SrCO_3$ to water contained, after reaction, <u>less</u> than 0.10 mole of bicarbonate ion (HCO_3^-). Since we have seen examples of the use of an excess of one reactant to improve product yield, we would not be surprised either to find that mixing 0.10 mole of $SrCO_3$ with 10 moles of water produced more HCO_3^- than was obtained from 0.10 mole <u>each</u> of $SrCO_3$ and H_2O.

Now let's look at a system prepared by adding 0.10 mole of $SrCO_3$ to 10 moles of water, sealing the system, and maintaining it at a constant temperature of $18^{\circ}C$. If we just consider the <u>stoichiometry</u>, as described by the equation given earlier, we <u>will</u> find some surprises! When all reaction has appeared to stop:

1. *The system contains about 1.3×10^{-5} mole of Sr^{2+}. (That's only <u>0.013%</u> of the theoretical amount!)*

2. *Instead of the expected equimolar amounts of CO_3^{2-}, HCO_3^-, and OH^- (as suggested by the equation), we find about 10^{-5} mole each of HCO_3^- and OH^- and only 3×10^{-6} mole of CO_3^{2-}.*

3. *Most of the $SrCO_3$ added is "just sitting there" and, unlike the addition of more water, an addition of more $SrCO_3$ produces <u>no more</u> products than we already had.*

Evidentally, some chemical systems are a bit more complicated than we might have expected simply from the aspects of stoichiometry we have studied thus far. The addition of strontium carbonate to water quickly forms a <u>saturated</u> <u>solution</u>, i.e., we reach a point at which further addition of $SrCO_3$ simply adds more solid to the heterogeneous mixture without changing the concentrations of species in the solution. This is one characteristic of an <u>equilibrium</u> <u>state</u>.

EQUILIBRIUM IS CHARACTERIZED BY THE CONSTANT COMPOSITION OF A CLOSED SYSTEM, AT CONSTANT TEMPERATURE.

If we had approached this equilibrium state in another way, by mixing solutions of Sr^{2+}, CO_3^{2-}, HCO_3^- and OH^- (in proper ratios) until a precipitate had formed, we would have found the same final composition of the system. Could it be that both the reactions "magically" stop at a certain point?

We can examine this possibility by an experiment using radioactive strontium-90. If solid $SrCO_3$ containing some radioactive $^{90}Sr^{2+}$ is added to a saturated aqueous solution of strontium carbonate it does not change the concentrations of Sr^{2+}, CO_3^{2-}, HCO_3^-, or OH^- in the solution. Initially, all the radioactivity is detected only in the undissolved $SrCO_3$. However, as time passes, radioactivity is detected in the solution itself. Evidently radioactive $^{90}Sr^{2+}$ ions are escaping from the solid into the solution. Since the concentration of strontium ion remains constant, some non-radioactive Sr^{2+} must be leaving the solution to replace $^{90}Sr^{2+}$ lost from the solid. In the saturated solution, an equilibrium system, the two "opposing" processes must be occurring at the same rate. Thus, the equilibrium state is dynamic.

CHEMICAL EQUILIBRIUM IS CHARACTERIZED BY EQUAL RATES OF OPPOSING REACTIONS WITHIN A CLOSED SYSTEM.

At equilibrium (and only for that state) we can represent a system by a single equation, using "double arrows" to indicate that both "forward" and "reverse" reactions are still going on, but at equal rates. The equilibrium system described as "saturated aqueous strontium carbonate" is then represented by either

$$2\ SrCO_{3(s)} + H_2O_{(\ell)} \rightleftarrows 2\ Sr^{2+}_{(aq)} + CO^{2-}_{3(aq)} + HCO^-_{3(aq)} + OH^-_{(aq)}$$

or

$$2\ Sr^{2+}_{(aq)} + CO^{2-}_{3(aq)} + HCO^-_{3(aq)} + OH^-_{(aq)} \rightleftarrows 2\ SrCO_{3(s)} + H_2O_{(\ell)}$$

Since both reactions (dissolving of $SrCO_3$ and precipitation of $SrCO_3$) are going on at the same time, the terms "reactants" and "products" as used with reference to a single reaction are inappropriate to the description of an equilibrium system. It is more meaningful to use "right hand species" and "left hand species", referring to a particular equilibrium equation.

The constant composition typical of an equilibrium system may be expressed, for mathematical purposes, by an equilibrium constant, related to the composition of the system through a conventional expression determined by the chemical equation written for the process. Without going into the derivation of such expressions, which we will leave to more advanced texts, we can point out the conventions used (and some simplified approximations) and illustrate some applications of equilibrium constants to the mathematical analysis of chemical equilibria.

Basic Conventions

1. The equilibrium constant is defined as the ratio of activities[1] of "right hand species" to "left hand species", each activity having the exponent shown as the coefficient of that species in the chemical equation. (As approximations of "activity", we will use pressure, in atm, for gases and molarity for species in solution.)

2. Pure solids or pure liquids have unit activities, i.e., their activities have the value of 1.

3. The chemical equation is written in the form most convenient for the process being described, or consistent with tabulated equilibrium constants. For example, the description of a saturated solution of silver chloride would be expressed as:

$$AgCl_{(s)} \rightleftharpoons Ag^+_{(aq)} + Cl^-_{(aq)}$$

In this unit we will deal with a particular type of heterogeneous equilibrium system, the saturated aqueous solution. For such a solution the equilibrium constant normally used is called the solubility product and symbolized as K_{sp}. An example will illustrate applications of solubility product expressions. Other examples will be given in the METHODS sections which follow.

Finding the K_{sp}-Value from an Experimental Solubility

Reference texts of chemical data, such as the Chemical Rubber Company's Handbook of Chemistry and Physics, typically report the solubility of a salt in grams (salt) per 100 ml (water). Since we wish to use molar concentrations, we must apply appropriate unity factor conversions (Unit 6). For example, suppose we wanted to calculate the approximate solubility product for silver sulfide [Ag_2S, exp. sol. = 2×10^{-5} g/100 ml (H_2O)]. A stepwise procedure might involve:

Step 1 - Writing the conventional chemical equation for the equilibrium system of saturated aqueous silver sulfide:

$$Ag_2S_{(s)} \rightleftharpoons 2Ag^+_{(aq)} + S^{2-}_{(aq)}$$

Step 2 - Writing the approximate K_{sp} - expression[2] (i.e., using molar concentrations, symbolized by C, in lieu of thermodynamic activities):

$$K_{sp} = C^2_{Ag^+} \times C_{S^{2-}}$$

---

[1] The "activity" of a chemical species is a thermodynamic term reasonably approximated for dilute aqueous solutions by molarity.

[2] Note: Since silver sulfide is a solid, its activity is unity, so:

$$K_{sp} = \frac{C^2_{Ag+} \times C_{S^{2-}}}{C_{Ag_2S}} = \frac{C^2_{Ag+} \times C_{S^{2-}}}{1} = C^2_{Ag+} \times C_{S^{2-}}$$

Step 3 - Evaluating molar concentration terms from solubility data[3] and mole ratios:

$$C_{Ag+} = \frac{2 \times 10^{-5} \text{ g(Ag}_2\text{S)}}{100 \text{ ml(soln)}} \times \frac{1 \text{ mole(Ag}_2\text{S)}}{248 \text{ g(Ag}_2\text{S)}} \times \frac{2 \text{ moles(Ag}^+)}{1 \text{ mole(Ag}_2\text{S)}} \times \frac{1000 \text{ ml}}{1 \text{ liter}}$$

$$C_{Ag+} = 1.6 \times 10^{-6}$$

$$C_S2- = \frac{2 \times 10^{-5} \text{ g(Ag}_2\text{S)}}{100 \text{ ml(soln)}} \times \frac{1 \text{ mole(Ag}_2\text{S)}}{248 \text{ g(Ag}_2\text{S)}} \times \frac{1 \text{ mole(S}^{2-})}{1 \text{ mole(Ag}_2\text{S)}} \times \frac{1000 \text{ ml}}{1 \text{ liter}}$$

$$C_S2- = 8 \times 10^{-7} \underline{M}$$

Step 4 - Evaluating K_{sp} from molarities:

$$K_{sp} = C_{Ag+}^2 \times C_S2- = (1.6 \times 10^{-6})^2 \times (8 \times 10^{-7})$$

$$K_{sp} = 2 \times 10^{-18}$$

(Since the use of thermodynamic activities would yield a "dimensionless" K-value, we shall omit units for our approximate constants, remembering that all concentrations used are in moles liter^{-1}.)

An investigation of solubility equilibria will provide a useful introduction to the general area of chemical equilibrium, a subject of considerable practical importance.

- References -

1. O'Connor, Rod. 1977. Fundamentals of Chemistry, second edition. New York: Harper & Row (Units 17, 18 and Excursion 2)

2. Brown, Theodore, and H. Eugene LeMay, Jr. 1977. Chemistry: The Central Science. Englewood Cliffs, N.J.: Prentice-Hall (Chapters 12, 14)

3. Masterton, William, and Emil Slowinski. 1977. Chemical Principles, fourth edition. Philadelphia: W.B. Saunders (Chapters 12, 18)

4. Nebergall, William, F.C. Schmidt, and H.F. Holtzclaw, Jr. 1976. College Chemistry, fifth edition. Lexington, MA: D.C. Heath (Chapters 13, 17, 19)

5. Hamm, Randall, and Carl Nyman. 1968. Chemical Equilibrium. Lexington, MA: Heath/Raytheon.

OBJECTIVES:

(1) Given the name or formula of a salt, be able to write the solubility product expression for a saturated aqueous solution of the salt.

---

[3] For very dilute solutions, ml(H_2O) = ml(solution).

(2) *Given the concentrations and volumes of two reactant solutions, be able to use tabulated K_{sp}-values (Appendix D) to predict whether or not a precipitate will form when the solutions are mixed and, if so, the chemical composition of the precipitate.*

*(3) *Given the description of a system involving a sparingly-soluble salt, be able to use tabulated K_{sp}-values (Appendix D) to calculate solubility of the salt.*

PRE-TEST:

```
Necessary Solubility Products:   (at 20 - 25°C)

BaSO₄     7.9 x 10⁻¹¹

CuS       8.0 x 10⁻³⁷

Mg(OH)₂   1.2 x 10⁻¹¹
```

Rendered as equations:

Necessary Solubility Products: (at 20 - 25°C)

$BaSO_4$ 7.9×10^{-11}

CuS 8.0×10^{-37}

$Mg(OH)_2$ 1.2×10^{-11}

(1) Most ions of the "heavy" metals are toxic to humans in moderate concentrations, often because of their action in damaging vital enzyme systems. All of the following salts are used in situations in which a soluble salt could furnish a toxic concentration of the metal ion, yet normal use conditions are safe because of the limited solubilities of the salts. For each salt, write the conventional solubility product expression.

(a) barium sulfate (K_{sp} = _____)
[Used for x-ray examinations of the gastrointestinal tract as an aqueous slurry of the salt.]

(b) mercury(I) chloride (K_{sp} - _____)
[*Under the trade name of* calomel, *used in medicine as a stimulant to bile secretion.*]

(Hint: Remember that the mercury(I) ion is the diatomic Hg_2^{2+}.)

(c) $Cu_2[Fe(CN)_6]$ (K_{sp} = _____)
[*Used in preparing synthetic membranes for water purification or desaliniza-tion by reverse osmosis. Note that both the copper(II) and hexacyanoferrate ions would be toxic in moderate concentrations.*]

(2) An environmental protection group surveying water contamination from industrial plants found that a metal plating industry was dumping waste copper(II) sulfate into a small local river. A few miles away, a paint manufacturer was dumping waste barium sulfide into another stream. These two streams eventually met to form part of a larger river system. If equal volumes of the two streams met at the junction, would a precipitate form when the $CuSO_4$ content of one stream was 2.0×10^{-5} M and the BaS content of the other was 4.0×10^{-8} M? If so, what

would be the composition of the precipitate?

*(3)   Americans consume an enormous quantity of "antacids" each year. Brand names such as TUMS, ROLAIDS, or ALKA-SELTZER are a familiar part of our national vocabulary. The production of chemicals for antacid preparations is a "big-money" business. Among such compounds are aluminum, magnesium, and calcium salts. Hydroxides, carbonates, and oxides are the most common anions employed for "neutralizing excess stomach acids". A particular company produces magnesium hydroxide for use in a well-known antacid preparation. The solid $Mg(OH)_2$, from precipitation by sodium hydroxide, must be washed thoroughly to remove residual traces of "lye", but not so thoroughly as to dissolve an excessive amount of the product. What maximum percentage of a 50.0 g $Mg(OH)_2$ precipitate might be lost by washing with 10.0 liters of water at 20-25°C?

Answers and Directions:

(1)   (a) $K_{sp} = C_{Ba^{2+}} \times C_{SO_4^{2-}}$

(b) $K_{sp} = C_{Hg_2^{2+}} \times C_{Cl^-}^2$

(c) $K_{sp} = C_{Cu^{2+}}^2 \times C_{[Fe(CN)_6]^{4-}}$

(2)   precipitate will form, CuS

*If all are correct, go on to question *3. If you missed any, study METHODS, sections 10.1 and 10.2.*

*(3)   0.17%

If your answer was correct, go on to RELEVANT PROBLEMS, Unit 10. If you missed it, study METHODS, section 10.3.

METHODS

10.1 Writing Solubility Product Expressions

Learning to write K_{sp}-expressions is simply a matter of becoming familiar with some standard conventions. A knowledge of common chemical names and formulas is assumed, but you can always find a formula, if you're uncertain, by looking up the name in some reference text, such as the Chemical Rubber Company's Handbook of Chemistry and Physics.

First of all, the chemical equation should be written as a description of the equilibrium system represented by the saturated solution of the salt. Conventionally, such an equation shows the complete formula of the solid salt at the left and the formulas or symbols of the ions formed when the salt dissolves at the right, with the coefficients required for a balanced equation. Double arrows are used to indicate an equilibrium condition. For example:

for lead iodide _____

$$PbI_{2(s)} \rightleftharpoons Pb^{2+}_{(aq)} + 2I^-_{(aq)}$$

for calcium phosphate _____

$$Ca_3(PO_4)_{2(s)} \rightleftharpoons 3Ca^{2+}_{(aq)} + 2\ PO_4^{3-}_{(aq)}$$

Once the proper equation is available, the solubility product expression (using the approximation of molar concentrations, represented by C) will be the product of the concentrations of the ions shown, with each concentration having an exponent equal to the coefficient of that ion in the balanced equation. For example:

for lead iodide _____

$$PbI_{2(s)} \rightleftharpoons Pb^{2+}_{(aq)} + \textcircled{2}I^-_{(aq)}$$

$$K_{sp} = C_{Pb^{2+}} \times C^2_{I^-}$$

for calcium phosphate _____

$$Ca_3(PO_4)_{2(s)} \rightleftharpoons \textcircled{3}Ca^{2+}_{(aq)} + \textcircled{2}PO_4^{3-}_{(aq)}$$

$$K_{sp} = C^3_{Ca^{2+}} \times C^2_{PO_4^{3-}}$$

EXAMPLE 1

The SOLUTION to Pre-Test question 1:

(a) for barium sulfate _____

equation: $BaSO_{4(s)} \rightleftharpoons Ba^{2+}_{(aq)} + SO_4^{2-}_{(aq)}$

from which: $K_{sp} = C_{Ba^{2+}} \times C_{SO_4^{2-}}$

(b) for mercury(I) chloride _____

equation: $Hg_2Cl_{2(s)} \rightleftharpoons Hg_2^{2+}_{(aq)} + \textcircled{2}Cl^-_{(aq)}$

from which: $K_{sp} = C_{Hg_2^{2+}} \times C^2_{Cl^-}$

(c) for $Cu_2[Fe(CN)_6]$

equation: $Cu_2[Fe(CN)_6]_{(s)} \rightleftharpoons \textcircled{2}Cu^{2+}_{(aq)} + [Fe(CN)_6]^{4-}$

from which: $K_{sp} = C^2_{Cu^{2+}} \times C_{[Fe(CN)_6]^{4-}}$

EXERCISE 1

"Washing soda", Na_2CO_3, can be used to precipitate most of the di- and tripositive ions responsible for water "hardness". Calcium ions, for example, form insoluble carbonates. In other cases, such as with Fe^{3+}, it is the hydroxide which is precipitated, using OH^- produced by the reaction of carbonate with water:

$$CO_{3(aq)}^{2-} + H_2O_{(\ell)} \rightleftarrows HCO_{3(aq)}^- + OH_{(aq)}^-$$

Write conventional solubility product expressions for calcium carbonate and iron(III) hydroxide.

(answers, page 248)

_ _

10.2 Predicting Precipitation

Since a saturated solution contains the <u>maximum</u>[4] concentrations of dissolved ions, the solubility product is the largest number that should be obtained from the products of concentration terms, as given by the appropriate solubility product expression. Any number exceeding the K_{sp} value for a hypothetical set of ion concentrations indicates that the solution for which the calculation was made is unstable with respect to ion concentrations and precipitation should occur, continuing until concentrations are reduced to those consistent with the K_{sp}-value.

In predicting precipitation for a case in which two different solutions are mixed, it is important to recognize that the mixing process <u>dilutes</u> both solutions. Thus, concentrations to be used for the ion-product calculation must be those resulting from dilution, rather than the original concentrations of the isolated solutions. An appropriate "concentration correction" can be made by multiplying the original concentration by the ratio of its volume to the total volume of the mixed solutions.

Once the proper ("diluted") concentrations are known, these are simply "plugged into" the appropriate ion-product expression for the salt being considered as a possible precipitate. If the ion-product calculated exceeds the K_{sp}-value for the salt, then a precipitate of that composition is expected. If, on the other hand, the ion-product is less than or equal[5] to K_{sp}, no precipitation should occur.

EXAMPLE 2

The <u>SOLUTION</u> to Pre-Test question 2:

A precipitate could occur only as the result of an exchange of ions between the two

---

[4] A condition of transient stability, called "supersaturation", may sometimes occur, but any disturbance is likely to induce rapid precipitation until concentrations are reduced to those of the saturated solution.
[5] An ion-product <u>equal</u> to the K_{sp} corresponds to the stable saturated solution.

salts mixed, since the original salts themselves are obviously soluble at the concentrations given. Thus, we wish to consider the possibility of forming saturated solutions (by precipitation) of CuS and $BaSO_4$.

For these salts (using K_{sp}-values given):

$$K_{sp}(CuS) = C_{Cu^{2+}} \times C_{S^{2-}} = 8.0 \times 10^{-37}$$

$$K_{sp}(BaSO_4) = C_{Ba^{2+}} \times C_{SO_4^{2-}} = 7.9 \times 10^{-11}$$

Ion concentrations, <u>after mixing</u> of equal volumes of the two solutions, are:

$$C_{Cu^{2+}} = 2 \times 10^{-5}\underline{M} \times \frac{1\ vol(Cu^{2+})}{2\ vol\ (total)} = 1 \times 10^{-5}\underline{M}$$

$$C_{SO_4^{2-}} = 2 \times 10^{-5}\underline{M} \times \frac{1\ vol(SO_4^{2-})}{2\ vol\ (total)} = 1 \times 10^{-5}\underline{M}$$

$$C_{Ba^{2+}} = 4 \times 10^{-8}\underline{M} \times \frac{1\ vol(Ba^{2+})}{2\ vol(total)} = 2 \times 10^{-8}\underline{M}$$

$$C_{S^{2-}} = 4 \times 10^{-8}\underline{M} \times \frac{1\ vol(S^{2-})}{2\ vol(total)} = 2 \times 10^{-8}\underline{M}$$

Then the hypothetical ion-products would be:

for CuS, $C_{Cu^{2+}} \times C_{S^{2-}} = (1 \times 10^{-5})(2 \times 10^{-8}) = 2 \times 10^{-13}$

[This exceeds $K_{sp}(8 \times 10^{-37})$, so CuS will precipitate.]

for $BaSO_4$, $C_{Ba^{2+}} \times C_{SO_4^{2-}} = (2 \times 10^{-8})(1 \times 10^{-5}) = 2 \times 10^{-13}$

[This is less than $K_{sp}(7.9 \times 10^{-11})$, so $BaSO_4$ will <u>not</u> precipitate.]

- -

EXERCISE 2

Because of its action as both an oxidizing agent and a strong acid, sulfuric acid is quite destructive to organic matter, such as hair and grease. The high density of the concentrated acid ($1.84\ g\ ml^{-1}$) adds to its utility as a "drain cleaner". The dense acid, when poured through standing water in a clogged sink trap, quickly sinks to the level of the blocking material, where it begins to attack any hair or grease. The extensive use of H_2SO_4 "drain cleaners" has resulted in the addition of sulfate ion in increasing concentrations to waste effluent streams. If a waste effluent stream from a group of vacation cottages (containing 3×10^{-3} mole liter^{-1} of SO_4^{2-}) runs into a mineral pool basin in which the dissolved minerals provide $8 \times 10^{-2}\underline{M}\ Ca^{2+}$, $7 \times 10^{-6}\underline{M}\ Sr^{2+}$, and $4 \times 10^{-8}\underline{M}\ Ba^{2+}$, will any precipitate form when the effluent volume equals 1/4 the mineral water volume? If so, what is the composition of the precipitate?

(answer, page 248)

[For K_{sp}-values, see Appendix D.]

Extra Practice

EXERCISE 3

Evidence suggests that fluoride ion, in a proper concentration range, is effective in reducing tooth decay, possibly by the deactivation of certain vital enzymes in oral bacteria by fluoride ion. This evidence is the basis for water fluoridation or the use of fluoride-containing toothpastes. When fluoride solutions come in contact with the teeth, some of the F^- ion is incorporated into the crystal lattice of the calcium salt (mainly calcium phosphate) which constitutes the tooth structure. Write the conventional solubility product expressions for two salts which may be formed by reaction of calcium phosphate with fluoride ion, CaF_2 and $Ca_2F(PO_4)$.

(answer, page 249)

EXERCISE 4

When fluoride is added to a municipal water supply, the process may involve addition of the amount of sodium fluoride calculated for producing the desired F^- concentration in the water reservoir being fluoridated. In cases where the water is quite "hard" (e.g., high Ca^{2+} concentration), it is necessary to consider possible fluoride loss by formation of insoluble fluorides. A stock solution of 0.10 $\underline{M}$ NaF was prepared for addition of 4.00 liters to a 4000 liter water supply which was $3 \times 10^{-4}\underline{M}$ in Ca^{2+} and $5 \times 10^{-6}\underline{M}$ in Mg^{2+}. Would any loss of F^- by precipitation be expected? If so, what would be the composition of the precipitate?

(answer, page 249)

[See Appendix D for K_{sp}-values.]

At this point you should try the competency level Self-Test questions on page 247 .

----

*10.3 Finding Solubilities from K_{sp}-Values

The use of molar concentrations in our approximate solubility product expressions enables us to use the K_{sp}-values to find the solubility of a salt in mole liter^{-1}. Once this has been established, it is a simple matter to determine the mass of a salt present in any particular volume of the saturated solution, using the familiar mass/mole unity factor (Unit 3).

*Proficiency Level

The key to finding the <u>molar</u> solubility lies in the recognition of the mole relation-ships among the salt and its aqueous ions as expressed by the chemical equation for the solubility equilibrium. Consider, for example, the case of a saturated solution of calcium phosphate as described by the equilibrium equation:

$$Ca_3(PO_4)_{(s)} \rightleftarrows 3Ca^{2+} + 2PO_4^{3-}$$

This equation tells us that for each mole of the salt <u>dissolved</u>, three moles of Ca^{2+} ion and two moles of PO_4^{3-} ion are produced. If we express the molar solubility (i.e., moles per liter of calcium phosphate dissolved) by S, then the molarities of the respective ions are given by:

$$C_{Ca^{2+}} = \frac{S[Ca_3(PO_4)]_2}{1} \times \frac{3[Ca^{2+}]}{1[Ca_3(PO_4)]_2} = 3S$$

$$C_{PO_4^{3-}} = \frac{S[Ca_3(PO_4)_2]}{1} \times \frac{2[PO_4^{3-}]}{1[Ca_3(PO_4)_2]} = 2S$$

Then a mathematical relationship can be established by use of the conventional K_{sp}-expres-sion:

$$K_{sp} [Ca_3(PO_4)_2] = C_{Ca^{2+}}^3 \times C_{PO_4^{3-}}^2 = (3S)^3 \times (2S)^2 = 108S^5$$

from which the <u>molar</u> solubility of the salt may be calculated:

$$S = \sqrt[5]{\frac{K_{sp}}{108}}$$

This approach is a general one and may be used, with appropriate mole ratio analysis, to find the solubility in water of any salt in mole liter^{-1} if the K_{sp}-value is known. It should be noted that the relationship among the component species of a solubility equili-brium can be found from the coefficients in the equilibrium equation <u>only</u> when the sole source of the ions involved is the salt itself. For a saturated solution of $Ca_3(PO_4)_2$ in 1.0 $\underline{M}$ Na_3PO_4, for example, the phosphate concentration at equilibrium would be <u>much</u> greater than that furnished just by the dissolved calcium phosphate. In addition, the approach described contains several approximations, not the least of which is the assumption that the only ions in solution are the same as those of the original salt. A more accurate treatment of solubility equilibria, including the reaction of ions from the salt with water or other species, is described in Unit 13.

The simple treatment discussed in this section is still a useful approximation for a number of practical situations.

*EXAMPLE 3

The SOLUTION to Pre-Test question 3:

For $Mg(OH)_2$ the solubility equilibrium is described by:

$$Mg(OH)_{2(s)} \rightleftarrows Mg^{2+}_{(aq)} + 2\ OH^-_{(aq)}$$

If we represent the molar solubility of $Mg(OH)_2$ by S, then:

$$C_{Mg^{2+}} = \frac{S[Mg(OH)_2]}{1} \times \frac{1[Mg^{2+}]}{1\cdot[Mg(OH)_2]} = S$$

$$C_{OH^-} = \frac{S[Mg(OH)_2]}{1} \times \frac{2[OH^-]}{1[Mg(OH)_2]} = 2S$$

Then:

$$K_{sp} = C_{Mg^{2+}} \times C^2_{OH^-} = (S) \times (2S)^2 = 4S^3$$

From which (using $K_{sp} = 1.2 \times 10^{-11}$, Appendix D):

$$S = \sqrt[3]{\frac{K_{sp}}{4}} = \sqrt[3]{\frac{1.2 \times 10^{-11}}{4}} = 1.44 \times 10^{-4} \text{ mole liter}^{-1}$$

The maximum loss by washing would be that from formation of the saturated solution, for which:

$$\frac{1.44 \times 10^{-4}\text{mole }[Mg(OH)_2]}{1 \text{ liter [solution]}} \times \frac{10 \text{ liters[solution]}}{1} \times \frac{58.3 \text{ g }[Mg(OH)_2]}{1 \text{ mole }[Mg(OH)_2]} =$$

$$0.084 \text{ g }[Mg(OH)_2]$$

The percentage loss, then, is:

$$\frac{0.084 \text{ g }[Mg(OH_2)] \text{ (lost)}}{50.0 \text{ g }[Mg(OH)_2]\text{(originally)}} \times 100\% = \underline{0.17\%}$$

- -

*EXERCISE 5

Lead poisoning is particularly insidious because it may result from a gradual accumulation of lead ion from an unexpected source. Symptoms may often be mistaken for those of less serious conditions until the lead concentration in the body has produced some permanent damage. Cases of slow poisoning from lead pigments in unglazed ceramic water vessels are not uncommon. How much lead (in mg) would be ingested by a person drinking 1.00 quart (946 ml) of water per day for 4 years from a ceramic water jug whose "chrome-yellow"

(PbCrO$_4$) pigment, unprotected by a glaze, has saturated all water stored in it?

[For the K$_{sp}$-value, see Appendix D]

(answer, page 250)

- -

Extra Practice

*EXERCISE 6

It is not unusual for those living on isolated farms and ranches to dump "junk" equipment somewhere out of sight, rather than to haul it to the closest municipal dumping grounds. Rusted ironware slowly reacts with water to form iron(III) hydroxide. What would be the concentration, in ppm[6](Fe^{3+}), of a ranch pond saturated with Fe(OH)$_3$ from rusted equipment thrown into it over a period of time?

[For the K$_{sp}$-value, see Appendix D]

(answer, page 250)

==

SELF-TEST (UNIT 10)   [answers, page 251]

10.1. A number of inorganic salts exhibit some brilliant colors which, along with their low solubilites, make them valuable commercial pigments for paints or glazes. Write the conventional solubility product expressions for the following common inorganic pigments:

(a) "Cobalt Yellow", K$_3$[Co(NO$_2$)$_2$]

(b) "Cobalt Violet", Co$_3$(PO$_4$)$_2$

(c) "Malachite Green", Cu$_2$(CO$_3$)(OH)$_2$

10.2. A waste effluent stream which is 2 x 10^{-3} M in fluoride, 3 x 10^{-7} M in sulfate, and 5 x 10^{-5} M in hydroxide empties into a natural lake having a calcium ion concentration of 4 x 10^{-3} M. At the stream/lake junction equal volumes of the two solutions mix and a white precipitate forms. What is the composition of the precipitate?

[For K$_{sp}$-values, see Appendix D.]

- -

If you completed Self-Test questions 10.1 and 10.2 correctly, you may go on to the proficiency level, try the RELEVANT PROBLEMS (Unit 10), or stop here. If not, you should consult your instructor for suggestions of further study aids.

- -

---

[6]ppm (part per million) $\simeq$ mg liter^{-1}

*10.3. A commercial pigment known as "Cobalt Yellow" has the scientific name potassium hexanitrocobaltate(III), for which the formula is $K_3[Co(NO_2)_6]$. The solubility product for this salt is 4.3×10^{-6}. The pigment is prepared by precipitation, after which it is washed with water to remove soluble impurities. What maximum percentage of the pigment might be lost by washing 250 g of the salt with 15 liters of water?

If you answered Self-Test question 10.3 correctly, you may go on to the RELEVANT PROBLEMS for Unit 10. If not, you should consult your instructor for suggestions of further study aids.

ANSWERS to EXERCISES, Unit 10

1. $[K_{sp} = C_{Ca^{2+}} \times C_{CO_3^{2-}} \ ; \ K_{sp} = C_{Fe^{3+}} \times C_{OH^-}^3]$ <u>Solution:</u>

   for calcium carbonate _____

   equation: $CaCO_{3(s)} \underset{\leftarrow}{\rightarrow} Ca^{2+}_{(aq)} + CO^{2-}_{3(aq)}$

   for which: $K_{sp} = C_{Ca^{2+}} \times C_{CO_3^{2-}}$

   for iron(III) hydroxide _____

   equation: $Fe(OH)_{3(s)} \underset{\leftarrow}{\rightarrow} Fe^{3+}_{(aq)} + ③ OH^-_{(aq)}$

   from which $K_{sp} = C_{Fe^{3+}} \times C_{OH^-}^3$

2. (Only $CaSO_4$ will be precipitated) <u>Solution:</u>
   From Appendix D:

   $K_{sp} (BaSO_4) = 7.9 \times 10^{-11}$

   $K_{sp} (CaSO_4) = 2.5 \times 10^{-5}$

   $K_{sp} (SrSO_4) = 3.2 \times 10^{-7}$

   Ion concentrations, after mixing of solutions:

   $C_{SO_4^{2-}} = 3 \times 10^{-3} \ \underline{M} \times \dfrac{1 \ \text{part} (SO_4^{2-})}{5 \ \text{parts} (\text{total})} = 6 \times 10^{-4} \ \underline{M}$

   $\hookleftarrow$ [1 part effluent + 4 parts "pool"]

   $C_{Ca^{2+}} = 8 \times 10^{-2} \ \underline{M} \times \dfrac{4 \ \text{parts} (Ca^{2+})}{5 \ \text{parts} (\text{total})} = 6.4 \times 10^{-2} \ \underline{M}$

   $C_{Sr^{2+}} = 7 \times 10^{-6} \ \underline{M} \times \dfrac{4 \ \text{parts} (Sr^{2+})}{5 \ \text{parts} (\text{total})} = 5.6 \times 10^{-6} \ \underline{M}$

$$C_{Ba^{2+}} = 4 \times 10^{-8} \ \underline{M} \times \frac{4 \text{ parts } (Ba^{2+})}{5 \text{ parts (total)}} = 3.2 \times 10^{-8} \ \underline{M}$$

Ion products for hypothetical solutions:

<u>for CaSO$_4$</u>: $C_{Ca^{2+}} \times C_{SO_4^{2-}} = (6.4 \times 10^{-2})(6 \times 10^{-4}) = 4 \times 10^{-5}$

[CaSO$_4$ <u>will</u> precipitate, $(4 \times 10^{-5}) > K_{sp}$.]

<u>for SrSO$_4$</u>: $C_{Sr^{2+}} \times C_{SO_4^{2-}} = (5.6 \times 10^{-6})(6 \times 10^{-4}) = 3 \times 10^{-9}$

[SrSO$_4$ will <u>not</u> precipitate, $(3 \times 10^{-9}) < K_{sp}$.]

<u>for BaSO$_4$</u>: $C_{Ba^{2+}} \times C_{SO_4^{2-}} = (3.2 \times 10^{-8})(6 \times 10^{-4}) = 2 \times 10^{-11}$

[BaSO$_4$ will <u>not</u> precipitate, $(2 \times 10^{-11}) < K_{sp}$.]

3. [K_{sp} for CaF$_2$ = $C_{Ca^{2+}} \times C_F^2$-; K_{sp} for Ca$_2$F(PO$_4$) = $C^2_{Ca^{2+}} \times C_{F^-} \times C_{PO_4^{3-}}$]

<u>Solution</u>:

for CaF$_2$ _____

equation: $CaF_{2(s)} \rightleftharpoons Ca^{2+}_{(aq)} + \textcircled{2} \ F^-_{(aq)}$

for which: $K_{sp} = C_{Ca^{2+}} \times C_{F^-}^2$

for Ca$_2$F(PO$_4$) _____

equation: $Ca_2F(PO_4)_{(s)} \rightleftharpoons \textcircled{2} \ Ca^{2+} + F^- + PO_4^{3-}$

for which: $K_{sp} = C^2_{Ca^{2+}} \times C_{F^-}^2 \times C_{PO_4^{3-}}$

4. (no fluorides precipitate) <u>Solution</u>:

From Appendix D:

$$K_{sp}(CaF_2) = 1.7 \times 10^{-10}$$

$$K_{sp}(MgF_2) = 6.8 \times 10^{-9}$$

Ion concentrations, after mixing of solutions:

$$C_{F^-} = 0.10 \ \underline{M} \times \frac{4 \text{ liters } (F^-)}{4001 \text{ liters(total)}} = 1.0 \times 10^{-4} \ \underline{M}$$

$$C_{Ca^{2+}} = 3 \times 10^{-4} \ \underline{M} \times \frac{1000 \text{ liters } (Ca^{2+})}{1001 \text{ liters(total)}} = 3 \times 10^{-4} \ \underline{M}$$

$$C_{Mg^{2+}} = 5 \times 10^{-6} \ \underline{M} \times \frac{1000 \text{ liters } (Mg^{2+})}{1001 \text{ liters(total)}} = 5 \times 10^{-6} \ \underline{M}$$

Ion products for hypothetical solutions:

<u>for CaF$_2$</u>: $K_{sp} = C_{Ca^{2+}} \times C_{F^-}^2 = (3 \times 10^{-4})(1 \times 10^{-4})^2 = 3 \times 10^{-12}$

[CaF$_2$ will <u>not</u> precipitate, $(3 \times 10^{-12}) < K_{sp}$]

for MgF_2: $K_{sp} = C_{Mg^{2+}} \times C_{F^-}^2 = (5 \times 10^{-6})(1 \times 10^{-4})^2 = 5 \times 10^{-14}$

$[MgF_2$ will <u>not</u> precipitate $(5 \times 10^{-14}) < K_{sp}.]$

*5. $[38 \text{ mg}(Pb^{2+})]$ *(This looks quite small, but a tiny amount of Pb^{2+} may cause significant biological damage.)*

<u>Solution:</u>

From Appendix D, K_{sp} for $PbCrO_4 = 1.8 \times 10^{-14}$

For $PbCrO_4$:

$$PbCrO_{4(s)} \rightleftharpoons Pb^{2+}_{(aq)} + CrO_4^{2-}{}_{(aq)}$$

Representing $C_{Pb^{2+}}$ by S, then:

$$C_{CrO_4^{2-}} = \frac{S(Pb^{2+})}{1} \times \frac{1(CrO_4^{2-})}{1(Pb^{2+})} = S$$

From which:

$$K_{sp} = C_{Pb^{2+}} \times C_{CrO_4^{2-}} = (S) \times (S) = S^2$$

$$S = \sqrt{K_{sp}} = \sqrt{1.8 \times 10^{-14}} = 1.34 \times 10^{-7} \text{ mole liter}^{-1}$$

Then the Pb^{2+} ingested will be:

$$\frac{1.34 \times 10^{-7} \text{ mole}(Pb^{2+})}{1 \text{ liter}} \times \frac{10^3 \text{ mmol}}{1 \text{ mole}} \times \frac{207.2 \text{ mg}(Pb^{2+})}{1 \text{ mmol}(Pb^{2+})} \times \frac{1 \text{ liter}}{10^3 \text{ ml}} \times \frac{946 \text{ ml}}{1 \text{ day}} \times$$

$$\frac{365.25 \text{ days}}{1 \text{ year}} \times \frac{4 \text{ years}}{1}$$

$[\text{mole}(Pb^{2+})/\text{liter} \rightarrow \text{mmol/liter} \rightarrow \text{mg/liter} \rightarrow \text{mg/ml} \rightarrow \text{mg/day} \rightarrow \text{mg/year} \rightarrow \text{mg}]$

*6. $[1.2 \times 10^{-5} \text{ ppm}(Fe^{3+})]$ <u>Solution:</u>

From Appendix D, K_{sp} for $Fe(OH)_3 = 6.0 \times 10^{-38}$

For $Fe(OH)_3$:

$$Fe(OH)_{3(s)} \rightleftharpoons Fe^{3+}_{(aq)} + 3 OH^-_{(aq)}$$

Representing $C_{Fe^{3+}}$ by S, then:

$$C_{OH^-} = \frac{S(Fe^{3+})}{1} \times \frac{3(OH^-)}{1(Fe^{3+})} = 3S$$

From which:

$$K_{sp} = C_{Fe^{3+}} \times C_{OH^-}^3 = (S) \times (3S)^3 = 27S^4$$

$$S = \sqrt[4]{\frac{K_{sp}}{27}} = \sqrt[4]{\frac{6.0 \times 10^{-38}}{27}} = 2.2 \times 10^{-10} \text{ mole liter}^{-1}$$

Then the Fe^{3+} concentration, in ppm will be:

$$\frac{2.2 \times 10^{-10} \text{ mole}(Fe^{3+})}{1 \text{ liter}} \times \frac{10^3 \text{mmol}}{1 \text{ mole}} \times \frac{55.85 \text{ mg}(Fe^{3+})}{1 \text{ mmol}(Fe^{3+})} \times \frac{1\text{ppm}}{1 \text{ mg liter}^{-1}}$$

===

ANSWERS *to* SELF-TEST, *Unit 10*

10.1. (a) $K_{sp} = C_{K^+}^3 \times C_{[Co(NO_2)_6]^{3-}}$

(b) $K_{sp} = C_{Co^{2+}}^3 \times C_{PO_4^{3-}}^2$

(c) $K_{sp} = C_{Cu^{2+}}^2 \times C_{CO_3^{2-}} \times C_{OH^-}^2$

10.2. CaF_2

*10.3. 54%

A.1. Several important mineral phases have been formed by reaction between bacterially produced sulfide and the marine sediment. A number of metal sulfides having very low solubility products have been formed in this manner. However, with the exception of iron, these metals are present in only trace amounts in the average marine sediment. Pyrite (FeS_2) is one of the most common minerals in recent marine sediments. Pyrite is an iron(II) polysulfide in which the sulfur atoms can be regarded as possessing a formal charge of negative one (S^{1-}) each, or alternatively, one atom may be divalent and the other zerovalent (S^{2-}, S^0). This implies that pyrite cannot be formed from iron(II) ion and hydrogen sulfide without the oxidation of some sulfide ion to elemental sulfur. Assuming that the sulfur in pyrite is simply described as S_2^{2-}, write the conventional solubility product expression for pyrite.

A.2. The availability of phosphorus to plants is determined largely by the ionic form of the element. The ionic form in turn is a function of the soil solution pH. Thus, in highly acid soils the $H_2PO_4^-$ species predominates. In acidic, organic soils that are deficient in iron, aluminum, and manganese, the dihydrogen phosphate ($H_2PO_4^-$) ions would be readily available for plant assimilation. When the same degree of acidity exists in a normal mineral soil, however, very different results occur since soluble iron, aluminum, and manganese are present in these soils. These metal ions react with the $H_2PO_4^-$ by forming insoluble salts such as $Al(OH)_2H_2PO_4$, and rendering the phosphorus unavailable for plant growth. Write the conventional solubility product expressions for aluminum hydroxyphosphate[1] and manganese(II) hydroxyphosphate[2].

A.3. Minerals in deep-sea sediments are of interest because they can be used to elucidate chemical processes operating on the ocean floor. Because they are formed <u>in situ</u> in the sediments, they can provide a record of physicochemical and biochemical reactions operating and having operated at their site of deposition during the period of their formation. For example, study of the layer-by-layer composition of ferro-manganese nodules might provide information on changes in the chemistry of the environment of deposition with time. There are a large variety of minerals formed in deep-sea sediments, most of them rare. Four groups are relatively abundant, however, and include barite, zeolites, iron oxides, and manganese oxides. Barite ($BaSO_4$) is the most common mineral of barium and is widespread in marine sediments. Marine barite generally has a high degree of purity, although it has been found to contain up to three percent strontium. If the water in an ocean trench (one volume, containing

---

[1]IUPAC name: aluminum dihydrogen hydroxide phosphate
[2]IUPAC name: manganese(II) dihydrogen hydroxide phosphate

9.7×10^{-3} mole liter^{-1} of SO_4^{2-}) is mixed, by tectonic activity, with water from another trench (two volumes) in which the dissolved minerals provide 4.0×10^{-4} $\underline{M}$ Ba^{2+} and 8.0×10^{-7} $\underline{M}$ Sr^{2+}, will any precipitate form? If so, what is the composition of the precipitate? [For K_{sp}-values, see Appendix D.]

A.4. The marine biomass dramatically affects chemical and physical processes in the oceans in a variety of ways, directly and indirectly. Sea water constituents are required as nutrients and they are critical for maintaining and propagating life. The marine biomass produces metabolic wastes, decaying organic matter, and animal excreta, all of which play an important role in recycling sea water components. A large segment of the marine biomass precipitates inorganic constituents in their hard parts. Biologic mineral fixation has a significant impact on chemical and physical processes in the sea. More than twenty different kinds of inorganic constituents have been firmly established as biologic precipitates in the hard parts of marine life, including the bacteria, plant, and animal phyla. Write the conventional solubility product expression for two salts, <u>Dahllite</u> ($Ca_7(CO_3)(PO_4)_4$), and <u>Whewellite</u> ($Ca[C_2O_4]H_2O$), examples of marine biologic precipitates.

A.5. Micronutrients become "limiting reagents" for crop growth when the crops are grown in highly leached acid sandy soils, muck soils, soils with high pH, or soils that have been intensively cultivated and heavily fertilized with macronutrients. The micronutrient cations, Fe^{2+}, Mn^{2+}, Zn^{2+}, Cu^{2+}, and Co^{2+}, are most soluble and available to plants under acid conditions. In very acid soils where there is a relative abundance of the micronutrient cations their concentrations may be high enough to be toxic to common plants. Liming soils increases the hydroxide (OH^-) ion concentration so that the micronutrient cations are precipitated as their hydroxides. This decrease in availability also reduces their toxicity to crops. A particular soil solution which was 3.0×10^{-4} $\underline{M}$ in Fe^{2+} and 5.0×10^{-5} $\underline{M}$ in Mn^{2+} had its pH raised until the OH^- concentration was 2.0×10^{-5} $\underline{M}$. Would any micronutrient loss by precipitation be expected? If so, what would be the composition of the precipitate? [See Appendix D for K_{sp}-values.]

*A.6. Precipitation and sedimentary deposits of calcium carbonate minerals in the world's oceans are essentially the product of biologic agents. A very diversified carbonate-secreting biomass is found in continental shelf areas. Magnesium and calcium carbonates are their precipitation products, with higher percentages of calcium carbonate being formed in the warm water regions. In the open oceans, pelagic micro-plankton, represented by <u>coccolithophorids</u> and <u>foraminifera</u>, constitute the principal precipitation agents of calcium carbonate in the calcite mineral form. A sample of skeletal calcite was obtained from a Project Mohole site, after which it was washed with water to remove soluble impurities. What maximum percentage of the calcite might be lost by washing 150 g of the skeletal remains with 20 liters of water? [The solubility product for calcite, $CaCO_3$, is 4.7×10^{-9}.]

*A.7. Sedimentary materials may be introduced or redistributed in oceans by mechanisms of ice transport. Ice in the marine environment can originate on land as ice caps or glaciers, or as sea ice, which is formed by the freezing of sea water. On land, ice in the form of sheets or glaciers is an effective agent for eroding and transporting the underlying terresterial surfaces. Powdered rock ("rock flour") is often produced by such activity. Thus, the icebergs formed when land ice breaks away from the parent material can contribute a variety of continental materials to the ocean sediments. A trench in the continental shelf has an exposed surface consisting of the mineral <u>celestite</u>, $SrSO_4$, which was deposited there by glacial activity. What would be the concentration in ppm $[Sr^{2+}]$, of the sea water saturated with $SrSO_4$ from the celestite in this submarine environment? [For the K_{sp}-value, see Appendix D.]

*A.8. Compared with the complex nutritional requirements of man and other animals the needs of plants are quite simple. Plants subsist entirely on inorganic materials, and even these make up a short list. The major plant nutrients are carbon dioxide, oxygen and water. Carbon dioxide and oxygen are, of course, universally available, and under field conditions they probably never fall to levels low enough to impair growth. Water is also an abundant resource, although it is less uniformly distributed. The relationship of water supply to plant growth has been understood since before the formal beginnings of agriculture. Water is often a factor limiting growth, and the correction of water deficiency by irrigation can yield spectacular results. If the waters in a farm reservoir used for irrigation are saturated with $Ca_3(PO_4)_2$ from the reservoir's walls, what would be the concentration of phosphate, an essential micronutrient, in the irrigation waters (in ppm $[PO_4^{3-}]$)? [See Appendix D for K_{sp}-value.]

*A.9. Animals can be classified into three groups based on their abilities to assimilate cellulose, a classification that is reflected in the structure of their digestive apparatus. Carnivores, such as dogs, and omnivores, such as man and the pig, have simple stomachs and have difficulty digesting cellulose. Nonruminant herbivores, such as the horse, rabbit, and the guinea pig, can derive sustenance from cellulose, but they assimilate it less effectively than the ruminants. Ruminant herbivores, such as cattle, sheep, goats, deer, elk, and buffalo, can efficiently break down cellulose and extract a large proportion of their dietary calories in the process. Moreover, the nutritional needs of most ruminants are exceptional in that they require relatively large quantities of Ca^{2+} for bones, milk, and antlers in some members of the group. How much calcium (in mg) would be ingested by a cow drinking 4.00 gallons (15.14 liters) of water per day from the farm reservoir, described in problem A.8, for one year?

*A.10. Iron oxides are an important constituent of deep sea sediments. They usually occur as amorphous, or poorly defined crystalline, reddish brown coatings on clay and other minerals. Occasionally, the mineral goethite, FeO(OH) is present. Most iron is present in sea water in the +3 oxidation state and occurs in the form of colloidal $Fe(OH)_3$. The slow settling of this phase probably accounts for the bulk of free iron oxides in deep sea sediments. Samples of sediments obtained from shallow Gulf coast estuaries have a top layer composed of iron(III) hydroxide. During a particular investigation, a 10 g sample of the iron(III) hydroxide sediment layer was washed with 12.0 liters of water to remove any soluble trace impurities. What maximum percentage of the iron(III) hydroxide might have been lost by this washing? [See Appendix D for K_{sp}-value.]

ANSWERS:

(A.1.) $K_{sp} = C_{Fe^{2+}} \times C_{S^{2-}}$, (A.2.) $K_{sp} = C_{Al^{3+}} \times C_{OH^-}^2 \times C_{H_2PO_4^-}$ and $K_{sp} = C_{Mn^{2+}} \times C_{OH^-} \times C_{H_2PO_4^-}$, (A.3.) Only $BaSO_4$ will precipitate., (A.4.) $K_{sp} = C_{Ca^{2+}}^7 \times C_{CO_3^{2-}} \times C_{PO_4^{3-}}^4$ and $K_{sp} = C_{Ca^{2+}} \times C_{C_2O_4^{2-}}$, (A.5.) Only $Fe(OH)_2$ will precipitate., (A.6.) 0.092%, (A.7.) 50 ppm [Sr^{2+}], (A.8.) 0.03 ppm [PO_4^{3-}], (A.9.) 109 mg yr^{-1}, (A.10.) 2.8×10^{-6}% (negligible)

RELEVANT
PROBLEMS

Unit 10: *Heterogeneous Equilibria*

Set B: *Biological & Medical Sciences*

B.1. Dystrophic calcification, a result of deranged calcium metabolism, is characterized by the deposit of calcium salts in injured, degenerating or dead tissues. It is encountered in any form of tissue injury, particularly in the presence of suppurative liquefaction or coagulation necrosis, or in fatty degeneration and enzymatic necrosis of fat. In these tissue alterations, phosphate-releasing enzymes may raise the level of phosphate in the local fluids to the point where the solubility product of calcium phosphate is exceeded. Dystrophic calcification is also prominent in the focal areas of fat necrosis caused by pancreatic disorders. The fatty acids that are released by the pancreatic lipases combine with serum calcium to form so-called insoluble "soaps". For each salt, write the conventional solubility product expression.

    a. Calcium phosphate

       [Formed in injured tissues and in kidney stones]

    b. Calcium palmitate, $Ca[CH_3(CH_2)_{14}CO_2]_2$

       [Formed in blood serum during fat necrosis]

    c. Magnesium ammonium phosphate

       [Formed in kidney stones]

B.2. Bone is produced by two totally distinct processes, membranous formation and endochondral formation. In membranous bone formation, an osteoid matrix is formed with connective tissue by specialized cells. This osteoid is then converted into bone by the mineralization of the matrix through the action of osteoblasts. Bone is thus formed without the intermediary development of cartilage. The mineralization process results from the deposition of a complex salt called <u>hydroxyapatite</u>, and having the composition, $Ca_5(PO_4)_3OH$. The membranous connective tissue, collagen, provides toughness and resiliency, while the <u>hydroxyapatite</u> provides hardness and rigidity. A similar mineralization process occurs with the formation of the dental enamel. By replacing the hydroxide ion in the <u>hydroxyapatite</u> structure with fluoride ion, it forms <u>fluorapatite</u>, $Ca_5(PO_4)_3F$. This <u>fluorapatite</u> is less soluble and more resistant to acid, hence generally reduces dental caries. For each salt write the conventional solubility product expression.

    a. <u>Hydroxyapatite</u>

    b. <u>Fluorapatite</u>

B.3. X-rays figure significantly in medicine as a diagnostic tool used to locate foreign bodies such as bullets, minute differences in tissues, such pathological conditions as tuberculosis, or broken bones. Pictures of the lungs are easy to interpret because air spaces are more transparent than lung tissues. Various other cavities in the body must be filled artificially with contrasting media, either more transparent or more opaque to x-rays than the surrounding tissue, so that a particular organ can be brought more sharply into view. Certain compounds are administered either by mouth or by injection into the blood stream for kidney or bladder examination. Stones in the kidney or gall bladder show up clearly as light rings against the darker background of the opaque dye. Barium sulfate, which is highly opaque to x-rays, is used for the x-ray examination of the gastrointestinal tract. To maintain a very low concentration of the toxic Ba^{2+} ion, suspensions of $BaSO_4$ may contain added sulfate ion.

In the preparation of a $BaSO_4$ suspension for a particular radiographic study of the G.I. tract, enough Epsom Salts ($MgSO_4 \cdot 7H_2O$) was added to saturated $BaSO_4$ to give a Mg^{2+} concentration of 0.01 M. What is the molar concentration of free Ba^{2+} in this suspension? What would it have been without the added Epsom Salts? [See Appendix D, for K_{sp}-value.]

B.4. Most of the fluoride of the body is present in the bones and teeth. The highest concentration is found in the enamel of the teeth. When an excessive level of fluoride in drinking water is ingested over a long period of time, a condition known as "mottled enamel" occurs. This enamel is discolored and brittle. In regions where the amount of fluoride in drinking water ranges between 0.1 and 1.2 parts per million, mottling does not occur, but the enamel seems to be more resistant than normal and dental caries are less likely to occur. The cause of dental caries is believed to be the erosion of enamel by organic acids formed from the carbohydrates of the food we eat by certain strains of bacteria present in the mouth. One explanation for the possible inhibiting action of fluoride ion on caries is that it acts as a bacterial enzyme inhibitor, thus interrupting the chain of fermentation reactions and preventing the formation of the organic acids in close proximity to the enamel. The addition of fluoride to the water supplies of communities is frequently used as a method of preventing dental caries. The method of choice is to add sufficient sodium fluoride to increase the fluoride concentration to 1 mg per liter. Would this fluoride ion concentration be possible in hard water containing 200 mg per liter of Ca^{2+} simply by adding the same amount of NaF as used for soft water? [See Appendix D, for K_{sp}-value.]

B.5. There are many conditions in which change in the color of the oral mucous membrane occurs. These color changes may result from the effects of nutritional disturbances, either deficiencies or excesses. For example, in nicotinic acid deficiency the tongue is "beefy red", glazed and painful. "Bleeding gums" are often presumed to be indicative of vitamin C deficiency. Changes in the oral tissue may also be caused by intoxication from metals such as arsenic, bismuth, mercury, and lead through industrial contact or ingestion of contaminated water or of certain medications. Metallic intoxication may result in ulceration of the mucous membrance and decomposition of the jaw bone. Spotty discoloration of the oral mucous membrane and linear pigmentation of the gums are prominent features. The latter feature is referred to as a "bismuth" or "arsenic line" (black), a "lead line" (blue-gray), or a "mercury line" (gray-violet). All pigmentation disappears upon removal of the local irritants responsible for the inflammation. The spotty discoloration or line pigmentation of the oral mucosa and gums, resulting from metallic intoxication, is caused by the deposition of metallic sulfides in the connective tissue. What sulfide ion concentration would be required to precipitate PbS and form the blue-gray "lead line" in a patient's gums if the lead ion concentration in these tissues has reached a level equivalent to 6.0×10^{-12} mole per liter of tissue fluid? [For K_{sp}-value, see Appendix D.]

*B.6. The formation of "stones" within the urinary tract occurs under a variety of conditions. It is apparent that in the formation of inorganic stones precipitation of mineral salts must occur. Highly concentrated urines, particularly those which are alkaline, predispose to the precipitation of alkaline salts. Excessive mobilization of minerals, particularly calcium phosphate from the bone, favor the formation of stones as the minerals are excreted in the urine. Excess mobilization of bone salts may occur as a result of such systemic problems as hypervitaminosis D, resorption of bone in old age, and immobilization of the skeletal system because of illness or injury. Hyperparathyroidism is well known to be associated with the formation of renal stones. A decrease in urinary circulation is an important factor in permitting the concentration and crystallization of salts with the formation of stones. Diets high in alkaline ash, high calcium content of the water supply, and certain hereditary factors also favor urinary stone formation. The chemical composition of renal stones is variable and it is uncommon to find a pure type. Calcium phosphate, $Ca_3(PO_4)_2$, is one of the most common components of an ordinary kidney stone. Other minerals frequently encountered are calcium carbonate, calcium oxalate, and magnesium ammonium phosphate. What is the approximate concentration, in ppm, of calcium phosphate in the urine of a patient that has several calcium phosphate stones forming in the kidney? [For the K_{sp}-value, see Appendix D.]

*B.7. Soluble salts of inorganic lead are strong protoplasmic poisons. They are stored in the body and by their progressive accumulation reach toxic levels and cause tissue damage. Lead poisoning may be acute from sudden absorption of large doses or it may be chronic and develop insidiously from the storage of small doses. These two forms of poisoning differ in their rate of development, but otherwise manifest themselves in identical morphologic and clinical changes. Sources of lead in the environment include volatilized and finely divided lead compounds from industry, old lead-base paints, lead salts, and lead halides (especially lead bromide) from the exhaust of internal combustion engines using leaded gasoline. Approximately 2.0×10^5 tons of lead enter the environment by the latter pathway each year. Lead poisoning may be manifest by lesions in the gastrointestinal and nervous systems. Functional disturbances are also observed in the liver, kidney and central nervous system. It is not unusual for lead to be found in the illegally produced whiskey sold in some parts of the United States. The lead comes from the lead solder used in the tubing of the distillation units and from the lead-containing automobile radiators used as condensers. The lead solder is generally used with a flux containing hydrochloric acid, which reacts with the lead to form lead(II) chloride, $PbCl_2$. What would be the concentration, in ppm $[Pb^{2+}]$, in a "moonshine" barrel filled with water and saturated with $PbCl_2$ from the distillation unit? [For the K_{sp}-value, see Appendix D.]

*B.8. Mercury and its compounds can enter the body via the lungs by inhalation, the gastrointestinal tract by ingestion, and the skin by contact exposure. The principal manifestations of acute inorganic mercury poisoning include pulmonary irritation after exposure to mercury vapor and kidney injury after exposure to mercury(II) salts. The human exposure to mercury has been intensified recently by its high production rate and the extensive use of its compounds in agriculture, electrical apparatus, industry, paints, and medicine. Elemental mercury is used in the preparation of dental amalgams and mercury(I) chloride, Hg_2Cl_2, has been used to prepare injections used in treating syphilis. How much mercury, as Hg_2^{2+}, (in mg) would be injected into the blood stream of a person required to take three 10 ml injections, each saturated with Hg_2Cl_2, per week for 26 weeks? [For the K_{sp}-value, see Appendix D.]

*B.9. The prolonged administration of silver compounds, usually as medications, may result in the accumulation of this metal in the tissues. It has been most commonly observed in individuals who have indulged in the excessive self-administration of "nose drops" containing silver salts. Gray-blue skin pigmentation results from the precipitation of a silver-protein complex, which yields a gray-blue metallic silver upon reduction. The most common site of accumulation is in the subcutaneous fibrous tissues or in the skin. Visceral deposits occur occasionally in the kidneys and liver. A patient accidentally ingested 100 ml of a silver salt solution having a silver ion concentration of 0.10 M. This was quickly treated by administering a 100 ml portion of 0.30 M salt water solution to precipitate the Ag^+ as $AgCl_{(s)}$. What is the concentration of silver ion remaining in the 200 ml mixture?

*B.10. The most important regressive change in cartilage precedes and is closely related to one type of bone development, intracartilaginous ossification. The cartilage cells become arranged in groups as in a vertebra, or in more or less irregular columns separated by wide, parallel bands of interstitial tissue, as in the end of a long bone. A high concentration of calcium salts accumulate in the zone of hypertrophic cartilage cells and minute granules of calcium carbonate are deposited in the interstitial substance, primarily in the vicinity of the cells. As these granules become larger and merge, the cartilage becomes opaque, hard and brittle. Most of the cells degenerate as vascular connective tissue invades the calcified cartilage preceding the formation of bone. What is the approximate concentration, in ppm, of calcium ion, Ca^{2+}, in the heterotrophic zone of a person that is experiencing calcification of the sternum? [For the K_{sp}-value, see Appendix D.]

ANSWERS:

(B.1.) (a) $K_{sp} = C_{Ca^{2+}}^3 \times C_{PO_4^{3-}}^2$, (b) $K_{sp} = C_{Ca^{2+}} \times C_{[CH_3(CH_2)_{14}CO_2]^-}^2$, (c) $K_{sp} = C_{Mg^{2+}} \times$

$C_{NH_4^+} \times C_{PO_4^{3-}}$, (B.2.) (a) $K_{sp} = C_{Ca^{2+}}^5 \times C_{PO_4^{3-}}^3 \times C_{OH^-}$, (b) $K_{sp} = C_{Ca^{2+}}^5 \times C_{PO_4^{3-}}^5 \times C_{F^-}$,

(B.3.) 7.9×10^{-9} $\underline{M}$ (in the Epsom Salts), 8.9×10^{-6} $\underline{M}$ (alone), (B.4.) yes [C_{F^-} may be 1.8×10^{-4} $\underline{M}$], (B.5.) 1.2×10^{-17} $\underline{M}$, (B.6.) 0.05 ppm [$Ca_3(PO_4)_2$], (B.7.) 3.3×10^3 ppm [Pb^{2+}], (B.8.) 0.21 mg, (B.9.) 2.4×10^{-9} $\underline{M}$, (B.10.) 2.7 ppm [Ca^{2+}]

RELEVANT
PROBLEMS

Unit 10: Heterogeneous Equilibria

Set I: Industrial Chemistry

I.1. Phosphoric acid, widely used in the manufacture of phosphate fertilizers, is made by treating crushed phosphate rock (largely $Ca_3(PO_4)_2$) with sulfuric acid. What is the solubility product expression for calcium phosphate?

I.2. In a hydrometallurgical (leaching) method of recovering copper, bacterial leaching aids apparently produce both iron(III) sulfate ($Fe_2(SO_4)_3$) and sulfuric acid. The iron(III) sulfate promotes the oxidation of copper sulfides to sulfates, which are more easily treated. What is the solubility product expression for iron(III) sulfate?

I.3. Gold is the backbone of international commerce and the effect of fluctuating gold prices over the past few years on the economies of industrial societies has been significant. Although most gold is found as the native metal, either pure or alloyed with silver, it does occur in commercially profitable deposits of petzite ore, Ag_3AuTe_2. What is the conventional solubility product for this mineral, assuming simple ions of silver, gold(I) and Te^{2-} are involved?

I.4. Most industrial operations require very large amounts of water as a coolant in heat-exchange processes. Muddy or cloudy water is generally unsatisfactory, since the dispersed solids may clog filters or deposit sediment in pipes and pumps. Murky water may be clarified on a large scale by the addition of flocculating agents to coagulate colloidal material, allowing it to settle out in holding tanks or ponds before the clarified water is sent to plant intakes. A recent flocculation method employs the addition of both calcium hydroxide and magnesium carbonate. If 56 g of $Ca(OH)_2$ and 45 g of $MgCO_3$ were added to 520 liters of water, would any chemical precipitation occur?

I.5. Hydrofluoric acid finds such diverse applications as the interior etching of "frosted" light bulbs, the catalysis of alkylation reactions in petroleum refineries, and the preparation of various fluoride additives for toothpastes. The commercial manufacture of hydrofluoric acid utilizes the action of sulfuric acid on the mineral fluorite, CaF_2. Will a precipitate form if 20 liters of waste 5.9 $\underline{M}$ H_2SO_4 is dumped into a settling pond containing 1100 liters of water saturated with fluorite (solubility: 0.02 g liter^{-1})?

*I.6. The Solvay process has been used for more than fifty years as the principal industrial source of sodium bicarbonate (baking soda). In this process, carbon dioxide gas is pumped into concentrated aqueous ammonia saturated with sodium chloride. Sodium bicarbonate precipitates and is removed by filtration. Care must be taken that the water used in the process is free of other ions that might form precipitates contaminating the baking soda, which is too soluble in water for purification by repeated washings. During the course of reaction, ammonia, CO_2, and Na^+ are consumed, but a high concentration of chloride ion remains. What is the maximum concentration of Ag^+ contaminant (in ppm) that could be tolerated without fear of precipitation of silver chloride from a solution 4.0 $\underline{M}$ in Cl^-? [See Appendix D for K_{sp}-data.]

*I.7. The phosphate rock mines in North Africa and in Florida are major resouces for a multimillion dollar industrial production of phosphorus. When a mixture of calcium phosphate, sand, and coke is heated in an electric furnace, elemental phosphorus is produced (as $P_{4(g)}$) along with carbon monoxide and a silicate "slag", $CaSiO_3$. The slag wastes are dumped in massive piles except in areas where it is economically feasible to utilize the slag in cement manufacture. What water "hardness", in ppm $[Ca^{2+}]$ would occur in rainwater pools saturated with $CaSiO_3$ ($K_{sp} = 6.7 \times 10^{-7}$)?

*I.8. Natural gas typically contains water vapor and hydrogen sulfide, both of which must be removed before the gas is pumped through pipelines. Hydrogen sulfide is corrosive to metal parts and its combustion to SO_2 would result in an undesirable odor for natural gas consumers. Water vapor is also somewhat of a corrosion problem and presents additional complications from possible ice crystal formation in pipelines and pumps during cold weather. Hydrogen sulfide can be removed by passing the natural gas over powdered iron(III) oxide, after which the water vapor can be removed by a desiccant. The iron(III) oxide is consumed and must be replaced periodically. In the waste oxide discarded is some Fe_2S_3. What molar concentration of sulfide ion would be found in water saturated with Fe_2S_3 ($K_{sp} = 1.0 \times 10^{-88}$)?

*I.9. Magnesium carbonate is used in the manufacture of a high density magnesite brick. This material is not well suited to general exterior use because of the erosive loss of magnesium carbonate. What percentage of a 28 g amount of surface-exposed $MgCO_3$ would be lost through the solvent action of 15 liters of rainwater, assuming sufficient contact time for saturation of the water with $MgCO_3$? [For K_{sp}-data, see Appendix D.]

*I.10. For many years zinc sulfide was of primary commercial value as the white pigment lithopone, but its industrial production was dramatically increased with the growth of the television industry. The fluorescence of zinc sulfide when bombarded with electrons was an ideal characteristic for a substance to coat television screens. What maximum percentage loss of zinc sulfide must be anticipated during the industrial purification of 25 kg ZnS by washing with 32 liters of water? [For K_{sp}-data, see Appendix D.]

ANSWERS:

(I.1.) $K_{sp} = C_{Ca^{2+}}^3 \times C_{PO_4^{3-}}^2$, (I.2.) $K_{sp} = C_{Fe^{3+}}^2 \times C_{SO_4^{2-}}^3$, (I.3.) $K_{sp} = C_{Ag^+}^3 \times C_{Au^+} \times C_{Te^{2-}}^2$,

(I.4.) $Mg(OH)_2$ and $CaCO_3$ will precipitate, (I.5.) $CaSO_4$ will precipitate, (I.6.) 6.5×10^{-6} ppm $[Ag^+]$, (I.7.) 33 ppm $[Ca^{2+}]$, (I.8.) 3×10^{-18} M, (I.9.) 29%, (I.10.) 5×10^{-11}%

RELEVANT
PROBLEMS

Unit 10: Heterogeneous Equilibria

Set E: Environmental Sciences

E.1. Many of the more common pesticides and insecticides are toxic to such a broad spectrum of living organisms as to be more properly called biocides. The effects of indiscriminate use of compounds such as DDT and parathion have now been generally recognized as potentially disastrous to the biosphere, and the use of such compounds has been sharply curtailed. As a result, many "more primitive" compounds have re-

turned to the market. One of these, lead arsenate, is 160 times less toxic than
<u>parathion</u>. However, both the Pb^{2+} and AsO_4^{3-} ions are harmful to humans, so care
must be exercised in avoiding environmental contamination by even the simple inorganic poisons. Write the conventional solubility product expression for lead
arsenate.

E.2. Commercial plants manufacturing phosphorus or phosphate fertilizers from <u>fluor-</u>
<u>apatite</u>, $Ca_5(PO_4)_3F$, are major sources of fluoride pollution, usually in the form
of fluorosilicates in the waste water effluent. What is the conventional solubility
product expression for <u>fluorapatite</u>?

E.3. Fluorides from industrial wastes (problem E.2) may be removed by addition of gypsum
to the aqueous wastes in settling ponds, resulting in the formation of calcium
fluoride. Will calcium fluoride precipitate in a settling pond containing 0.020 <u>M</u>
F^- if sufficient gypsum is added to introduce Ca^{2+} to the extent of 0.0010 mole
$liter^{-1}$? [See Appendix D for K_{sp}-data.]

E.4. Tannery wastes are rich in sodium and calcium sulfides. Such wastes may deplete the
oxygen of the aquatic ecosystem because of the ready susceptibility of S^{2-} to oxidation. Pretreatment of the sulfide solution with atmospheric oxygen in the presence
of an iron(II)/iron(III) redox catalyst system can be used to convert the sulfide to
the more innocuous SO_4^{2-}. Will calcium sulfate precipitate in such a system if the
water contains 85 ppm Ca^{2+} and the sulfate concentration reaches 0.015 <u>M</u>? [See
Appendix D for K_{sp}-data.]

E.5. Phosphate additives to detergents have now been banned or sharply curtailed because
of the effect of high phosphate concentrations on the growth of algae and other undesirable aquatic plants using phosphate as a nutrient. Many municipal and industrial wastes, however, are still rich in phosphate. Waste water treatment with
Al^{3+} can reduce the phosphate concentration significantly. Will aluminum phosphate
precipitate when 12 liters of 4.2 <u>M</u> Al^{3+} solution are added to 25,000 liters of
waste water 25 ppm in PO_4^{3-}? [K_{sp} for $AlPO_4$ = 8.9 x 10^{-22}]

*E.6. Waste effluent from many mining and metallurgical operations is often quite acidic,
containing a number of inorganic acids. The direct addition of such wastes to
natural streams or lakes can produce a hydrogen-ion concentration high enough to be
destructive to several species of aquatic life. If the acidic wastes are exposed
to limestone, hydrogen ion is removed by reaction with the carbonate of limestone.
Under certain conditions, however, sulfuric acid in the waste effluent can produce
an insoluble coating of calcium sulfate on the surface of the limestone, preventing
any further reaction with hydrogen ion. What is the maximum concentration (in mole

liter^{-1}) of SO_4^{2-} that could be present in a saturated solution of $CaCO_3$ without precipitation of $CaSO_4$? [See Appendix D for K_{sp}-data.]

*E.7. Phosphates play many vital roles in life processes, ranging from the participation of simple phosphate species in biological buffers to the exceedingly complex organophosphates of the nucleic acids involved in the "genetic code". Phosphate is a very stable chemical species and circulates along with other biogenic species throughout the biosphere. The production and use of phophates for agriculture, industry, and the home have increased enormously over the past thirty years. As a result, the flow of phosphates into natural waters has increased accordingly and high phosphate concentrations are of serious concern. Phosphates have always been leached from the soil by natural waters, but this has been a slow process because of the low solubilities of most phosphate minerals. What is the concentration of phosphate ion in water that is saturated with calcium phosphate, a common "phosphate rock"? [For K_{sp}-data, see Appendix D.]

*E.8. What is the concentration of calcium ion, in ppm, in the saturated solution (E.7) if the density of the water is 1.08 g ml^{-1}? [For K_{sp}-data, see Appendix D.]

*E.9. What is the approximate concentration of Pb^{2+} (in ppm) in the waters of a smelter "holding pond" containing a settled slurry of $PbCO_3$? [K_{sp} for $PbCO_3$ = 1.5 x 10^{-13}.]

*E.10. The average adult human can excrete about 2 mg of Pb^{2+} per day. Any intake in excess of this amount accumulates in the body and, if sufficient continued accumulation occurs, chronic lead poisoning can result, with associated damage to the brain and nervous system. In streams near heavily travelled highways, the concentration of Pb^{2+} from compounds emitted from the exhausts of automobiles still using "leaded" gasoline may be relatively high. How many milligrams of Pb^{2+} would be ingested daily by an individual drinking 1.5 liters day^{-1} of water saturated with lead sulfate (K_{sp} = 1.3 x 10^{-8})?

ANSWERS:

(E.1.) $K_{sp} = C_{Pb^{2+}}^3 \times C_{AsO_4^{3-}}^2$, (E.2.) $C_{Ca^{2+}}^5 \times C_{PO_4^{3-}}^3 \times C_{F^-}$, (E.3.) yes, (E.4.) yes,

(E.5.) yes, (E.6.) 0.36 $\underline{M}$, (E.7.) 3.3 x 10^{-7} $\underline{M}$, (E.8.) 2.0 x 10^{-2} ppm [Ca^{2+}], (E.9.)

0.08 ppm [Pb^{2+}], (E.10.) 35 mg

UNIT 11: ACID/BASE EQUILIBRIA

There are a number of different ways of looking at acid/base chemistry. We shall restrict our considerations to aqueous systems, in which the Brønsted-Lowry concepts are particularly useful:

AN ACID IS ANY SPECIES ACTING AS A PROTON (H^+) DONOR.

A BASE IS ANY SPECIES ACTING AS A PROTON ACCEPTOR.

The key to these concepts is the word acting. This implies that we need to know more about a substance than just its chemical formula in order to classify it as an acid or a base (or, perhaps, as neither). Water, for example, is ordinarily thought of as a "neutral" substance, but because of the highly polarized O-H bond and the presence of two lone pairs of electrons, water may function as either an acid or a base:

$$HCl_{(g)} + H_2O_{(\ell)} \rightleftharpoons H_3O^+_{(aq)} + Cl^-_{(aq)}$$

Water acts here as a base, accepting a proton from an HCl molecule.

$$[H \!:\! \overset{\cdot\cdot}{\underset{\cdot\cdot}{Cl}} \!:\! + \; H \!:\! \overset{\cdot\cdot}{\underset{\cdot\cdot}{O}} \!:\! H \rightleftharpoons H \!:\! \overset{\cdot\cdot}{\underset{H}{O}} \!:\! H^+ + \; \!:\! \overset{\cdot\cdot}{\underset{\cdot\cdot}{Cl}} \!:\! ^-]$$

Alternatively, water may furnish a proton, acting as an acid.

$$NH_{3(g)} + H_2O_{(\ell)} \rightleftharpoons NH^+_{4(aq)} + OH^-_{(aq)}$$

$$[H \!:\! \overset{\cdot\cdot}{\underset{H}{N}} \!:\! H + \; H \!:\! \overset{\cdot\cdot}{\underset{\cdot\cdot}{O}} \!:\! H \rightleftharpoons H \!:\! \overset{H}{\underset{H}{N}} \!:\! H^+ + \; \!:\! \overset{\cdot\cdot}{\underset{\cdot\cdot}{O}} \!:\! H^-]$$

Note that both of these processes have been formulated by equilibrium equations ("double arrows"). The Brønsted-Lowry system considers acid/base behavior to involve a dynamic competition between two bases (e.g., H_2O, Cl^- or NH_3, OH^-) for a proton.

This approach enables us to treat aqueous acids and bases by a mathematical analysis similar to that developed for solubility equilibria. To illustrate this approach, we will consider some typical aqueous solutions of acids and bases as examples of homogeneous[1] equilibrium systems.

When hydrogen chloride is dissolved in water, the resulting solution is found experimentally to contain very little molecular HCl. Nearly all the acid has ionized:

$$HCl_{(aq)} + H_2O_{(\ell)} \rightleftharpoons H_3O^+_{(aq)} + Cl^-_{(aq)}$$

---

[1]Heterogeneous equilibria involve two or more phases of matter (e.g., solid/gas, solid/solution), while homogeneous systems are those in which all components are in the same phase (gas or liquid solution).

Such acids are said to be "strong" and their solutions are not very accurately described by simple equilibrium calculations. A similar situation occurs with "strong" bases, such as the methoxide ion obtained by dissolving sodium methoxide in water:

$$CH_3O^-_{(aq)} + H_2O_{(\ell)} \rightleftharpoons CH_3OH_{(aq)} + OH^-_{(aq)}$$

We shall, therefore, restrict our discussion to equilibrium systems more amenable to a simple mathematical analysis, the weak acids and bases. The reaction of these species with water is appreciably less than 100%, so an equilibrium analysis is quite applicable.

Since we will be dealing entirely with relatively dilute aqueous solutions, the concentration of liquid water will remain essentially constant. We may think of the constant c_{H_2O} as omitted[2] from the equilibrium constant expressions for weak acids and bases. For convenience, we will not even bother to write H_2O in equations for acid dissociation. The H_3O^+ species is actually only one of a number of species containing H^+ and its use, while helping to visualize the competitive nature of an acid/base system, does not add any more "reality" to our formulations since all the ions and molecules in the system are associated with one or more water molecules in solution. We must use H_2O in equations representing aqueous weak bases in order to have a balanced equation, but we will still omit c_{H_2O} from the mathematical expressions employed.

The conventions introduced in Unit 10 for writing equilibrium expressions apply to all equilibrium systems. As examples, consider the cases of aqueous ammonia and aqueous hydrofluoric acid:

aq. ammonia $\qquad NH_{3(aq)} + H_2O_{(\ell)} \rightleftharpoons NH^+_{4(aq)} + OH^-_{(aq)}$

for which, by convention, symbolizing the basic behavior of ammonia by a subscript b:

$$K_b = \frac{c_{NH_4^+} \times c_{OH^-}}{c_{NH_3}}$$

aq. hydrofluoric acid $\qquad HF_{(aq)} \rightleftharpoons H^+_{(aq)} + F^-_{(aq)}$

for which, by convention, symbolizing the acidic behavior of HF by a subscript a:

$$K_a = \frac{c_{H^+} \times c_{F^-}}{c_{HF}}$$

Our primary interest in aqueous acids and bases will be in their influence on the H^+-ion or OH^--ion concentration of the solution. When these concentrations are relatively small (as in aqueous solutions of the weak acids or bases), they are often expressed by a

[2]In "true" equilibrium constants, based on activities (Unit 10), the activity of a pure liquid, like that of a pure solid, is defined as unity (1).

pH or pOH notation, defined by:

$$pH = -\log C_{H^+}$$

$$pOH = -\log C_{OH^-}$$

These are related[3], in an aqueous system around $20\text{-}25^0C$, by:

$$pH + pOH = 14.00$$

The following sections will illustrate applications of pH or pOH notation and the mathematical analysis of simple equilibria to some common acid/base systems.

- References -

1. O'Connor, Rod. 1977. Fundamentals of Chemistry, second edition. New York: Harper & Row (Units 19, 20)

2. Brown, Theodore, and H. Eugene LeMay, Jr. 1977. Chemistry: The Central Science. Englewood Cliffs, N.J.: Prentice-Hall (Chapters 16, 17)

3. Masterton, William, and Emil Slowinski. 1977. Chemical Principles, fourth edition. Philadelphia: W.B. Saunders (Chapters 19, 20)

4. Nebergall, William, F.C. Schmidt, and H.F. Holtzclaw, Jr. 1976. College Chemistry, fifth edition. Lexington, MA: D.C. Heath (Chapters 15, 18)

5. van der Werf, C.A. 1961. Acids, Bases, and the Chemistry of the Covalent Bond. New York: Van Nostrand/Reinhold.

OBJECTIVES:

   (1) Given the name or formula of a weak acid, the label concentration of its aqueous solution, and its dissociation constant (or access to Appendix D), be able to calculate the pH of the solution.

   (2) Given the name or formula of a weak base, the label concentration of its aqueous solution, and its dissociation constant (or access to Appendix D), be able to calculate the pH of the solution.

   (3) Given the composition of a conjugate-pair buffer and the dissociation constant for one of its components (or access to Appendix D), be able to calculate the pH of the buffer solution.

---

[3]Water itself is slightly ionized, in pure water to the extent of about 10^{-7} mole per liter each of H^+ and OH^-:

$$H_2O_{(\ell)} \rightleftharpoons H^+_{(aq)} + OH^-_{(aq)}$$

$$K_w \text{ (ion-product for water)} = C_{H^+} \times C_{OH^-} = 10^{-7} \times 10^{-7} = 10^{-14}$$

*(4) *Given the description of a conjugate-pair buffer, the dissociation constant for one of its components (or access to Appendix D), and the volume and concentration of a strong acid or base added to a specified volume of the buffer, be able to calculate the resulting pH change of the solution.*

PRE-TEST:

Necessary Dissociation Constants:	
$HCN_{(aq)}$	4.8×10^{-10}
$NH_{3(aq)}$	1.8×10^{-5}
$H_2PO_4^-{}_{(aq)}$	6.2×10^{-8}

(1) Hydrogen cyanide is a very toxic gas having the faint odor of bitter almonds. In some states the death sentence of a convicted felon was executed by use of the "gas chamber", in which HCN gas was used. Some insects also use HCN for protection against their enemies. Certain centipedes, for example, excrete a dilute aqueous solution of HCN when disturbed. What is the approximate pH of a 0.20 $\underline{M}$ solution of hydrocyanic acid? _____

(2) "Household ammonia" is often used in cleaning solutions for porcelain or glass. Television advertising has even used the pungent odor of ammonia vapors to "plug" a product as "so strong you can smell it". What is the approximate pH of an aqueous solution labeled 3.0 $\underline{M}$ ammonia? _____

(3) A "phosphate" buffer system helps control the pH of the blood. Many carbonated beverages also use this system, although at a significantly higher concentration than that found in the bloodstream. What is the approximate pH of a buffer made by dissolving 15 g each of sodium dihydrogen phosphate and sodium hydrogen phosphate in sufficient water to form 500 ml of solution? _____

*(4) A number of buffer preparations are sold for the control of gastric acidity, among them a "phosphate" buffer which is rendered more palatable by the addition of sugar and artificial flavoring agents. The buffer solution is initially at a pH of about 7.4, maintained by 0.5 $\underline{M}$ $H_2PO_4^-$ and 0.8 $\underline{M}$ HPO_4^{2-} concentrations. If 20 ml of this solution was taken as a medication for "upset stomach" under conditions corresponding to addition of the buffer solution to 100 ml of 0.10 $\underline{M}$ HCl (a stomach pH of 1.0), what would be the resulting pH of the 120 ml mixture?

_____

---

*Proficiency Level

Answers and Directions:

(1) 5.0   (2) 11.9   (3) 7.1

If all are correct, go on to question *4. If you missed any, study METHODS, sections 11.1 - 11.3.

*(4) 6.7

If your answer was correct, go on to RELEVANT PROBLEMS, UNIT 11. If you missed it, study METHODS, section 11.4.

METHODS

11.1 Finding the pH of a Weak Acid Solution

If we limit our considerations to the simple case of a weak monoprotic[4] acid, we can represent the equilibrium system by the general equation:

$$HX_{(aq)} \rightleftharpoons H^+_{(aq)} + X^-_{(aq)}$$

The equilibrium constant for the acid dissociation, symbolized by K_a, is then:

$$K_a = \frac{C_{H^+} \times C_{X^-}}{C_{HX}}$$

(in which the C terms represent <u>equilibrium</u> concentrations, in moles per liter.) If we are primarily interested in the hydrogen-ion concentration of the solution, i.e., the solution pH, then we can make some useful simplifying approximations. These will be valid <u>only</u> under the conditions indicated:

1. If the acid is sufficiently strong <u>and</u> sufficiently concentrated so that it is the main source of H^+ (i.e., $H_2O \rightleftharpoons H^+ + OH^-$ is a negligible contribution), and the <u>only</u> source of the conjugate[5] base (X^-), then:

$$C_{H^+} \simeq C_{X^-}$$

2. If the acid is sufficiently weak[6] so that the <u>equilibrium</u> concentration of the undissociated acid is quite close to that indicated by the "label concentration",

---

[4]An acid is said to be <u>monoprotic</u> if it acts to furnish only one proton per acid species, e.g., $HF \rightleftharpoons H^+ + F^-$, $HPO_4^{2-} \rightleftharpoons H^+ + PO_4^{3-}$. The dissociations of diprotic ($2H^+$) or triprotic ($3H^+$) acids are sequential, with each successive acid being weaker. For example, for $H_3PO_4 \rightleftharpoons H^+ + H_2PO_4^-$, $K_a = 7.5 \times 10^{-3}$; for $H_2PO_4^- \rightleftharpoons H^+ + HPO_4^{2-}$, $K_a = 6.2 \times 10^{-8}$; for $HPO_4^{2-} \rightleftharpoons H^+ + PO_4^{3-}$, $K_a = 2.0 \times 10^{-13}$.

[5]A conjugate pair is represented by any two species differing only by a proton (H^+). For example, NH_3/NH_4^+ and HF/F^- are two conjugate pairs. We say that ammonia is the conjugate base of ammonium ion, or that HF is the conjugate acid of fluoride ion, etc.

[6]For our purposes, this will correspond to a dissociation constant of around 10^{-4} or smaller.

then we may safely approximate the equilibrium concentration by the molarity indicated on the label for the acid solution:

$$c_{HX} \simeq \underline{M} \text{ (acid)}$$

Now, introducing these approximations into our general equilibrium constant expression, we have

$$K_a = \frac{c_{H^+} \times c_{X^-}}{c_{HX}} \simeq \frac{c_{H^+} \times c_{H^+}}{\underline{M} \text{ (acid)}}$$

Solution of this approximate equation gives:

$$\frac{c_{H^+}^2}{\underline{M}(\text{acid})} \simeq K_a$$

$$c_{H^+}^2 \simeq K_a \times \underline{M} \text{ (acid)}$$

$$c_{H^+} \simeq \sqrt{K_a \times \underline{M} \text{ (acid)}}$$

From the definition of pH and the appropriate use of logarithms (Appendix B):

$$pH = -\log c_{H^+} \simeq -\log \sqrt{K_a \times \underline{M} \text{ (acid)}}$$

$$pH \simeq -\tfrac{1}{2} \log \left[K_a \times \underline{M} \text{ (acid)} \right]$$

This is a very handy equation for making routine calculations of pH for solutions of weak acids. It must be remembered, of course, that this expression is an approximation whose validity requires the conditions specified in its derivation. It is still useful for a wide variety of simple applications.

It is not necessary, and perhaps not even desirable, that this equation be memorized. If its derivation is understood, then that general approach can be applied to any suitable system. In any case, it is most convenient to delay actual calculations to the final step, which will correspond to that of the equation derived.

EXAMPLE 1

The SOLUTION to Pre-Test question 1:
The K_a-value for HCN is 4.8×10^{-10}.
The equilibrium system in aqueous HCN may be represented by:

$$HCN_{(aq)} \rightleftharpoons H^+_{(aq)} + CN^-_{(aq)}$$

for which:

$$K_a = \frac{c_{H^+} \times c_{CN^-}}{c_{HCN}} \simeq \frac{c_{H^+}^2}{\underline{M}(\text{HCN})}$$

$$4.8 \times 10^{-10} \approx \frac{c_{H^+}^2}{0.20}$$

$$c_{H^+} \approx \sqrt{(4.8 \times 10^{-10}) \times (0.20)}$$

$$pH = -\log c_{H^+} \approx -\tfrac{1}{2} \log (9.6 \times 10^{-11}) \approx \underline{5.0}$$

EXERCISE 1

Boric acid solutions are very weakly acidic and their concentrations are limited by the relatively low solubility of boric acid in water. Such solutions are frequently employed in "eyewash" preparations. Commercial "eye drops", such as MURINE, contain boric acid as one of the ingredients, although these solutions also contain several other substances affecting the pH. What is the approximate pH of a laboratory stock solution prepared for emergency eyewash use, if the label reads 0.18 $\underline{M}$ Boric Acid?

[See Appendix D for the K_a-value.]

(answer, page 279)

11.2 Finding the pH of a Weak Base Solution

If we follow a general approach similar to that developed for a weak monoprotic acid, then our equilibrium system will be given by:

$$B_{(aq)} + H_2O_{(aq)} \rightleftarrows BH^+_{(aq)} + OH^-_{(aq)}$$
$$\text{(weak base)}$$

for which the conventional constant, symbolized for a base by K_b, is:

$$K_b = \frac{c_{BH^+} \times c_{OH^-}}{c_B}$$

Note that c_{H_2O} is "omitted" from the expression for reasons previously discussed.

We can make the same kinds of simplifying approximations used with weak acids, recognizing the limitations these impose, and derive:

$$c_{OH^-} \approx \sqrt{K_b \times \underline{M} \text{ (base)}}$$

The pOH can then be found:

$$pOH \approx -\tfrac{1}{2} \log [K_b \times \underline{M} \text{ (base)}]$$

after which, from the K_w-expression for water:
$$pH = 14.00 - pOH$$

271

EXAMPLE 2

The *SOLUTION* to Pre-Test question 2:

The K_a-value for NH_3 is 1.8×10^{-5}.

$$NH_{3(aq)} + H_2O_{(\ell)} \rightleftarrows NH_{4(aq)}^+ + OH_{(aq)}^-$$

$$K_b = \frac{c_{NH_4^+} \times c_{OH^-}}{c_{NH_3}} \approx \frac{c_{OH^-}^2}{\underline{M} \text{ (base)}}$$

$$c_{OH^-} \approx \sqrt{K_b \times \underline{M} \text{ (base)}}$$

$$pOH \approx -\tfrac{1}{2} \log \ [(1.8 \times 10^{-5}) \times 3.0 \ \underline{M}] \approx 2.1$$

$$pH = 14.00 - pOH \approx 14.00 - 2.1 \approx \underline{11.9}$$

- -

EXERCISE 2

Most marine organisms, unlike land mammals and insects, eliminate waste nitrogen in the form of ammonia. What is the pH of an aquarium, assuming that the only species of concern is waste ammonia at a concentration of 0.023 $\underline{M}$?

(answer, page 279)

- -

11.3 Finding the pH of a Simple Buffer Solution

By definition, a buffer solution is one which is able to maintain a nearly constant pH on dilution or on limited addition of acid or base. The simplest type of buffer is the so-called "conjugate-pair" system, prepared by mixing a weak acid and one of its salts (e.g., acetic acid with sodium acetate) or a weak base and one of its salts (e.g., ammonia with ammonium chloride). Since the buffer will contain significant amounts of both members of an acid/base conjugate pair, the equilibrium system for the buffer may be represented by an equilibrium equation using either member of the pair as the "left hand" species. The equation which is most convenient for the particular calculation is used.

It is fairly easy to see why a buffer should resist pH change when acid or base is added if we consider a "generalized conjugate-pair buffer" of the weak acid HX and one of its salts, furnishing the X^- ion. If extra strong acid (represented by H^+) is added, it is consumed by the base in the buffer:

$$X^- + H^+ \rightarrow HX$$

while added strong base (OH^-) will react with the acid component of the buffer:

$$HX + OH^- \rightarrow H_2O + X^-$$

In either case, accumulation of extra H^+ or OH^- is avoided by action of buffer components, so pH change is prevented, at least until some significant fraction of the necessary buffer component is consumed.

The resistence of a buffer to pH change on dilution can also be illustrated by our "generalized case". If we choose to represent the buffer by:

$$HX_{(aq)} \rightleftarrows H^+_{(aq)} + X^-_{(aq)}$$

then:

$$K_a = \frac{C_{H^+} \times C_{X^-}}{C_{HX}}$$

from which[7]:

$$C_{H^+} = K_a \left[\frac{C_{HX}}{C_{X^-}} \right]$$

Thus it is apparent that the hydrogen-ion concentration (hence, the pH) of a buffer depends only on the magnitude of the dissociation constant and the ratio of concentrations of the buffer components. Since both components are in the same solution, addition of more water (dilution) will change the values of the concentrations, but not the ratio. This is, per-haps, even easier to see if we write the concentrations with their proper units:

$$C_{H^+} = K_a \times \frac{moles\ (HX)}{liters(buffer)} \times \frac{liters(buffer)}{moles\ (X^-)}$$

$$C_{H^+} = K_a \times \frac{moles\ (HX)}{moles\ (X^-)}$$

Thus, within the limits of ideal behavior of the buffer components, the pH of the solution is independent of the buffer volume.

A similar derivation could be made from the use of an equilibrium equation consistent with K_b for the basic component of the buffer:

$$C_{OH^-} = K_b \times \frac{moles\ (X)}{moles(XH^+)}$$

From these two relationships, by using the definitions of pH and pOH (including pH = 14.00 - pOH), we may obtain, for dealing with conjugate-pair buffer solutions:

When K_a is readily available:

$$pH = -\log \left[K_a \times \frac{moles(acid)}{moles(base)} \right]$$

---

[7]Note that for a buffer, $C_{H^+} \neq C_{X^-}$, since the salt provides an "extra" source of X^- ion.

When K_b is more readily available:

$$pOH = -log \left[K_b \times \frac{moles(base)}{moles(acid)} \right]$$

(then pH = 14.00 - pOH)

Although these equations do introduce some approximation, they are generally reliable for most common conjugate-pair buffers. If molar concentrations of the buffer components are known, these may obviously be used to form the buffer component ratios. The equations need not be memorized, but their forms are such as to be easily remembered and their regular use will prove quite convenient.

EXAMPLE 3

The SOLUTION to Pre-Test question 3:

The K_a-value for $H_2PO_4^-$ is available (6.2×10^{-8}), while the K_b for HPO_4^{2-} is not, so we will describe the buffer system by:

$$H_2PO_4^-{}_{(aq)} \underset{\leftarrow}{\rightarrow} H^+{}_{(aq)} + HPO_4^{2-}{}_{(aq)}$$

for which

$$pH = -log \left[(6.2 \times 10^{-8}) \times \frac{moles(H_2PO_4^-)}{moles(HPO_4^{2-})} \right]$$

We must convert the __mass__ composition data to __moles__, using the method introduced in Unit 3. The volume of the solution is not needed for our calculation.

$$\frac{15 \ g(NaH_2PO_4)}{1} \times \frac{1 \ mole(NaH_2PO_4)}{120 \ g(NaH_2PO_4)} \times \frac{1 \ mole(H_2PO_4^-)}{1 \ mole(NaH_2PO_4)} = 0.125 \ mole(H_2PO_4^-)$$

$$\frac{15 \ g(Na_2HPO_4)}{1} \times \frac{1 \ mole(Na_2HPO_4)}{142 \ g(Na_2HPO_4)} \times \frac{1 \ mole(HPO_4^{2-})}{1 \ mole(Na_2HPO_4)} = 0.106 \ mole(HPO_4^{2-})$$

then:

$$pH = -log \left[(6.2 \times 10^{-8}) \times \frac{0.125}{0.106} \right] = \underline{7.1}$$

EXERCISE 3

Nitrogen is an essential element for plant growth and the production of "agrichemicals" containing nitrogen, for use as fertilizers, is a massive industrial venture. A number of such soil additives are available. Anhydrous ammonia is a simple choice. If the soil is acidic, the ammonia will be bound as a salt, but alkaline soils cannot do this and much of the ammonia is lost by vaporization. For alkaline soils, acidic ammonium salts are preferred, such as ammonium nitrate or ammonium sulfate. As an ammonium salt dissolves and percolates through an alkaline soil, some of the ammonium ion is converted to ammonia by loss of protons to soil bases. What is the pH of a ground water sample buffered by a mixture of ammonia and ammonium nitrate resulting from conversion of 26% of residual NH_4NO_3 additive to ammonia?

See Appendix D for dissociation constants.

(answer, page 280)

==

Extra Practice

EXERCISE 4

Formic acid solutions can cause painful skin burns. Certain ants use formic acid in their defense mechanisms and it was once suggested that bee venom was mainly formic acid, although we know now that this was incorrect. What is the approximate pH of 0.15 M formic acid?

[For the K_a-value, see Appendix D.]

(answer, page 280)

EXERCISE 5

Ethanolamine, $HOCH_2CH_2NH_2$, is found among the degradation products of complex organic compounds called cephalins. The cephalins are believed to have an important function in the clotting of blood. The base dissociation constant for ethanolamine is 3.2×10^{-5}. What is the approximate pH of a 0.50 M solution of ethanolamine?

(answer, page 281)

EXERCISE 6

Bacteria play a suprisingly large role in the modern chemical industry. Bacterial fermentations, familiar in the production of alcohol, are also used in large scale processes for making acetone and a number of other important organic compounds. Catalysis by bacterial enzymes is used extensively in the pharmaceutical chemicals industry and the analysis of microbial byproducts has opened new fields in antibiotic research. Bacterial cultures must be maintained under carefully controlled conditions, including pH control by buffers. Buffers for such purposes cannot be selected only on the basis of their pH characteristics. Buffer components must not be toxic to the microorganisms in the cultures, nor, if the buffer action is to be maintained, may the buffers contain chemicals metabolized by the organisms. For a particular culture medium, a buffer composed of 75 g of acetic acid and 90 g of sodium acetate per 25 liters has been found suitable. What is the approximate pH of this buffer?

[See Appendix D for dissociation constants.]

(answer, page 281)

==

At this point you should try the competency level Self-Test questions on page 278.

==

METHODS

*11.4 Buffer pH Change

The pH of a conjugate pair buffer is a function of the dissociation constant for one component and the mole ratio of the two components, as derived in Section 11.3:

when K_a is available:

$$pH = -\log \left[K_a \times \frac{moles(acid)}{moles(base)} \right]$$

when K_b is more readily available:

$$pOH = -\log \left[K_b \times \frac{moles(base)}{moles(acid)} \right]$$

(then pH = 14.00 - pOH)

To determine the change in pH resulting from the addition of strong acid or base, it is necessary to calculate the pH of the original buffer and then to repeat the calculation for the buffer component mole ratio resulting from the reaction with added base or acid. The latter calculation requires some simple solution stoichiometry (Unit 6), along with a recognition of the nature of the reaction taking place. It should be apparent that added acid will consume an equivalent amount of the base component of the buffer, while producing an equivalent amount of the conjugate acid in the buffer. Added base should have the reverse affect. Since buffer pH is independent of solution volume, within limits of ideal behavior, the dilution of the buffer by addition of a reacting solution has no significant effect on pH.

*EXAMPLE 4

The SOLUTION to Pre-Test question 4:

Since K_a for $H_2PO_4^-$ is available (6.2×10^{-8}), and K_b for HPO_4^{2-} is not, we will describe the buffer equilibrium by:

$$H_2PO_4^- \rightleftharpoons H^+_{(aq)} + HPO_4^{2-}_{(aq)}$$

for which

$$pH = -\log \left[K_a \times \frac{moles(H_2PO_4^-)}{moles(HPO_4^{2-})} \right]$$

We already know the pH of the original buffer (7.1), so we are concerned with what happens when the buffer is mixed with 0.10 $\underline{M}$ HCl. The expected reaction <u>consumes</u> HPO_4^{2-} and <u>pro-</u><u>duces</u> $H_2PO_4^-$:

$$HPO_4^{2-} + H^+ \rightarrow H_2PO_4^-$$
$$\text{(from HCl)}$$
10 mmol

The <u>*quantitative*</u> *change depends on the stoichiometric consumption of the added acid:*

$$\text{mmols }(H^+)\text{ added} = \frac{0.10 \text{ mmol}(H^+)}{1 \text{ ml}} \times \frac{100 \text{ ml}}{1} = 10 \text{ mmol}(H^+)$$

Thus, when equilibrium is reestablished:

<u>for $H_2PO_4^-$:</u>

$$\left[\frac{0.5 \text{ mmol}}{1 \text{ ml}} \times \frac{20 \text{ ml}}{1}\right] + \left[10 \text{ mmol}\right] = 20 \text{ mmol}(H_2PO_4^-)$$
$$\text{(original)} \qquad \text{(newly formed)}$$

<u>for HPO_4^{2-}:</u>

$$\left[\frac{0.8 \text{ mmol}}{1 \text{ ml}} \times \frac{20 \text{ ml}}{1}\right] - \left[10 \text{ mmol}\right] = 6 \text{ mmol}(HPO_4^{2-})$$
$$\text{(original)} \qquad \text{(consumed)}$$

The pH of the final solution, then, using mmol ratios (equivalent to mole ratios) is:

$$\text{pH} = -\log\left[(6.2 \times 10^{-8}) \times \frac{20}{6}\right] = \underline{6.7}$$

---

*EXERCISE 7

   Although the <u>pH</u> of a buffer is independent of both the volume and dilution of the buffer, the <u>buffer capacity</u> is not. <u>Buffer capacity</u> is a measure of how much acid or base a buffer can accomodate without significant pH change. Estimate the pH that would have resulted for the situation described in EXAMPLE 4 if an economy-minded person with a limited knowledge of buffers had diluted his phosphate buffer 1:100 with distilled water and then swallowed 20 ml of the <u>diluted</u> solution to neutralize the 100 ml of stomach acid.

(answer, page 281)

===

<u>Extra Practice</u>
*EXERCISE 8

   The bicarbonate/"carbonic acid" buffer system is of the utmost importance in the control of blood pH. Evidence suggests that "carbonic acid" (H_2CO_3) is actually a negligible component, as such. The acid properties of this buffer system are more accurately attributed to dissolved carbon dioxide:

$$CO_{2(aq)} + H_2O_{(\ell)} \rightleftharpoons H^+_{(aq)} + HCO^-_{3(aq)}$$

What would be the approximate pH __change__ if 15 ml of 0.10 $\underline{M}$ NaOH were added to 50 ml of a buffer in which the concentration of HCO_3^- was 0.30 $\underline{M}$ and the concentration of dissolved CO_2 was 0.15 $\underline{M}$?

For the dissociation constants, see Appendix D.

(answer, page 282)

==

SELF-TEST (UNIT 11)   [answers, page 283]

11.1. The tart taste of soured milk is due primarily to the formation of lactic acid by a bacterial fermentation of milk sugar.

$$CH_3CHCO_2H \qquad \text{(lactic acid, } K_a = 1.4 \times 10^{-4}\text{)}$$
$$\quad | \quad$$
$$\quad OH$$

What is the approximate pH of 0.25 $\underline{M}$ lactic acid?

11.2. Nicotine ($C_{10}H_{14}N_2$), the toxic alkaloid from tobacco, has been used in dilute solutions as an insecticide. The K_b-value for the group accounting for most of the basic character of nicotine is 1.0×10^{-6}. What is the approximate pH of a solution containing 10.0 g of nicotine per liter?

11.3. Deuterium is a nonradioactive isotope of hydrogen, having an atomic mass twice as great as that of "normal" hydrogen atom. In an experiment to study the effects of "heavy water" (D_2O) on living organisms, the solution to be used for a bacterial culture was prepared using D_2O, rather than H_2O, and buffered by a mixture of 15 g of sodium acetate and 10 g of deuteroacetic acid, CH_3CO_2D (K_a in $D_2O = 5.5 \times 10^{-6}$), per 100 ml of solution. What was the approximate pH of this buffer? [Perhaps we should call it, in this case, the pD.]

If you completed Self-Test questions 11.1 - 11.3 correctly, you may go on to the _proficiency level, try the RELEVANT PROBLEMS (Unit 11), or stop here. If not, you should_ _consult your instructor for suggestions of further study aids._

----

*11.4. A bacterial culture was maintained in an aqueous medium buffered by a system which was 0.25 $\underline{M}$ in ammonia and 0.35 $\underline{M}$ in ammonium chloride. Portions of the solution were withdrawn periodically and treated with picric acid, to stain the bacteria for examination. Assuming that the monoprotic picric acid can be approximated for our purposes as essentially 100% ionized, what would be the pH of a solution made by mixing equal volumes of the buffered solution and 0.050 $\underline{M}$

picric acid? [K_b for ammonia $= 1.8 \times 10^{-5}$.]

If you answered this question correctly, you may go on to the RELEVANT PROBLEMS (Unit 11). If not, you should consult your instructor for suggestions of further study aids.

ANSWERS to EXERCISES, Unit 11

1. (5.0) <u>Solution</u>:

From Appendix D, K_a for boric[8] acid $= 6.0 \times 10^{-10}$

$$H_3BO_{3(aq)} \rightleftharpoons H^+_{(aq)} + H_2BO^-_{3(aq)}$$

$$K_a = \frac{C_{H^+} \times C_{H_2BO_3^-}}{C_{H_3BO_3}} \simeq \frac{C^2_{H^+}}{\underline{M}(H_3BO_3)}$$

$$C_{H^+} \simeq \sqrt{K_a \times \underline{M}(H_3BO_3)}$$

$$pH \simeq -\tfrac{1}{2} \log [(6.0 \times 10^{-10}) \times 0.18]$$

2. (10.8) <u>Solution</u>:

$$NH_{3(aq)} + H_2O_{(\ell)} \rightleftharpoons NH^+_{4(aq)} + OH^-_{(aq)}$$

$$K_b = \frac{C_{NH_4^+} \times C_{OH^-}}{C_{NH_3}} \simeq \frac{C^2_{OH^-}}{\underline{M}(base)}$$

$$C_{OH^-} \simeq \sqrt{K_b \times \underline{M}(base)}$$

$$pOH \simeq -\tfrac{1}{2} \log [(1.8 \times 10^{-5}) \times 0.023] \qquad 3.2$$

$$pH = 14.00 - pOH \simeq 14.00 - 3.2$$

---

[8]The acidity of boric acid appears, from careful investigation, to be more accurately attributed to the system:

$$B(OH)_3 + H_2O_{(\ell)} \rightleftharpoons H^+_{(aq)} + [B(OH)_4]^-_{(aq)}$$

However, both formulations are simplifications of a fairly complex equilibrium system and they are mathematically equivalent.

3. (8.8) Solution:

Since we need the mole ratio of buffer components, we'll start that calculation first. If 26% of the NH_4NO_3 was converted to NH_3, then 74% was left as NH_4NO_3. Using appropriate unity factors:

$$\frac{74 \text{ g}(NH_4NO_3) \text{ [left]}}{100 \text{ g}(NH_4NO_3) \text{ [orig]}} \times \frac{1 \text{ mole}(NH_4NO_3)}{80 \text{ g}(NH_4NO_3)} = 0.925 \text{ mole}(NH_4NO_3)/100 \text{ g(orig)}$$

$$\frac{26 \text{ g}(NH_4NO_3) \text{ [converted]}}{100 \text{ g}(NH_4NO_3) \text{ [orig]}} \times \frac{1 \text{ mole}(NH_4NO_3)}{80 \text{ g}(NH_4NO_3)} \times \frac{1 \text{ mole}(NH_3)}{1 \text{ mole}(NH_4NO_3)} =$$

$$0.325 \text{ mole}(NH_3)/100 \text{ g(orig)}$$

Since K_b for NH_3 is available (Appendix D) and K_a for NH_4^+ is not, we will use:

$$NH_{3(aq)} + H_2O_{(\ell)} \underset{\leftarrow}{\rightarrow} NH_{4(aq)}^+ + OH_{(aq)}^-$$

for which:

$$pOH = -\log \left[K_b \times \frac{\text{moles}(NH_3)}{\text{moles}(NH_4NO_3)} \right]$$

$$= -\log \left[(1.8 \times 10^{-5}) \times \frac{0.325/100}{0.925/100} \right] = 5.2$$

then:

$$pH = 14.00 - pOH = 14.00 - 5.2$$

4. (2.2) Solution:

From Appendix D, K_a for formic acid is 2.0×10^{-4}. The equilibrium system in aqueous formic acid may be represented by:

$$HCO_2H_{(aq)} \underset{\leftarrow}{\rightarrow} H_{(aq)}^+ + HCO_{2(aq)}^-$$

for which:

$$K_a = \frac{C_{H^+} \times C_{HCO_2^-}}{C_{HCO_2H}} \approx \frac{C_{H^+}^2}{\underline{M}(HCO_2H)}$$

$$2.0 \times 10^{-4} \approx \frac{C_{H^+}^2}{0.15}$$

$$C_{H^+} \approx \sqrt{(2.0 \times 10^{-4}) \times 0.15}$$

$$pH = -\log C_H+ \approx -\tfrac{1}{2} \log [(2.0 \times 10^{-4}) \times 0.15]$$
$$\approx -\tfrac{1}{2} \log (3.0 \times 10^{-5})$$
$$\approx -\tfrac{1}{2} \log [\log 3.0 + \log 10^{-5}]$$
$$\approx -\tfrac{1}{2} [(0.477) + (-5)]$$
$$\approx \underline{2.2}$$

5. (11.6) Solution:

$$K_b \approx \frac{C_{OH^-}^2}{\underline{M}(base)}$$

$$C_{OH^-} \approx \sqrt{K_b \times \underline{M}(base)}$$

$$pOH \approx -\tfrac{1}{2} \log[K_b \times \underline{M}(base)]$$
$$\approx -\tfrac{1}{2} \log[(3.2 \times 10^{-5}) \times 0.50] = 2.4$$
$$pH = 14.00 - pOH \approx 14.00 - 2.4$$

6. (4.7) Solution:

Since K_a for acetic acid (1.8×10^{-5}) is available:

$$CH_3CO_2H_{(aq)} \rightleftarrows H^+_{(aq)} + CH_3CO_{2(aq)}^-$$

for which:

$$pH = -\log \left[K_a \times \frac{moles(CH_3CO_2H)}{moles(CH_3CO_2^-)} \right]$$

The mole ratio may be obtained from the mass ratio, using unity factors:

$$\frac{75 \text{ g}(CH_3CO_2H)}{90 \text{ g}(CH_3CO_2Na)} \times \frac{1 \text{ mole}(CH_3CO_2H)}{60 \text{ g}(CH_3CO_2H)} \times \frac{82 \text{ g}(CH_3CO_2Na)}{1 \text{ mole}(CH_3CO_2Na)} \times \frac{1 \text{ mole}(CH_3CO_2Na)}{1 \text{ mole}(CH_3CO_2^-)}$$

$$= 1.14 \text{ mole}(CH_3CO_2H)/\text{mole}(CH_3CO_2^-)$$

Then:

$$pH = -\log [(1.8 \times 10^{-5}) \times 1.14]$$

===

*7. (~1.1) Solution:

$$HPO_4^{2-} + H^+ \rightarrow H_2PO_4^-$$

HPO_4^{2-} available:

$$\frac{0.8 \text{ mmol}}{1 \text{ ml}} \times \frac{1 \text{ ml}}{100 \text{ ml}} \times \frac{20 \text{ ml}}{1} = 0.16 \text{ mmol}$$

("dilution correction")

H^+ to be neutralized:

$$\frac{0.1 \text{ mmol}}{\text{ml}} \times \frac{100 \text{ ml}}{1} = 10 \text{ mmol}$$

$H_2PO_4^-$ originally present:

$$\frac{0.5 \text{ mmol}}{1 \text{ ml}} \times \frac{1 \text{ ml}}{100 \text{ ml}} \times \frac{20 \text{ ml}}{1} = 0.10 \text{ mmol}$$

("*dilution correction*")

The "stomach acid" will consume essentially <u>all</u> of the HPO_4^{2-}, so:

HPO_4^{2-}: approx. <u>zero</u> (at least, very small)

$H_2PO_4^-$: $0.10 + 0.16 = 0.26$ mmol

H^+ : $10 - 0.16 = 9.84$ mmol

Now, the "parent" acid, H_3PO_4, is also weak (K_a, Appendix D, $= 7.5 \times 10^{-3}$), so the excess hydrogen ion will convert most of the $H_2PO_4^-$ to H_3PO_4:

$$H_2PO_4^- + H^+ \rightleftarrows H_3PO_4$$

$H_2PO_4^-$: approx. <u>zero</u> (at least, very small)

H_3PO_4 : approx. 0.26 mmol

H^+ : $9.84 - 0.26 = 9.58$ mmol

This is still such a large excess that the acidity will be due mainly to H^+ from the "stomach acid", rather than from the small amount produced by ionization of phosphate species, so:

$$c_{H^+} \simeq \frac{9.6 \text{ mmol}}{120 \text{ ml}} = 0.08 \ \underline{M}$$

$$pH \simeq - \log (0.08) \qquad 1.1$$

By diluting the buffer, our "patient" may have saved some money, but he exceeded the buffer capacity to the extent that he now has a "fuller" stomach, but one just about as acidic as before.

*8. (0.14 pH unit) <u>Solution</u>:

From Appendix D, K_a for "carbonic acid" is 4.2×10^{-7}

For the system:

$$CO_{2(aq)} + H_2O_{(\ell)} \rightleftarrows H^+_{(aq)} + HCO^-_{3(aq)}$$

the pH of the original buffer is:

$$pH = - \log \left[(4.2 \times 10^{-7}) \times \frac{0.15}{0.30} \right] = 6.68$$

The effect of added NaOH is:

$$CO_{2(aq)} + OH^-_{(aq)} \rightleftarrows HCO^-_{3(aq)}$$

so, after addition of the base:

$\underline{CO_2}$:

$$\left[\frac{0.15 \text{ mmol}}{1 \text{ ml}} \times \frac{50 \text{ ml}}{1}\right] - \left[\frac{0.1 \text{ mmol}}{1 \text{ ml}} \times \frac{15 \text{ ml}}{1}\right] = 6.0 \text{ mmol}(CO_2)$$

      (original)   (consumed)

$\underline{HCO_3^-}$:

$$\left[\frac{0.30 \text{ mmol}}{1 \text{ ml}} \times \frac{50 \text{ ml}}{1}\right] - \left[\frac{0.1 \text{ mmol}}{1 \text{ ml}} \times \frac{15 \text{ ml}}{1}\right] = 16.5 \text{ mmol } (HCO_3^-)$$

      (original)   (newly formed)

Then, after the "new" equilibrium is established:

$$pH = -\log\left[(4.2 \times 10^{-7}) \times \frac{6.0}{16.5}\right] = 6.82$$

The pH change is then given by:

$$6.82 - 6.68$$

(final) (orig.)

===

ANSWERS *to* SELF-TEST, *Unit 11*

      11.1.   (2.2)

      11.2.   (10.4)

      11.3.   (5.3)

   *11.4.   (8.95)

A.1.   The composition of interstitial waters associated with marine sediments is influenced by the hydrochemical activity of the particular area, sediment-water interactions, interstitial water convection, and diffusion of dissolved species. In the average ocean area, interstitial sodium and chloride concentrations remain constant, similar to their concentrations in the ocean water. Interstitial potassium and magnesium are depleted rapidly due to uptake by clays and also carbonate precipitation for magnesium. Sulfate is depleted through bacterial processes, especially in organic-rich areas. Calcium and strontium are frequently enriched by biogenic activity. In a particular parcel of porewater, all of the sulfate had been reduced and the principal species remaining in solution was the ammonia, NH_3. What is the approximate pH of an aliquot of this porewater, if the ammonia concentration is 0.008 $\underline{M}$? [See Appendix D for K_b-value.]

A.2.   The availability and utilization of boron in soils is largely determined by the pH. Boron occurs in acid soils as boric acid, H_3BO_3, which is readily available to plants. In acid sandy soils, soluble boron nutrients are leached downward. In heavier soils, rapid leaching does not occur. At higher pH values, boron is less available due to its fixation by clays and other minerals. Boron is frequently held by organic complexes from which it may be released for crop use. The amount of boron held in the topsoil is usually much larger than in the subsoils. Consequently, boron deficiency is more noticeable in periods of dry weather. What is the approximate pH of a soil solution in which the acidity is attributed to a concentration of boric acid of 2.3×10^{-3} $\underline{M}$? [See Appendix D for K_a-value.]

A.3.   The dissolved and dispersed constituents in sea water are moved about by a complicated mass transportation system which includes gravity, diffusion, and biological transportation; physical factors such as currents, evaporation, and sedimentation; and geochemical factors such as volcanic activity and chemical reactions. Organisms selectively use many sea water constituents, hence these are subject to a "biological fractionation". Thus, organisms affect the components of sea water both by synthesis and decomposition. The average concentration of dissolved organic compounds in the sea ranges between 0.1 ppm and 10.0 ppm. These organic substances are produced by the biomass and range in composition from nonpolar aliphatic hydrocarbons to complex highly polar acids, alcohols, and carbohydrates. The decomposition of a particular type of marine phytoplankton produces propionic acid such that the surrounding waters have the acid present in a concentration of 0.05 $\underline{M}$. What is the pH of these waters

284

with respect to the propionic acid? [K_a for $CH_3CH_2CO_2H = 1.4 \times 10^{-5}$.]

A.4. The major source of phosphorus in the world's oceans is from land runoff, which has been estimated to provide approximately 2×10^6 metric tons of this valuable nutrient annually. This input is small when compared to the 1.7×10^9 metric tons already in the ocean system. Phosphorus is structurally important as a skeletal component, a buffer component in the marine environment, and an indispensable component for energy conversion processes in marine organisms. A "phosphate" buffer system helps control the pH of sea water. The blood and many carbonated beverages also use a phosphate mixture as a part of their buffer system. What is the pH of an artificial sea water matrix made by dissolving 20 g each of sodium dihydrogen phosphate and sodium hydrogen phosphate in sufficient water to form 1.00 liter of solution? [K_a for $H_2PO_4^- = 6.2 \times 10^{-8}$.]

A.5. Fecal matter from marine organisms and sea birds releases nitrogen compounds to the hydrosphere with relatively little accumulation of nitrogen in the sediments. The total combined nitrogen in the marine environment remains virtually constant because nitrogen fixation balances denitrification. Generally, nitrogen is released from its organic form as ammonia. In the sea water, some of the ammonia is converted to ammonium ion, which is subsequently oxidized to nitrite and nitrate. What is the pH of a sea water sample buffered by a mixture consisting of 15% ammonia and 85% ammonium nitrate, by weight? [K_b for $NH_3 = 1.8 \times 10^{-5}$.]

*A.6. In the marine euphotic zone, most of the ammonium ion originates from microbial decomposition and excretion, but some also enters surface waters through rainfall and runoff from land. In some regions of the world the upper layers of coastal waters are isolated during summer months from the deep waters. The warm surface waters are less dense than the deep waters, hence they stay on top of the colder deep waters so there is negligible mixing. Since the nitrate level in the euphotic zone is close to zero, the ammonium ion is that layer's principal form of nitrogen. With the decreasing solar radiation in the fall the surface waters become cooler and more turbulent. As mixing occurs many phytoplankton are carried below the photosynthetic zone, where they die. They release ammonia, which is oxidized to nitrate. The remaining phytoplankton population is so small they cannot utilize all the available nitrate. Therefore, nitrate is returned to the marine system in considerable quantities. If a marine buffer system is initially at a pH of 9.17, maintained by 0.9 $\underline{M}$ NH_3 and 1.1 $\underline{M}$ NH_4NO_3 concentrations, what is the new pH obtained by adding 10 ml of 0.10 $\underline{M}$ NaOH to 50 ml of the buffer system. [K_b for $NH_3 = 1.8 \times 10^{-5}$.]

*Proficiency Level

*A.7.   The chief constituents of the carbon cycle are methane (CH_4), carbon monoxide (CO), carbon dioxide (CO_2), carbonates (such as $CaCO_3$), and organic matter. The rate of CO_2 production by combustion of fossil fuels is approximately 3.7×10^{14} moles annually. This should cause an annual increase of 1.3 ppm in the atmospheric CO_2 concentration if all this CO_2 remained in the atmosphere. Since the observed annual increment is only about half of this value, it is obvious that a portion of the fossil fuel CO_2 is being absorbed by other carbon reservoirs. One of these reservoirs is the sea. Here the fossil fuel carbon dioxide dissolves in the water and subsequently participates in several important reactions. The dissolved carbon dioxide may be used by phytoplankton to produce organic matter. It may react with water to form "carbonic acid" which functions in concert with the bicarbonate species to form one of the most important buffer systems in the sea. It may also be precipitated as calcium carbonate in the open ocean, largely in the form of the tiny shells of foraminifera. What would be the approximate pH change if 20 ml of 0.10 M NaOH were added to 50 ml of a sea water sample buffered by the H_2CO_3/HCO_3^- system in which the concentration of HCO_3^- was 0.50 M and the concentration of H_2CO_3 (principally aqueous CO_2) was 0.30 M? [K_a for H_2CO_3 = 4.2×10^{-7}.]

*A.8.   What would be the pH change if 40 ml of 0.25 M HCl were added to the original buffer system described in problem A.7?

*A.9.   The distribution of the various forms of phosphorus in the sea is under the broad control of biological and physical agencies that are similar to those which control the marine chemistry of nitrogen. There are many phosphorite deposits on the continental shelves around the world. Unlike the regions of ferromanganese nodule development, where oxidizing conditions prevail, the regions of phosphorite accumulation are reducing environments. The phosphorites have as their general formula: $Ca_{10}(PO_4)_2(CO_3)_6F_2$. A water sample collected over a phosphorite deposit was found to have a buffer system which was 0.003 M in $H_2PO_4^-$ and 0.10 M in HPO_4^{2-}. What would be the pH of a 25.0 ml aliquot of this solution that was treated with 10.0 ml of 1.0×10^{-3} M HCl?

*A.10. What would be the pH change if 25.0 ml of 0.001 M NaOH were added to the 25.0 ml aliquot of the original buffer system described in problem A.9?

ANSWERS:

(A.1.) 10.6, (A.2.) 5.9, (A.3.) 3.1, (A.4.) 7.13, (A.5.) 9.17, (A.6.) 9.19, (A.7.) increase of 0.09 pH unit, (A.8.) decrease of 0.45 pH unit, (A.9.) 8.67, (A.10.) increase of 0.18 pH unit

RELEVANT
PROBLEMS

Unit 11: Acid/Base Equilibria

Set B: Biological & Medical Sciences

B.1. The urine is by far the most important excretory product of the animal body and it is the medium through which the end-products of nitrogeneous metabolism and soluble mineral salts are almost exclusively eliminated under normal conditions. All these substances are found in the urine in aqueous solution or suspension. The catabolism of purines from nucleoproteins leads to the formation of the uric acid that is excreted in the urine. What is the approximate pH of a urine sample, based on a uric acid concentration of 2.0×10^{-3} $\underline{M}$? [K_a for uric acid = 1.3×10^{-4}.]

B.2. Lemon juice occupies a unique position among the citrus juices produced commercially. Because of its acidity, smaller quantities are used at any one time and for many more purposes than may be true for other fruit juices. This accounts for the general desire of consumers to keep some lemon fruit or juice on hand for various applications as a flavorful and healthful source of a "household" acid. Except for sugar and salt, lemon juice has probably been used more extensively to enhance and develop inherent food flavor than any other food item. Most commercial lemon juice goes into concentrated juices and compounded juice products where flavor and standardization are primary considerations. In addition to citric acid, the most important single constituent of the lemon, other constituents of the natural juice are desirable for their effects on flavor and for their health values. The elements aluminum, copper, lithium, titanium, manganese, cobalt, zinc, strontium, boron, tin, nickel, chromium, vanadium, zirconium, molybdenum, and barium are all present in lemon juice, in trace amounts. Citric acid is a polyprotic acid but only the dissociation of one proton is very significant.

$$(HC_6H_7O_7 \rightleftharpoons H^+ + C_6H_7O_7^- , \quad K_a = 8.2 \times 10^{-4})$$

What is the approximate pH of lemon juice if the citric acid concentration is

8.5×10^{-2} $\underline{M}$? (Assume that all the acid present is citric acid.)

B.3. The carbon dioxide arising from the oxidation of nutrients is chiefly an excretory product, transported from the tissues to the lungs by the blood. As the CO_2 diffuses into the blood, it reacts quickly with water to establish the equilibrium system:

$$CO_{2(aq)} + H_2O_{(\ell)} \overset{\rightarrow}{\leftarrow} H_2CO_{3(aq)} \overset{\rightarrow}{\leftarrow} H^+_{(aq)} + HCO^-_{3(aq)}$$

The further ionization of bicarbonate to carbonate is negligible. One of the principal buffers in the blood may be thought of as the carbonic acid/bicarbonate ion conjugate pair. What would be the approximate pH of a blood sample in which the concentration of HCO^-_3 was 0.032 $\underline{M}$ and the concentration of H_2CO_3 (dissolved CO_2) was 0.0016 $\underline{M}$? [For the dissociation constant, see Appendix D.]

B.4. An additional buffer in the blood consists of the $H_2PO_4^-/HPO_4^{2-}$ system. The combination of this and the H_2CO_3/HCO_3^- buffer system maintains the pH of normal blood around 7.4. The maintenance of a nearly constant pH is essential, since many body cells are quickly destroyed by either excess acid or excess alkali. The blood buffers must be able to counteract sudden addition of acids, such as lactic acid produced by strenuous muscle activity, or bases such as some of the nitrogenous waste products from amino acid and nucleoprotein metabolism. The phosphate buffer system in the blood helps to lower the blood pH below that afforded by the H_2CO_3/HCO_3^- itself. What would be the approximate pH of the phosphate buffer system alone in a sample in which the concentration of HPO_4^{2-} was 0.136 $\underline{M}$ and the concentration of $H_2PO_4^-$ was 0.133 $\underline{M}$? [For the dissociation constant, see Appendix D.]

B.5. The phenolic group of organic corrosives includes phenol itself and cresol. Each of these compounds act both as local corrosives and as systemic poisons. These compounds may be absorbed either through the skin or intestinal routes. When spilled upon the skin, they cause immediate necrosis of tissues and large chemical burns which ulcerate and become superficially infected. When phenols are ingested in significant amounts, immediate coagulation of the mucosa of the upper alimentary tract occurs, followed by sloughing necrosis. Through either pathway of contact, these compounds may be absorbed systemically and cause central nervous system damage and vascular collapse. Hepatic and renal necrosis ensues within 24 to 48 hours. As phenol is dissolved in water, it establishes the equilibrium system:

$$\bigcirc\!\!-OH_{(aq)} \overset{\rightarrow}{\leftarrow} \bigcirc\!\!-O^-_{(aq)} + H^+_{(aq)}$$

Such solutions are frequently used as general disinfectants. What is the approximate pH of an aqueous solution in which the phenol concentration is 0.20 $\underline{M}$? [K_a for phenol = 1.3×10^{-10}.]

B.6. Amines are important derivatives of ammonia in which one or more of the hydrogens on a molecule of ammonia are replaced by aliphatic or aromatic substituents. In general, the aliphatic amines are stronger bases than ammonia, while the aromatic amines are weaker. Many heterocyclic amines of biochemical importance consist, at least in part, of ring systems in which one of the ring atoms is nitrogen. Some of the important heterocyclic amines are indole, nicotine, niacin, pyrimidine, and purine. The pyrimidine and purine systems are particularly important in biochemical genetics, the chemistry of heredity. Aniline, $C_6H_5-NH_2$, has been used in making sun-tan preparations to prevent ultraviolet radiation from reaching the skin. The base dissociation constant for aniline is 3.8×10^{-10}. What is the approximate pH of a 0.40 M solution of aniline?

*B.7. Maintenance of the normal water content of cells is of critical importance for the preservation of their health and function. The intracellular water is a closely guarded parameter that varies only in conditions of extreme water imbalance. Any alteration, such as a decrease in the intracellular fluid, is immediately compensated for by the interstitial fluid which bathes the cell. Osmotic force is the most important influence that regulates this water balance. Since sodium is the principal ion in the interstitial fluid, it is the preponderant influence in establishing the osmotic pressure of the interstitial fluid. The regulation of sodium concentration in the body fluids helps to maintain the normal fluid partition in the body. Metabolic alkalosis is caused by an increase in the blood pH, resulting from very high concentrations of bicarbonate ion. In this case, the kidney excretes alkaline urine to make the interstitial fluids more acidic. Sodium ions are excreted along with the bicarbonate ions. One blood buffer system is initially at a pH of about 7.68, maintained by 0.0080 M H_2CO_3 and 0.16 M HCO_3^- concentrations. If 20 ml of this solution is treated by addition of 5.0 ml of 0.10 M HCl, what would be the resulting pH of the 25 ml mixture?

*B.8. The pH of venous plasma is normally maintained in the very narrow range of 7.38 to 7.41. A drop in this pH, that is, a shift toward the acid side, is called metabolic acidosis. Acidosis is fairly common because ordinary metabolism produces acids. The body has two agents, the kidneys and the buffers of the blood, to remove this excess acid and maintain a steady pH. The kidneys "pump" hydrogen ions from the blood to the urine for excretion. Sodium ions are reabsorbed by the blood to maintain charge balance. In addition to the H_2CO_3/HCO_3^- buffer, the blood also contains the $H_2PO_4^-/HPO_4^{2-}$ buffer system. What would be the approximate pH change if 20 ml of 0.20 M KOH were added to 60 ml of a buffer in which the concentration of HPO_4^{2-} was 0.40 M and the concentration of $H_2PO_4^-$ was 0.25 M? [K_a for $H_2PO_4^-$ = 6.2×10^{-8}.]

*B.9.   Lactic acid acidosis may develop during strenuous muscular work. Glycolysis, an important source of ATP under anaerobic conditions, produces the lactic acid. When the lactic acid forms faster than it can be removed, its level in the blood will rise, and the level of the bicarbonate ion will drop, as will the pH. The level of lactic acid in blood may vary from 1-2 millimoles per liter during periods of rest to 10-12 millimoles per liter after vigorous exercise. When the lactic acid concentration reaches the upper level, further exercise is almost impossible because of pain and muscle cramping. To recover from lactic acid acidosis, the overexerted person simply needs to rest. What would be the approximate pH <u>change</u> if 25 ml of 0.15 <u>M</u> NaOH were added to 45 ml of a buffer in which the concentration of sodium lactate, $CH_3CH(OH)CO_2Na$, was 0.50 <u>M</u> and the concentration of lactic acid, $CH_3CH(OH)CO_2H$, was 0.30 <u>M</u>? [K_a for lactic acid = 1.4×10^{-4}.]

*B.10. For nearly every chemical reaction in living organisms, enzymes are present which greatly speed up the rate of the reaction. The enzymes are not consumed during these reactions and are considered to be organic catalysts. For example, hydrogen peroxide is a highly active chemical, often used for bleaching or for cleansing minor wounds. It is also formed continually as a byproduct of certain chemical reactions in living cells. It is toxic to tissue, and if it were not immediately removed or broken down by the cells it would destroy them. In the presence of a proper catalyst, hydrogen peroxide is converted into two harmless substances, water and oxygen. Most enzymes operate inside the living cell, but a few, especially digestive enzymes, function outside the cell. These enzymes are particularly convenient for laboratory work, since they can function in the "test tube". In order to determine what factors affect enzyme activity, experiments are designed so as to limit the number of parameters, under consideration. Control of pH in laboratory situations is frequently accomplished by using the acetic acid/acetate buffer system. What would be the approximate pH <u>change</u> if 5.0 ml of 0.10 <u>M</u> HCl were added to 50 ml of a buffer in which the concentration of CH_3CO_2Na was 0.50 <u>M</u> and the concentration of CH_3CO_2H was 0.10 <u>M</u>? [K_a for $CH_3CO_2H = 1.8 \times 10^{-5}$.]

ANSWERS:

(B.1.) 3.3,   (B.2.) 2.1,   (B.3.) 7.7,   (B.4.) 7.2,   (B.5.) 5.3,   (B.6.) 9.1,   (B.7.) 6.99,
(B.8.) increase of 0.20 pH unit,   (B.9.) increase of 0.21 pH unit,   (B.10.) decrease of 0.05
pH unit

RELEVANT

PROBLEMS

Unit 11: Acid/Base Equilibria

Set I: Industrial Chemistry

I.1.   Persons with metabolic disturbances requiring reduced carbohydrate intake, such as
diabetics, require some sweetening agent other than the typical sugars. Although
these needs offer an appreciable market for artificial sweeteners, an even larger
demand exists from the millions of Americans desiring to lose weight for health or
cosmetic reasons. The first artificial sweetener to capture a large market was
saccharin, which gained wide popularity during the sugar rationing days of World
War II. In dilute solutions, saccharin is about 450 times "sweeter" than table sugar
and its taste is detectable in concentrations as low as 10 ppm. In the pure form,
or in concentrated solution, saccharin is sour-to-bitter and, even in the concentrations recommended for sweetening, the discriminating palate can detect a "non-sugar"
flavor. In addition, saccharin decomposes when heated so that it cannot be used
effectively for sweetening foods before cooking or baking. The cyclamates became
serious competitors for the artificial sweetener market because of their better flavor and greater heat stability. However, evidence that large dosages of cyclamates
might induce cancer in test animals removed them from production and the saccharin
industry alone had the field, an annual market exceeding two million kilograms.
Recently, large doses of saccharin were shown to produce cancer in test animals and
the saccharin industry has been severely curtailed. The ionization constant is
2.5×10^{-12} for:

(saccharin, $C_7H_5SNO_3$)

What is the approximate pH of a sweetening solution prepared by dissolving 18 g of
saccharin in 250 ml of water?

I.2.   The pharmaceuticals industry manufactures about 15,000 metric tons of aspirin annually for the U.S. market, making acetylsalicylic acid one of the most profitable chemicals in the industry. In addition to "simple" aspirin tablets, the drug is sold in
powdered form and as a component of such widely known preparations as ALKA-SELTZER,

BUFFERIN, and ANACIN. The advertising investments alone for aspirin and aspirin-containing drugs represent significant costs, ultimately paid by the consumer. One of the arguments employed by proponents of so-called "buffered" aspirin is that aspirin alone is an acid. Calculate the pH of a 0.10 $\underline{M}$ solution of acetylsalicylic acid and compare this with the pH of 0.10 $\underline{M}$ HCl, the acid normally present in the stomach.

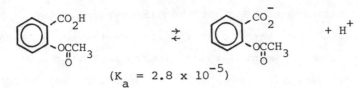

$$(K_a = 2.8 \times 10^{-5})$$

I.3. Lactic acid is one of the oldest known organic acids. It occurs naturally as the primary acid constituent of sour milk. Commercially, lactic acid is produced by the controlled fermentation of the hexose sugars from molasses, corn, or milk, although procedures have been developed that allow its production by synthetic methods. In the fermentation method, the fermenter is charged with whey, a by-product in the manufacture of cheese from whole milk, and inoculated with a culture of $\underline{Lactobacillus}$ $\underline{bulgaricus}$. During the fermentation period, calcium carbonate is added periodically to maintain a proper pH. Therefore, the product is isolated as calcium lactate and is treated with dilute sulfuric acid to regenerate the free acid. The principal uses of lactic acid are in the food industry, primarily as an acidulant and for manufacture of bread additives. Considerable technical grade lactic acid is used as a chemical intermediate and in the tanning industry for decalcification of hides. One of the markets for lactic acid is for the preparation of an infant feeding formula. At one stage in the formula preparation, a solution is prepared containing 8.0 g of lactic acid per liter of water. What is the approximate pH of this solution?

$$\underset{\underset{CH_3CHCO_2H}{|}}{OH} \rightleftarrows \underset{\underset{CH_3CHCO_2^-}{|}}{OH} + H^+ \qquad (K_a = 1.4 \times 10^{-4})$$

I.4. Among the more profitable products of the pharmaceuticals industry are two local anesthetics widely employed in dentistry, $\underline{Novocaine}$ and $\underline{Lidocaine}$. Both are effective in blocking nerve impulses to the brain, preventing the transmission of "pain messages". The latter is currently the drug of choice by most dentists, since it is only one-eighth as toxic as the older $\underline{Novocaine}$ (first prepared in 1905). What is the approximate pH of a solution labeled "1.5% LIDOCAINE, by weight. Density 1.00 g ml^{-1}"?

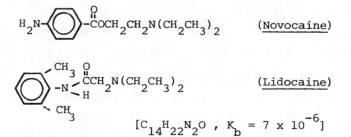

$[C_{14}H_{22}N_2O$, $K_b = 7 \times 10^{-6}]$

I.5. Lidocaine (problem 1.4) is not very soluble in water and its aqueous solution is
slightly alkaline. Commercial preparations can be made with a wider range of concentrations and with a pH approximately the same as that of human blood by using a mixture of Lidocaine and its "hydrochloride" salt:

(Lidocaine hydrochloride)

What mole ratio of lidocaine/lidocaine hydrochloride would provide a pH approximating
that of blood (7.4)?

I.6. Although instrumental techniques have replaced many of the older "wet chemical" analyses in industrial research and quality control laboratories, there are still many
uses for "classical" separations and analysis procedures. In the workup of samples of
euxenite (a uranium ore), for example, many of the ore components must be removed by
chemical processes before final spectroscopic assays can be made. At one stage in a
typical procedure, an ammonia/ammonium chloride buffer is employed in the selective
precipitation of certain metal carbonates. What is the pH of a buffer made by adding
25 ml of 15 M ammonia and 12 g of ammonium chloride to sufficient distilled water to
form 500 ml of solution? [For K_b-data, see Appendix D.]

*I.7. What change in pH would result from the addition of 5.0 ml of 0.30 M sodium hydroxide
to 45 ml of the buffer described in problem I.6?

*I.8. Hypochlorite bleaches are used as household laundry aids such as CHLOROX. In addition,
these chemicals find many industrial uses, as in the bleaching of recycled paper.
Sodium hypochlorite, one of the more common bleaching agents, is manufactured industrially by the electrolysis of brine (aqueous NaCl). The chlorine generated at the
anode is mixed with sodium hydroxide formed during the cathode reaction. Since
hypochlorite ion is decomposed by acids, these bleaching agents must be used in
slightly alkaline solutions. In industrial applications requiring the mixing of
hypochlorite solutions with acidic solutions, a buffer must be employed to maintain
a high pH. Ammonia/ammonium ion buffers cannot be used because of the generation of

*Proficiency Level   293

toxic compounds from reaction of OCl^- with NH_3. A carbonate/bicarbonate buffer is, however, quite satisfactory. What pH change would occur on mixing equal volumes of 0.1010 $\underline{M}$ HCl and a 0.25 $\underline{M}$ CO_3^{2-}/0.15 $\underline{M}$ HCO_3^- solution? [For K_a-data see Appendix D.]

*I.9. In most industrial operations requiring pH control, solutions are routinely monitored by the use of pH meters. In certain apple processing plants, however, the fairly concentrated sodium hydroxide solution used to "chemically peel" the apples is highly corrosive to the glass electrodes typically used with pH meters. In such cases, chemical indicators may be employed to test the pH of samples periodically withdrawn from the NaOH solutions. A standard buffer of pH 12.0 is used to provide a comparison of the indicator color required to show that the NaOH solution is still adequately "strong" for effective removal of the apple "skins". What relative volume of 0.30 $\underline{M}$ PO_4^{3-} and 0.50 $\underline{M}$ HPO_4^{2-} solutions could be mixed to provide a standard buffer of pH 12.0? [See Appendix D for K_a-data.]

*I.10. Strongly alkaline buffers must be stored in plastic bottles, since most types of glass are attacked by OH^- ion. In addition, repeated use of the buffer tends to result in a gradual decrease in its pH by the reaction of carbon dioxide absorbed from the air:

$$CO_2 + OH^- \rightarrow HCO_3^-$$

If the buffer described in problem I.9 had, through regular use, changed to pH 11.3, what volume of 0.10 $\underline{M}$ NaOH could be added to 250 ml of the buffer to restore the solution to pH 12.0?

ANSWERS:

(I.1.) 6.0, (I.2.) aspirin: 2.8, HCl: 1.0, (I.3.) 2.5, (I.4.) 10.8, (I.5.) 3.5×10^{-2} mole <u>lidocaine</u> per mole <u>lidocaine</u> hydrochloride, (I.6.) 9.48, (I.7.) increase of 0.05 pH unit, (I.8.) decrease of ~0.55 pH unit, (I.9.) 1 vol. $[PO_4^{3-}]$ to 3 vol. $[HPO_4^{2-}]$, (I.10.) ~140 ml

---

RELEVANT
PROBLEMS

Unit 11: Acid/Base Equilibria

Set E: Environmental Sciences

E.1. A lagoon is a shallow body of salt water, partially isolated from the ocean by a sand barrier or coral reef. Aqueous communication with the ocean is maintained through inlets, passes, or percolation through the barrier. The composition of lagoon water is partly dependent upon the amount of seawater entering the lagoon as a result of tides, winds, or percolation. Continental waters carry minerals from their drainage basins into the lagoon by rivers or by water table discharge. The fresh water intake is a function of rainfall. Evaporation of lagoon water, especially in arid climates where evaporation exceeds fresh water discharge into the lagoon, produces a condition of hypersalinity. The pH of lagoon water is controlled by several factors according to the local environment. The pH of seawater is usually between 7.5 and 8.4 and this contributes to the alkalinity of lagoon water. Continental waters discharging into lagoons often show low pH values due to humic and tannic acid drainage from swamps. In a tropical lagoon that receives a large amount of sunshine the rate of photosynthesis and carbon dioxide consumption are correspondingly large, and this tends to raise the pH. The aquatic ecosystem of a natural lagoon is dependent on maintenance of a fairly constant pH. A 10 ml sample of a natural lagoon water was found to contain 3.6×10^{-11} mole of H^{+}. What was the pH of the sample?

E.2. A manufacturing company was established on the shore of the lagoon (problem E.1) to convert coconut oil to soap. Waste sodium hydroxide was dumped into the lagoon at a rate such that the OH^{-} concentration of the lagoon water reached a steady level of 1.6×10^{-5} $\underline{M}$. What was the <u>change</u> of pH of the lagoon water?

E.3. Chemical fertilizers are commonly used to introduce nitrogen and other essential elements into the soil. Depending upon the soil pH and the needs for other elements, nitrogenous fertilizers may be anhydrous ammonia, ammonium nitrate, ammonium sulfate, or ammonium phosphate. All of these are quite soluble in water and may be washed from the soil by natural rains or by artificial irrigation systems. The "run-off" waters from freshly-fertilized agricultural land may thus contain relatively high concentrations of ammonia or ammonium salts which affect the pH of the water. Ammonium nitrate and ammonium sulfate are acidic salts (K_a for NH_4^{+} > K_b for NO_3^{-} or SO_4^{2-}), while ammonia or ammonium phosphate (K_b for PO_4^{3-} > K_a for NH_4^{+}) yield alkaline solutions. What is the approximate pH of agricultural "run-off" water whose alkalinity is equivalent to 5×10^{-4} $\underline{M}$ NH_3? [$K_b = 1.8 \times 10^{-5}$]

E.4. Most natural waters contain detectable amounts of nitrate ion, some of which is present as part of the natural "nitrogen cycle" of the biosphere. Man has added to this, and continues to do so in ever-increasing amounts. The results have been noted in two major problem areas. Aquatic plant life, including algae, utilize nitrogen in varying forms. Nitrate is one such form and its increasing concentration, along with that of phosphates, is linked to the wide-spread eutrophication (nutrient enrichment) of lakes and streams, with the resulting "bloom" of algae and other water plants. In addition, nitrate poses a significant health problem to both humans and other animals. In the stomach, nitrate may be reduced to nitrite. This, in turn, may react with the iron of hemoglobin to form methemoglobin, which is incapable of acting as an oxygen carrier. Infants are particularly susceptible to this problem and a large number of "blue baby" deaths have been traced to nitrates in drinking water. As a result, the U.S. Public Health Service has set a limit of 45 ppm NO_3^- in drinking water for adults and a 10 ppm limit in drinking water for infants. Few water-treatment plants are capable of removing nitrate from drinking water supplies, so the increasing nitrate concentrations of natural waters represent very real public health problems in many localities. A solution used as a stock reagent in biological testing of the effects of nitrite ion on small animals was 0.10 $\underline{M}$ in nitrous acid and 0.30 $\underline{M}$ in sodium nitrite. What was the pH of this solution? [For K_a-data, see Appendix D.]

E.5. Agricultural fertilizers furnish enormous amounts of nitrate, either directly (e.g., as NH_4NO_3) or indirectly (e.g., from oxidation of ammonia or ammonium ion). Although the intent of fertilizer use is to increase soil nutrients needed for plant growth, a large percentage of chemical fertilizers is washed from the soil into neighboring water systems. Cations (e.g., NH_4^+ or K^+) from fertilizers are readily adsorbed by clays in the soil, but the nitrate anion is not. When you consider that chemical fertilizer use is expected to grow from 63 million metric tons (1970) to 115 million metric tons (1980) over a ten year period, it is not difficult to see why water pollution by agricultural run-off is an increasing problem. Nearly a third of this run-off will probably be nitrate-contaminated. In some localities, the drainage from feedlots and barnyards represents a larger source of nitrate (from decomposing animal wastes) than does the elution of nitrate fertilizers. In most cases, however, this is a fairly localized problem. What is the pH of a feedlot pond in which the C_{H^+} is determined by the contaminants ammonia (3.8 g liter^{-1}) and ammonium nitrate (5.4 g liter^{-1})? [See Appendix D for K_b-data.]

*E.6. What pH change would result from the addition of 1.00 lb (454 g) of lye (NaOH) to the feedlot pond (problem E.5), if the total pond water volume is 5300 liters?

*E.7. The pH of ocean water is generally maintained in the range 7.4-8.6 by the buffering action of the dissolved CO_2 system:

$$CO_{2(aq)} + H_2O_{(\ell)} \; \underset{\leftarrow}{\rightarrow} \; HCO_{3(aq)}^- + H_{(aq)}^+$$
$$\downarrow\uparrow$$
$$CO_{3(aq)}^{2-} + H_{(aq)}^+$$

The marine ecosystem has developed for optimum function in the "normal" pH range and any major pH change resulting from acidic or alkaline pollutants can have a disastrous affect on a number of interdependent marine organisms. What pH change would result from the addition of 1500 liters of an acidic waste effluent 0.05 $\underline{M}$ in H^+ to the 52,000 liters of isolated waters in a small inlet buffered by a system whose principal components are 2.4×10^{-3} $\underline{M}$ CO_2 and 1.7×10^{-4} $\underline{M}$ HCO_3^-? [See Appendix D for K_a-data.] (Assume negligible CO_3^{2-}.)

*E.8. A few years ago "phosphate" additives to detergents were widely promoted because of the utility of the tripolyphosphate ion as an agent for complexing with Mg^{2+} and Ca^{2+} ions, the primary species responsible for water "hardness". Thousands of metric tons of "phosphates" poured into municipal waste water plants and thence into rivers, lakes, and streams. Tripolyphosphate eventually hydrolyzes to phosphate, shown to be the limiting factor in eutrophication (problem E.4):

$$P_3O_{10}^{5-} + H_2O \rightarrow 2 \; HPO_4^{2-} + H_2PO_4^-$$

In 1970, the Environmental Protection Agency forced a drastic reduction in the phosphate content of detergents. A number of substitutes were tested, but most were eventually shown to be potential health hazards and, in 1972, the U.S. Surgeon General strongly suggested a return to phosphate additives. Many of the more common detergents today still contain phosphates, although, in most cases, these constitute a much smaller percentage of the detergent mixture than in pre-1970 days. What would be the pH of the hydrogen phosphate/dihydrogen phosphate mixture resulting from the hydrolysis of 18 g liter^{-1} of tripolyphosphate, according to the equation previously shown? [For K_a-data, see Appendix D.]

*E.9. What pH change would result from the mixing of equal volumes of the tripolyphosphate hydrolysis mixture (problem E.8) and 0.015 $\underline{M}$ NaOH?

*E.10. Although fluoride is added in small amounts to many municipal water supplies to reduce the incidence of dental decay, fairly low concentrations of F^- can be toxic to many aquatic organisms. In moderate concentrations, fluoride can be a cumulative poison to humans, acting to disrupt the activity of certain vital enzyme systems. Fluoride wastes are a particular problem with aluminum producing industries and with phosphate plants using fluorapatite rock. To prevent fluoride pollution, principally as HF, of the air, many plants pass the contaminated exhaust air through a solution of caustic soda. The resulting mixture of aqueous HF and NaF is then expelled as waste effluent. What is the pH of a fluoride waste solution resulting from absorption of 328 kg of HF by 2000 liters of 4.0 $\underline{M}$ NaOH? [For K_a-data, see Appendix D.]

ANSWERS:

(E.1.) 8.4,   (E.2.) increase of 0.8 pH unit,   (E.3.) 10.0,   (E.4.) 3.82,   (E.5.) 9.77,
(E.6.) increase of 0.02 pH unit,   (E.7.) decrease of ~1.4 pH units,   (E.8.) 7.51,   (E.9.)
increase by 0.56 pH unit,   (E.10.) 3.14

UNIT 12: COMPLEX IONS

Let's think back briefly to our earlier discussion of acids and bases, in particular to the behavior of ammonia as a proton acceptor:

$$H:N:H \quad + \quad H^+ \quad \rightleftharpoons \quad H:N:H \quad +$$

Two factors are noted. As far as the <u>ammonia</u> molecule was concerned, its ability to act as a base was due to the presence of a lone pair of valence electrons. The <u>proton</u>, on the other hand, was <u>seeking</u> an electron pair for formation of a covalent bond. There are many other species which, like ammonia, can furnish a pair of electrons for bond formation:

$$H:\overset{..}{\underset{..}{O}}:^- \quad + H^+ \quad \rightleftharpoons \quad H:\overset{..}{\underset{..}{O}}:H$$
hydroxide

$$:\overset{..}{\underset{..}{F}}:^- \quad + \quad H^+ \quad \rightleftharpoons \quad :\overset{..}{\underset{..}{F}}:H$$
fluoride

$$:N:::C:^- \quad + \quad H^+ \quad \rightleftharpoons \quad :N:::C:H$$
cyanide

Are there, perhaps, species other than proton which might seek a lone pair of electrons for bond formation?

There are, in fact, very many species which, like the proton, act to accept a pair of electrons to form a covalent bond:

$$H:\overset{:O:}{\underset{:O:}{O}}:Al \quad + \quad :\overset{..}{\underset{..}{O}}:H^- \quad \rightleftharpoons \quad \left[H:\overset{:O:}{\underset{:O:}{O}}:Al:\overset{..}{\underset{..}{O}}:H \right]^-$$

[Aluminum hydroxide will dissolve in excess sodium hydroxide, forming the tetra-hydroxoaluminate(III) ion.]

$$Cu^{2+} + :\overset{H}{\underset{H}{N}}:H \quad \rightleftharpoons \quad \left[Cu:\overset{H}{\underset{H}{N}}:H \right]^{2+}$$

[The Copper(II) ion reacts with ammonia. If sufficient ammonia is available, reaction continues until the deep blue $[Cu(NH_3)_4]^{2+}$ complex is formed.]

The wide variety of reactions of this type led G. N. Lewis to suggest that the proton is involved in acid/base chemistry only as a special case of a much more general phenomenon, the competition for an electron-pair. In the Lewis concepts, then:

AN ACID IS ANY SPECIES ACTING TO ACCEPT A PAIR OF ELECTRONS TO FORM A COVALENT BOND.

A BASE IS ANY SPECIES ACTING TO FURNISH A PAIR OF ELECTRONS TO FORM A COVALENT BOND.

We shall concentrate on a particular area of Lewis acid/base chemistry, the formation or instability of complex ions in aqueous solutions. The electron-pair competition, like that in the simple case involving protons (Unit 11), may result in a dynàmic equilibrium, for which an equilibrium constant may be written and used in a mathematical analysis. Most complex ions, however, present a much more complicated equilibrium system than that of a simple monoprotic acid.

Metal ions in water are bound in more or less regular clusters of water molecules. These hydrates require that a Lewis base, such as NH_3 or CN^-, must replace a water molecule. The anion or neutral molecule bonding to the metal ion in a complex is called a ligand. Rarely do we find complexes involving only one ligand per metal ion and the equilibrium system often contains a relatively large number of components. Consider, for example, a more accurate description of the equilibrium system involving ammonia and aqueous copper(II) ions:

$$[Cu(H_2O)_6]^{2+} + NH_3 \; \rightleftharpoons \; [Cu(H_2O)_5(NH_3)]^{2+} + H_2O$$

$$[Cu(H_2O)_5(NH_3)]^{2+} + NH_3 \; \rightleftharpoons \; [Cu(H_2O)_4(NH_3)_2]^{2+} + H_2O$$

$$[Cu(H_2O)_4(NH_3)_2]^{2+} + NH_3 \; \rightleftharpoons \; [Cu(H_2O)_3(NH_3)_3]^{2+} + H_2O$$

$$[Cu(H_2O)_3(NH_3)_3]^{2+} + NH_3 \; \rightleftharpoons \; [Cu(H_2O)_2(NH_3)_4]^{2+} + H_2O$$

$$H_2O + NH_3 \; \rightleftharpoons \; NH_4^+ + OH^-$$

Any solution containing Cu^{2+}, NH_3, and H_2O will contain some of every species shown. A mathematical analysis of such a complicated, interrelated mixture would appear almost hopeless, at least without the aid of a rather sophisticated computer program. It isn't really "hopeless", but it is "pretty tricky". Fortunately, we can illustrate some aspects of complex ion equilibria without the necessity of tracing through a detailed analysis.

Le Chatelier's principle suggests that any addition of a reactant to a system at equilibrium will disturb the equilibrium in such a way as to consume more of that reactant until a new equilibrium state is achieved. This means, for us, that the addition of a large excess of the ligand will favor consumption of the ligand so that the final equilibrium system will contain the "complete" complex as the major component containing the metal ion. Under the conditions of excess complexing agent, then, we can deal with a simplified net equation. In the case of the copper/ammonia system, for example, the use of a large excess of ammonia permits us to work with the single equation:

$$\left[Cu(H_2O)_6\right]^{2+} + 4NH_3 \;\rightleftharpoons\; \left[Cu(H_2O)_2(NH_3)_4\right]^{2+} + 4H_2O$$

We can simplify the equation still further by omitting "H_2O", recognizing that maximum hydration of a metal ion occurs in aqueous solution even if we choose not to include it in an equation:

$$Cu^{2+} + 4NH_3 \;\rightleftharpoons\; \left[Cu(NH_3)_4\right]^{2+}$$

Using such an equation, it is not difficult to estimate the concentration of "free" metal ion:

e.g., for the case of Cu^{2+} in excess aq. NH_3:

$$K_{eq} = \frac{C_{\left[Cu(NH_3)_4\right]^{2+}}}{C_{Cu^{2+}} \times C_{NH_3}^4}$$

$$C_{Cu^{2+}} = \frac{C_{\left[Cu(NH_3)_4\right]^{2+}}}{K_{eq} \times C_{NH_3}^4}$$

in which the concentrations are those at equilibrium.

The equilibrium constants for complex ions may be tabulated in various ways. Formation constants refer to the equation as written with the complex as the "right hand" species, while instability constants refer to the case of the complex written as the "left hand" species. When equilibrium constants for stepwise formation of a complete complex are given, these may be combined to obtain the net formation constant:

$$K_{(net)} = K_1 \times K_2 \times K_3 \times \dots.$$

Appendix D contains formation constants for a number of the more common complex ions. It should be obvious from the conventions used in writing equilibrium constant expressions (Unit 10) that an inverse relationship exists between the formation constant and the instability constant for a given complex:

$$K_{instability} = \frac{1}{K_{formation}}$$

We will limit our study of complex ions to the relatively simple cases of writing appropriate equilibrium equations and calculating concentrations of "free" metal ions in solutions containing excess complexing agent.

- References -

1. O'Connor, Rod. 1977. Fundamentals of Chemistry, second edition. New York: Harper & Row (Units 21, 27 and Excursion 2)

2. Brown, Theodore, and H. Eugene LeMay, Jr. 1977. Chemistry: The Central Science. Englewood Cliffs, N.J.: Prentice-Hall (Chapters 17, 23)

3. Masterton, William, and Emil Slowinski. 1977. Chemical Principles, fourth edition. Philadelphia: W.B. Saunders (Chapter 21)

4. Nebergall, William, F.C. Schmidt, and H.F. Holtzclaw, Jr. 1976. College Chemistry, fifth edition. Lexington, MA: D.C. Heath (Chapters 19, 32)

5. Quagliano, J.V. and L.J. Vallarino. 1969. Coordination Chemistry. Indianapolis, IN: Heath/Raytheon.

OBJECTIVES:

(1) *Given the formula and instability constant for a complex ion and the equilibrium concentrations of the complex and the excess complexing agent, be able to calculate the equilibrium concentration of the "free" metal ion.*

*(2) *Given the formula of a complex ion, the net or stepwise formation constants for the complex (or access to Appendix D), and the concentrations and volumes of a metal ion solution and a solution of the complexing agent, be able to calculate the approximate equilibrium concentration of the "free" metal ion in the solution resulting from mixing the two reactants (with the complexing agent in excess.)*

PRE-TEST:

> Necessary Equilibrium Constants:
>
> for $[Ag(S_2O_3)_2]^{3-}$, $(net)K_{instability} = 3.4 \times 10^{-14}$
>
> for $[Fe(CN)_6]^{4-}$, $(net)K_{formation} = 1.0 \times 10^{24}$

(1) A solution of sodium thiosulfate (photographer's "hypo") is used in the "fixing bath" in photographic developing laboratories. The silver bromide in a black-and-white film emulsion is "activated" when exposed to light. The activated salt is selectively reduced to black metallic silver in the work-up of a photographic negative. The function of the "hypo" is to dissolve any residual silver bromide from the film, in regions not exposed to light, before the developed film can be taken from the dark room. Although silver bromide is only very slightly soluble in water, it can be washed from the film by the aqueous thiosulfate because of the formation of the stable dithiosulfatoargentate(I) complex:

*Proficiency Level

$$AgBr_{(s)} + 2S_2O_3^{2-}{}_{(aq)} \rightleftharpoons \left[Ag(S_2O_3)_2\right]^{3-}_{(aq)} + Br^-_{(aq)}$$

What is the approximate molarity of "free" silver ion in a used "fixing bath" in which the concentration of the $\left[Ag(S_2O_3)_2\right]^{3-}$ complex is 0.15 $\underline{M}$ and the concentration of free thiosulfate ion is 2.65 $\underline{M}$? _____

*(2) Hemoglobin is a complex organic molecule (mol. wt. ~68,000) containing four iron(II) ions per molecule, each bound to the ligand groups of a porphyrin segment of hemoglobin. The oxygen-carrying properties of the red blood cells are due to the ability of the hemoglobin in these cells to complex with molecular oxygen. The toxicity of cyanides is due, to a major extent, to the fact that cyanide ion is bound more tightly by the Fe^{2+} of hemoglobin than is O_2, so cyanides effectively tie up the hemoglobin needed for oxygen transport in the body. To illustrate the affinity of Fe^{2+} for cyanide, estimate the concentration of "free" Fe^{2+} that remains after mixing 45 ml of 0.010 $\underline{M}$ $FeSO_4$ with 5.0 ml of 2.0 $\underline{M}$ KCN.

Answers and Directions:

   (1) 7.3×10^{-16} $\underline{M}$ (Ag^+)

*If you answered this question correctly, go on to question *2. If you missed it, study METHODS, section 12.1.*

   *(2) 9×10^{-22} $\underline{M}$ (Fe^{2+})

If you answered this question correctly, go on to RELEVANT PROBLEMS, UNIT 12. If you missed it, study METHODS, section 12.2.

METHODS

12.1 "Free" Metal Ion in a Solution of Known Complex Concentration

If we can write a suitable equation for the equilibrium system involving a complex ion and its "free" components in aqueous solution, then the calculation of the metal ion concentration is relatively simple. To avoid the complications of "intermediate" complexes, we will deal only with systems in which the complexing agent is present in large excess. Under these conditions, we may assume that the concentrations of "intermediate" complexes are negligible.

By convention, the net instability constant refers to an equation in which the complex is written as the "left-hand" species and the free metal ion and free complexing agent are written as "right-hand" species, with the coefficients necessary for a balanced equation. For the instability equation describing the tetraaminecopper(II) complex system, for example, we would write:

$$[Cu(NH_3)_4]^{2+} \rightleftharpoons Cu^{2+} + 4NH_3$$

Once the appropriate equation has been written, the equilibrium constant ("instability constant") expression is derived in the same way as that for any other equilibrium system (Unit 10). For the tetraaminecopper(II), for example:

$$[Cu(NH_3)_4]^{2+} \rightleftharpoons Cu^{2+} + 4NH_3$$

$$K_{instability} = \frac{C_{Cu^{2+}} \times C_{NH_3}^4}{C_{[Cu(NH_3)_4]^{2+}}}$$

When the equilibrium concentrations are known for the complex and the excess "free" complexing agent, these are substituted in the $K_{instability}$-expression and, taking careful note of exponents, the resulting equation is solved for the equilibrium concentration of the "free" metal ion.

EXAMPLE 1

The *SOLUTION* to Pre-Test question 1:

$$K_{instability} = 3.4 \times 10^{-14} \text{ for}$$

$$[Ag(S_2O_3)_2]^{3-} \rightleftharpoons Ag^+ + ②S_2O_3^{2-}$$

$$3.4 \times 10^{-14} = \frac{C_{Ag^+} \times C_{S_2O_3^{2-}}^2}{C_{[Ag(S_2O_3)_2]^{3-}}} = \frac{C_{Ag^+} \times (2.65)^2}{0.15}$$

$$C_{Ag^+} = \frac{(3.4 \times 10^{-14}) \times 0.15}{(2.65)^2} = 7.3 \times 10^{-16} \underline{M}$$

= =

EXERCISE 1

Silver is found in nature in a number of different forms, including ores containing the free metal. Its principal ores, however, are "horn silver" (containing silver chloride) and "argentite" (containing silver sulfide). A method widely used for isolating silver from these ores is the cyanide process, in which the crushed ore is stirred with aqueous sodium cyanide. The silver is solubilized as the dicyanoargentate complex. This is then reduced with zinc to obtain the metallic silver. For "argentite" ore the process is described by:

$$Ag_2S_{(s)} + 4CN^-_{(aq)} \rightarrow 2[Ag(CN)_2]^-_{(aq)} + S^{2-}_{(aq)}$$
(crude)

$$2[Ag(CN)_2]^- + Zn \rightarrow \underline{2Ag} + [Zn(CN)_4]^{2-}$$

The net instability constant for dicyanoargentate(I) is 1.8×10^{-19}. What is the molarity of "free" silver ion in a solution in which the concentration of the complex is 0.20 $\underline{M}$ and the concentration of the excess cyanide ion is 1.2 $\underline{M}$?

(answer, page 308)

==

Extra Practice
EXERCISE 2

In the <u>cyanide process</u> for silver ores (Exercise 1) the second step involves formation of the tetracyanozincate(II) ion. What is the molarity of "free" Zn^{2+} ion in a solution in which the concentration of the complex is 0.10 $\underline{M}$ and the concentration of the excess cyanide ion is 1.1 $\underline{M}$? The instability constant for $[Zn(CN)_4]^{2-}$ is 1.1×10^{-18}.

(answer, page 308)

==

At this point you should try the <u>competency</u> level Self-Test question on page 307.

==

*12.2 "Free" Metal Ion From Pre-equilibrium Data

The analysis of a problem involving the formation of a complex ion from specified amounts of reactants is similar to that employed in calculating the pH change of a buffer (Unit 11). That is, both equilibrium and stoichiometric considerations are required.

First, the equation consistent with the <u>net</u>[1] formation constant must be written. Then the number of moles (or mmols) of the metal ion and of the complexing agent must be calculated (from available volumes and molarities, Unit 6). The assumption is made that, in the presence of a large excess of the complexing agent, nearly all the original metal ion is converted to the complete complex. It is at this point that stoichiometry is used to determine the number of moles of complex formed <u>and</u> of complexing agent remaining (original - amount consumed by complex formation). These quantities are then converted to concentrations, for use in the $K_{formation}$-expression to calculate the approximate molarity of

---

[1] This constant will either be available or stepwise formation constants will be available, from which the net constant may be calculated:

$$K_{net} = K_1 \times K_2 \times K_3 \times \ldots.$$

*Proficiency Level

"free" metal ion remaining at equilibrium.

EXAMPLE 2

    The *SOLUTION* to Pre-Test question 2:

$$K_{formation} \text{ for } [Fe(CN)_6]^{4-} = 1.0 \times 10^{24}$$

for which the equilibrium equation is:

$$Fe^{2+} + 6 \; CN^- \rightleftharpoons [Fe(CN)_6]^{4-}$$

and the formation constant expression is:

$$K_{formation} = \frac{c_{[Fe(CN)_6]^{4-}}}{c_{Fe^{2+}} \times c_{CN^-}^6}$$

The original reactants available are:

Fe^{2+}:

$$\frac{0.010 \text{ mmol}}{1 \text{ ml}} \times \frac{45 \text{ ml}}{1} = 0.45 \text{ mmol}(Fe^{2+})$$

CN^-:

$$\frac{2.0 \text{ mmol}}{1 \text{ ml}} \times \frac{5.0 \text{ ml}}{1} = 10 \text{ mmol}(CN^-)$$

Assuming that, in a large excess of CN^-, nearly all Fe^{2+} is converted to $[Fe(CN)_6]^{4-}$, then at equilibrium:

$$c_{[Fe(CN)_6]^{4-}} = \frac{0.45 \text{ mmol}(Fe^{2+})}{50 \text{ ml (solution)}} \times \frac{1 \text{ mmol } [Fe(CN)_6]^{4-}}{1 \text{ mmol } (Fe^{2+})} = 9.0 \times 10^{-3} \; \underline{M}$$

$$[45 + 5]$$

$$c_{CN^-} = \left[\frac{10 \text{ mmol}(CN^-)}{50 \text{ ml(solution)}} \right] - \left[\frac{0.45 \text{ mmol}(Fe^{2+})}{50 \text{ ml(solution)}} \times \frac{6 \text{ mmol}(CN^-)}{1 \text{ mmol}(Fe^{2+})} \right] = 0.146 \; \underline{M}$$

                (original)   (consumed)

Then, substituting in the K-expression gives:

$$1.0 \times 10^{24} = \frac{(9.0 \times 10^{-3})}{c_{Fe^{2+}} \times (0.146)^6}$$

from which:

$$c_{Fe^{2+}} = \frac{(9.0 \times 10^{-3})}{(1.0 \times 10^{24}) \times (0.146)^6} = 9 \times 10^{-22} \; \underline{M}$$

*EXERCISE 3

Cadmium has found extensive use as an electroplated coating for metals, both because of the protection if affords the metal from oxidation and for the decorative appearance of the blue-grey cadmium coating. Quality electroplating never uses a concentrated solution of the free metal ion because of the irregular electrodeposition which occurs. Typically a solution of some complex ion of the metal is employed so that a large quantity of the metal ion is available, but the "free" metal ion is maintained at a low concentration, approaching that of the complex equilibrium. Most cadmium-plating operations use the tetra-cyano complex. What would be the approximate concentration of "free" Cd^{2+} ion in an electrolyte solution prepared by adding 10.0 liters of 6.0 $\underline{M}$ NaCN to 90.0 liters of 0.10 $\underline{M}$ $CdCl_2$?

For necessary formation constants, see Appendix D.

(answer, page 309)

===

Extra Practice

*EXERCISE 4

Most commercial silverplating operations use an electrolyte solution of dicyanoargentate(I), to maintain a low concentration of free Ag^+ for uniform electrodeposition. [See Exercise 3.] What would be the approximate molarity of "free" silver ion in a solution prepared by adding 5.0 liters of 6.0 $\underline{M}$ NaCN to 95.0 liters of 0.10 $\underline{M}$ $AgNO_3$? (The net formation constant for $[Ag(CN)_2]^-$ is 5.6×10^{20}.)

(answer, page 310)

===

SELF-TEST (UNIT 12)   [answers, page 311]

12.1. Potassium hexacyanoferrate(III) (older name: potassium ferricyanide) is used in the "blueprinting" process. The paper is impregnated with a solution containing the $K_3[Fe(CN)_6]$ and iron(III) citrate. When a section of the dried paper is exposed to light, the iron(III) is reduced to iron(II) by reaction with the citrate ion. Any unexposed areas are washed to remove the unreacted components. The exposed regions turn blue through formation of the colored salt, $KFe[Fe(CN)_6]$. What is the approximate molarity of "free" Fe^{3+} in a solution having an excess cyanide concentration of 2.0 $\underline{M}$ and a hexacyanoferrate (III) concentration of 0.20 $\underline{M}$? [The net instability constant for this complex is 1.0×10^{-31}.]

If you answered this question correctly, you may go to the proficiency level, try the

RELEVANT PROBLEMS (Unit 12), or stop here. If not, you should consult your instructor for suggestions of further study aids.

- -

*12.2. Copper is one of the most important metals of our modern technological society. The depletion of high-grade ores has made it necessary to locate and use low-grade ores and to recycle waste copper products. Nearly half of our present copper production is from recycling "scrap" copper. In geologic surveys for new ore deposits, a quick "field-test" for copper involves dissolving a small amount of ore sample in nitric acid, then adding excess ammonia. The formation of the deep blue tetraamminecopper(II) complex is a sensitive test capable of detecting copper in ore samples of commercial interest. What is the approximate molarity of "free" Cu^{2+} in a solution prepared by adding 1.0 ml of 10 $\underline{M}$ ammonia to 4.0 ml of 0.20 $\underline{M}$ $Cu(NO_3)_2$? [For the stepwise formation constants see Appendix D.]

- -

If you answered this question correctly, you may go on to the RELEVANT PROBLEMS for Unit 12. If not, you should consult your instructor for suggestions of further study aids.

- -

ANSWERS to EXERCISES, Unit 12

1. [$2.5 \times 10^{-20} \underline{M}$ (Ag^+)] Solution:

$$[Ag(CN)_2]^- \overset{\rightarrow}{\leftarrow} Ag^+ + 2 CN^-$$

$$K_{instability} = \frac{C_{Ag^+} \times C^2_{CN^-}}{C_{[Ag(CN)_2]^-}}$$

$$1.8 \times 10^{-19} = \frac{C_{Ag^+} \times (1.2)^2}{0.20}$$

$$C_{Ag^+} = \frac{(1.8 \times 10^{-19}) \times 0.20}{(1.2)^2}$$

2. [7.5×10^{-20} $\underline{M}$ (Zn^{2+})] Solution:

$$[Zn(CN)_4]^{2-} \overset{\rightarrow}{\leftarrow} Zn^{2+} + 4 CN^-$$

$$K_{instability} = \frac{C_{Zn^{2+}} \times (1.1)^4}{C_{[Zn(CN)_4]^{2-}}}$$

$$1.1 \times 10^{-18} = \frac{C_{Zn^{2+}} \times (1.1)^4}{0.10}$$

$$C_{Zn^{2+}} = \frac{(1.1 \times 10^{-18}) \times 0.10}{(1.1)^4}$$

- -

*3. [4.6×10^{-18} M (Cd^{2+})] Solution:

To find the net formation constant (using stepwise formation constants from Appendix D.):

$$K_{net} = (3.0 \times 10^5) \times (1.3 \times 10^5) \times (4.3 \times 10^4) \times (3.5 \times 10^3) = 5.9 \times 10^{18}$$

For which, the equilibrium equation is:

$$Cd^{2+} + 4\ CN^- \rightleftharpoons [Cd(CN)_4]^{2-}$$

and the formation constant expression is:

$$K_{formation} = \frac{C_{[Cd(CN)_4]^{2-}}}{C_{Cd^{2+}} \times C_{CN^-}^4}$$

The original reactants available are:

Cd^{2+}:

$$\frac{0.10\ mole}{1\ liter} \times \frac{90.0\ liters}{1} = 9.0\ moles\ (Cd^{2+})$$

CN$^-$:

$$\frac{6.0\ mole}{1\ liter} \times \frac{10.0\ liters}{1} = 60\ moles\ (CN^-)$$

Assuming that the large excess of CN$^-$ converts nearly all the Cd^{2+} to [Cd(CN)$_4$]$^{2-}$, then at equilibrium:

$$C_{[Cd(CN)_4]^{2-}} = \frac{9.0\ moles\ (Cd^{2+})}{100\ liters} \times \frac{1\ mole\ [Cd(CN)_4]^{2-}}{1\ mole\ (Cd^{2+})} = 9.0 \times 10^{-2}\ M$$

$$[90 + 10]\nearrow$$

$$C_{CN^-} = \left[\frac{60\ moles\ (CN^-)}{100\ liters}\right] - \left[\frac{9.0\ moles\ (Cd^{2+})}{100\ liters} \times \frac{4\ moles\ (CN^-)}{1\ mole\ (Cd^{2+})}\right] = 0.24\ M$$

(original)   (consumed)

Then, substituting in the K-expression gives:

$$5.9 \times 10^{18} = \frac{(9.0 \times 10^{-2})}{C_{Cd}^{2+} \times (0.24)^4}$$

from which:

$$C_{Cd}^{2+} = \frac{(9.0 \times 10^{-2})}{(5.9 \times 10^{18}) \times (0.24)^4}$$

*4.   [1.4×10^{-20} M (Ag$^+$)] <u>Solution</u>:

The net formation constant is 5.6×10^{20}, for which:

$$Ag^+ + 2CN^- \rightleftharpoons [Ag(CN)_2]^-$$

and the formation constant expression is:

$$K_{formation} = \frac{C_{[Ag(CN)_2]^-}}{C_{Ag}^+ \times C_{CN^-}^2}$$

The original reactants available are:

<u>Ag$^+$</u>:

$$\frac{0.10 \text{ mole}}{1 \text{ liter}} \times \frac{95.0 \text{ liters}}{1} = 9.5 \text{ moles (Ag}^+)$$

<u>CN$^-$</u>:

$$\frac{6.0 \text{ mole}}{1 \text{ liter}} \times \frac{5.0 \text{ liters}}{1} = 30 \text{ moles}$$

Assuming nearly complete conversion by excess CN$^-$ of Ag$^+$ to [Ag(CN)$_2$]$^-$, then at equilibrium:

$$C_{[Ag(CN)_2]^-} = \frac{9.5 \text{ moles (Ag}^+)}{100 \text{ liters}} \times \frac{1 \text{ mole [Ag(CN)}_2]^-}{1 \text{ mole (Ag}^+)} = 9.5 \times 10^{-2} \text{ M}$$

$$[95 + 5]$$

$$C_{CN^-} = \left[\frac{30 \text{ moles (CN}^-)}{100 \text{ liters}}\right] - \left[\frac{9.5 \text{ moles (Ag}^+)}{100 \text{ liters}} \times \frac{2 \text{ moles (CN}^-)}{1 \text{ mole (Ag}^+)}\right] = 0.11 \text{ M}$$

(original)   (consumed)

Then, substituting in the K-expression gives:

$$5.6 \times 10^{20} = \frac{(9.5 \times 10^{-2})}{C_{Ag}^+ \times (0.11)^2}$$

from which:

$$C_{Ag}^+ = \frac{(9.5 \times 10^{-2})}{(5.6 \times 10^{20}) \times (0.11)^2}$$

==

12.1. 3.1×10^{-34} $\underline{M}$ (Fe^{3+})

*12.2. 9.6×10^{-15} $\underline{M}$ (Cu^{2+})

A.1. Iron is one of the major components of the lithosphere. Phosphorus is a minor constituent. These elements rarely occur as dissolved species at concentrations greater
than 10^{-4} $\underline{M}$ but they play very important roles in oceanography. Phosphate is found
dissolved in natural waters as a result of weathering and dissolution of phosphate
minerals, soil erosion and transport, soil fertilization, biological transfer, and
the use of soluble phosphate compounds in detergent manufacture, water treatment and
industrial wastes. Phosphate, pyrophosphate, and the higher polyphosphates are
known to form complexes and insoluble salts with a variety of metal ions such as
Fe^{3+}, Ca^{2+}, and Mg^{2+}. The extent of complexing between various phosphates and metal
ions will depend upon the relative concentrations of the ions involved. What is the
approximate concentration (molarity) of "free" iron(III) ion in a sea water sample
in which the concentration of $[Fe(HP_2O_7)_2]^{3-}$ complex is 9.9×10^{-6} $\underline{M}$ and the concentration of free pyrophosphate ion is 9.9×10^{-12} $\underline{M}$? [For $[Fe(HP_2O_7)_2]^{3-}$, (net)
$K_{inst} = 1.0 \times 10^{-22}$]

A.2. High concentrations of cadmium have been found in marine organisms along industrialized coastal areas. The cadmium concentration in these coastal waters is in the
range of 0.05 to 1.0 ppb (parts per billion). In the open seas its concentration is
usually much lower. The cadmium level in deep sea sediments averages 225 ppb, with
much lower levels in areas rich in shell debris. Cadmium forms complexes with a
variety of organic compounds and it is frequently concentrated in the kidneys of
several kinds of marine animals. Levels as high as 500 ppm have been observed in
sea otters. The $[CdCl]^+$ ion pair is the predominant species of cadmium in sea water.
The instability constant for chlorocadmium(II) is 3.16×10^{-2}. What is the molarity
of "free" cadmium in sea water in which the concentration of the ion-pair complex is
1.26×10^{-9} $\underline{M}$ and the concentration of excess chloride is 0.50 $\underline{M}$?

A.3. Mining operations in the United States discharge large amounts of iron into the marine environment. Compared to the natural flow from weathering and land runoff the
man made contribution is small. The natural transport of iron involves release from
minerals on the land mass, transport with surface water runoff to the oceans, solu-
bilization as complex ions in the marine environment, or precipitation with mineral
particles on the continental shelves. Biogenic processes provide for vertical
transport in dead organisms and fecal material. In the sediment zone, released iron
is usually precipitated as an oxide. During the course of iron precipitation, heavy
metals such as mercury, zinc, cadmium, and copper may be coprecipitated. The $[FeF]^{2+}$

ion pair is one of the predominant soluble iron complexes found in sea water. The instability constant for fluoroiron(III) is 6.3×10^{-6}. What is the molarity of "free" iron(III) in sea water in which the concentration of the ion-pair complex is 5.38×10^{-8} $\underline{M}$ and the concentration of excess fluoride is 6.31×10^{-5} $\underline{M}$?

A.4. Calcareous shell deposits are mined from a number of offshore deposits, particularly in the Gulf of Mexico and off the Southwest coast of Iceland. These shells are used in a variety of chemical processes and as a substitute for gravel in the building of highways. Specific shell deposits from some parts of the world are used in the manufacture of portland cement. Most calcareous shell deposits are renewed by a complex chain of events. For example, the marine animals extract calcium from the sea water to build their shells. The waves break the shells free, crushing and grinding them. Tidal currents sweep the particles into bays where they are deposited. The influx of shell material far exceeds the rate of extraction, thus providing a replenishing mineral deposit. The ion-pair complex $[CaHCO_3]^+$ provides an important source of biogenic calcium for marine animals. What is the approximate molarity of "free" Ca^{2+} in a sea water sample in which the concentration of $[CaHCO_3]^+$ complex is 1.0×10^{-4} $\underline{M}$ and the concentration of free HCO_3^- ion is 1.67×10^{-3} $\underline{M}$? [For $[CaHCO_3]^+$, $K_{inst} = 5.5 \times 10^{-2}$]

A.5. The availability of various micronutrients in farm land has a significant effect on the land's productivity. Several of the so-called trace elements regulate the important enzyme controlled reactions essential to plant development and reproduction. For example, cobalt is essential for the symbiotic fixation of nitrogen, chloride controls both root and top growth, zinc is concerned with the formation of growth hormones, and copper is involved in respiration and utilization of iron. As might be expected, soil composition regulates the availability of micronutrients for plants. For example, soil solutions containing polyphosphates in the form of pyrophosphate, $P_2O_7^{4-}$, have a mechanism for maintaining a soluble copper(II) species available for plant utilization. The process is described by:

$$Cu^{2+}_{(aq)} + P_2O_7^{4-}_{(aq)} \rightleftharpoons [CuP_2O_7]^{2-}$$

The formation constant for the $[CuP_2O_7]^{2-}$ ion is 6.3×10^8. What is the molarity of "free" copper(II) ion in a soil solution in which the concentration of the complex is 4.5×10^{-6} $\underline{M}$ and the concentration of the excess pyrophosphate ion is 8.6×10^{-7} $\underline{M}$?

*A.6. The availability of the Cu^{2+} ion for plants is often limited, in certain types of alkaline soils, by formation of insoluble hydroxides. In recent years attention has been directed to the use of synthetic complexes as a method of protecting the heavy metal cations from precipitation in the soil. In some cases, the mobilization of the native micronutrients can be accomplished by addition of fertilizers containing a substance that can serve both as a nutrient and as a ligand in complex formation. To illustrate the mobilization of Cu^{2+} by adding a fertilizer containing ammonia, estimate the concentration of "free" Cu^{2+} that remains after mixing 50 ml of 0.0010 $\underline{M}$ $CuSO_4$ with 10.0 ml of 2.0 $\underline{M}$ NH_3. [See Appendix D for formation constants.]

*A.7. When Cu^{2+} nutrient solutions are to be used in a greenhouse, it may be more convenient to prepare a solution by using a strong complexing agent such as the chelating ligand ethylenediaminetetraacetate, $[EDTA]^{4-}$, instead of ammonia, which is more volatile and more difficult to handle and store. As an example, consider a solution containing Cu^{2+} and EDTA. If the concentration of free EDTA anion, $EDTA^{4-}$, is the same as that of the free ammonia given in Problem A.6, (i.e., 2.0 $\underline{M}$ $EDTA^{4-}$), estimate the concentration of "free" Cu^{2+} that remains after mixing 50 ml of 0.0010 $\underline{M}$ $CuSO_4$ with 10.0 ml of the $EDTA^{4-}$. [For $[Cu(EDTA)]^{2-}$, $K_{formation} = 6.3 \times 10^{18}$.]

*A.8. Amino acids are formed in ocean waters by the breakdown of proteins from plant and animal sources. These compounds constitute a separate class of naturally occuring complexing agents. Amino acids such as glycine, $H_3\overset{+}{N}CH_2CO_2^-$, form very stable complexes with some metal ions. For example, the net formation constant of the monoglycinate complex of Cu^{2+} is 1.3 x 10^8. Thus, the naturally occuring amino acids may play a significant role in the transport of heavy metal ions in the hydrosphere. What would be the approximate molarity of "free" copper(II) ion in a solution prepared by adding 3.0 liters of 5.0 $\underline{M}$ sodium glycinate to 47.0 liters of 0.010 $\underline{M}$ $Cu(NO_3)_2$? (The complex is formulated as $[Cu(H_2NCH_2CO_2)]^+$.)

*A.9. Seaweed has been used in animal feedstuffs for many years. However, the novel practice of using seaweed and its extract as a fertilizer was first developed in 1950. Seaweed contains all known trace elements that are essential for plant nutrition. Moreover, these trace elements are present in a form acceptable to plants. Many trace elements, iron in particular, cannot be absorbed by plants or animals in their common natural form because they are precipitated by competitive reactions. For example, iron is precipitated by calcium carbonate in high-lime soils, so that fruit trees growing in these soils suffer from an iron deficiency known as chlorosis. The technique of chelating makes it possible to provide the necessary trace elements for plants in readily acceptable forms. Such chelating properties are possessed by the starches, sugars, and proteins in seaweed and seaweed products. Consequently, these constituents are complexed with the iron, cobalt, manganese, copper, zinc and other trace elements found naturally in seaweed. That is why the trace elements in seaweed and seaweed extracts will not precipitate, even in alkaline soils, but remain available to plants which need them. In a greenhouse study to illustrate the affinity of seaweed chelates for nutrient manganese, a "control" solution was prepared using the diethylenetriaminepentaacetate ligand, $DTPA^{5-}$, according to the reaction:

$$Mn^{2+} + DTPA^{5-} \rightleftharpoons [MnDTPA]^{3-}$$

Estimate the concentration of "free" Mn^{2+} that remains after mixing 30 ml of 0.050 $\underline{M}$ $MnSO_4$ with 10 ml of 2.0 $\underline{M}$ $DTPA^{5-}$. [For $[MnDTPA]^{3-}$, $K_{formation} = 5.0 \times 10^{16}$]

*A.10. Animals have eaten seaweed in its natural state since the beginning of recorded history. At times grazing animals will turn to seaweed because of a shortage of other foods, but both domestic and wild animals will eat seaweed even when ample supplies of grass and other herbage are available. Seaweed is a valuable supplement for dairy cows, in low concentrations, because it provides many naturally chelated trace elements in usable forms. These improve the milk quality, including butterfat content. The effectiveness of the seaweed chelates depend upon their stability constants. For maximum mineral absorption the amino acid or other ligand must produce a chelate with a stability constant that is sufficiently favorable to compete with other substances that might render the element unavailable, without holding the metal so tightly bound that the metal ion cannot be released at the intestinal wall or in the blood stream. Some highly stable, nonabsorbable metal chelates are formed in feedstuffs and in the intestinal tract. Some of these have no useful biological purpose or value and some, such as the phytic acid - Zn^{2+} chelate, may interfere with normal metabolism by preventing the essential mineral from being absorbed, and thereby creating a deficiency. Studies reveal that the availability of Zn^{2+} in purified diets containing protein is greatly enhanced by the addition of $EDTA^{4-}$. Seaweed contains natural chelates that provide similar aid in the utilization of Zn^{2+} and many other nutrient elements. What would be the approximate molarity of "free" Zn^{2+} in a synthetic feedstuff solution prepared by adding 2.0 liters of 4.0 $\underline{M}$ $EDTA^{4-}$ to 30.0 liters of 0.010 $\underline{M}$ $ZnSO_4$? [For $[ZnEDTA]^{2-}$, $K_{formation} = 1.58 \times 10^{17}$.]

ANSWERS:

(A.1.) 1.0×10^{-5} $\underline{M}$, (A.2.) 8.0×10^{-11} $\underline{M}$, (A.3.) 5.4×10^{-9} $\underline{M}$, (A.4.) 3.3×10^{-3} $\underline{M}$, (A.5.) 8.3×10^{-9} $\underline{M}$, (A.6.) 1.4×10^{-14} $\underline{M}$, (A.7.) 4.0×10^{-22} $\underline{M}$, (A.8.) 2.5×10^{-10} $\underline{M}$, (A.9.) 1.6×10^{-18} $\underline{M}$, (A.10.) 2.5×10^{-19} $\underline{M}$.

RELEVANT
PROBLEMS

B.1. Chelation is a chemical process wherein ions of a metal are seized in a "clawlike" structure by a chemical complexing agent. The medical importance of chelating agents hinges on the fact that metals play many critical roles in the living organism. Metabolic processes depend on the presence of magnesium, iron, sodium, potassium, calcium and traces of such metals as zinc, manganese, copper, and cobalt. However, certain other metals, even in micro-quantities, are extremely toxic. Consequently, chelate drugs with appropriate properties can play several therapeutic roles. For example, chelating agents have been designed that will bind toxic metals, forming a stable complex that can be excreted. Ethylenediaminetetraacetate ion ($EDTA^{4-}$) is usually the treatment of choice for lead poisoning. It increases the urinary excretion of lead dramatically, according to the following reaction:

$$Pb^{2+} + EDTA^{4-} \rightleftharpoons [Pb(EDTA)]^{2-}$$

What is the approximate molarity of "free" lead ion in a urine sample in which the concentration of the $[Pb(EDTA)]^{2-}$ complex is 1.3×10^{-5} M and the concentration of excess $EDTA^{4-}$ is 0.50 M? [The instability constant for $[Pb(EDTA)]^{2-}$ is 9.1×10^{-19}]

B.2. Some chelates are designed to bind essential trace elements and deliver them to appropriate tissue sites. Other chelates are capable of deactivating bacteria or viruses by depriving them of metals they need for their metabolism or delivering metals to them that are toxic. One of the very important factors in chelate drug design is a knowledge of the relative affinity that a chelating agent has for a particular metal ion. For example, to remove a toxic metal the chelate must have a greater affinity for the metal ion than that of any associated biological substance. To deliver metals to tissues the chelate must have a relatively low affinity for the metal, compared to that of the desired "host". The instability or formation constant of the metal-chelate complex is important in assessing this relative affinity. One of the functions of the antipyretic and analgesic drug aspirin is to scavenge Cu^{2+} ion from the blood and return it to the cell sites from which it was lost. The salicylate ion from the original aspirin acts as the chelating agent:

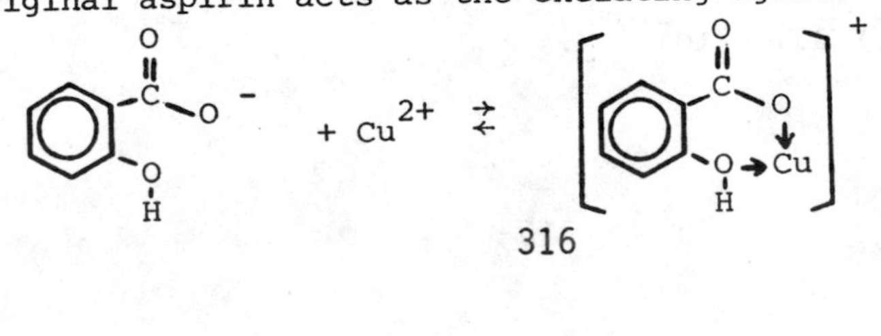

A 100 ml sample of human blood, taken from a patient on aspirin therapy, was analyzed and found to contain at equilibrium 98 micrograms of Cu^{2+}, 8.1 mg of salicylate ion, and 6.8×10^{-6} mole liter^{-1} of the $[Cu(C_7H_5O_3)]^+$ complex. Calculate the approximate formation constant for this complex.

B.3. Recently the toxicology of selenium has received considerable attention, largely as a result of the discovery that selenium occurring naturally in certain soils is absorbed by some of the plants used as food by animals and man. Another reason for this growing interest is that selenium and its compounds are being produced and consumed industrially and in research in ever increasing quantities. The toxic effects depend on the type of selenium compound absorbed, the characteristics of the animal which absorbs the compound, and the conditions under which the absorption occurs. Seleniferous materials can produce either acute or chronic poisoning in animals. Mild cases of selenium poisoning are characterized by a noticeable increase in the incidence of dental caries. This cariogenic effect is one that increases the solubility of tooth enamel in acid media. Selenium in the form of biselenite ion ($HSeO_3^-$), complexes with Ca^{2+} ion and thus reduces the stability of the calcium fluorophosphate lattice of the dental enamel:

$$Ca_2F(PO_4) \;\rightleftarrows\; 2\ Ca^{2+} + F^- + PO_4^{3-}$$

$$Ca^{2+} + HSeO_3^- \;\rightleftarrows\; [CaHSeO_3]^+$$

A 10 ml sample of saliva taken from a patient living in a seleniferous soil region contained 1.4×10^{-8} mg of the $[CaHSeO_3]^+$ complex and had a biselenite concentration of 1.5×10^{-5} $\underline{M}$. What is the instability constant for this complex if the concentration of "free" Ca^{2+} ion in the saliva sample was 1.63×10^{-12} $\underline{M}$?

B.4. To function properly most biological systems require a number of different enzymes (complex organic catalysts), substrates (the chemicals acted on by the enzymes), and enzyme cofactors. These cofactors may be organic molecules (called coenzymes), simple metal ions (called activators), or a metal ion complex. Fifteen different metal ions, in addition to ammonium ion, have been identified as activators for various enzyme systems. These are invariably ions of the lighter metals (atomic numbers 11-55). Heavy metals are often highly toxic to living organisms because of the irreversible damage they may cause to vital enzyme systems. Mercury(II) and lead ion poisoning, for example, can be traced directly to their destruction of enzymes. ATP (adenosine triphosphate) has many functions in the body. As a coenzyme its activity is enhanced by complexing with the magnesium ion:

$$(ATP)^{4-} + Mg^{2+} \rightleftharpoons [Mg(ATP)]^{2-}$$

The formation constant for this complex is approximately 9.0×10^4. If a physiological fluid was found to contain, at equilibrium, 5.0×10^{-4} $\underline{M}$ Mg^{2+} and 3.0×10^{-3} $\underline{M}$ $(ATP)^{4-}$, what is the approximate concentration of the complex $[Mg(ATP)]^{2-}$?

B.5. The biological effects of radiation, although seldom seen in ordinary clinical practice, are one of the most important concerns of mankind in our atomic age. When high energy radiation penetrates living cells, it leaves in its wake a mixture of "strange", unstable inorganic and organic species. As these species seek stability, covalent bonds within them may break permanently. The fragments may recombine either with themselves or with their neighbors. New molecules foreign to the cell are formed. When these disruptions occur in a nucleic acid molecule within a cell nucleus the genetic code will be altered, hence cell mutation will occur. At the least the cell may be reproductively dead. Efforts to minimize the damaging effects of radiation have been instrumental in the identification of certain types of substances that may be developed as effective radioprotective agents. A wide variety of chemicals exhibit some potential as radioprotective drugs. Among these are several components of honey bee venom. An analog of some bee components shows unusual potential. The mechanism of the protective action of this analog, glycylhistamine, is not yet understood but many radioprotective drugs are recognized copper(II) chelating agents. Glycylhistamine also forms a copper(II) complex according to the reaction:

$$GLY-HIS^- + Cu^{2+} \rightleftharpoons [Cu(GLY-HIS)]^+$$

What is the approximate molarity of "free" copper(II) ion in a radioprotective drug solution in which the concentration of the $[Cu(GLY-HIS)]^+$ complex is 2.1×10^{-6} $\underline{M}$ and the concentration of $(GLY-HIS)^-$ is 0.45 $\underline{M}$? [The formation constant for $[Cu(GLY-HIS)]^+$ is 5.2×10^7.]

*B.6. Medical diagnosis is both a science and an art, requiring sensitive observation as well as medical knowledge. The x-ray instrument is one of the physician's most valuable diagnostic tools. The x-ray machine is an extraordinary camera that allows the physician to view a patient's internal parts. Each of the patient's organs or bones absorbs a certain quantity of the x-rays, allowing the rest to pass through body tissue and "expose" the x-ray film. The photographic film is covered with a silver bromide/gelatin matrix. The x-rays striking the AgBr particles form "excited" Ag^+ ions. This only occurs in places the x-rays can reach. During "development" the excited ions are selectively reduced to black metallic silver. The film "fixing" process requires washing away the excess or unused AgBr with "hypo" (a solution of sodium thiosulfate). The insoluble AgBr forms a soluble silver complex in this process according to the reaction:

$$AgBr + 2 S_2O_3^{2-} \rightarrow [Ag(S_2O_3)_2]^{3-} + Br^-$$

Very often, an interesting biological phenomenon can be observed in the "hypo" solution. Platelets or communities of the aerobic colorless sulfur bacteria, Thiobacillus thiooxidans, can be seen floating on the surface of the "hypo" solution. These bacteria utilize the thiosulfate ion as an energy source according to the reaction:

$$S_2O_3^{2-} + H_2O + 2 O_2 \rightarrow 2 H^+ + 2 SO_4^{2-}$$

Used "hypo" solutions are not infected by these bacteria because of the toxicity of silver in the complex, $[Ag(S_2O_3)_2]^{3-}$. What is the approximate molarity of "free" Ag^+ ion in a solution formed by dissolving 15.0 g of AgBr in 4.0 liters of 3.5 $\underline{M}$ $Na_2S_2O_3$? [For $[Ag(S_2O_3)_2]^{3-}$, the $K_{instability} = 3.4 \times 10^{-14}$.]

*B.7. Symptoms of manganese deficiency in different mammals are similar, but not identical. The nature and severity of symptoms depend partly on the previous nutritional history of the animal. Manganese deficiency is usually characterized by reduced growth, slightly reduced mineralization, defective bone structure, and decreased reproductive performance in both males and females. Congenital manganese deficiency also produces animals that withstand stress poorly, have decreased resistance to infection, and suffer pancreatic pathology that is manifest in abnormal utilization of glucose. The most dramatic manganese deficiency syndrome occurs in poultry, especially young chicks. The disease, termed perosis, is characterized by gross enlargement and malformation of legs. What would be the approximate molarity of "free" manganese(II) ion in a fortified drinking water solution prepared for chickens by adding 10.0 liters of 4.0 $\underline{M}$ EDTA^{4-} to 40.0 liters of 0.10 $\underline{M}$ MnSO$_4$? [The net formation constant for $[Mn(EDTA)]^{2-}$ is 1.1×10^{14}.]

*B.8. The exact biochemical role of zinc is not completely understood. However, pathologic symptoms of zinc deficiency, skin lesions and corneal damage, are similar to those occurring in animals deprived of other nutrients such as riboflavin, Vitamin A, biotin, pantothenic acid, pyridoxine, or essential fatty acids. Thus, zinc may be concerned in the metabolism of one or more of these nutrients. Carbonic anhydrase, a zinc-containing enzyme, plays an important role in the acid-base equilibrium of the body and in the release of carbon dioxide in the lungs. It is also involved in the hydration of CO_2 in gastric mucosa, a necessary reaction for the neutralization of the excess alkalinity remaining from the secretion of H^+ in the production of gastric HCl. Carbonic anhydrase also plays a role in the calcification of bone and the formation of eggshells in birds. What would be the approximate molarity of the free Zn^{2+} ion in a nutrient solution prepared by adding 10.0 liters of 3.5 $\underline{M}$ sodium citrate to 70.0 liters of 0.10 $\underline{M}$ $ZnSO_4$? [The net formation constant for $[Zn(CIT)]^-$ is 3.2 x 10^5.]

*B.9. The magnesium activation of enzymes occurs through the formation of loosely bound metal-enzyme complexes. One of the most important functions of magnesium is in the activation of the numerous enzymes that split and transfer phosphate groups. These are the phosphatases and the enzymes concerned in the reactions involving adenosine triphosphate (ATP). Several of the magnesium-activated enzymes also require potassium as an additional cofactor. Potassium forms a loosely bound complex with ATP according to the reaction:

$$K^+ + ATP^{4-} \rightleftharpoons [K(ATP)]^{3-}$$

What would be the approximate molarity of the "free" K^+ ion in a solution prepared by mixing 2.0 liters of 2.50 $\underline{M}$ $(ATP)^{4-}$ with 100 ml of 0.010 $\underline{M}$ KCl? [The net formation constant for $[K(ATP)]^{3-}$ is 200.]

*B.10. Chemical poisons are divided into categories on the basis of their specific modes of action. Some are violent tissue poisons and exert a profound effect on any organ or structure contacted. Others only act systemically and exert their principal effect on vulnerable organ sites. Some chemicals appear relatively harmless by themselves but exhibit harmful synergistic effects with other chemicals. Nitrilo-triacetic acid (NTA), $N(CH_2CO_2H)_3$, is an organic compound capable of chelating the cations that cause water hardness. Under pressure to reduce phosphate pollution in natural waters, the detergent industry turned to NTA as a phosphate substitute. Following a short-term use, a better understanding of its biological impact began to emerge. The NTA was capable of forming soluble heavy metal complexes with mercury, cadmium, and lead. Tests indicated that these NTA-heavy metal complexes cause fetal malformations and birth defects. Moreover, the biodegradation of NTA may form compounds such as nitrates and nitrosamines which can lead to other pathological conditions. What would be the approximate molarity of free Cd^{2+} ion in a solution prepared by adding 5.0 liters of 8.0 $\underline{M}$ NTA^{3-} to 95.0 liters of 0.20 $\underline{M}$ $Cd(NO_3)_2$? [The net formation constant for $[Cd(NTA)]^-$ is 3.2 x 10^{10}.]

ANSWERS:
(B.1.) 2.4 x 10^{-23} $\underline{M}$, (B.2.) ∿7.5 x 10^2, (B.3.) 2.9 x 10^{-6}, (B.4.) 0.14 $\underline{M}$, (B.5.) 9.0 x 10^{-14} $\underline{M}$, (B.6.) 5.7 x 10^{-17} $\underline{M}$, (B.7.) 1.0 x 10^{-15} $\underline{M}$, (B.8.) 7.8 x 10^{-7} $\underline{M}$,

(B.9.) 1.0×10^{-6} $\underline{M}$,   (B.10.) 2.8×10^{-11} $\underline{M}$.

RELEVANT
PROBLEMS

Unit 12:   Complex Ions

Set I: Industrial Chemistry

I.1.   The commercial electrolytic method for production of aluminum was developed in 1886 by a student at Oberlin College, Charles Hall. Two complex ions are involved in the total process, tetrahydroxoaluminate and hexafluoroaluminate. Bauxite, an impure aluminum oxide hydrate, is treated with hot aqueous sodium hydroxide, forming a solution of $[Al(OH)_4]^-$, and leaving behind other metal oxide impurities such as Fe_2O_3. When the pH is adjusted properly, $Al(OH)_3$ precipitates from the complex ion solution and the solid aluminum hydroxide is heated to form Al_2O_3. This, in turn, is fused with cryolite, $Na_3[AlF_6]$, and the molten mixture is electrolyzed to produce aluminum metal. Bauxite deposits are rapidly being depleted and intensive research is underway to develop economical alternates to the Hall Process. To demonstrate the strong affinity of F^- for Al^{3+} in cryolite, calculate the concentration of "free" Al^{3+} in a solution 0.050 $\underline{M}$ in cryolite and 0.40 $\underline{M}$ in F^-. [The formation constant for hexafluoroaluminate is 6.9×10^{19}.]

I.2.   Hydrometallurgy is concerned with methods of producing metals, or some of their compounds, by reactions which take place in water. The first hydrometallurgical methods were used for the recovery of gold and copper. Recently, the methods of hydrometallurgy have been extended to include other elements, especially uranium. The principal operations used in hydrometallurgy include leaching, solution purification, and precipitation. Leaching is a process of dissolving the desired mineral by the chosen solvent. The problem in leaching is to set up the mechanical, chemical, and economic conditions that will permit a maximum dissolution of the desired metal, at a reasonable profit. Solution purification involves getting rid of the barren gangue from which the valuable mineral has been leached. This may vary from simple draining and washing to more elegant techniques such as countercurrent distribution. Precipitation is the final step, in which the metal or compound is produced as such and needs only refining to be ready for use. Silver is obtained from its sulfide ore, Ag_2S, by a process which starts by leaching with dilute cyanide solution, according to the equation:

$$Ag_2S_{(s)} + 4\ CN^- + H_2O \rightleftharpoons 2[Ag(CN)_2]^- + SH^- + OH^-$$

After solution purification, the equilibrium concentration of $[Ag(CN)_2]^-$ is 0.8 $\underline{M}$ and the excess cyanide concentration is 0.2 $\underline{M}$. What is the molar concentration of free Ag^+ in the solution? [For $[Ag(CN)_2]^- \rightleftharpoons Ag^+ + 2\ CN^-$, $K = 1.8 \times 10^{-19}$.]

I.3.   The exposure of a film to light activates silver bromide in the photographic emulsion so that treatment with "developer" (mild reducing agents), forms black metallic silver by the preferential reduction of activated silver bromide. Subsequently, black areas appear on the film where the light exposure was greatest. If the developed negative were now exposed to light, the light sensitive silver bromide still in the emulsion would be reduced and the entire film would turn black. Therefore, before the negative can be used, it must be fixed, i.e., the unexposed silver bromide must be removed. This is accomplished by placing the negative in a solution of sodium thiosulfate, $Na_2S_2O_3$, commonly called hypo. The hypo solution is able to dissolve the silver bromide by the reaction:

$$AgBr_{(s)} + 2\ S_2O_3^{2-} \rightleftharpoons [Ag(S_2O_3)_2]^{3-} + Br^-$$

After all the unchanged silver bromide on the negative has been dissolved in the hypo bath, the negative is washed with water to remove all traces of hypo, and then dried. What is the approximate concentration of free Ag^+ in a hypo wash solution containing 0.15 mole liter^{-1} of dithiosulfatoargentate(I) and 1.2 mole liter^{-1} of thiosulfate ion? [For formation constants, see Appendix D.]

I.4.   Metal-organic complexes are of great commercial interest. They are especially important in the areas of biology, catalysis, and analytical chemistry and in the technology of organic dyes and pigments. The commercial value of metal complex dyes is due especially to their resistance to fading on exposure to light and their improved washing "fastness". These improvements are brought about by converting dyes with coordinating groups to their metal complexes. The most important synthetic metallized organic pigments are used as the coloring components in printing inks and paints, for the coloring of rubber and plastics, in the dying of aqueous dispersions of paper pulp, and for the printing of textiles by the pigment printing method, in which the insoluble pigment in anchored to a fabric by heat-curable resins. The most useful chelating groups, which must be located in favorable positions in the dye molecule in order to form cyclic chelates, are the -OH, -COOH, -N=, -NH and -NH$_2$ groups. Pyridine-2- azo -p-dimethylaniline (PAD) reacts with Ni^{2+} to form the colored metal complex, [Ni(PAD)]$^{2+}$, that has an absorption maximum at 550 mμ. What is the concentration of the complex in an equilibrium system in which C_{Ni2+} = 5.0 $\times 10^{-3}$ $\underline{M}$ and C_{PAD} = 1.11 $\times 10^{-5}$ $\underline{M}$? [$K_{formation}$ = 1.72 $\times 10^4$]

I.5. Early gold mining techniques were both primitive and inefficient, involving primarily a simple mechanical separation of the heavy gold from lighter silicate minerals. The discarded wastes ("mine tailings") were usually still quite rich in gold content. In the early twentieth century, a chemical technique was developed that soon earned enormous profits from the waste dumps of abandoned gold mines. A dilute solution of sodium cyanide percolated through gold-containing wastes in contact with air extracts the gold by a process represented by:

$$4\ Au_{(s)} + 2\ H_2O_{(\ell)} + O_{2(g)} + 8\ CN^-_{(aq)} \rightarrow 4\ [Au(CN)_2]^-_{(aq)} + 4\ OH^-_{(aq)}$$

Metallic gold is then recovered by adding powdered zinc:

$$2\ [Au(CN)_2]^-_{(aq)} + Zn_{(s)} \rightarrow [Zn(CN)_4]^{2-}_{(aq)} + 2\ Au_{(s)}$$

Since most gold deposits contain some silver, which is also extracted by the cyanide process, the precipitated gold must be further purified by treatment with concentrated sulfuric acid, which reacts only with the silver. What is the concentration of "free" Zn^{2+} in a solution 0.050 M in tetracyanozincate and 0.15 M in cyanide ion? [The formation constant for tetracyanozincate ion is 1.1×10^{18}.]

*I.6. The development of new applications for lasers in science, technology, and medicine has generated a demand for lasers with greater power, efficiency, and stability. Most lasers currently use either gases or solids as the active medium. One essential property for an active medium is a high degree of optical perfection, i.e., freedom from localized irregularities. This is an inherent property in gases at low pressures, but in solids it is a major problem. Crystals and glasses are usually formed at high temperatures through processes requiring considerable effort and expense. A liquid is not subject to such irregularities as crystal defects and is, therefore, a promising active medium. Liquids have the additional advantages of not cracking or shattering, as solids will, and of simple production on a large scale at relatively low cost. In typical solid-state lasers, the rare-earth ions are employed in the active medium. When they are dissolved in water to use in liquid lasers their fluorescence efficiency is very low due to the agitation of solvent molecules. This competitive radiative dissipation of energy has been overcome by incorporating the ions into cages by using chelates. An active medium was prepared by an industrial research laboratory for use in a liquid laser that was 0.5 M in $[Nd(BzAc)_3]$. Estimate the concentration (molarity) of free Nd^{3+} ion in this solution, assuming that intermediate complexes are negligible. [K_{inst} for the complex is 5.0×10^{-19}.]

*I.7. What is the approximate concentration of "free" Ag^+ in a photographer's "hypo" solution (problem I.3) in which 0.25 g of AgBr has been dissolved, if the concentration of $S_2O_3^{2-}$ in the 500 ml of solution used was 0.15 $\underline{M}$ before any AgBr was added?

*I.8. The familiar industrial blueprint is a "negative-image" in which the blue color is due to the formation of insoluble potassium iron(II) hexacyanoferrate(III). In manufacturing the paper used in preparing blueprints, the paper is coated with a solution of $K_3[Fe(CN)_6]$ and iron(III) ammonium citrate. When a black-and-white image is placed over the paper and exposed to light, the light penetrating the white areas excites Fe^{3+} ions for selective reduction by citrate to Fe^{2+}. The iron(II) ions produced form the blue $KFe[Fe(CN)_6]$ precipitate. In unexposed areas, corresponding to lines on the original black-and-white "master", Fe^{3+} remains unreduced and the chemicals in these areas can be washed out with water, leaving white lines on a blue background. What is the approximate concentration of "free" Fe^{3+} in a solution used for blueprint reagent preparation, for which 250 ml of 0.10 $\underline{M}$ Fe^{3+} were mixed with 750 ml of 0.50 $\underline{M}$ KCN? [For formation constants, see Appendix D.]

*I.9. Platinum is an extremely valuable metal, used in making expensive jewelry, as well as metal parts and equipment requiring a high resistance to corrosion. The most important use of platinum, however, is based on the special surface characteristics of the metal that make it uniquely valuable as a contact catalyst. Platinum catalysts are used in most of the "catalytic converters" installed as part of the exhaust emission control systems in modern automobiles. The industrial applications of platinum catalysts are so important that modern industry would essentially disappear if platinum were to suddenly become unavailable. Industrial uses include such catalytic operations as the oxidation of SO_2 to SO_3 for sulfuric acid production, the hydrogenation of vegetable oils for preparation of margerine and shortening, the isomerization and cyclization of hydrocarbons in the petrochemicals industry, and the Haber Process synthesis of ammonia from H_2 and N_2 gases. Pure platinum can be obtained from the crude metal and various ores by a series of reactions starting with a treatment with aqua regia (a mixture of HCl and HNO_3) to form a solution of chloroplatinic acid, H_2PtCl_6. Impurities are precipitated by making the solution alkaline with sodium hydroxide, with the soluble $Na_2[PtCl_6]$ then isolated and treated with NH_4Cl. A yellow precipitate, $(NH_4)_2[PtCl_6]$, is recovered and purified by successive recrystallizations. Pure platinum is then obtained by thermal decomposition of the ammonium hexachloroplatinate(IV):

$$3\ (NH_4)_2[PtCl_6]_{(s)} \rightarrow 3\ Pt_{(s)} + 2\ N_{2(g)} + 16\ HCl_{(g)} + 2\ NH_4Cl_{(g)}$$

The net formation constant for hexachloroplatinate(IV) is for all practical purposes a meaningless term, since no detectable "free" Pt^{4+} ions exist in aqueous systems. Some idea of the affinity of platinum for chloride can be obtained, however, from the tetrachloroplatinate(II) complex:

$$Pt^{2+} + 4\ Cl^- \rightleftarrows [PtCl_4]^{2-}$$

$$(K_{formation} = 1.0 \times 10^{16})$$

What concentration of "free" Pt^{2+} would exist in a solution prepared by mixing equal volumes of 0.10 $\underline{M}$ Pt^{2+} and 4.0 $\underline{M}$ Cl^-?

*I.10. The tetrachloroplatinate(II) ion (problem I.9) finds industrial application in the electroplating of thin platinum coatings on more reactive metals for corrosion protection. It is the "free" Pt^{2+} from the complex that is reduced at the cathode in the electroplating operation. If intermediate complexes are negligible, what is the "free" Pt^{2+} concentration prepared by dissolving 25 kg of $Na_2[PtCl_4]$ in sufficient water to form 50 liters of solution?

ANSWERS:
(I.1.) 1.8×10^{-19} $\underline{M}$, (I.2.) 3.6×10^{-18} $\underline{M}$, (I.3.) 3.6×10^{-15} $\underline{M}$, (I.4.) 9.5×10^{-4} $\underline{M}$, (I.5.) 9.0×10^{-17} $\underline{M}$, (I.6.) 1×10^{-5} $\underline{M}$, (I.7.) 4.4×10^{-15} $\underline{M}$, (I.8.) 1.9×10^{-29} $\underline{M}$, [Note: This corresponds to about eleven Fe^{3+} ions per million liters.], (I.9.) 4.8×10^{-19} $\underline{M}$, (I.10.) 2.2×10^{-4} $\underline{M}$

RELEVANT PROBLEMS

Unit 12: Complex Ions

Set E: Environmental Sciences

E.1. The organic matter of marine sediments and terrestrial soils consists mainly of a series of complex, high - molecular weight, brown, nitrogenous polymers, collectively referred to as "humus". One fraction of the humus is a family of compounds known as "fulvic acid", which can help to transport iron into plants via a series of soluble complexes. Thus, humic materials provide a natural soil additive for prevention of iron chlorosis. Although "fulvic acid" is a family of compounds, rather than a single species, we can represent the complexing action by a simplified equation:

$$Fe^{3+} + HFV \rightleftarrows [Fe(FV)]^{2+} + H^+ \qquad (K_{formation} = 2.0 \times 10^5)$$

What is the molar concentration of this complex in a natural equilibrium system having a pH of 5.0, a "free" Fe^{3+} concentration of 3.6×10^{-8} $\underline{M}$, and a free "fulvic acid" concentration of 5.0×10^{-5} $\underline{M}$?

E.2. The sea is an enormous and extremely complicated aqueous system. All of the natural elements are found in sea water. Fresh water in lakes and streams is a dilute solution and most of the ionic solutes are present as their simple hydrates. Seawater, on the other hand, is much more concentrated and offers many opportunities for ionic interactions. Thus, many components of seawater are complex ions, usually of simple ligands such as chloride, hydroxide, carbonate, bicarbonate, or sulfate. These natural complexes play important roles in the ecosystems of the sea. Contamination of the sea by other complexing agents, or by metal ions competing for natural ligands, may have significant effects on aquatic life. A one liter aliquot of a seawater sample was analyzed and found to contain, at equilibrium, 40 mg of free Ca^{2+}, 0.1464 g of HCO_3^-, and 1.45×10^{-6} mole of the $[CaHCO_3]^+$ complex. Calculate the formation constant for this complex.

$$Ca^{2+} + HCO_3^- \rightleftharpoons [CaHCO_3]^+$$

E.3. Although much attention has been focused on the health hazards of tars and nicotine in tobacco smoke, two gaseous pollutants have received relatively little attention. Both carbon monoxide and hydrogen cyanide are among the products of the incomplete combustion of tobacco. Both these gases are highly toxic and, even in the small amounts found in tobacco smoke, complex strongly with the iron(II) in hemoglobin to reduce the oxygen-transport capability of the blood. To illustrate the affinity of cyanide for iron(II), estimate the concentration of "free" Fe^{2+} at equilibrium in a solution 0.30 M in hexacyanoferrate(II) and 0.010 M in cyanide. [For formation constants, see Appendix D.]

E.4. Water pollution from mercury compounds used as fungicides for treatment of seeds or for preventing slime growth during the pulping process in papermills has been recognized as a serious health problem. A number of other industrial operations also produce mercury-containing wastes. It is now known that certain aquatic bacteria can convert mercury and most mercury compounds to organomercurials, such as dimethylmercury, that are concentrated in successive stages of various food chains. As a result, some food fish have been identified as potential mercury-carriers. Mercury is a cumulative poison acting to destroy neurons in brain tissue. Mercury pollution is generally a localized problem in fresh water, since most mercury compounds are quite insoluble. In salt waters, however, stable water-soluble chloro complexes help maintain mercury pollutants in solution for widespread distribution. Calculate the net formation constant for tetrachloromercurate(II) assuming negligible intermediate complexes, on the basis of an analysis of an equilibrium system shown to contain 0.050 M $[HgCl_4]^{2-}$, 0.20 M Cl^-, and 2.7×10^{-14} M Hg^{2+}.

E.5. When the Federal Water Quality Commission imposed major restrictions on phosphates in detergents, in 1970, a number of compounds were tested as alternatives to tri-polyphosphate as a detergent additive for complexing with the Ca^{2+} and Mg^{2+} of "hard" water. One of these was the sodium salt of nitrilotriacetic acid (NTA), furnishing the nitrilotriacetate ion:

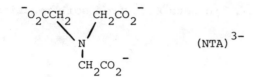

For a few months, this appeared to be a satisfactory substitute for tripolyphosphate, but studies soon identified NTA as a major health hazard. Unlike the tripolyphosphate, which is a fairly selective complexing agent, NTA formed soluble complexes with a wide variety of cations, including such toxic ions as Pb^{2+}, Hg^{2+}, and Cd^{2+}. These ions, usually precipitated as insoluble salts, could be widely spread through rivers and lakes as the NTA complexes. Both Hg^{2+} and Pb^{2+} cause irreversible neurological damage and Cd^{2+} leads to a bone deterioration by loss of Ca^{2+}. The formation constant for $[Cd(NTA)_2]^{4-}$ is 1.6×10^{15}. What is the concentration of "free" Cd^{2+} in a solution 0.035 $\underline{M}$ in $[Cd(NTA)_2]^{4-}$ and 0.012 $\underline{M}$ in $(NTA)^{3-}$?

*E.6. If a soluble cadmium salt is added to an aqueous system at pH 8.5, $Cd(OH)_2$ will precipitate if the concentration of Cd^{2+} exceeds 0.002 $\underline{M}$. What concentration of "free" Cd^{2+} would result from mixing equal volumes of 0.20 $\underline{M}$ Cd^{2+} and 1.4 $\underline{M}$ $(NTA)^{3-}$? Does this calculation suggest that the "total available" Cd^{2+} could significantly exceed the "precipitation limit" at pH 8.5 if sufficient $(NTA)^{3-}$ were present? [See Problem I.5. for additional information.] ("available" = "free" + "complexed")

*E.7. Most natural iron salts are insoluble within the usual pH range of soils and waters. The addition of ammonia or ammonium phosphate fertilizers further reduces the chance for Fe^{3+} to be dissolved and carried into plant systems, where it is essential to chlorophyll function. When insufficient Fe^{3+} is available for plant use, chlorophyll function is retarded and plants exhibit a condition known as iron chlorosis. This condition is usually noticed first in young leaves, which appear abnormally light green to almost white. Iron chlorosis may affect any type of green plant, but grasses, fruit trees, and bean plants are particularly susceptible. Land flooded by waters contaminated by alkaline waste materials is becoming an increasing problem in terms of iron chlorosis. Most soils in which the iron chlorosis occurs do contain sufficient iron, but not in a form available for plant use. Corrective measures require either the addition of soluble iron salts or the treatment

*Proficiency Level

of the soil with some iron-solubilizing agent. Certain iron chelates have proved quite useful by protecting the iron against soil reactions while maintaining it in a form easily assimilated by plants. Once the chelate is assimilated, it provides a continuous source of Fe^{3+} for the needs of the plant's chlorophyll system. "Sequestrene 220Fe", a commercial iron chelate micronutrient containing sodium diethylenetriaminepentaacetatoferrate(III), protects Fe^{3+} from soil reactions as the chelate system, formulated as:

$$[Fe(HDTPA)]^- \rightleftharpoons Fe^{3+} + (HDTPA)^{4-}$$

$$(K_{instability} = 3.3 \times 10^{-20})$$

The recommended application of "Sequestrene 220Fe" for a particular fruit orchard was 1.00 kg per 40 liters of water. What was the equilibrium concentration (mole liter^{-1}) of free Fe^{3+} ion in this solution? [The formula weight of the complex salt is 468.]

*E.8. The plants in a bean field were found to be suffering from iron chlorosis attributed to the loss of "available" iron caused by the periodic flooding of the field by the alkaline waters of a nearby river polluted by industrial wastes. The soil water pH of the field was 8.67. Under these conditions a concentration of Fe^{3+} exceeding 6×10^{-22} M will result in precipitation of $Fe(OH)_3$. Would application of a 15 g liter^{-1} "Sequestrene 220Fe" to this field provide only "available iron" without exceeding the "precipitation limits" of $Fe(OH)_3$? [Assume negligible pH change and negligible dilution. See Problem E.7. for additional information.]

*E.9. The presence of chloride ion in seawater helps to maintain a mercury(II) pollutant level well above that of the "precipitation limit" of $Hg(OH)_2$. What concentration of "free" Hg^{2+} would result from the dumping of 1500 liters of 0.30 M Hg^{2+} into a small bay containing 28,000 liters of seawater 0.55 M in Cl^-? [See Problem E.4. for additional information.]

*E.10. Plutonium is a radioactive element used in nuclear power plants. Within recent years, increasing concern has been expressed over the possibility of accidental plutonium release to the environment from nuclear plants. The element is extremely poisonous and the ingestion of even small amounts presents the additional hazard of an "internal radiation source". Although most plutonium compounds are only sparingly soluble in water, the distribution of plutonium(III) in an aquatic system is aided by fluoride ion, due to formation of the soluble $[PuF]^{2+}$ complex. What concentration of "free" Pu^{3+} would result from the addition of 250 liters of 1.3×10^{-5} M Pu^{3+} to 3000 liters of "fluoridated" water, 1.0 ppm in F^-? [The instability constant for fluoroplutonium(III) is 3.6×10^{-7}.]

328

ANSWERS:

(E.1.) 3.6×10^{-2} $\underline{M}$,   (E.2.) 0.61,   (E.3.) 3.0×10^{-20} $\underline{M}$,   (E.4.) 1.2×10^{15},   (E.5.) 1.5×10^{-13} $\underline{M}$,   (E.6.) 2.5×10^{-16}, yes,   (E.7.) 4.2×10^{-11} $\underline{M}$,   (E.8.) no [$C_{Fe^{3+}}$ of 3.3×10^{-11} $\underline{M}$ exceeds "limits".],   (E.9.) 2.8×10^{-16} $\underline{M}$,   (E.10.) 7.6×10^{-9} $\underline{M}$.

EQUILIBRIUM

UNIT 13: SIMULTANEOUS EQUILIBRIA

We have already seen, in the case of complex ions (Unit 12), examples of multicomponent equilibrium systems in which some species, such as the ammonia in the $[Cu(NH_3)_4]^{2+}$ system, participates simultaneously in a number of competing reactions. Two additional cases are of practical importance: the so-called "hydrolysis" of salts and the pH - control of precipitations. Both of these are amenable to a fairly simple treatment, using certain approximations, within the concentration range of a number of useful applications. A third case, that of the polyprotic acids (such as aqueous H_3PO_4), requires a more detailed analysis, which we will leave to more advanced texts.

Pure water has a pH of 7 and is said to be "neutral". If we dissolve a salt such as sodium nitrate in the water, the pH remains about 7, but addition of ammonium nitrate cause the pH to drop sharply into the acid range. Sodium fluoride has the opposite effect, producing an alkaline solution (pH > 7). Since sodium nitrate had no appreciable effect on the pH, we might suspect that the acidity of ammonium nitrate was due to the NH_4^+ ion, while the basic character of a sodium fluoride solution is probably attributable to the fluoride ion. Such is, indeed, the case.

Remember (Unit 11) that water may _act_ as an acid or a base, provided some other species is present to serve as a proton acceptor or donor, respectively. The pH changes resulting from adding NH_4NO_3 or NaF to water might then be described by equations for "hydrolysis" reactions, i.e., reactions in which water _participates_ as one of the reactants:

pH decrease when NH_4NO_3 is added to water:

$$NH_4^+{}_{(aq)} + H_2O_{(\ell)} \; \overset{\rightarrow}{\leftarrow} \; NH_{3(aq)} + H_3O^+{}_{(aq)}$$

pH increase when NaF is added to water:

$$F^-{}_{(aq)} + H_2O_{(\ell)} \; \overset{\rightarrow}{\leftarrow} \; HF_{(aq)} + OH^-{}_{(aq)}$$

Note that, in both cases, the products of the "hydrolysis" reaction include the "parent" weak acid or base of the ion reacting with water. This is a general observation which may be summarized as:

THE CONJUGATE[1] ION OF A WEAK ACID OR BASE REACTS WITH WATER TO ESTABLISH AN EQUILIBRIUM SYSTEM CONTAINING THE ORIGINAL WEAK ACID OR BASE.

The calculation of pH for a solution of a salt from a weak acid or base follows exactly the same procedures as were used for the simple cases described in Unit 11. However, most tabulations of dissociation constants contain the K_a or K_b-values for the

---

[1]Remember (Unit 11), that members of a conjugate pair are species differing only by a proton (H^+).

weak acids and bases, rather than for their conjugate ions. To find the constants needed for "hydrolysis" equilibria, we need to consider two <u>simultaneous equilibria</u> in aqueous solutions. Let's trace through an example, using the equilibrium system represented by an aqueous solution of sodium fluoride. Two interactive equilibria may be represented.

$$F^-_{(aq)} + H_2O_{(\ell)} \;\rightleftharpoons\; HF_{(aq)} + OH^-_{(aq)}$$

and (in any system containing liquid water):

$$H_2O_{(\ell)} \;\rightleftharpoons\; H^+_{(aq)} + OH^-_{(aq)}$$

For the hydrolysis equilibrium,

$$K_{b(F^-)} = \frac{C_{HF} \times C_{OH^-}}{C_{F^-}}$$

and the ionization of water is given by (Unit 11):

$$K_w = C_{H^+} \times C_{OH^-}$$

If we look in a table of dissociation constants (e.g., Appendix D), we do not find a K_b-value of F^-. The most closely related constant available is K_a for HF. With a little ingenuity, we may be able to use this K_a.

From the ion-product for water:

$$C_{OH^-} = \frac{K_w}{C_{H^+}}$$

Substituting this for C_{OH^-} in the K_b-expression gives:

$$K_{b(F^-)} = \frac{\boxed{C_{HF}} \times K_w}{\boxed{C_{F^-} \times C_{H^+}}}$$

in which the encircled portion is recognized as the inverse of the K_a-expression of HF, so that:

$$K_{b(F^-)} = \frac{K_w}{K_{a(HF)}}$$

This approach is general for any "hydrolysis" of a monoprotic species:

$$K_{a(acid)} = \frac{K_w}{K_{b(conj.\ base)}} \quad \text{or} \quad K_{b(base)} = \frac{K_w}{K_{a(conj.\ acid)}}$$

Thus, a tabulated value of K_a or K_b for a weak acid or weak base can be used with K_w for water (1.0×10^{-14}) to calculate the constant needed for an equilibrium solution established by dissolving in water a salt of the weak acid or base. This system is useful as long as the other ion of the salt does not react significantly with water. Salts containing cations of the Group IA or IIA elements or anions of strong acids (e.g., Cl^-, Br^-, I^-, NO_3^-, ClO_4^-) are amenable to this simplified treatment.

Anions such as hydroxide or those from weak acids (e.g., CO_3^{2-} or S^{2-}) are often used as precipitating agents for cations. In our introductory treatment of solubility equilibria (Unit 10) we worked with the approximation that only the ions of the insoluble salt needed to be considered. For many situations this approximation is perfectly satisfactory. However, the use of "simultaneous equilibria" analysis permits us to consider some cases on a more accurate bases, and to see how the control of pH (perhaps by a buffer, Unit 11) will allow us to choose conditions to induce or prevent precipitation or to selectively precipitate some particular salt from a mixture of cations. The easiest case to see is that of the metal hydroxides, for example copper(II) hydroxide:

$$Cu(OH)_{2(s)} \rightleftharpoons Cu^{2+}_{(aq)} + 2OH^-_{(aq)}$$

If we consider also the equilibrium of water itself ($H_2O \rightleftharpoons H^+ + OH^-$), we see that we can make the following calculation:

$$K_{sp} = C_{Cu^{2+}} \times C_{OH^-}^2$$

$$K_{sp} = C_{Cu^{2+}} \times \left[\frac{K_w}{C_{H^+}}\right]^2$$

$$\log K_{sp} = \log C_{Cu^{2+}} + 2 \log K_w - 2 \log C_{H^+}$$

$$\log C_{Cu^{2+}} = -2 \log K_w + \log K_{sp} + 2 \log C_{H^+}$$

But, for $Cu(OH)_2$, both K_{sp} and K_w are constants, and ($- \log C_{H^+} = $ pH), so:

$$\log C_{Cu^{2+}} = \text{[a constant]} - 2pH$$

This relationship shows that the concentration of Cu^{2+} ion in a saturated solution of $Cu(OH)_2$ can be varied simply by varying the pH.

Further discussion of pH-control of precipitation is given in the METHODS sections which follow.

- References -

1. O'Connor, Rod. 1977. Fundamentals of Chemistry, second edition. New York: Harper & Row (Units 19, 21 and Excursion 2)

2. Brown, Theodore, and H. Eugene LeMay, Jr. 1977. Chemistry: The Central Science. Englewood Cliffs, N.J.: Prentice-Hall (Chapters 16, 17)

3. Masterton, William, and Emil Slowinski. 1977. Chemical Principles, fourth edition. Philadelphia: W.B. Saunders (Chapters 18, 19)

4. Nebergall, William, F.C. Schmidt, and H.F. Holtzclaw, Jr. 1976. College Chemistry, fifth edition. Lexington, MA: D.C. Heath (Chapters 18, 19)

5. Moeller, Therald and Rod O'Connor. 1972. Ions in Aqueous Systems. New York: McGraw-Hill.

OBJECTIVES:

(1) *Given a description of an aqueous solution of the salt of a weak acid or a weak base and access to appropriate dissociation constants (e.g., Appendix D), be able to calculate the approximate pH of the solution.*

(2) *Given a description of a solution containing metal ions and access to K_{sp}-values for their hydroxides (e.g., Appendix D), be able to calculate the pH necessary for selective precipitation of specified hydroxides.*

*(3) *Given descriptions of a solution containing metal ions, the precipitating agent available, and necessary equilibrium constants (e.g., Appendix D), be able to make calculations of appropriate buffer component ratios to avoid or selectively induce specified precipitations.*

PRE-TEST:

Necessary Equilibrium Constants

for NH_3, $K_b = 1.8 \times 10^{-5}$

for $Ca(OH)_2$, $K_{sp} = 4.0 \times 10^{-6}$

for $Mg(OH)_2$, $K_{sp} = 1.2 \times 10^{-11}$

for FeS, $K_{sp} = 4.0 \times 10^{-19}$

for CdS, $K_{sp} = 1.0 \times 10^{-28}$

for CH_3CO_2H, $K_a = 1.8 \times 10^{-5}$

(1) Ammonium chloride is used in medicine as an expectorant and as an aid to certain antibiotics administered for kidney or bladder infections, in which a low pH favors antibiotic activity. What would be the approximate pH of a solution prepared by dissolving a 100 mg tablet of ammonium chloride in 50.0 ml of pure water? _____

(2) A number of the commercial drain-cleaners contain either solid or aqueous NaOH or KOH because of the effectiveness of these in solubilizing the animal greases

often responsible for clogged drains. Partly as a result of "high pressure" advertising techniques urging the regular use of some of these drain-cleaners, waste effluent from many high-density housing areas is strongly alkaline. This is, of course, diluted when the effluent (perhaps "treated" to some extent in a municipal sewage plant) reaches a river, lake, or sea. What is the minimum pH to which an alkaline waste stream could convert a natural lake 0.05 $\underline{M}$ in Ca^{2+} and 0.005 $\underline{M}$ in Mg^{2+} so that only $Mg(OH)_2$ would precipitate? _____

At what pH would $Ca(OH)_2$ also precipitate? _____

*(3) Cadmium sulfide is a valuable commercial pigment, sold under the name of <u>cadmium yellow</u>. It is prepared commercially by saturating a slightly acidic solution of cadmium ion with hydrogen sulfide gas. Cadmium solutions are normally contaminated by small amounts of Fe^{2+} ion. If FeS (black) is significantly precipitated along with the CdS, the color of the pigment is ruined. What would be the minimum mole ratio of acetic acid to acetate ion for a buffer to be used for precipitation of only CdS from a solution which was 0.15 $\underline{M}$ Cd^{2+} and 2×10^{-4} $\underline{M}$ Fe^{2+}? _____

In saturated aq. H_2S, in acidic solution, a reasonable approximation of sulfide-ion concentration is given by ($C_{H^+}^2 \times C_{S^{2-}} = 1.3 \times 10^{-21}$).

Answers and Directions:

(1) 5.3,　(2) 9.7 [$Ca(OH)_2$ at pH 12.0]

If both are correct, go on to question *3. If you missed any, study METHODS, sections 13.1 and 13.2.

*(3) moles (CH_3CO_2H): moles ($CH_3CO_2^-$) = 45:1

If your answer was correct, go on to RELEVANT PROBLEMS, Unit 13. If you missed it, study METHODS, section 13.3.

METHODS

13.1　Salt "Hydrolysis" and pH

There is no <u>real</u> difference between the behavior of an "acidic or basic salt" in water and that of a molecular weak acid or weak base, because the ions formed from these <u>are</u> <u>themselves</u> weak bases or acids. The only difference between the two equilibrium systems:

$$NH_{3(aq)} + H_2O_{(\ell)} \rightleftarrows NH_{4(aq)}^+ + OH_{(aq)}^-$$

and

*Proficiency Level

$$CN^-_{(aq)} + H_2O_{(\ell)} \; \overset{\rightarrow}{\leftarrow} \; HCN_{(aq)} + OH^-_{(aq)}$$

is that we can find the K_b for ammonia easily (Appendix D) and we have to calculate that for cyanide ion from the tabulated K_a-value for its conjugate acid (page 331).

$$K_{b(CN^-)} = \frac{K_w}{K_{a(HCN)}}$$

Once we have the necessary dissociation constant, the remaining procedure is identical to that introduced in Unit 11.

EXAMPLE 1

The SOLUTION to Pre-Test question 1:

When NH_4Cl is dissolved in water, the solution consists mainly of NH_4^+ and Cl^- ions. Chloride ion is the conjugate base of a strong acid (HCl), so it will not react significantly with water to recapture a proton. Ammonium ion, on the other hand, is the conjugate acid of the weak base NH_3. The equilibrium system, then, is represented as (omitting H_2O for simplicity):

$$NH_4^+{}_{(aq)} \; \overset{\rightarrow}{\leftarrow} \; NH_3{}_{(aq)} + H^+_{(aq)}$$

for which:

$$K_{a(NH_4^+)} = \frac{K_w}{K_{b(NH_3)}}$$

Using $K_w = 1.0 \times 10^{-14}$ and $K_b = 1.8 \times 10^{-5}$:

$$K_{a(NH_4^+)} = \frac{1.0 \times 10^{-14}}{1.8 \times 10^{-5}} = 5.6 \times 10^{-10}$$

Then, by the method developed in Unit 11:

$$K_a = \frac{C_{NH_3} \times C_{H^+}}{C_{NH_4^+}} \simeq \frac{C^2_{H^+}}{\underline{M}\,(NH_4^+)}$$

and, from the information given and the formula weight of NH_4Cl:

$$\underline{M}(NH_4^+) = \frac{100\ mg(NH_4Cl)}{50\ ml} \times \frac{1\ mmol(NH_4Cl)}{53.5\ mg(NH_4Cl)} \times \frac{1\ mmol(NH_4^+)}{1\ mmol(NH_4Cl)} = 3.7 \times 10^{-2}\ \underline{M}$$

so that:

$$C_{H^+} = \sqrt{(5.6 \times 10^{-10}) \times (3.7 \times 10^{-2})}$$

and

$$pH = -\tfrac{1}{2} \log[(5.6 \times 10^{-10}) \times (3.7 \times 10^{-2})] = \underline{5.3}$$

- -

EXERCISE 1

Washing soda ($Na_2CO_3 \cdot 10H_2O$) is used to "soften" water by removal of ions such as Ca^{2+} or Fe^{3+} which would otherwise form insoluble salts with soaps needed for cleaning. Although calcium carbonate is precipitated, it is the hydroxide precipitate, $Fe(OH)_3$, which removes iron(III) ion from the solution. The principal source of the OH^- needed is the hydrolysis of carbonate ion:

$$CO_{3(aq)}^{2-} + H_2O_{(\ell)} \;\rightleftharpoons\; HCO_{3(aq)}^{-} + OH_{(aq)}^{-}$$

What is the approximate pH of a solution made by dissolving one tablespoon (21.3 g) of washing soda in 1.00 pint (473 ml) of pure water?
[For dissociation constants, see Appendix D.]

(answer, page 342)

- -

13.2 pH-Control of Hydroxide Precipitations

If we know the solubility product (Unit 10) for a metal hydroxide and the molarity of the metal ion in solution, then we may easily calculate the hydroxide ion concentration of a saturated solution of the compound from the conventional K_{sp}-expression. For precipitation to occur, the hydroxide-ion concentration must exceed[2] the value calculated for the saturated solution. To avoid precipitation, the hydroxide-ion concentration must be equal to or less than that calculated for the saturated solution. Thus, for selective precipitation of metal hydroxides, the upper limit of OH^- concentration (and, hence, the "maximum pH")[3] is the C_{OH^-} calculated from the K_{sp} of the metal hydroxide which is not to be precipitated. The lower limit of OH^- concentration (minimum pH) is that which just exceeds the C_{OH^-} calculated from the K_{sp} of the metal hydroxide which is to be precipitated. Of course, if more than two metal ions are involved, the calculations must be made for each to determine which ones set the maximum and minimum pH limits.

EXAMPLE 2

The SOLUTION to Pre-Test question 2:

for $Mg(OH)_2$: $\quad K_{sp} = C_{Mg^{2+}} \times C_{OH^-}^2 = 1.2 \times 10^{-11}$

---

[2]Remember that the OH^- concentration calculated from K_{sp} is for the equilibrium solution, i.e., no further precipitation occurs at that concentration.
[3]Remember, since pH = 14 - pOH and pOH is the negative log of C_{OH^-}, pH increases as C_{OH^-} increases.

for Ca(OH)$_2$: K$_{sp}$ = C$_{Ca^{2+}}$ x C$^2_{OH^-}$ = 4.0 x 10^{-6}

For a saturated solution of Mg(OH)$_2$:

$$C_{OH^-} = \sqrt{\frac{K_{sp}}{C_{Mg^{2+}}}} = \sqrt{\frac{1.2 \times 10^{-11}}{0.005}}$$

pOH = $-\frac{1}{2}$ log (2.4 x 10^{-9}) = 4.31

pH = 14.00 - 4.31 = 9.69

So, the <u>minimum</u> pH at which Mg(OH)$_2$ would precipitate from a solution 0.005 <u>M</u> in Mg^{2+} must <u>just exceed</u> 9.69, say ~<u>9.7</u>.

For a saturated solution of Ca(OH)$_2$

$$C_{OH^-} = \sqrt{\frac{K_{sp}}{C_{Ca^{2+}}}} = \sqrt{\frac{4.0 \times 10^{-6}}{0.05}}$$

pOH = $-\frac{1}{2}$ log (8.0 x 10^{-5}) = 2.05

pH = 14.00 - 2.05 = 11.95

Thus, Ca(OH)$_2$ would not begin to precipitate until the pH just exceeded 11.95, say ~12.0.

EXERCISE 2

Many natural waters obtain iron salts from various minerals through which the water has percolated. In most cases, natural water supplies are slightly alkaline. Iron(III) hydroxide is so slightly soluble that, even at very low concentrations, the iron(III) in natural waters is usually present as "colloidal" (dispersed solid) Fe(OH)$_3$. In regions where the iron content of the water is moderately high, it is not uncommon for well water to exhibit a distinct yellow turbidity from the colloidal Fe(OH)$_3$. What would be the maximum pH for a natural water containing 33 ppm[4] of iron(III) to <u>avoid</u> formation of any solid Fe(OH)$_3$?

[For K$_{sp}$-values, see Appendix D.]

(answer, page 343)

Extra Practice
EXERCISE 3

A municipal water plant uses a stock solution of 1.5 <u>M</u> NaF for introducing fluoride into the water, since it is easier to control the addition and mixing of a solution than it

---

[4]ppm (parts per million) $\simeq$ mg liter^{-1}

is of a solid additive. What is the approximate pH of the stock solution used for this water fluoridation process?

[See Appendix D for dissociation constants.]

(answer, page 343)

EXERCISE 4

Paints containing the pigment <u>white lead</u> ($PbSO_4$) now find rather limited markets, partly because of the problems of lead poisoning with children who have ingested such paints from chewing on painted objects. Quite recently it was discovered that certain pencils available to school children, who often engage in "pencil gnawing", were painted with dangerous "lead-based" paints. An alternative white pigment, <u>lithopone</u>, is much safer. It is prepared by mixing equimolar amounts of aqueous zinc sulfate and aqueous barium sulfide, forming a mixed precipitate of the white salts $BaSO_4$ and ZnS. Although barium hydroxide is moderately soluble, zinc hydroxide is not.[5] If the pH of the reaction mixture is too high, zinc hydroxide will coprecipitate and such contamination of the lithopone produces undesirable mixing and "covering" characteristics of the pigment. The barium sulfide solution used in the preparation of <u>lithopone</u> is alkaline, due to "hydrolysis" of sulfide ion, while the zinc sulfate solution is slightly acidic.[6] If a <u>lithopone</u> production is planned to employ mixing equal volumes of 0.20 <u>M</u> BaS and 0.20 <u>M</u> $ZnSO_4$, what is the maximum pH allowable for the resulting mixture if coprecipitation of zinc hydroxide is to be avoided?

(answer, page 344)

===

At this point you should try the <u>competency</u> level Self-Test questions on page 342.

===

*13.3 Buffer-Control of Precipitation

Many of the anions used in preparing common insoluble salts undergo "hydrolysis" reactions (Section 13.1) in water, for example:

Sulfide - $\quad S^{2-}_{(aq)} + H_2O_{(l)} \rightleftarrows HS^-_{(aq)} + OH^-_{(aq)}$

Carbonate - $\quad CO_3^{2-}_{(aq)} + H_2O_{(l)} \rightleftarrows HCO_3^-_{(aq)} + OH^-_{(aq)}$

Sulfate - $\quad SO_4^{2-}_{(aq)} + H_2O_{(l)} \rightleftarrows HSO_4^-_{(aq)} + OH^-_{(aq)}$

Phosphate - $\quad PO_4^{3-}_{(aq)} + H_2O_{(l)} \rightleftarrows HPO_4^{2-}_{(aq)} + OH^-_{(aq)}$

---

[5] For $Zn(OH)_2$, $K_{sp} = 2.5 \times 10^{-10}$. Zinc hydroxide, like $Al(OH)_3$ [Unit 12] will react with excess OH^- to form a soluble complex, $[Zn(OH)_4]^{2-}$.
[6] The <u>hydrated</u> zinc ion acts as a proton donor.

*Proficiency Level   338

An analysis of these equilibria in terms of Le Chatelier's principle (Unit 12) suggests that the concentration of the "precipitating anion" (e.g., S^{2-}, CO_3^{2-}, SO_4^{2-}, PO_4^{3-}) can be varied by changing the hydroxide ion concentration of the solution. This is the basis for the pH-control of precipitation reactions, and this control may be exercised by use of an appropriate buffer (Unit 11).

The problem of selecting an appropriate buffer for a desired pH control is one involving a consideration of the simultaneous equilibria of saturated solutions, anion "hydrolysis", and buffer systems. A stepwise approach is probably the most efficient way of handling these problems:

1. The concentration limits of a "precipitating anion" are determined from appropriate solubility product expressions (Unit 10).

2. The concentration limits of hydroxide ion (and, hence, the pH limits) are determined from the "hydrolysis" equilibrium of the "precipitating anion" (Section 13.1). [Alternatively, pH limits may be found from other information, depending on the conditions of a particular process. See EXAMPLE 3.]

3. The buffer composition limits are determined from the analysis of the buffer equilibrium system (Unit 11).

Each of these "steps" has been discussed earlier, so our problem is simply one of "putting them all together" for the appropriate analysis of buffer-controlled precipitation.

*EXAMPLE 3

The SOLUTION to Pre-Test question *3:

Step 1 - Calculating limits for $C_{S^{2-}}$:

For CdS, $K_{sp} = 1.0 \times 10^{-28}$, so precipitation of CdS requires a "free" S^{2-} concentration greater than:

$$C_{S^{2-}} = \frac{K_{sp}}{C_{Cd^{2+}}} = \frac{1.0 \times 10^{-28}}{0.15} = 6.7 \times 10^{-28} \underline{M}$$

For FeS, $K_{sp} = 4.0 \times 10^{-19}$, so avoiding the precipitation of FeS requires that the concentration of "free" S^{2-} may not exceed:

$$C_{S^{2-}} = \frac{K_{sp}}{C_{Fe^{2+}}} = \frac{4 \times 10^{-19}}{2 \times 10^{-4}} = 2 \times 10^{-15} \underline{M}$$

Thus, the concentration of "free" S^{2-} must be maintained with the range of

$\leq 2 \times 10^{-15}$ to $> 6.7 \times 10^{-28}$

    (maximum)   (minimum)

Step 2 - Calculating pH (C_{H^+}) limits:

Since, in saturated H_2S in an acidic medium, we can approximate the pH limits from:

$$C_{H^+}^2 \times C_{S^{2-}} = 1.3 \times 10^{-21}$$

the maximum C_{H^+} permitting CdS to precipitate[7] is:

$$C_{H^+} = \sqrt{\frac{1.3 \times 10^{-21}}{6.7 \times 10^{-28}}} \quad \text{[minimum } C_{S^{2-}} \text{ needed]} = 1.4 \times 10^3$$

and the minimum C_{H^+} to __avoid__ precipitation of FeS is:

$$C_{H^+} = \sqrt{\frac{1.3 \times 10^{-21}}{2 \times 10^{-15}}} \quad \text{[maximum allowable } C_{S^{2-}}] = 8.1 \times 10^{-4}$$

Thus, C_{H^+} must be maintained at or above 8.1×10^{-4} __M__ to avoid precipitation of the FeS.

Step 3 - Calculating buffer limits:

For an acetic acid/acetate buffer (K_a for $CH_3CO_2H = 1.8 \times 10^{-5}$), the C_{H^+} is given by (Unit 11):

$$C_{H^+} = K_a \times \frac{\text{moles } (CH_3CO_2H)}{\text{moles } (CH_3CO_2^-)}$$

Then for a minimum C_{H^+} of 8.1×10^{-4}

$$\frac{\text{moles } (CH_3CO_2H)}{\text{moles}(CH_3CO_2^-)} = \frac{C_{H^+}}{K_a} = \frac{8.1 \times 10^{-4}}{1.8 \times 10^{-5}} = 45$$

Thus, the buffer must contain a mole ratio of $CH_3CO_2H : CH_3CO_2^-$ of 45:1.

---

[7]This was not requested in the original question, but it is necessary information for designing the real process. Note that no __real__ solution could have a hydrogen-ion concentration as high as 1.4×10^3 __M__. This indicates that there is, in fact, no way of using pH control to prevent the precipitation of CdS from 0.15 __M__ Cd^{2+} solution saturated by H_2S.

*EXERCISE 5

"Limestone" is a mineral rich in calcium carbonate, but also containing silicates and carbonates of magnesium and other metals. Lime (CaO) is prepared commercially on a massive scale by the thermal decomposition of the calcium carbonate in "limestone". Lime is sold as a soil additive (for acidic soils), as a component of plaster for the construction industry, and as a reagent for a wide variety of industrial chemical processes. "Slaked lime", $Ca(OH)_2$, prepared by adding water to lime, is mixed with sand and water to form mortar for the building trade. Since "limestone" from different sources varies in composition, lime plants must maintain analytical laboratories. Although almost all of the natural "limestones" are commercially valuable, the composition of the "limestone" determines to some extent the market potential. Lime containing more than 5% MgO, for example, cannot be used in the production of "slaked lime" for mortar because of the very slow reaction of magnesium oxide with water.

One analytical scheme employed in limestone assay uses the selective precipitation of $CaCO_3$ from a buffered solution of Ca^{2+} and Mg^{2+} by the addition of aqueous sodium carbonate. What mole ratio of ammonia to ammonium chloride would prepare a buffer whose pH is the maximum possible for the selective precipitation of $CaCO_3$ from a solution which is 0.10 $\underline{M}$ in Ca^{2+} and 0.16 $\underline{M}$ in Mg^{2+}, by addition of an equal volume of 0.20 $\underline{M}$ Na_2CO_3 ?
[For necessary equilibrium constants, see Appendix D.]

(answer, page 345)

Extra Practice

*EXERCISE 6

Iron smelters maintain facilities for ore analysis to help in planning the routine economic and engineering aspects of smelter operation. Smaller companies unable to justify the expense of sophisticated instrumental laboratories often rely on a gravimetric (mass) assay for iron content of ores. The iron from the ore, along with other metals, is dissolved by addition of nitric acid to a pulverized sample of the ore. Since nitric acid is a good oxidizing agent, the iron in the resulting solution is present as Fe^{3+}. To the acidic solution is added sufficient ammonia/ammonium chloride buffer to selectively precipitate $Fe(OH)_3$. This is collected by filtration, heated to a high temperature for conversion to Fe_2O_3, and the resulting mass (along with that of the original ore sample) is used to calculate the % iron in the ore.

What is the maximum mole ratio of NH_3 to NH_4^+ for a buffer to be used for the selective precipitation of $Fe(OH)_3$ from a solution which is, after addition of the buffer, 0.20 $\underline{M}$ in Fe^{3+} and 0.0070 $\underline{M}$ in Mn^{2+} ?
[For necessary equilibrium constants, see Appendix D.]   (answer, page 346)

==

SELF TEST (UNIT 13)   [answers, page 347]

13.1. A solution of calcium propionate, $Ca(CH_3CH_2CO_2)_2$, is used by most commercial bakeries as a dough additive to retard the growth of mold on bread or other bakery products. What is the approximate pH of a solution containing 80 g of calcium propionate per liter? [For dissociation constants, see Appendix D.]

13.2. Copper(II) sulfate is an effective algaecide and is sometimes added to swimming pools or decorative ponds to prevent the growth of algae. What must be the maximum pH limit for pool water to avoid the precipitation of copper(II) hydroxide when $CuSO_4 \cdot 5H_2O$ is added to the pool in the amount of 1.5 g per gallon (3.79 liters) of water?   [For K_{sp}-values, see Appendix D.]

If you completed Self-Test questions 13.1 and 13.2 correctly, you may go on to the proficiency level, try the RELEVANT PROBLEMS (Unit 13), or stop here. If not, you should consult your instructor for suggestions of further study aids.

----

*13.3. Pollution of natural waters by mercury is a problem of increasing concern, having gained international attention in the 1960's when a number of persons died after eating mercury-contaminated fish in such widely-separated areas as Japan and Guatemala. Federal restrictions now limit the dumping of industrial mercury-containing wastes and have set the "safe" level of mercury in water at 5 parts per billion. The mercury(II) ion may be assayed, even in fairly dilute solution, by precipitating and weighing the very insoluble mercury(II) sulfide. What minimum mole ratio of acetic acid to acetate ion would provide a buffer of pH sufficient to permit the precipitation of only HgS from a solution which is 5.4×10^{-3} M in Mn^{2+} and 3.0×10^{-5} M in Hg^{2+} when the solution is saturated with H_2S gas?

[Assume for the saturated H_2S system that $c_{H^+}^2 \times c_{S^{2-}} = 1.3 \times 10^{-21}$. For other equilibrium constants, see Appendix D.]

If you answered this question correctly, you may go on to the RELEVANT PROBLEMS for Unit 13. If not, you should consult your instructor for suggestions of further study aids.

ANSWERS to EXERCISES, Unit 13

1. (11.8) Solution: (following the method introduced in Unit 11):

for $CO_3^{2-} + H_2O \rightleftharpoons HCO_3^- + OH^-$

$$K_{b(CO_3^{2-})} = \frac{C_{HCO_3^-} \times C_{OH^-}}{C_{CO_3^{2-}}} \simeq \frac{C_{OH^-}^2}{\underline{M}(CO_3^{2-})}$$

$$K_{b(CO_3^{2-})} = \frac{K_w}{K_{a(HCO_3^-)}} = \frac{1.0 \times 10^{-14}}{4.8 \times 10^{-11}} \quad \text{(Appendix D)}$$

$$\underline{M}(CO_3^{2-}) = \frac{21.3 \text{ g}(NaCO_3 \cdot 10H_2O)}{0.473 \text{ liter}} \times \frac{1 \text{ mole}(Na_2CO_3 \cdot 10H_2O)}{286 \text{ g}(Na_2CO_3 \cdot 10H_2O)} \times \frac{1 \text{ mole}(CO_3^{2-})}{1 \text{ mole}(Na_2CO_3 \cdot 10H_2O)}$$

$$C_{OH^-} = \sqrt{K_b \times \underline{M}(CO_3^{2-})} = \sqrt{(2.1 \times 10^{-4}) \times 0.157}$$

$$pOH = - \log (3.3 \times 10^{-5}) = 2.2$$

$$pH = 14.00 - pOH = 14.00 - 2.2$$

2. (~2.6) Solution:

From Appendix D, K_{sp} for $Fe(OH)_3 = 6.0 \times 10^{-38}$

For the saturated solution, represented by:

$$Fe(OH)_{3(s)} \overset{\rightarrow}{\underset{\leftarrow}{}} Fe^{3+}_{(aq)} + 3 \, OH^-_{(aq)}$$

$$K_{sp} = C_{Fe^{3+}} \times C_{OH^-}^3$$

from which:

$$C_{OH^-} = \sqrt[3]{\frac{K_{sp}}{C_{Fe^{3+}}}}$$

If the iron(III) content of the water is 35 ppm, then:

$$C_{Fe^{3+}} = \frac{35 \text{ mg}(Fe^{3+})}{1 \text{ liter}} \times \frac{1 \text{ liter}}{1000 \text{ ml}} \times \frac{1 \text{ mmol}(Fe^{3+})}{55.85 \text{ mg}(Fe^{3+})} = 6.3 \times 10^{-4} \, \underline{M}$$

Then, for the saturated solution:

$$C_{OH^-} = \sqrt[3]{\frac{6.0 \times 10^{-38}}{6.3 \times 10^{-4}}}$$

$$pOH = - \frac{1}{3} \log (9.5 \times 10^{-35}) = 11.34$$

$$pH = 14.00 - pOH = 2.66$$

To avoid precipitation, then, the minimum pH must be just lower than 2.66, say ~2.6.

3. (8.7) Solution:

From Appendix D, $K_{a(HF)} = 6.9 \times 10^{-4}$.

For the equilibrium system:

$$F^-_{(aq)} + H_2O_{(\ell)} \rightleftharpoons HF_{(aq)} + OH^-_{(aq)}$$

$$K_{b(F^-)} = \frac{C_{HF} \times C_{OH^-}}{C_{F^-}} \simeq \frac{C^2_{OH^-}}{\underline{M}(F^-)}$$

from which:

$$C_{OH^-} \simeq \sqrt{K_b \times \underline{M}(F^-)}$$

Since $K_{b(F^-)} = \dfrac{K_w}{K_{a(HF)}}$:

$$K_{b(F^-)} = \frac{1.0 \times 10^{-14}}{6.9 \times 10^{-4}} = 1.4 \times 10^{-11}$$

Then:

$$pOH = -\tfrac{1}{2} \log[(1.4 \times 10^{-11}) \times 1.5] = 5.3$$

$$pH = 14.00 - pOH = 14.00 - 5.3$$

4. (~8.6) Solution:

For $Zn(OH)_2$, $K_{sp} = 2.5 \times 10^{-10}$.

if a saturated solution is formed:

$$Zn(OH)_{2(s)} \rightleftharpoons Zn^{2+}_{(aq)} + 2OH^-_{(aq)}$$

for which:

$$K_{sp} = C_{Zn^{2+}} \times C^2_{OH^-}$$

$$C_{OH^-} = \sqrt{\frac{K_{sp}}{C_{Zn^{2+}}}}$$

Remember (Unit 10) that mixing different solutions dilutes both, so after mixing (and before ZnS begins to precipitate):

$$C_{Zn^{2+}} = \frac{0.20\ \underline{M}}{1} \times \frac{1}{2} = 0.10\ \underline{M}$$

To avoid precipitation then, C_{OH^-} must be less than:

$$C_{OH^-} = \sqrt{\frac{2.5 \times 10^{-10}}{0.10}}$$

And the pH must be less than that calculated from:

$$pOH = -\tfrac{1}{2} \log(2.5 \times 10^{-11}) = 5.30$$

$$pH = 14.00 - pOH = 14.00 - 5.30 = 8.70$$

A pH < 8.70, say ~8.6, would avoid coprecipitation of $Zn(OH)_2$.

- -

*5. [moles (NH_3): moles $(NH_4Cl) \simeq 56:1$] <u>Solution</u>:

From Appendix D:

$$K_{sp}(CaCO_3) = 4.7 \times 10^{-9}$$

$$K_{sp}(MgCO_3) = 4.0 \times 10^{-5}$$

$$K_a(HCO_3^-) = 4.8 \times 10^{-11}$$

After mixing equal volumes, before any reaction occurs:

$$C_{CO_3^{2-}} = 0.20 \underline{M} \times \tfrac{1}{2} = 0.10 \underline{M}$$

$$C_{Ca^{2+}} = 0.10 \underline{M} \times \tfrac{1}{2} = 0.050 \underline{M}$$

$$C_{Mg^{2+}} = 0.016 \underline{M} \times \tfrac{1}{2} = 0.0080 \underline{M}$$

Step 1 - *Carbonate concentration limits:*

To precipitate $CaCO_3$, $C_{CO_3^{2-}}$ must exceed:

$$C_{CO_3^{2-}} = \frac{K_{sp}}{C_{Ca^{2+}}} = \frac{4.7 \times 10^{-9}}{0.050} = 9.4 \times 10^{-8} \underline{M}$$

and, to <u>avoid</u> precipitation of $MgCO_3$, $C_{CO_3^{2-}}$ must <u>not</u> exceed:

$$C_{CO_3^{2-}} = \frac{K_{sp}}{C_{Mg^{2+}}} = \frac{4.0 \times 10^{-5}}{0.0080} = 5.0 \times 10^{-3} \underline{M}$$

Step 2 - *Hydroxide concentration limits:*

For $CO_3^{2-} + H_2O \rightleftarrows HCO_3^- + OH^-$

$$K_b(CO_3^{2-}) = \frac{K_w}{K_a(HCO_3^-)} = \frac{1.0 \times 10^{-14}}{4.8 \times 10^{-11}} = 2.1 \times 10^{-4}$$

$$2.1 \times 10^{-4} = \frac{C_{HCO_3^-} \times C_{OH^-}}{C_{CO_3^{2-}}} \simeq \frac{C_{OH^-}^2}{C_{CO_3^{2-}}}$$

To precipitate $CaCO_3$, then, C_{OH^-} must exceed:

$$C_{OH^-} = \sqrt{(2.1 \times 10^{-4}) \times (9.4 \times 10^{-8})} = 4.4 \times 10^{-6} \underline{M}$$
$$\text{↳[necessary for } CaCO_3]$$

but, to <u>avoid</u> precipitation of $MgCO_3$, C_{OH^-} may not exceed:

$$C_{OH^-} = \sqrt{(2.1 \times 10^{-4}) \times (5.0 \times 10^{-3})} = 1.0 \times 10^{-3} \; \underline{M}$$

[maximum to avoid $MgCO_3$]

Step 3 - Buffer composition:

For $NH_3 + H_2O \rightleftharpoons NH_4^+ + OH^-$

$$C_{OH^-} = K_b \times \frac{\text{moles } (NH_3)}{\text{moles } (NH_4^+)}$$

So, to avoid precipitation of $MgCO_3$, the maximum buffer component ratio is:

$$\frac{\text{moles } (NH_3)}{\text{moles } (NH_4^+)} = \frac{C_{OH^-}}{K_b} = \frac{1.0 \times 10^{-3}}{1.8 \times 10^{-5}} = \underline{56}$$

(added as NH_4Cl)

Note that this ratio still provides a more than sufficient hydroxide-ion concentration for precipitation of $CaCO_3$:

$$C_{OH^-} = 56 \times K_b = 1.0 \times 10^{-3}$$

(only 4.4×10^{-6} necessary)

*6. [moles (NH_3): moles $(NH_4^+) \simeq 0.29:1$] Solution:

From Appendix D:

$$K_{sp} \; [Fe(OH)_3] = 6.0 \times 10^{-38}$$

$$K_{sp} \; [Mn(OH)_2] = 2.0 \times 10^{-13}$$

$$K_b \; [NH_3] = 1.8 \times 10^{-5}$$

Steps 1 and 2 - (Combined, since OH^- is the precipitating anion):

for $Fe(OH)_3$, C_{OH^-} must exceed:

$$C_{OH^-} = \sqrt[3]{\frac{K_{sp}}{C_{Fe^{3+}}}} = \sqrt[3]{\frac{6.0 \times 10^{-38}}{0.20}} = 6.7 \times 10^{-13} \; \underline{M}$$

for $Mn(OH)_2$, C_{OH^-} must not exceed:

$$C_{OH^-} = \sqrt{\frac{K_{sp}}{C_{Mn^{2+}}}} = \sqrt{\frac{2.0 \times 10^{-13}}{0.0070}} = 5.3 \times 10^{-6} \; \underline{M}$$

Step 3 -

Then, for an NH_3/NH_4^+ buffer:

$$C_{OH^-} = K_b \times \frac{\text{moles (NH}_3)}{\text{moles (NH}_4^+)}$$

and the maximum component ratio is:

$$\frac{\text{moles(NH}_3)}{\text{moles(NH}_4^+)} = \frac{C_{OH^-}}{K_b} = \frac{5.3 \times 10^{-6}}{1.8 \times 10^{-5}} = 0.29$$

==

ANSWERS to SELF-TEST, Unit 13

13.1. (9.5)

13.2. (~6.0)

*13.3. moles (CH_3CO_2H) : moles $(CH_3CO_2^-) \approx 5.6 : 1$

A.1. Copper deficiencies in crops are treated by the application of copper fertilizers which are available in both organic and inorganic forms. Applications can be made directly to the soil, through foliar sprays, or by seed treatment prior to planting. One typical organic source utilizes the chelate, $[Cu(EDTA)]^{2-}$. Copper deficiencies in various crops produce abnormal coloring and development, lowered quality in fruit and grain, and lowered yields. The availability of copper is dependent upon soil pH. In alkaline systems the copper(II) ion forms the insoluble $Cu(OH)_2$. What would be the maximum pH for a soil solution containing 45 ppm of uncomplexed copper(II) to avoid formation of any solid $Cu(OH)_2$? [For K_{sp}-values, see Appendix D.]

A.2. The production of fluid fertilizers has increased significantly in recent years. Popularity of such fertilizers is justified because they offer an excellent opportunity for custom-type methods of preparation. Micronutrient sources that are used in fluid fertilizers include oxides, inorganic salts, and organic derivatives such as synthetic chelates and natural organic complexes. Most micronutrients are applied with solutions of mixed nitrogen-phosphorus-potassium fertilizers. Solubilities of the micronutrients are frequently enhanced in these fluid preparations by addition of carrier solutions such as ammonium nitrate. This salt improves the solubility of boron, copper, manganese, and zinc. What would be the approximate pH of a carrier solution prepared by dissolving 2.0 g of ammonium nitrate per 50.0 ml of pure water?

A.3. Pasture grasses and grazing animals both have delicately balanced needs for trace elements. Too little may produce deficiency symptoms; too much may be poisonous. In the case of many trace elements, nutritional levels and toxic levels are close together. Thus, inattention, or difficulties during manufacture, may lead to unsatisfactory mixing, in which case toxic concentrations may appear in certain batches. A case of fluorine poisoning in a dairy herd was caused by adding a commercial mineral supplement to the cows' diet. Analysis of the mineral supplement showed a fluorine content of 2,846 ppm of F^- ion. What is the approximate pH of the solution used to prepare this supplement? [For K_a-value, see Appendix D.]

A.4. One of the outstanding characteristics of sea water is the presence and concentration of acids and bases. Their composition is governed by the physical and chemical processes that occur in the environment. Physical processes include the evaporation of water, solute concentration by freezing, or addition of more water by precipitation. Chemical processes include addition or removal of carbon dioxide at the air-water interface, precipitation or dissolution of calcium carbonate or other compounds, and ion-exchange processes between sea water and its constituents and suspended materials or sediments. Of the major anions in sea water, only carbonate and dihydrogen borate make a significant contribution to the alkalinity of sea water. Their effects are summarized according to the equations:

$$CO_3^{2-} + H_2O \rightleftharpoons HCO_3^- + OH^-$$

$$H_2BO_3^- + H_2O \rightleftharpoons H_2BO_3 + OH^-$$

The vertical distribution of carbonate components vary as a function of the depth of oceanic waters. What is the approximate pH of sea water, at a depth of 1562 meters, resulting from a carbonate ion concentration of 2.23×10^{-3} mole liter^{-1}? [See Appendix D for ionization constants.]

A.5. Cadmium, a very dangerous trace element in the marine environment, is of concern because of its tendency to accumulate in animals. The average concentration of cadmium in sea water is approximately 0.20 mg liter^{-1}. Attention has been focused on the effects of cadmium in marine organisms because of the ever increasing number of sources and quantities of cadmium finding its way into the sea. Since cadmium bears a close geochemical kinship to zinc it is associated with zinc and often enters the environment because of incomplete technical separation. Large quantities also enter the marine environment from aerosols that were emitted in the manufacture of batteries, plastics, pigments, alloys, and fertilizers. In sea water, cadmium may be precipitated as $Cd(OH)Cl$. What minimum pH would be required to precipitate $Cd(OH)Cl$ in a marine system having a chloride ion concentration of 0.535 $\underline{M}$ and a cadmium concentration of 30 mg liter^{-1}? [K_{sp} for $Cd(OH)Cl$ is 3.2×10^{-11}.]

A.6. One measurement of the reactivities of elements in sea water is based on the degree of undersaturation for the elements in sea water, especially in the case of cations and dissolved gases. The maximum concentration that a cation could attain would conceivably be regulated by the solubility of its least soluble compound, as determined by the ionic species present, if no other chemical reaction reduced its abundance. As a first approximation, it might be deduced that those elements with the lowest concentrations would be the most reactive, while those at or near saturation would be essentially inert in the marine environment. What is the minimum pH that would have to be attained in the sea, containing 2.0 µg liter^{-1} of Ni^{2+} and 8.0 µg liter^{-1} of Mn^{2+}, so that only $Ni(OH)_2$ would precipitate? [K_{sp} values for $Ni(OH)_2 = 1.6 \times 10^{-16}$; $Mn(OH)_2 = 2.0 \times 10^{-13}$.]

*A.7. The stability of carbon dioxide and its distribution among the gaseous, liquid, and solid phases of the earth makes this gas an essential link in the carbon cycle. Carbon dioxide gas from the atmosphere is continually cycled into the biosphere by photosynthesis but is, simultaneously, replenished by the oxidation of organic matter and by respiration in animals. Atmospheric carbon dioxide is interfaced, by molecular exchange, with the dissolved carbon dioxide of the hydrosphere. The dissolved hydrospheric carbon dioxide may have been transferred from the atmosphere, having been produced from the oxidation of organic carbon by biogenic and non-biogenic processes, or having originated from the dissolving of carbonates. In some marine systems, the selective precipitation of calcium carbonate from naturally buffered solutions may occur at certain carbonate levels. What mole ratio of ammonia to ammonium ion would provide a buffer whose pH is the maximum possible for the selective precipitation of $CaCO_3$ from a solution initially 0.01 $\underline{M}$ in Ca^{2+}, 0.06 $\underline{M}$ in Mg^{2+}, and 0.40 $\underline{M}$ in CO_3^{2-}? [For necessary equilibrium constants, see Appendix D.]

*A.8. The addition of micronutrients to standard fertilizer preparations offers a practical method for applying micronutrients to crops or pasture land. Several methods have been developed for incorporating micronutrients in fertilizers. These include adding the micronutrients to fertilizers by incorporation during granulation of the fertilizer, dry-blending with solid fertilizers, coating fertilizer granules with powdered materials, dissolving them in liquid fertilizer, and adding them to suspension products. The mode of addition of micronutrient materials to fertilizers is a factor that can have a significant effect on both the physical and chemical properties of the mixture. When adding micronutrient materials, consideration must be given to such factors as the chemical compatibility of the ingredients, the segregation characteristics of the mixture, and the solubility in liquid systems. One common analytical scheme employed in micronutrient assay uses the method of selective hydroxide precipitation to separate Fe(III) from the bivalent Mn(II) ion. What is the maximum mole ratio of NH_3 to NH_4^+ for a buffer to be used for the selective precipitation of $Fe(OH)_3$ from a solution which is, after addition of the buffer, 0.15 $\underline{M}$ in Fe^{3+} and 0.005 $\underline{M}$ in Mn^{2+}? [For necessary equilibrium constants, see Appendix D.]

*A.9. The proportions of micronutrients added to fertilizers are generally small, but vary widely, depending on specific crops and soils and on whether or not the micronutrient enriched fertilizer is to be used in direct application or is to be mixed first with other fertilizer materials. When micronutrients are incorporated in fertilizer carriers, the chemical interactions that may occur during formulation, in storage, or during the initial dissolution in soil embrace a wide variety of reaction mechanisms. Conditions of formulation and storage determine the rate and extent of these reactions, and most of the reactions produce compounds of much lower water solubility than the original micronutrient source. A classical scheme used to separate several heavy metal ions in micronutrient assay uses the method of selective sulfide precipitation. What would be the minimum mole ratio of acetic acid to acetate ion for a buffer to be used for precipitation of only ZnS from a solution which was 0.20 $\underline{M}$ Zn^{2+} and 4.0 x 10^{-4} $\underline{M}$ Fe^{2+}? [For equilibrium constants, see Appendix D and page 334.]

*A.10. The six elements nitrogen, phosphorus, potassium, calcium, magnesium, and sulfur are obtained from the soil by plants in such large quantities that they are designated as macronutrients. Plant growth may be retarded by their deficiencies in the soil. The total calcium in the soil shows great variation but is usually more available than potassium. When it is lacking, soils tend to be acid. This deficiency is corrected by addition of lime or dolomite, [CaMg(CO$_3$)$_2$]. Magnesium, besides its nutrient value, functions in the soil much as does calcium. Where liming is practiced, deficiency in magnesium is automatically corrected. What mole ratio of NH$_3$/NH$_4^+$ would a soil chemist use to prepare a buffer whose pH is the maximum possible for the selective precipitation of CaCO$_3$ from a solution which is 0.20 $\underline{M}$ in Ca^{2+} and 0.13 $\underline{M}$ in Mg^{2+}, by addition of an equal volume of 0.25 $\underline{M}$ Na$_2$CO$_3$? [For necessary equilibrium constants, see Appendix D.]

ANSWERS:

(A.1.) 6.2, (A.2.) 4.8, (A.3.) 8.2, (A.4.) 10.8, (A.5.) 7.34, (A.6.) At pH ~9.8, only Ni(OH)$_2$ precipitates., (A.7.) [NH$_3$]:[NH$_4^+$] $\simeq$ 20 : 1, (A.8.) [NH$_3$]:[NH$_4^+$] $\simeq$ 0.35 : 1, (A.9.) [CH$_3$CO$_2$H]:[CH$_3$CO$_2^-$] $\simeq$ 60 : 1, (A.10.) [NH$_3$]:[NH$_4^+$] $\simeq$ 0.06 : 1

RELEVANT
PROBLEMS

Unit 13: Simultaneous Equilibria

Set B: Biological & Medical Sciences

B.1. The skeletal retention of fluoride ion is significantly increased when drinking water contains traces of molybdenum. Moreover, molybdenum also increases fluoride absorption and retention in muscle tissue and in the brain. The increases in deposition of fluoride, however, are not always proportional to the amount of molybdenum. This is of interest because the fluoride ion probably plays an important role in strengthening bone and tooth structures. However, as is the case with most substances, excess fluoride ion is poisonous. Toxic levels can be assimilated in a number of ways. A particularly hazardous source is the smoke from aluminum factories, containing a high content of fluoride which enters the food chain through vegetation of surrounding pasture lands. The diseased cattle grazing such a pasture showed fluoride ion levels of 38 parts per million (ppm). What would be the approximate pH of a solution made by dissolving 122 mg of sodium fluoride in 1.0 liter of water? [For dissociation constant, see Appendix D.]

B.2. The potential effect of a chemical agent upon a plant or animal depends upon the vulnerability of individual tissues, the mode of action of the agent, and its concentration. Tissues and cells vary widely in their susceptibility to specific forms of chemical injury. For example, benzene damages principally the liver and bone marrow cells, although this toxin is usually absorbed through the lungs and carried by the blood stream to all tissues of the body. On the other hand, cyanide is a highly toxic protoplasmic poison which inhibits the cell oxidative respiratory enzymes. In some cases toxins are produced _in situ_ under anaerobic conditions. For example, ethylene is produced in wet soils and inhibits root growth while cyanide may be formed in the roots of peach trees, causing their death. What is the approximate pH of a solution that is 0.10 $\underline{M}$ in cyanide ion (CN^-)? [For dissociation constant, see Appendix D.]

B.3. Almost all organisms are afflicted periodically by a malady described as "altered tissue reactivity", allergy. The causes of allergies are as varied as the chemical substances in the environment. It has long been recognized that what is one man's food may be another man's poison. One third of the population is credited with some idiosyncrasy to food. The chief dietary offenders include wheat, milk, egg, chocolate, fish, potato, and tomato. Botanically related foods may give what is known as "group sensitization". If one is sensitive to oranges he is likely to be sensitive to grapefruit, lemons, and other citrus fruits. Other causes of allergy are less common in occurrence, but are none the less potent when they are involved. Bacteria may become antigens, and bacterial allergy is one of the oldest and best-known examples of the allergic process. Fungus allergy is also encountered, along with allergy to molds and smuts. "Mycoban", a topical fungicide, is prepared by using a stock solution of 2.0 $\underline{M}$ sodium propionate, $CH_3CH_2CO_2Na$, for introducing sodium propionate into an ointment or gel. What is the approximate pH of the stock solution used for this fungicide preparation process? [See Appendix D, for dissociation constants.]

B.4. Insect allergy is often a very serious problem. Many persons are sensitive to chemicals that the insects produce and use as weapons in their defense mechanisms. The most common offenders include mosquitoes, fleas, ants, and bees. Occasionally an individual is so sensitive to certain insect attack that he goes immediately into collapse and may die unless treated promptly with a specific remedy, such as epinephrine (adrenalin) by hypodermic injection. Formic acid was first observed to be among the distillation products of certain ants in 1670. What would be the approximate pH of a solution prepared by dissolving 17.0 g of sodium formate in 50.0 ml of pure water? [See Appendix D for dissociation constants.]

B.5. Early studies of the functions of magnesium in animals were concerned with the neuromuscular paralysis that may occur when magnesium salts are administered intravenously. In the 1920's magnesium was discovered to activate alkaline phosphatase. Since then hundreds of enzymes have been found to be activated by magnesium. Magnesium is concentrated within the cells of the tissues. The liver, striated muscle, kidney, and brain contain about 450 to 500 parts per million, while the blood serum contains only about 10% of that amount, or approximately 50 mg liter^{-1}. Like calcium, some of the serum magnesium is bound to protein and, although the total blood magnesium varies from species to species, all warm-blooded animals have about the same amount of magnesium bound to protein. What is the minimum pH that a physiological solution 0.03 $\underline{M}$ in Ca^{2+} and 0.02 $\underline{M}$ in Mg^{2+} could sustain so that only $Mg(OH)_2$ would precipitate? [For equilibrium constants, see Appendix D.]

B.6. The human adult contains approximately four to five grams of iron. This represents about 50 ppm of the body weight. About 92% of the iron is present in complex forms bound to porphyrinc. Bone marrow is one of the last iron reservoirs to be depleted of iron, and also one of the last to be restored during recovery from iron depletion. Thus, bone marrow iron is one of the best clinical criteria of body iron reserves. In the body the globular protein apoferritin is used in the storage of iron by formation of an iron-protein complex called ferritin. The iron of ferritin is in the Fe^{3+} state in the form of $Fe(OH)_3$ micelles. What would be the minimum pH for a physiological fluid containing 50 ppm of iron(III) to form solid $Fe(OH)_3$? [For K_{sp} values, see Appendix D.]

B.7. Copper deficiency in animals is manifest by a variety of clinical syndromes in which different species exhibit different symptoms. Anemia is a general symptom for all species, but bone disorders, fibrosis of the myocardium, and gastrointestinal disorders have all been observed in copper deficient animals of several species. All have responded to treatment by administration of adequate amounts of copper. Bone defects in grazing cattle and sheep on copper deficient pastures are characterized by spontaneous fractures and a condition very similar to rickets. In preparing solutions to be used for treating copper deficiency, the Cu^{2+} will precipitate as $Cu(OH)_2$ if the pH is too high. What would be the maximum pH for a solution containing 75.0 ppm of Cu^{2+} to avoid formation of any solid $Cu(OH)_2$? [For K_{sp}-values, see Appendix D.]

353

*B.8.   Radiant energy, such as x-ray, is capable of injuring or killing cells. The morphologic effects of radiant energy on cells are by no means unique and resemble alterations produced by a variety of other injurious agents, such as many of the chemical poisons, bacterial toxins and other lethal physical agents. However, it may be possible to identify radiation reaction by the pattern of tissue injury produced. Certain generalizations have been made with respect to the effect of radiant energy upon specific body tissues. First, embryonic, immature, undifferentiated, non-specialized cells are in general more susceptible to radiant injury than mature, well-differentiated adult cells. Secondly, all cells have an increased susceptibility or vulnerability to radiant injury while in mitosis. An analytical scheme employed in a research laboratory to assay skeletal radiation damage uses the selective precipitation of $CaCO_3$ from a buffered solution of Ca^{2+} and Mg^{2+} by addition of aqueous sodium carbonate. What mole ratio of NH_3/NH_4^+ would prepare a buffer whose pH is the maximum possible for the selective precipitation of $CaCO_3$ from a solution which is 0.05 $\underline{M}$ in Ca^{2+} and 0.01 $\underline{M}$ in Mg^{2+}, by addition of an equal volume of 0.15 $\underline{M}$ Na_2CO_3? [For necessary equilibrium constants, see Appendix D.]

*B.9.   Manganese is distributed in practically all of the body tissues. Bone is the richest source of manganese (~3.5 ppm) and may serve as a storage reservoir for the element, as for calcium and magnesium. Manganese is believed to be involved in amino acid metabolism, not only because of its activation of some of the hydrolyzing enzymes, but also because it forms chelates with many amino acids. The complexes of amino acids and manganese are transported in the body more rapidly than the amino acids alone. This links protein metabolism to manganese turnover and establishes the role of amino acid complexes in transport of the metal. In a particular blood assay the NH_3/NH_4^+ buffer was added to a serum solution, containing Fe^{3+} and Mn^{2+}, to selectively precipitate $Fe(OH)_3$. What was the maximum mole ratio of NH_3 to NH_4^+ for the buffer that was used to selectively precipitate $Fe(OH)_3$ from the serum solution which was, after addition of the buffer, 0.10 $\underline{M}$ in Fe^{3+} and 0.005 $\underline{M}$ in Mn^{2+}? [For necessary equilibrium constants, see Appendix D.]

*B.10. Several areas in the eastern United States have exceedingly low soil cobalt. Because of these low levels of cobalt, neither grasses nor legumes have sufficient cobalt to meet animal nutritional needs. Surveillance of cobalt levels in soils is easily monitored by observing the growth patterns of the cobalt-indicator plant, Nyassa sylvatica. There is no direct relationship between cobalt deficiency in soils and plants of a particular area and cobalt deficiency in humans or monogastric animals. Monogastric animals require cobalt specifically, and only in the form of vitamin B_{12}.

Synthesis of this vitamin by microorganisms in the rumen, and transfer of B_{12} from ruminants to monogastric animals in meat and milk, is the major route by which monogastric animal requirements for this vitamin are met. A biochemical assay of the leaves of a cobalt-indicator plant, Nyassa sylvatica, requires a scheme for the separation of Fe^{2+} and Co^{2+} ions. The separation was made by the selective precipitation of CoS from a solution of leaf extract which was 0.10 M in Co^{2+} and 3.0×10^{-4} M in Fe^{2+}. What would be the minimum mole ratio of acetic acid to acetate ion required to buffer the selective precipitation of CoS? [For necessary equilibrium constants, see Appendix D and page 334.]

ANSWERS:

(B.1.) 7.3, (B.2.) 11.2, (B.3.) 9.6, (B.4.) 9.2, (B.5.) 9.4, (B.6.) ~2.6, (B.7.) ~6.1, (B.8.) $[NH_3]:[NH_4^+]$ ~ 1.4 : 1, (B.9.) $[NH_3]:[NH_4^+]$ ~ 0.35 : 1, (B.10.) $[CH_3CO_2H]:[CH_3CO_2^-]$ ~ 56 : 1

I.1. Monochloroacetic acid, $ClCH_2COOH$, is manufactured by passing chlorine into glacial acetic acid in the presence of a sulfur or red phosphorus catalyst. Chloroacetic acid is commonly known industrially as MCA. It is used almost entirely as an intermediate in the manufacture of other chemicals, specifically 2,4-D, 2,4,5-T (both important herbicides) and carboxymethyl cellulose, which is manufactured by the reaction of alkali cellulose with sodium chloroacetate. Most large manufacturers of these important chemicals usually manufacture their own MCA and sodium-MCA. Smaller amounts of MCA are required in the manufacture of ethyl chloroacetate, glycine (a pharmaceutical used in buffered aspirin) and thioglycolic acid, of which the most important derivative is ammonium thioglycolate (used primarily in permanent wave solutions). The carboxymethyl cellulose, which is manufactured from the sodium-MCA, is physiologically inert and is employed as a protective coating for textiles and paper, sizing, ice cream and other emulsion stabilizers, and as an additive to impart strength to sausage casings and other films. Calculate the pH of a 0.2 $\underline{M}$ solution of sodium chloroacetate. [K_a for MCA = 1.4×10^{-3}.]

I.2. Ammonium chloride is generally produced by the reaction of ammonium sulfate and sodium chloride solutions. The ammonium chloride is subsequently removed by crystallization. Alternately, ammonium chloride is produced as a by-product of manufacture of sodium carbonate. The waste liquor from the carbonate process contains primarily calcium chloride and sodium chloride. This may be partially evaporated to recover sodium chloride, leaving a 50% solution of calcium chloride. This solution is charged with ammonia and carbon dioxide gases. The reaction is represented by the equation:

$$CaCl_{2(aq)} + 2\ NH_{3(g)} + CO_{2(g)} + H_2O_{(\ell)} \rightarrow 2\ NH_4Cl_{(aq)} + CaCO_{3(s)}$$

The precipitated calcium carbonate is filtered off and the ammonium chloride is crystallized from solution. The manufacture of ammonium chloride is an excellent example of a situation where the obvious manufacturing process, neutralization of hydrochloric acid with ammonia, is not economical. Approximately 50% of all ammonium chloride produced is used in the production of "dry" cells. The remainder is divided among a variety of applications, including uses as a mordant, in soldering flux, and as a pharmaceutical acidulant. The solubility of NH_4Cl in water is 41 g per 100 ml at room temperature. What is the approximate pH of saturated ammonium chloride solution? [See Appendix D for ionization constants.]

I.3. Propionic acid, $CH_3CH_2CO_2H$, may be prepared on an industrial scale by the catalytic oxidation of 1-propanol ($CH_3CH_2CH_2OH$). The acid is also prepared commercially by the reaction between ethyl alcohol and carbon monoxide or by the fermentation of hexoses, pentoses and lactic acid. Propionic acid or its salts occurs naturally in various dairy products, Swiss cheese containing as much as one percent of propionates. The sodium and calcium salts are marketed commercially. These salts are prepared by the neutralization of propionic acid with sodium or calcium hydroxide. Both salts require a nearly neutral medium for effective utilization of their fungistatic and fungicidal properties. Calcium and sodium propionates are used by the baking and dairy industries as inhibitors of molds. Both substances are available commercially under the name <u>Mycoban</u>. A variety of pharmaceutical preparations are available which contain sodium or calcium propionate blended with other chemicals in a jelly base or vehicle. These pharmaceuticals have a wide variety of uses ranging from ophthalmic ointments to topical dressings for treatment of superficial fungus infections. The solubility of calcium propionate in water is about 48 g per 100 ml at room temperature. What is the approximate pH of saturated calcium propionate solution? (K_a for propionic acid is 1.4×10^{-5}.)

I.4. Sodium sulfide is an inorganic compound that has attained a very important position in the <u>organic</u> chemical industry. It is consumed as a reducing agent in the manufacture of amino compounds and it is used extensively in the preparation of dyes. It is used in the leather industry as a depilatory, i.e., for the removal of hair from hides. Sodium polysulfide, a derivative of sodium sulfide, is one of the necessary reactants for making Thiokol synthetic rubber. Other industries where its use is important are the rayon, metallurgical, photographic, and engraving fields. Sodium sulfide is produced by treating barium sulfide with sodium carbonate, according to the equation:

$$BaS + Na_2CO_3 \rightarrow Na_2S + BaCO_3$$

Since both sulfide and carbonate ions hydrolyze to form OH^-, the individual reactant solutions are fairly alkaline. What maximum concentration of Ba^{2+} could be used for a solution of pH 12.5 if the precipitation of barium hydroxide is to be avoided? [K_{sp} for $Ba(OH)_2 \approx 5 \times 10^{-3}$]

I.5. Economic considerations are obviously important in any commercial process. Most calcium salts are considerably less expensive than the corresponding barium salts, yet BaS is the reagent of choice for preparation of sodium sulfide (problem I.4). A possible reason for this choice can be seen by comparing the relative concentration limits for Ca^{2+} and Ba^{2+} ions in alkaline solution. What is the maximum Ca^{2+} concentration to avoid precipitation of $Ca(OH)_2$ from a solution of pH 12.5? [See Appendix D for K_{sp}-data.]

I.6.   In any chemical industry using pipelines and pumps for distribution of alkaline fluids, a potential problem (in addition to the obvious corrosion problems) is the formation of metal hydroxide sediments that can clog lines and valves. This is a particular difficulty with ions forming very insoluble hydroxides. Even when the amount of initial precipitate is small, the accumulation of successive precipitates at crucial locations is a matter of considerable concern to plant engineers. Iron(III) ion is invariably present in solutions contacting iron or steel equipment and $Fe(OH)_3$ is extremely insoluble ($K_{sp} = 6.0 \times 10^{-38}$). What is the maximum concentration of Fe^{3+}, in parts per billion, if $Fe(OH)_3$ precipitation is to be avoided in an alkaline stream of pH 8.0?

*I.7. Cobalt is a relatively rare metal that finds important uses in such specialty alloys as ALNICO (used in making high-grade permanent magnets) and CARBALOY (for the cutting edges of special drill bits). Cobalt ores also contain nickel and copper, usually as sulfides. For the production of cobalt pure enough for most alloying purposes, the ore is oxidized and dissolved in acid, after which lime is added for "selective" precipitation of cobalt(III) oxide. The oxide, in turn, is reduced by carbon to form metallic cobalt. The small amounts of copper and nickel impurities present no significant problems for most typical uses. When ultrapure cobalt, or its compounds, is required, additional stages of production are required. In one method tested by a metallurgical laboratory, cobalt(II) hydroxide was precipitated from a solution buffered to permit selective precipitation of $Co(OH)_2$ in the presence of trace amounts of Cu^{2+} and Ni^{2+} without coprecipitation of $Cu(OH)_2$ and $Ni(OH)_2$. The ammonia/ammonium ion buffer system is not suitable for this use because of the complexing of Co^{2+} with NH_3. A trimethylamine/trimethylammonium chloride buffer is preferred. What mole ratio of $(CH_3)_3N/(CH_3)_3NH^+$ would provide a buffer pH satisfactory for precipitation of $Co(OH)_2$ from 0.30 $\underline{M}$ Co^{2+}, while providing the maximum possible hydroxide concentration to avoid precipitation of $Cu(OH)_2$ when the solution is 5×10^{-9} $\underline{M}$ in Cu^{2+}? [For K_b and K_{sp}-data, see Appendix D.]

*I.8. The solubility product for $Ni(OH)_2$ is 1.6×10^{-14}. Would the buffer system employed in problem I.8 prevent precipitation of nickel hydroxide from a cobalt(II) solution contaminated by 0.020 $\underline{M}$ Ni^{2+}?

*I.9. A few years ago lead-producing companies had one of the most diverse markets of any of the metallurgical industries. Metallic lead was used, either pure or in a number of different alloys, in diverse applications ranging from automobile batteries to fishing "sinkers". Lead compounds were found in a host of commercial products.

*Proficiency Level

Tetraethyl lead was the "anti-knock" component in the gasoline recommended for most high power automobiles. "Red lead" (Pb_3O_4) was used in corrosion-resistant paints. Lead(IV) oxide formed one of the electrodes in "storage" batteries. Various white compounds, such as $PbSO_4$, $PbCO_3$, and $Pb_3(CO_3)_2(OH)_2$ ["white lead"], were valuable paint pigments, as was $PbCrO_4$ ["chrome yellow"]. With the increasing concern over lead as an environmental pollutant, many of these applications have disappeared or been sharply curtailed. "Unleaded" gasoline seems assured of the future automobile market and many of the lead-based pigments have been restricted to certain "low-hazard" uses. What mole ratio of ammonia to ammonium ion would provide the maximum pH for the selective precipitation of $PbCO_3$ from 0.20 $\underline{M}$ Pb^{2+} by addition of an equal volume of 0.20 $\underline{M}$ CO_3^{2-} if $Pb(OH)_2$ precipitation is to be avoided? [For $Pb(OH)_2$, $K_{sp} = 3.0 \times 10^{-16}$, For $PbCO_3$, $K_{sp} = 1.5 \times 10^{-13}$.]

*I.10. Zinc sulfide finds many industrial uses. The paint industry employs it as a white, relatively nontoxic pigment (lithopone). In the television and electronics indus-try, it is used as the coating for TV screens and oscilloscope screens because of its fluorescence when bombarded with electrons. In the area of medical instrumenta-tion, the fluorescence of ZnS to x-rays makes it an ideal coating for fluoroscope screens. What mole ratio of acetic acid to sodium acetate would provide a buffer of minimum pH for the selective precipitation of ZnS from a solution 0.20 $\underline{M}$ in Zn^{2+} and 5×10^{-14} $\underline{M}$ in Cd^{2+}, when the solution is saturated with H_2S gas (solubility 0.10 mole liter^{-1})? [For K_a and K_{sp}-data, see Appendix D.]

ANSWERS:

(I.1.) 8.1, (I.2.) 4.2, (I.3.) 9.78, (I.4.) ~5 $\underline{M}$, (I.5.) 0.004 $\underline{M}$, (I.6.) 3.4 x 10^{-12}ppb,
(I.7.) $[(CH_3)_3N]:[(CH_3)_3NH^+] \simeq 0.11 : 1$,   (I.8.) No, (I.9.) $[NH_3]:[NH_4^+] \simeq 0.003 : 1$,
(I.10.) $[CH_3CO_2H]:[CH_3CO_2^-] \simeq 45 : 1$

RELEVANT
PROBLEMS

Unit 13: Simultaneous Equilibria

Set E: Environmental Sciences

E.1. Ammonium salts are the most common chemical fertilizers. Such compounds as ammonium nitrate, ammonium sulfate, and ammonium phosphate are used in the millions of metric tons for the addition of essential plant nutrients to agricultural lands. Ammonium salts are, in general, quite soluble in water and the run-off from agricultural areas can introduce significant amounts of these chemicals into rivers and lakes. In moderate concentrations, ammonium salts can increase the growth of algae and other aquatic plants. In higher concentrations, the pH of the aquatic ecosystem may be unfavorably altered. What is the pH of agricultural run-off water whose acidity is equivalent to 1.8 g liter^{-1} of ammonium nitrate? [See Appendix D for ionization constants.]

E.2. Soaps, unlike most synthetic detergents, may be precipitated as insoluble calcium salts in natural "hard" waters. However, in "soft" waters, such as streams and lakes formed by the melting of snow packs in mountainous areas, soaps can accumulate in sufficient concentrations to be considered significant pollutants. The biological degradation of soaps by aquatic microorganisms is an oxygen-consuming process and can deplete the available dissolved oxygen. In addition, the hydrolysis of the organic anion of the soap increases the alkalinity of the system. What is the pH of a solution 0.30 $\underline{M}$ in sodium stearate, $CH_3(CH_2)_{16}CO_2Na$? [The K_a for stearic acid is ~1.4 x 10^{-5}.]

E.3. Fluoride concentrations around 1 ppm are common in most municipal water supplies as a result of fluoridation to help in retarding tooth decay. At concentrations only slightly higher (2-6 ppm), fluoride is considered a pollutant, since regular ingestion of such a solution leads to the formation of brown spots on teeth. As the fluoride concentration increases, the adverse effects of fluoride become more pronounced. High concentrations can damage vital enzyme systems and blood levels in excess of 100 ppm can disturb the $Na^+/K^+/Ca^{2+}$ ionic balance by precipitation of calcium fluoride. A 3-4 g amount of fluoride can be lethal to an adult human. Fluoride pollution is a significant problem from aluminum plants and certain other chemical industries if care is not exercised to remove fluoride from waste effluent streams. What is the pH of a solution containing 4.0 g liter^{-1} of sodium fluoride? [For ionization constants, see Appendix D.]

E.4. One of the ways of removing colloidal particles and some bacteria in water purification systems involves the use of flocculating agents that form precipitates "trapping" dispersed particles. Aluminum sulfate is a common flocculating agent employed in many municipal water plants. When added along with "slaked lime" (aqueous calcium hydroxide), the $Al_2(SO_4)_3$ reacts to form a gelatinous precipitate of $Al(OH)_3$. What is the maximum pH for a "limed" solution that is 1.2 x 10^{-5} $\underline{M}$ in Mg^{2+} if the precipitation of $Mg(OH)_2$ is to be avoided? [See Appendix D for K_{sp}-data.]

E.5. In chrome-plating industries, both cyanide and chromate ions are present in the initial aqueous wastes. Until recently, many industries dumped these wastes directly into lakes and streams. With improved attention to water quality standards, methods have been developed to eliminate all or most of the toxic cyanides and chromates. Treatment of the waste solutions with chlorine gas serves to oxidize cyanide to harmless bicarbonate and subsequent treatment with sulfur dioxide reduces CrO_4^{2-} to Cr^{3+}. However, the latter treatment simply exchanges one pollutant for another if the waste water is too acidic to permit precipitation of $Cr(OH)_3$, which could be "captured" in "settling ponds". What is the minimum pH required to induce precipitation of $Cr(OH)_3$ [K_{sp} = 7.0 x 10^{-31}] from a 0.045 $\underline{M}$ Cr^{3+} solution?

E.6. It is unrealistic to expect that we will ever have "zero" water pollution. The cost of reducing most pollutants to below certain "reasonable" levels would be prohibitive and, with most chemical purification methods, the excessive treatments required could result in "pollution from the anti-pollutants". What minimum pH would be required to initiate precipitation of $Cr(OH)_3$ from a 7 x 10^{-22} $\underline{M}$ Cr^{3+} solution (about 10 Cr^{3+} ions per liter)? What mass of sodium hydroxide would have to be added to 25,000 liters of water to produce this pH, assuming no other pH-affecting species are present?

*E.7. In investigating ways of removing Cr^{3+} from industrial wastes (problem E.5), a water quality research laboratory considered the use of buffers for controlling the selective precipitation of $Cr(OH)_3$. What mole ratio of acetic acid to acetate ion would provide a pH high enough to induce the precipitation of chromium hydroxide alone from a solution 0.10 $\underline{M}$ in Cr^{3+} and 6 x 10^{-11} $\underline{M}$ in Fe^{3+}? [For K_a and K_{sp}-data, see Appendix D.]

*E.8. The composition of natural waters depends on substances leached from the soil by rain, absorbed from the atmosphere, or added - either accidentally or deliberately - from neighboring ecosystems. One of the most important component sets is the calcium carbonate (calcite) - carbon dioxide system:

$$CO_{2(g)} \rightleftarrows CO_{2(aq)} \qquad [K = 3.4 \times 10^{-2}]$$

$$CO_{2(aq)} + H_2O_{(\ell)} \rightleftarrows H^+_{(aq)} + HCO^-_{3(aq)} \qquad [K = 4.2 \times 10^{-7}]$$

$$HCO^-_{3(aq)} \rightleftarrows H^+_{(aq)} + CO^{2-}_{3(aq)} \qquad [K = 4.8 \times 10^{-11}]$$

$$CaCO_{3(s)} \rightleftarrows Ca^{2+}_{(aq)} + CO^{2-}_{3(aq)} \qquad [K_{sp} = 4.7 \times 10^{-9}]$$

The composition of this system, which is also closely related to the pH of the aqueous medium, is controlled by the partial pressure of CO_2 in the atmosphere and by the presence of $CaCO_3$ in the sedimentary deposits. Man's combustion of millions of tons of fossil fuels annually places a significant stress on the carbon dioxide balance of nature. Most of this excess CO_2 is absorbed in the sea, causing a decrease in the pH of the upper layers of seawater and corresponding changes in the chemical and ecological systems. A natural lagoon of pH 8.3 was saturated with calcite and in equilibrium with atmospheric CO_2 so that the "total dissolved inorganic carbon" (CO_2, HCO_3^-, CO_3^{2-}) was 4.0×10^{-2} $\underline{M}$. What was the molar concentration of Ca^{2+} in the lagoon water?

*E.9. An oil-burning power plant was built near the lagoon (problem E.8) and atmospheric CO_2 from the plant's exhaust stacks was dissolved in the lagoon water so that the "total dissolved inorganic carbon" content was increased to 9.0×10^{-2} $\underline{M}$ and the pH was decreased to 8.1. What was the approximate equilibrium concentration of Ca^{2+} in the new system?

*E.10. Lead from gasoline additives and "lead-based" paints has become an increasing problem in water supplies. The Pb^{2+} ion is capable of deactivating a number of essential enzyme systems and has been shown to produce irreversible brain damage when ingested in high concentrations or over a prolonged interval. Lead poisoning can occur from drinking contaminated water or from inhaling air containing gaseous lead compounds. A number of cases of poisoning have been attributed to the use of imported water pots containing lead pigments unprotected by glaze coating. Water supplies near heavily traveled highways have been found to contain hazardous concentrations of lead compounds. Lead chromate is used as a pigment, called "chrome yellow", in various paints and ceramics. It is only sparingly soluble in alkaline solution ($K_{sp} = 1.8 \times 10^{-14}$), but its solubility is increased in slightly acidic solution by the removal of chromate ion:

$$PbCrO_{4(s)} \rightleftarrows Pb^{2+}_{(aq)} + CrO^{2-}_{4(aq)}$$
$$CrO^{2-}_{4(aq)} + H^+ \rightleftarrows HCrO^-_{4(aq)}$$
$$(K_a \text{ for } HCrO_4^- = 3.2 \times 10^{-7})$$

What mole ratio of acetic acid to acetate ion would provide a maximum pH suitable for the selective precipitation of lead chromate from a solution 3.0×10^{-9} $\underline{M}$ in Pb^{2+} and 0.71 $\underline{M}$ in Ca^{2+} when the total "chromate species" concentration is 0.010 $\underline{M}$? [See Appendix D for additional data.]

ANSWERS:
(E.1.) 5.54, (E.2.) 9.2, (E.3.) 8.1, (E.4.) ~11.0, (E.5.) ~4.4, (E.6.) ~11.0, 1.0 kg, (E.7.) $[CH_3CO_2H]:[CH_3CO_2^-] \simeq 2.9 : 1$, (E.8.) 1.2×10^{-5} $\underline{M}$, (E.9.) ~9×10^{-6} $\underline{M}$, (E.10.) $[CH_3CO_2H]:[CH_3CO_2^-] \simeq 0.16 : 1$

EQUILIBRIUM

UNIT 14: OXIDATION-REDUCTION PROCESSES

Two aspects of electrochemical cells are of considerable practical importance. The first of these is <u>current</u>, which tells us how fast we can transfer electrons to run a motor, light a bulb, or electroplate a metal. A measurement of current, when coupled with time, also permits us to determine the quantity of chemical changes taking place during an electrochemical process. Current is related to a <u>continuing</u> chemical change and, as such, is treated by the principles of stoichiometry, not equilibrium, in Unit 8.

The other aspect, of equal importance, is <u>voltage</u>. The voltage of a cell may be measured under conditions of negligible current flow, that is, under conditions approaching zero <u>net</u> chemical change. This is one of the characteristics of chemical equilibrium. The other characteristic is the <u>dynamic</u> nature of the system, in which two opposing chemical processes occur at the same time and at the same rate (Unit 10). Let's consider a simplified way of looking at a copper/zinc electrochemical cell to see how voltage may be related to chemical equilibrium.

A useful model for a crystalline metal suggests that the crystal lattice consists of a regular array of positive metal ions in a loose "cloud" of mobile valence electrons. These cations are highly attractive to polar water molecules. Thus, we might conceive of a situation in which a piece of metal placed in water might "dissolve" a little bit as some of its cations are pulled from the crystal lattice by water molecules. This would give us a saturated solution of the "metal" in water and saturated solutions represent dynamic equilibrium systems (Unit 10). Now, this solution process shouldn't contain very much metal ion since, unlike the case of the electrically neutral salts, the solid metal is now negatively charged due to the "extra" electrons left behind when cations are removed. The piece of metal, being negative, should strongly attract the dissolved positive ions.

Different metals exhibit differences in their cation-electron attractions, roughly parallel to differences in the ionization potentials of the isolated metal atoms. It would not be surprising, then, to find that equilibrium constants for metal/metal-ion systems vary appreciably. For zinc and copper, for example, the equilibrium constant is larger for the zinc/zinc-ion system:

$$Zn_{(s)} \;\overset{\rightarrow}{\leftarrow}\; Zn^{2+}_{(aq)} + 2e^-$$

$$Cu_{(s)} \;\overset{\rightarrow}{\leftarrow}\; Cu^{2+}_{(aq)} + 2e^-$$

$$K_{Zn} > K_{Cu}$$

Each equilibrium could be described by an equilibrium constant or, since surplus negative charge is accumulated on the solid metal, by an <u>electrode</u> <u>potential</u> expressing (in simplistic terms) the availability of extra electrons on a piece of metal if someone should want them to travel someplace as an electric current.

Unfortunately, we don't have any way of measuring that electron accumulation directly. We do, however, have a way for comparing the potentials on two different metals. A voltmeter measures potential difference (that's basically what "voltage" refers to) and, if we set up the proper experimental conditions, we can measure the voltage as a comparison of two electrode potentials.

Since the activity (Unit 10) of the solid metal is unity, the electrode potential should be independent of the size of the metal piece used. Le Chatelier's principle (Unit 10) suggests, however, that adding more metal ion (e.g., by adding a soluble salt of the metal) should result in a new equilibrium in which more metal ion is returned to the solid metal, "neutralizing" some of the excess negative charge. Thus, the electrode potential should be a function of the concentration of metal ion in the solution. Such is, indeed, the case.

In would be handy to have some way of comparing electrode potentials without the necessity of constructing all possible combinations for voltage measurement. Since we can't easily determine a single electrode potential, we can do the next best thing by selecting some electrode system as a reference standard and tabulating all other potentials on a comparative basis. The universally-accepted reference is the Standard Hydrogen Electrode, a system in which hydrogen gas at 1.00 atm pressure is bubbled over the surface of an inert platinum electrode immersed in an aqueous solution of H^+ ions at unit activity (approximately 1.0 $\underline{M}$ H^+). This system is assigned a Standard Electrode Potential of ZERO, so that the voltage (potential difference) measured for a cell using the hydrogen system[1] as one electrode would represent the potential of the other electrode (on a comparative scale). A table of Standard Electrode Potentials (referring to electrodes in which all components are in their Thermodynamic Standard States, Unit 7) is given in Appendix E.

Electrode potentials can be determined for any oxidation-reduction half-reaction (Unit 2), i.e., the electrodes are not limited to metal/metal-ion systems.

The relationship between electrode potential (or cell potential as "voltage") and solution concentration or gas pressure is expressed mathematically by the Nernst Equation. The derivation of this equation is beyond the scope of our discussions[2], but we can see how the equation may be applied to some practical situations concerned with cell voltage.

The Nernst Equation	$E = E^o - \dfrac{0.0592}{n} \log Q$

---

[1]The "Standard Hydrogen Electrode" is, in reality, a hypothetical electrode. The mathematical treatment is, however, quite valid.

[2]A derivation of the Nernst Equation may be found in most standard textbooks of Physical Chemistry.

For our use of this equation, the value of 0.0592 may be regarded simply as a "constant" at 25°. _E_ represents the actual cell voltage, _E°_ represents the difference[3] between the Standard Electrode Potentials for the two electrodes of the cell, _n_ is the number of electrons transferred (net loss _or_ net gain) for the process as represented by the balanced equation for the cell discharge reaction, and _Q_ is approximated by the concentration[4] ratio of the cell components in the _form_ of an equilibrium constant formulated for the balanced equation for the cell discharge reaction.

[It should be noted that the Nernst Equation can be used for other calculations, such as the variation of single electrode potentials with concentration, and that we have defined its terms in a limited sense as appropriate to our particular applications.]

In addition to the use of the Nernst equation for calculations related to cell voltage, it can provide us with some interesting thermodynamic data. We have mentioned earlier (Units 7 and 8) that we can "tap" the energy of an oxidation-reduction process for useful work more efficiently by an electrochemical process than by a direct combustion process. A calculation of enthalpy difference (ΔH) could only reveal the total heat content difference and, although we know that heat connot all be converted to useful work, ΔH calculations told us nothing about what fraction of the enthalpy change we might expect to be useful to us. Another thermodynamic quantity, the free energy difference (symbolized by ΔG for a constant pressure process), does tell us the maximum amount of energy available for useful work in a cyclic process. The free energy difference can be determined from a measurement or calculation of a cell potential:

$$\Delta G = -23nE$$

The constant employed (23) permits us to express ΔG in kilocalories. Both _n_ and _E_ represent the same quantities as in the Nernst Equation. For a more detailed treatment of this relationship, see "References" (page 366).

The value calculated for ΔG sets an "upper limit" on the energy we can expect to utilize for work. In reality, of course, there are further practical limitations, such as friction, with which we must contend.

We will limit our problem situations in this unit to those involving cell concentrations and voltage, as related by the Nernst Equation. For examples of ΔG calculations, see the "References" listed.

---

[3]For our purposes, this difference is always obtained by subtracting $E°$ for the oxidation (electron loss) half-reaction from $E°$ for the reduction (electron gain) half-reaction. For a description of how to determine the equations for these half-reactions from the net equation for the complete cell process, see Unit 2. Note that all processes in Appendix E are shown as reduction. The sign of the potential is, however, invariant.

[4]We will express concentrations of dissolved species as molarities and use gas pressures in units of atmospheres (1.00 atm = 760 torr).

1. O'Connor, Rod. 1977. Fundamentals of Chemistry, second edition. New York: Harper & Row (Units 23, 24)

2. Brown,Theodore, and H. Eugene LeMay, Jr. 1977. Chemistry: The Central Science. Englewood Cliffs, N.J.: Prentice-Hall (Chapters 15, 19)

3. Masterton, William, and Emil Slowinski. 1977. Chemical Principles, fourth edition. Philadelphia: W.B. Saunders (Chapters 14, 22, 23)

4. Nebergall, William, F.C. Schmidt, and H.F. Holtzclaw, Jr. 1976. College Chemistry, fifth edition. Lexington, MA: D.C. Heath (Chapters 20, 22)

5. Lyons, Ernest H., Jr. 1967. Introduction to Electrochemistry. Lexington, MA: Heath/Raytheon.

OBJECTIVES:

(1) *Given the balanced equation for a cell discharge reaction, the concentrations of cell components, and access to Standard Electrode Potentials (Appendix E), be able to use the Nernst Equation to calculate cell voltage.*

*(2) *Given the description of a cell discharge process, appropriate voltage and concentration data, and access to Standard Electrode Potentials (Appendix E), be able to use the Nernst Equation to calculate the concentration of a specified component of the cell.*

PRE-TEST:

The Nernst Equation: $E = E° - \dfrac{0.0592}{n} \log Q$

Necessary Electrode Potentials: (Simplified Formulation)

$E° = -0.36$ for: $PbSO_{4(s)} + 2H^{+}_{(aq)} + 2e^{-} \rightleftarrows Pb_{(s)} + H_2SO_{4(aq)}$

$E° = +1.68$ for: $PbO_{2(s)} + H_2SO_{4(aq)} + 2H^{+}_{(aq)} + 2e^{-} \rightleftarrows PbSO_{4(s)} + 2H_2O_{(\ell)}$

The formula weight of H_2SO_4 is 98.1.

---

(1) A rather unusual way of measuring "voltage" is employed when the "gas station" attendant checks an automobile battery "charge" by withdrawing a sample of the electrolyte with a device called a hydrometer, which simply measures the density of the electrolyte solution. The interpretation of electrolyte density in terms of battery voltage is possible because both the density of the solution and the voltage of the battery are functions of the sulfuric acid content of the electrolyte. In the lead storage cell, one electrode is lead and the other is lead

366

dioxide. During discharge of the cell, sulphuric acid from the electrolyte is consumed and <u>both</u> electrodes are coated with the lead sulfate formed:

$$Pb_{(s)} + PbO_{2(s)} + 2H_2SO_{4(aq)} \rightleftarrows 2PbSO_{4(s)} + 2H_2O_{(\ell)}$$

Since all components of the cell are pure solids or liquids except for the aqueous sulfuric acid, H_2SO_4 is the only variable in the Nernst Equation. The concentration of H_2SO_4 is easily determined from a measurement of solution density and the hydrometer is simply calibrated to relate density-to-concentration-to-cell "charge" (voltage). In a "fresh" automobile battery, the electrolyte is 38% H_2SO_4 by weight and its density is 1.286 g ml^{-1}. Under these conditions, each cell produces about 2.08 volts. By connecting three, six, or twelve cells in series, a battery of "6", "12", or "24" volts, respectively is formed.

What is the approximate voltage of a lead storage cell in which the electrolyte density is 1.018 g ml^{-1}, corresponding to a solution containing 30.2 g H_2SO_4 per liter? _____

*(2) For a lead storage battery, as described in question 1, the voltage of a <u>used</u> "6.0 volt" battery was found to be 5.73 volts. What was the approximate concentration of H_2SO_4 in the electrolyte, in grams per liter? _____

Answers and Directions:

(1) 2.01 volts *(Note that the lead cell maintains a nearly constant voltage over a wide range of electrolyte concentrations.)*

*If you answered this question correctly, go on to question *2. If you missed it, study METHODS, section 14.1.*

*(2) 0.62 g liter^{-1}

If you answered this question correctly, go on to RELEVANT PROBLEMS, Unit 14. If you missed it, study METHODS, section 14.2.

METHODS

14.1 Cell Voltage and the Nernst Equation

To use the Nernst Equation for calculation of cell voltage, we must be able to evaluate the terms represented in the equation by E^o, n, and Q. Since the voltage of the cell <u>must</u> be positive[5], if it is a real cell, our final calculation must give us a positive number. This gives us a check on the "reasonableness" of our calculation. To evaluate

---

[5]If we measure a cell potential with a voltmeter having both positive and negative deflection, a negative reading means only that we connected the voltmeter wires to the wrong terminals of the cell.

the terms of the Nernst Equation:

1. $E°$ is found by subtracting the standard electrode potential for the half-reaction corresponding to <u>oxidation</u> for the cell reaction, <u>as</u> <u>expressed</u> <u>by</u> <u>the</u> <u>chemical</u> <u>equation</u> <u>given</u>, from that for the other half-reaction. To identify the <u>oxidation</u> half-reaction, it is only necessary to find which of the equations for tabulated standard electrode potentials is the <u>reverse</u> of the change involving those species as shown in the equation for cell discharge.

   for example:

   cell equation—$Cu^{2+} + Zn \rightleftarrows Cu + Zn^{2+}$

   standard electrode potentials—

   $Cu^{2+} + 2e^- \rightleftarrows Cu$, $E° = +0.34$
   $Zn^{2+} + 2e^- \rightleftarrows Zn$, $E° = -0.76$

   Since the equation shows conversion of Zn to Zn^{2+} (<u>reverse</u> of $Zn^{2+} + 2e^- \rightleftarrows Zn$), $E° = (+0.34) - (-0.76) = 1.10$ volts

   [reduction] [oxidation]

2. The value of <u>n</u> is found by multiplying the number of electrons shown in <u>either</u> electrode potential equation by the coefficient of the species in the cell equation.

   for example:

   cell equation—$Al + 3Ag^+ \rightleftarrows Al^{3+} + 3Ag$

   standard electrode potentials—

   $Al^{3+} + 3e^- \rightleftarrows Al$, $E° = -1.66$
   $Ag^+ + e^- \rightleftarrows Ag$, $E° = +0.80$

   Since the equation shows <u>one</u> Al, $n = 1 \times 3e^- = 3$. [Or, for three Ag^+, $n = 3 \times 1e^- = 3$.]

3. The Q term is written in the <u>form</u> of an equilibrium constant. That means "right hand species" over "left hand species", with exponents of concentration terms equal to corresponding coefficients in the chemical equation. As in the case of equilibrium constants (Unit 10), only dissolved species or gases are included.

   for example:

   for the Al/0.30 M Al^{3+}, 0.50 M Ag^+/Ag cell whose discharge reaction is formulated as:

   $$Al_{(s)} + 3Ag^+_{(aq)} \rightleftarrows Al^{3+}_{(aq)} + 3Ag_{(s)}$$

   $$Q = \frac{c_{Al^{3+}}}{c^3_{Ag^+}} = \frac{(0.30)}{(0.50)^3}$$

EXAMPLE 1

The _SOLUTION_ to Pre-Test question 1:

For the cell discharge reaction:

$$Pb_{(s)} + PbO_{2(s)} + H_2SO_{4(aq)} \rightleftarrows 2PbSO_{4(s)} + 2H_2O_{(\ell)}$$

corresponding to the simplified standard electrode potentials:

$$PbSO_4 + 2H^+ + 2e^- \rightleftarrows Pb + H_2SO_4, \quad E° = -0.36$$

$$PbO_2 + H_2SO_4 + 2H^+ + 2e^- \rightleftarrows PbSO_4 + 2H_2O, \quad E° = +1.68$$

1. Since the cell equation shows conversion of _lead_ to _lead sulfate_, the _reverse_ of the equation for the lead electrode; it is the _lead_ which is oxidized, so:

$$E° = (+1.68) - (-0.36) = 2.04 \text{ volts}$$
$$[\text{for } PbO_2] \quad [\text{for } Pb]$$

2. Since all species in the cell equation have the same coefficients (unity) as in the electrode equations:

$$n = 1 \times 2e^- = \underline{2}$$

3. Since the only dissolved species (there are no gases involved) is $H_2SO_{4(aq)}$, a "left-hand species" with a coefficient of 2:

$$Q = \frac{1}{c_{H_2SO_4}^2} = \frac{1}{[M(H_2SO_4)]^2}$$

and, from the concentration data provided, using a mole/mass unity factor:

$$\frac{30.2 \text{ g}(H_2SO_4)}{1 \text{ liter}} \times \frac{1 \text{ mole}(H_2SO_4)}{98.1 \text{ g}(H_2SO_4)} = 0.308 \underline{M}$$

so that:

$$Q = \frac{1}{(0.308)^2} = \underline{10.5}$$

Then, substituting in the Nernst Equation:

$$E = E° - \frac{0.0592}{n} \log Q$$

$$E = (+2.04) - \frac{0.0592}{2} \log(10.5) = \underline{2.01 \text{ volts}}$$

EXERCISE 1

In the methane/oxygen fuel cell (page 374), the cell voltage is a function of both the gas pressures (of CH_4 and O_2) and the solution concentrations (of OH^- and CO_3^{2-}). This cell, unlike the lead storage cell, cannot be recharged under practical conditions. In addition, although the gases could be maintained almost indefinitely, by periodically refilling their storage cylinders, the hydroxide content of the electrolyte is depleted during cell discharge and carbonate is accumulated. As a result, this cell is not practical for prolonged use at even moderate current rates because of the necessity for frequently draining and replacing the electrolyte solution. This is one of the reasons why such fuel cells have not found more extensive use.

The cell discharge reaction is formulated as:

$$CH_{4(g)} + 2\ O_{2(g)} + 2\ OH^-_{(aq)} \rightleftarrows CO^{2-}_{3(aq)} + 3\ H_2O_{(\ell)}$$

and the standard electrode potentials are tabulated as:

$$O_2 + 2\ H_2O + 4\ e^- \rightleftarrows 4\ OH^-, \quad E° = +0.40$$
$$CO_3^{2-} + 7\ H_2O + 8\ e^- \rightleftarrows CH_4 + 10\ OH^-, \quad E° = -1.06$$

What voltage would be produced by a methane/oxygen fuel cell when the gas pressures are 1.00 atm each and the electrolyte is 0.10 $\underline{M}$ in KOH and 0.45 $\underline{M}$ in K_2CO_3 ?

(answer, page 374)

Extra Practice

EXERCISE 2

What would be the voltage of a partially-depleted methane/oxygen fuel cell (EXERCISE 1) in which the gas pressures are 0.90 atm each and the electrolyte is 0.010 $\underline{M}$ in KOH and 0.495 $\underline{M}$ in K_2CO_3 ?

(answer, page 375)

At this point you should try the competency level Self-Test question on page 373.

*14.2 Concentrations from Cell Voltage

The relationship between concentration and cell potential expressed by the Nernst Equation permits us to use a voltage measurement to determine the concentration of a reactant or product species in an electrochemical cell. This is the basis for the familiar

*Proficiency Level

pH meter[6], which is just a sensitive voltmeter that has been calibrated for direct reading of pH units.

The application of a Nernst Equation calculation to the problem of finding a cell-component concentration is really no different, except for the "arithmetic", from using the equation to calculate cell potential. The determination of $\underline{E^\circ}$ and $\underline{n}$ is made as described in Section 14.1 and the $\underline{form}$ of the $\underline{Q}$ term is established by the same method discussed. All we need to know, then, are the experimentally-determined concentrations (or gas pressures) for all other cell components and the measured value of the cell potential. The problem then becomes one of solving a logarithmic equation.

*EXAMPLE 2

The SOLUTION to Pre-Test question *2:

Since this is a "6.0" volt battery, it consists, when new, of three 2.08 volt cells (page 367). We will assume[7] that in the "used" battery all cells have been equally discharged, so that the cell potential for each is:

$$E = \frac{5.73}{3} = 1.91 \text{ volts}$$

Then, by the method outlined in Section 14.1:

1. $E^\circ = 2.04$ volts (page   )

2. From either electrode potential, since both Pb and PbO_2 have coefficients of unity in the cell equation:

$$n = 2$$

3. Since all species other than H_2SO_4 are solids or liquid and H_2SO_4 (a "left-hand" species) has a coefficient of 2 in the balanced cell equation:

$$Q = \frac{1}{c_{H_2SO_4}^2}$$

Then, substituting in the Nernst Equation:

$$E = E^\circ - \frac{0.0592}{n} \log Q$$

$$1.91 = 2.04 - \frac{0.0592}{2} \log \left[\frac{1}{c_{H_2SO_4}^2} \right]$$

from which:

---

[6]Note that the definition of pH ($-\log c_{H^+}$) is obviously related to the $\underline{\log Q}$ term of the Nernst Equation.

[7]It is not unusual in a real automobile battery to find that some design or operation problem results in unequal cell discharge.

$$\log \left[\frac{1}{c_{H_2SO_4}^2} \right] = -2 \log c_{H_2SO_4} = \frac{2(2.04 - 1.91)}{0.0592}$$

$$\log c_{H_2SO_4} = \frac{2(0.13)}{(-2) \times (0.0592)} = -2.2 = (0.8 - 3)$$

$$c_{H_2SO_4} = \text{antilog}(0.8 - 3) = 6.3 \times 10^{-3} \underline{M}$$

Conversion to grams per liter, using a mass/mole unity factor gives:

$$\frac{6.3 \times 10^{-3} \text{ moles}(H_2SO_4)}{1 \text{ liter}} \times \frac{98.1 \text{ g}(H_2SO_4)}{1 \text{ mole}(H_2SO_4)} = 0.62 \text{ g liter}^{-1}$$

----

*EXERCISE 3

Modern pH meters use rather sophisticated electrode systems for which it is difficult to write a simple equation for the cell process. Electrode technology has advanced to the stage of developing a "swallowable" duel-electrode probe for measuring the pH of the gastric fluid, or a "micro-electrode" system so tiny that it could be inserted into the venom sac of a hornet to measure the pH of the aqueous venom. The hydrogen ion is not the only species for which such measurements can be made. So-called "ion-specific electrodes" can be used to monitor the fluoride-ion concentration in a municipal water-fluoridation operation, the potassium-ion concentration of a neurological fluid sample, or the chloride ion concentration of perspiration on the skin.

A more primitive, but easier to describe, "pH meter" can be assembled using a hydrogen electrode (in which hydrogen gas at a measured pressure is allowed to bubble through an aqueous solution and across the surface of an inert platinum electrode immersed in the solution). Any other electrode desired can be used and a measurement of the potential difference between the two electrodes can be used to calculate the pH of the solution in the hydrogen electrode compartment. The two electrode compartments must be connected, by a porous membrane for example, in such a way as to minimize electrolyte mixing while permitting sufficient ion migration to maintain the electroneutrality of both compartments of the cell.

A "primitive" pH meter, as described in the preceding paragraph, was used to measure the pH of a gastric fluid sample. The sample was placed in contact with a platinum electrode over which a stream of hydrogen was bubbled at 1.1 atm pressure. The other electrode compartment consisted of a zinc strip immersed in a 0.10 $\underline{M}$ Zn^{2+} solution. When the cell was connected, its potential was found to be 0.70 volt. What was the approximate pH of the gastric fluid if the cell discharge reaction was represented by:

$$Zn_{(s)} + 2H^{+}_{(aq)} \rightleftarrows Zn^{2+}_{(aq)} + H_{2(g)}$$

[For Standard Electrode Potentials, see Appendix E.]

(answer, page 375)

Extra Practice

*EXERCISE 4

Chlorine is an extremely dangerous gas, which can be appreciated if you recall reading about the dreadful results of its use as a "poison gas" during World War I. Most chlorine prepared commercially is obtained by the electrolysis of brine. The gas, or liquified chlorine, is sold extensively to water purification plants and chemical industries producing chlorine compounds. The hazards of working with chlorine require a careful monitoring system to detect leakage of the gas into the atmosphere of the plant or the surrounding area.

A young engineer once proposed the design for a chlorine-monitoring system in which the atmosphere was continuously assayed by sweeping a measured volume of air over an inert electrode immersed in a 0.50 $\underline{M}$ sodium chloride solution connected to an aluminum/0.30 $\underline{M}$ Al^{3+} electrode compartment. A recording voltmeter was to maintain a record of chlorine concentration in the air by plotting the voltage of the Al/Al^{3+}, Cl_2/Cl^- electrochemical cell. Assuming that the planned cell would work as conceived, what partial pressure (in atm) of chlorine in the atmosphere would correspond to a cell potential of 2.94 volts? The cell reaction under these conditions may be represented by:

$$2Al_{(s)} + 3Cl_{2(g)} \rightleftarrows 2Al^{3+}_{(aq)} + 6Cl^-_{(aq)}$$

[For the Standard Aluminum Electrode Potential, see Appendix E. The $E°$-value for $Cl_{2(g)} + 2e^- \rightleftarrows 2Cl^-$ is +1.33 volts.]

(answer, page 376)

SELF-TEST (UNIT 14)   [answers, page 377]

14.1. A "primitive" pH meter was constructed in which a platinum electrode was immersed in a solution of pH 4.30, through which hydrogen gas was bubbled at a pressure of 1.10 atm. This electrode compartment was connected, to form an electrochemical cell, with a second compartment containing aluminum metal immersed in 0.25 $\underline{M}$ Al^{3+} solution. Calculate the potential of the cell, if the discharge reaction is given by:

$$2Al_{(s)} + 6H^+_{(aq)} \rightarrow 2Al^{3+}_{(aq)} + 3H_{2(g)}$$

[For Standard Electrode Potentials, see Appendix E.]

If you answered this question correctly, you may go on to the __proficiency__ level, try the RELEVANT PROBLEMS for Unit 14, or stop here. If not, you should consult your instructor for suggestions of further study aids.

==

*14.2. A methane/oxygen fuel cell registered a potential of 1.40 volts under the conditions of gas pressure at 0.90 atm each and carbonate concentration of 0.25 M. The discharge reaction of this cell is given by:

$$CH_{4(g)} + 2\ O_{2(g)} + 2\ OH^-_{(aq)} \rightarrow CO^{2-}_{3(aq)} + 3H_2O_{(\ell)}$$

What was the approximate pH of the electrolyte of this cell?

$$CO_3^{2-} + 7H_2O + 8e^- \rightarrow CH_4 + 10\ OH^-, \quad E° = -1.06$$

$$O_2 + 2H_2O + 4e^- \rightarrow 4\ OH^-, \quad E° = +0.40$$

==

If you answered this question correctly, you may go on to the RELEVANT PROBLEMS for Unit 14. If not, you should consult your instructor for suggestions of further study aids.

ANSWERS to EXERCISES, Unit 14

1. (1.45 volts) __Solution:__

   1. *Since the cell equation shows conversion of methane to carbonate, the __reverse__ of the carbonate electrode equation:*

      $$E° = (+0.40) - (-1.06) = \underline{1.46\ volts}$$

      $$[O_2] \qquad [CO_3^{2-}]$$

   2. *Since the cell equation shows a coefficient of one for methane (and two for oxygen):*

      $$n = 1 \times 8e^- = \underline{8}$$
      $$(or\ n = 2 \times 4e^- = 8)$$

   3. *From the balanced cell equation, using pressure (in atm) for gases:*

*Proficiency Level

$$Q = \frac{C_{CO_3^{2-}}}{P_{CH_4} \times P_{O_2}^2 \times C_{OH^-}^2} = \frac{0.45}{(1) \times (1)^2 \times (0.10)^2} = 45$$

Then, substituting in the Nernst Equation:

$$E = E° - \frac{0.0592}{n} \log Q$$

$$E = 1.46 - \frac{0.0592}{8} \log (45)$$

2. (1.43 volts) <u>Solution</u>:

Steps 1 and 2 are the same as in Exercise 1.

3. $$Q = \frac{C_{CO_3^{2-}}}{P_{CH_4} \times P_{O_2}^2 \times C_{OH^-}^2} = \frac{0.45}{(0.9) \times (0.9)^2 \times (0.010)^2} = 6.8 \times 10^3$$

Then, substituting in the Nernst Equation:

$$E = E° - \frac{0.0592}{n} \log Q$$

$$E = 1.46 - \frac{0.0592}{8} \log (6.8 \times 10^3)$$

- -

*3. (~1.5) <u>Solution</u>:

From Appendix E:

$$Zn^{2+} + 2e^- \rightleftharpoons Zn, \quad E° = -0.76$$

$$2H^+ + 2e^- \rightleftharpoons H_2, \quad E° = \text{zero (by definition)}$$

For the cell process as described by the equation, since <u>zinc is oxidized</u>:

$$E° = (0) - (-0.76) = \underline{+0.76 \text{ volt}}$$

From the cell equation coefficients and either electrode equation:

$$n = \underline{2}$$

From the cell equation and electrode descriptions:

$$Q = \frac{C_{Zn^{2+}} \times P_{H_2}}{C_{H^+}^2} = \frac{(0.10) \times (1.1)}{C_{H^+}^2}$$

Then, substituting in the Nernst Equation:

$$E = E° - \frac{0.0592}{n} \log Q$$

$$0.70 = 0.76 - \frac{0.0592}{2} \log\left[\frac{0.11}{c_{H^+}^2}\right]$$

And solving this logarithmic equation:

$$\log\left[\frac{0.11}{c_{H^+}^2}\right] = \frac{2(0.76 - 0.70)}{0.0592} = 2.03$$

$$\log (0.11) - 2 \log c_{H^+} = 2.03$$

$$- 2 \log c_{H^+} = 2.03 - \log (0.11)$$

$$- \log c_{H^+} = \frac{2.03 - \log (0.11)}{2}$$

And, by definition, $- \log c_{H^+} = pH$.

*4. (2.3×10^{-3}) Solution:

From Appendix E:

$$Al^{3+} + 3e^- \rightleftarrows Al, \quad E° = -1.66$$

(given) $Cl_2 + 2e^- \rightleftarrows 2Cl^-, \quad E° = +1.33$

For the cell equation as given, since aluminum is oxidized:

$$E° = (+1.33) - (-1.66) = \underline{2.99 \text{ volts}}$$

From the cell equation coefficients and the electrode equations:

$$n = 2 \times 3e^- = \underline{6}$$

$$(or, n = 3 \times 2e^- = 6)$$

From the cell equation and electrode descriptions:

$$Q = \frac{c_{Al^{3+}}^2 \times c_{Cl^-}^6}{P_{Cl_2}^3} = \frac{(0.30)^2 \times (0.50)^6}{P_{Cl_2}^3} = \frac{(1.4 \times 10^{-3})}{P_{Cl_2}^3}$$

Then, substituting in the Nernst Equation:

$$E = E° - \frac{0.0592}{n} \log Q$$

$$2.94 = 2.99 - \frac{0.0592}{6} \log\left[\frac{(1.4 \times 10^{-3})}{P_{Cl_2}^3}\right]$$

And, solving this logarithmic equation:

$$\log\left[\frac{(1.4 \times 10^{-3})}{P_{Cl_2}^3}\right] = \frac{6(2.99 - 2.94)}{0.0592} = 5.07$$

$$\log (1.4 \times 10^{-3}) - 3 \log (P_{Cl_2}) = 5.07$$

$$\log (P_{Cl_2}) = \frac{\log(1.4 \times 10^{-3}) - 5.07}{3} = -2.64$$

$$P_{Cl_2} = \text{antilog} (-2.64) = \text{antilog} (0.36 - 3)$$

===

ANSWERS to SELF-TEST, Unit 14

    14.1.   1.41 volts

    *14.2.   ~9.7

A.1. Oxygen is one of several dissolved gases in the sea that has biological significance. Overall, the biological processes in the sea can be viewed in terms of photosynthesis for oxygen production and respiration for oxygen utilization. Biochemically, these processes are quite complicated. The chemical reactions involved and the metabolic activities of plants, animals, and microorganisms yield a wide variety of end products that result in the consumption of oxygen or the restoration of oxygen to the life cycle. End products of oxygen consumption in the sea are mainly carbon dioxide or carbonates, water, sulfates, and nitrates. In the aerobic conversion of ammonium ion to nitrite, an example of oxygen consumption in the marine environment, the biological cell voltage is a function of the partial pressure of oxygen and the solution concentrations (of NH_4^+, NO_2^- and H^+). The overall biological process, mediated by microorganisms, can be formulated as:

$$2\ NH_{4\,(aq)}^+ + 3\ O_{2\,(g)} \rightleftarrows 2\ NO_{2\,(aq)}^- + 4\ H_{(aq)}^+ + 2\ H_2O_{(\ell)}$$

and the standard electrode potentials are tabulated as:

$$O_2 + 4H^+ + 4e^- \rightleftarrows 2H_2O \qquad E° = 1.23\ v$$

$$NO_2^- + 8H^+ + 6e^- \rightleftarrows NH_4^+ + 2H_2O \qquad E° = 0.90\ v$$

What voltage would be produced by this "biochemical cell" when the O_2 gas pressure is 1.0 atm and the electrolyte is 0.20 $\underline{M}$ in NH_4^+, 0.50 $\underline{M}$ in NO_2^- and 1.0×10^{-5} $\underline{M}$ in H^+?

A.2. Oxygen restoration in the euphotic or photosynthetic zone of the sea is primarily accomplished by photosynthesis. However, some oxygen restoration occurs indirectly as plants and other organisms consume such nutrients as nitrates or sulfates and release oxygen in the combined form to the life cycle by either aerobic or anaerobic processes. Therefore, nitrate and sulfate-reducing bacteria help to regulate the quantity of oxygen in sea water. For example, acids produced by sedimentary bacteria cause oxygen in limestone and other insoluble carbonates to be released to the sea as carbon dioxide. The carbon dioxide is then ready for assimilation by phytoplankton, to be used in photosynthetic processes which produce "free" oxygen. The concentration of carbon dioxide in sea water is also an important factor in pH control. Denitrification, an important natural process of nitrate reduction, is the mechanism by which fixed nitrogen in the sea is returned to the atmosphere and simultaneously, helps in maintaining the availability of oxygen and carbon for use in the life cycle. This biochemical process, using methanol as "the organic compound", can be formulated as:

$$5CH_3OH_{(aq)} + 6NO_3^-{}_{(aq)} + 6H^+{}_{(aq)} \rightleftarrows 5CO_{2(g)} + 3N_{2(g)} + 13H_2O_{(\ell)}$$

and the electrode potentials are tabulated as:

$$2NO_3^- + 12H^+ + 10e^- \rightleftarrows N_2 + 6H_2O \qquad E^o = 1.244 \text{ v}$$

$$CO_2 + 6H^+ + 6e^- \rightleftarrows CH_3OH + H_2O \qquad E^o = 0.442 \text{ v}$$

What voltage would be produced by this "biochemical cell" when the gas pressures are 1.00 atm each and the electrolyte is 0.50 $\underline{M}$ in CH_3OH, 0.75 $\underline{M}$ in NO_3^-, and 0.020 $\underline{M}$ in H^+?

A.3. Organic chemical pollutants can exert specific adverse effects on sea water quality in addition to the consumption of dissolved oxygen. Such pollutants may affect human and aquatic-life physiology, as well as industrial applications, esthetic, and recreational uses of ocean or bay areas. The current levels of dissolved organic compounds in the sea range from 10 ppm near coastal areas to 0.10 ppm in deep-sea water. In the lower regions of sea water methane-forming bacteria, such as <u>Methano-ccus</u> and <u>Methanobacterium</u>, anaerobically degrade much of the dissolved organic matter, thus eliminating a part of the biological oxygen demand. If the methane-forming process does not occur, the biological oxygen demand might be filled later when the organic pollutant comes into contact with dissolved oxygen. The overall reaction for the anaerobic degradation of organic matter (formulated as "carbohydrate", CH_2O) can be expressed as:

$$2 CH_2O_{(aq)} \rightleftarrows CO_{2(g)} + CH_{4(g)}$$
(carbohydrate)

and the standard electrode potentials are tabulated as:

$$CH_2O + 4H^+ + 4e^- \rightleftarrows CH_4 + H_2O \qquad E^o = 0.410 \text{ v}$$

$$CO_2 + 4H^+ + 4e^- \rightleftarrows CH_2O + H_2O \qquad E^o \ -0.071 \text{ v}$$

What voltage would be produced by this "biochemical cell" when the gas pressures are 1.50 atm each and the carbohydrate is 0.30 $\underline{M}$?

A.4. The formation of hydrogen sulfide occurs in marine basins that have such poor circulation that waters are totally anoxic. The only living organisms that can exist in this environment are specialized bacteria, such as <u>Desulphovibris desulphuricum</u>. These bacteria utilize oxygen from sulfate ions, rather than dissolved oxygen, for the oxidative processes. In the process, sulfate ions are converted to hydrogen sulfide. Concentrations as high as 3.4×10^{-4} $\underline{M}$ in H_2S are sometimes reached. This anoxic region is separated from oxygenated waters by a zone in which dissolved oxygen reacts with the hydrogen sulfide. The overall chemical reaction can be formulated as:

$$HS^-_{(aq)} + 2 O_{2(g)} \rightleftarrows SO^{2-}_{4(aq)} + H^+_{(aq)}$$

and the standard electrode potentials are tabulated as:

$$O_2 + 4H^+ + 4e^- \rightleftarrows 2 H_2O \qquad E^0 = 1.23 \text{ v}$$

$$SO_4^{2-} + 9H^+ + 8e^- \rightleftarrows HS^- + 4H_2O \qquad E^0 = 0.244 \text{ v}$$

What "biochemical voltage" would be produced in this system when the partial pressure of oxygen is 0.05 atm and the electrolyte is 3.4×10^{-4} $\underline{M}$ in HS^-, 0.030 $\underline{M}$ in SO_4^{2-} and 1.0×10^{-5} $\underline{M}$ in H^+?

A.5. What voltage would be produced by the "biochemical cell" described in problem A.4 when the gas pressure is 0.10 atm and the electrolyte is 1.0×10^{-3} $\underline{M}$ in HS^-, 0.080 $\underline{M}$ in SO_4^{2-}, and 1.0×10^{-6} $\underline{M}$ in H^+?

*A.6. Soils are remarkably complex in their mineral and organic content and in their microbial composition. Although no two soils are alike, all good topsoils have several beneficial qualities in common. They are able to store water and plant nutrients, releasing them to plant roots upon demand. The essential macronutrients include nitrogen, phosphorus, potassium and magnesium. The micronutrients, substances required in trace levels, include iron, copper, zinc, manganese, boron, and molybdenum. One of the principal natural reservoirs of the micronutrient iron is its silicate mineral forms. The chemical weathering of the iron silicates, provided that free access of atmospheric oxygen is limited, forms mainly Fe^{2+} in solution. As long as the solution remains nonoxidizing and slightly acidic, Fe^{2+} is stable, probably in the form of complexes. In oxygenated systems, Fe^{3+} is formed, which may be precipitated as $Fe(OH)_3$ or more complex oxides. The overall process for direct oxidation of Fe^{2+} to Fe^{3+} may be formulated as:

$$4Fe^{2+}_{(aq)} + O_{2(g)} + 4H^+_{(aq)} \rightleftarrows 4Fe^{3+}_{(aq)} + 2H_2O_{(\ell)}$$

A particular soil solution which is 1.0×10^{-5} $\underline{M}$ in Fe^{3+}, 1.0×10^{-3} $\underline{M}$ in Fe^{2+} with the $P_{O_2} = 0.21$ atm, corresponds to a cell potential of 0.316 volt. What is the approximate pH of the solution? [For Standard Electrode Potentials, see Appendix E.]

*A.7. Corrosion is the destructive attack of a metal by chemical or electrochemical reaction with its environment. Corrosive destruction takes many forms depending on the temperature, the nature of the metal or alloy, the presence of foreign matter, the nature of the corrosive medium, the presence of oxygen, oxide scales, built-in crevices, galvanic effects between dissimilar metals, and the presence of "stray" electrical currents from external sources. While corrosion can take any one of several forms, the mechanism of attack in aqueous systems generally involves some aspect of electrochemistry. There is a flow of electricity from certain areas of a metal's surface to other areas through an electrolyte, such as sea water. The anodic part of the metal is that portion which is corroded and from which the "ion current" leaves the metal to enter the solution. The "ion current" leaves the solution and returns to the cathodic region. The circuit is completed outside the electrolyte by a union between the two metallic electrode regions. Products of the anodic and cathodic processes migrate through the electrolyte and may enter into further reactions that yield many of the common visible corrosion products. The corrosion of nickel in sea water can be formulated as:

$$O_{2(g)} + 2H_2O_{(\ell)} + 2Ni_{(s)} \rightleftharpoons 2Ni(OH)_{2(s)}$$

and the standard electrode potentials are tabulated as:

$$Ni(OH)_2 + 2e^- \rightleftharpoons Ni + 2OH^- \qquad\qquad E^o = -0.72 \text{ v}$$

$$O_2 + 2H_2O + 4e^- \rightleftharpoons 4OH^- \qquad\qquad E^o = 0.40 \text{ v}$$

During a particular corrosion process, the cell potential was found to be 1.09 volts. What was the approximate partial pressure of oxygen in this system?

*A.8.   The effects of certain types of stress on the performance of materials in the marine environment take the form of stress corrosion cracking and corrosion fatigue. Stress corrosion cracking is a result of the combined effects of corrosion and tensile stresses. The later may result from corrosion processes developing in or around welded joints. Susceptibility to stress corrosion effects is the principal factor limiting the level of stress and tensile properties that can be used safely with high strength steels in the design of vessels and other structures for service in sea water and sea air. Certain types of chromium steels have failed by stress corrosion cracking in a "salt atmosphere". The corrosion of chromium in salt water can be formulated as:

$$2Cr_{(s)} + H_2O_{(\ell)} + 3\,O_{2(g)} \rightleftharpoons 2H^+_{(aq)} + Cr_2O_7^{2-}{}_{(aq)}$$

and the standard electrode potentials are tabulated as:

$$Cr_2O_7^{2-} + 14H^+ + 12e^- \rightleftharpoons 2Cr + 7H_2O \qquad E^o = 0.30 \text{ v}$$

$$O_2 + 4H^+ + 4e^- \rightleftharpoons 2H_2O \qquad E^o = 1.23 \text{ v}$$

A sea water system, around a corroded drilling platform, is 0.50 $\underline{M}$ in $Cr_2O_7{}^{2-}$ with the P_{O_2} = 0.030 atm. The system has a potential of 0.91 volt. What is the approximate hydrogen ion concentration?

*A.9. Iron deficiency in plants is often difficult to correct. An iron-containing spray may not enter a chlorotic leaf, or may enter only small areas giving a spotty green appearance to a partially recovered leaf. In such cases there is need for better coverage of existing foliage and repeated spraying to expose new leaves. In some instances, injection of salts into the stem has been used with success. However, this method is more expensive and may cause salt injury and infections in the stem. The most widely used foliar spray is a 5.0% "ferrous sulfate" ($FeSO_4$) solution. A common method used for the volumetric determination of iron(II) in plant and soil samples uses an oxidation-reduction reaction that can be formulated as:

$$MnO_4{}^-_{(aq)} + 8H^+_{(aq)} + 5Fe^{2+}_{(aq)} \rightleftarrows 5Fe^{3+}_{(aq)} + 4H_2O_{(\ell)} + Mn^{2+}_{(aq)}$$

for which the standard electrode potentials are:

$$MnO_4{}^- + 8H^+ + 5e^- \rightleftarrows Mn^{2+} + 4H_2O \qquad E^o = 1.491 \text{ v}$$

$$Fe^{3+} + e^- \rightleftarrows Fe^{2+} \qquad E^o = 0.770 \text{ v}$$

A particular titration mixture which is 0.10 $\underline{M}$ in Fe^{3+}, 0.02 $\underline{M}$ in Mn^{2+}, 0.10 $\underline{M}$ in $MnO_4{}^-$ and 0.10 $\underline{M}$ in Fe^{2+} has a cell potential of 0.65 volt. What is the approximate pH of the solution?

*A.10. The sulfur cycle is very common in the marine environment, especially along certain fault lines where hydrogen sulfide enriched waters emerge. In certain areas, sulfur deposition occurs on a noticeable scale and may become sufficiently important to produce elemental sulfur in quantities suitable for recovery. Within sulfate-rich waters, sulfate reduction produces hydrogen sulfide gas in the lower water mass. The sulfide may then be oxidized to free sulfur. The overall process can be formulated as:

$$SO_4{}^{2-}_{(aq)} + 2H^+_{(aq)} + 3H_2S_{(g)} \rightleftarrows 4S_{(s)} + 4H_2O_{(\ell)}$$

for which the standard electrode potentials are:

$$SO_4{}^{2-} + 8H^+ + 6e^- \rightleftarrows S + 4H_2O \qquad E^o = 0.36 \text{ v}$$

$$S + 2H^+ + 2e^- \rightleftarrows H_2S \qquad E^o = 0.17 \text{ v}$$

A Gulf Coast sulfate-rich system is 0.018 $\underline{M}$ in $SO_4{}^{2-}$ and has a P_{H_2S} = 0.09 atm. The system has a measured cell potential of 0.010 volt. What is the approximate pH of the solution?

ANSWERS:

(A.1.) 0.42 v,   (A.2.) 0.78 v,   (A.3.) 0.46 v,   (A.4.) 0.99 v,   (A.5.) 1.0 v,   (A.6.) ~ 4.3,
(A.7.) ~ 9 x 10^{-3} atm,   (A.8.) ~ 0.8 $\underline{M}$,   (A.9.) ~ 0.8,   (A.10.) ~ 6.7

RELEVANT
PROBLEMS

Unit 14:   Oxidation-Reduction Processes

Set B: Biological & Medical Sciences

B.1. One type of bacteria involved in iron corrosion, Ferrobacillus ferrooxidans, is associated with the oxidation of "ferrous" ion (Fe^{2+}) to "ferric" ion (Fe^{3+}). As a
result, masses of iron(III) hydroxide accumulate on or near the cells. The Ferro-
bacillus is a short, motile, rod-shaped cell which is a strict autotroph, deriving
its energy from the Fe^{2+} to Fe^{3+} oxidation process. These bacteria obtain carbon
from carbon dioxide and nitrogen from dissolved inorganic nitrogen salts such as
nitrates. The overall energy producing reaction for these bacteria can be formulated
as:

$$4Fe^{2+}_{(aq)} + O_{2(g)} + 4H^+_{(aq)} \rightleftharpoons 4Fe^{3+}_{(aq)} + 2H_2O_{(\ell)}$$

and the standard electrode potentials are tabulated as:

$$O_2 + 4H^+ + 4e^- \rightleftharpoons 2H_2O \qquad E^o = 1.23 \text{ v}$$

$$Fe^{3+} + e^- \rightleftharpoons Fe^{2+} \qquad E^o = 0.77 \text{ v}$$

What voltage would be produced in this "biological cell" when the O_2 gas pressure is
0.21 atm and the electrolyte is 0.010 $\underline{M}$ in Fe^{3+}, 1.0 x 10^{-6} $\underline{M}$ in H^+, and 0.10 $\underline{M}$ in
Fe^{2+}?

B.2. All living organisms have to do work to survive. Man does muscular work and trans-
mits nerve impulses to perform his body functions. All organisms synthesize various
compounds and many of the syntheses involve large activation energies. Electric eels
can do impressive electrical work and some plants and animals even produce light
energy. In all physiological systems, biological substrates undergo degradations
which convert chemical energy into work. The physiological conversion of chemical
energy to work has been resolved into many parts so that each step is closely approximated by an equilibrium system. This allows the organism to most effectively utilize the energy produced by the chemical reactions in each of the metabolic steps.

A substantial part of the energy released by digestion is unavailable for useful work because these hydrolytic reactions are not coupled to other processes. Consequently, most of this energy is dissipated as body heat. The reaction for the oxidation of lactate to pyruvate can be expressed by the following equation:

$$
\begin{array}{cc}
CH_3 & CH_3 \\
| & | \\
H\text{-}C\text{-}OH & \rightleftharpoons \quad C\text{=}O \quad + 2H^+ + 2e^- \\
| & | \\
COO^- & COO^-
\end{array}
$$

$$\quad \text{(lactate)} \qquad\qquad \text{(pyruvate)}$$

Inside muscle cells, when lactate accumulates, nicotinamide adenine dinucleotide (NAD) participates as indicated by the equation:

$$\text{lactate}^- + NAD^+ \rightleftharpoons \text{pyruvate}^- + NADH + H^+$$

for which the standard electrode potentials are:

$$\text{pyruvate}^- + 2H^+ + 2e^- \rightleftharpoons \text{lactate}^- \qquad E^o = -0.19 \text{ v}$$

$$NAD^+ + H^+ + 2e^- \rightleftharpoons NADH \qquad E^o = -0.32 \text{ v}$$

What would be the potential for this "biological cell" when the electrolyte is 0.020 $\underline{M}$ in pyruvate, 0.010 $\underline{M}$ in NADH, 5.0×10^{-8} $\underline{M}$ in H^+, 2.0×10^{-4} $\underline{M}$ in lactate, and 0.0080 $\underline{M}$ in NAD^+?

B.3. In starvation or in certain diabetic states, in which the metabolism of carbohydrate is impaired, the organism must rely on fatty acid oxidation for an energy source. One of the important steps in this process is the conversion of β-hydroxybutyrate to acetoacetate, ultimately utilizing the reduction of molecular oxygen. In simplistic terms, the net sequence may be formulated by:

$$
\begin{array}{cc}
OH & O \\
| & \| \\
2 \; CH_3CHCH_2CO_2^- + O_2 & \rightleftharpoons \quad 2 \; CH_3CCH_2CO_2^- + H_2O
\end{array}
$$

$$\text{(β-hydroxybutyrate)} \qquad\qquad \text{(acetoacetate)}$$

for which the standard electrode potentials are:

$$
\begin{array}{cc}
O & OH \\
\| & | \\
CH_3CCH_2CO_2^- + 2H^+ + 2e^- \rightleftharpoons CH_3CHCH_2CO_2^- & \qquad E^o = -0.27 \text{ v}
\end{array}
$$

$$O_2 + 4H^+ + 4e^- \rightleftharpoons 2H_2O \qquad E^o = +1.23 \text{ v}$$

What would be the potential in this biochemical system when the O_2 gas pressure is 0.05 atm and the electrolyte is 2.0×10^{-3} $\underline{M}$ in acetoacetate and 5.8×10^{-4} $\underline{M}$ in β-hydroxybutyrate?

B.4. The catabolism of glucose in the body is characterized by a complex series of an-aerobic and aerobic reactions, through which the glucose is metabolized to carbon dioxide and water. The types of chemical reactions that can be identified include hydrolysis, condensation, hydration, oxidation, dehydration, fragmentation, and de-hydrogenation. Although some of these reactions are energy consuming, the overall process has a net gain in energy. Some of this energy is fixed as ATP, the body's "energy reservoir", and the remainder is manifest as body heat. One of these re-actions, the oxidation of malic acid to oxaloacetic, can be formulated as:

$$
\underset{\text{(malic acid)}}{\overset{HO}{H-\underset{CH_2CO_2H}{\overset{|}{\underset{|}{C}}}-CO_2H}} \rightleftharpoons \underset{\text{(oxaloacetic acid)}}{\overset{O}{\underset{CH_2CO_2H}{\overset{\|}{\underset{|}{C}}}-CO_2H}} + 2H^+ + 2e^-
$$

This reaction produces about 20.0% of the body's ATP. When malic acid accumulates, nicotinamide adenine dinucleotide (NAD) participates in hydrogen transport as in-dicated by the reaction:

$$malate^- + NAD^+ \rightleftharpoons oxaloacetate^- + NADH + H^+$$

for which the standard electrode potentials are:

$$oxaloacetate^- + 2H^+ + 2e^- \rightleftharpoons malate^- \qquad E^o = -0.17 \text{ v}$$
$$NAD^+ + H^+ + 2e^- \rightleftharpoons NADH \qquad E^o = -0.32 \text{ v}$$

What would be the potential for this biological reaction when the electrolyte is 0.030 $\underline{M}$ in oxaloacetate, 0.020 $\underline{M}$ in NADH, 0.0050 $\underline{M}$ in NAD^+, 3.0×10^{-4} $\underline{M}$ in malate, and 1.0×10^{-7} $\underline{M}$ in H^+?

B.5. Photosynthesis is one of the fundamental chemical processes for sustaining life in the biosphere. It is accomplished by both land and aquatic plants. Utilizing solar energy, plants convert the components of water and carbon dioxide to some form of carbohydrate, such as glucose, releasing oxygen to the environment. The light-induced movement of electrons via electron-carrying molecules from water to various electron acceptors is called "photosynthetic electron transport" [Discussed in Unit 8, problem B.3]. Normally, electrons would tend to flow from NADPH to oxygen, but in photosynthetic electron transport the electrons are induced to flow in the reverse direction, from water to $NADP^+$. The overall reaction is formulated as:

$$2H_2O_{(\ell)} + 2NADP^+_{(aq)} \xrightarrow{\text{(light)}} 2NADPH_{(aq)} + 2H^+_{(aq)} + O_{2(g)}$$

for which the standard electrode potentials are:

$$NADP^+ + H^+ + 2e^- \rightleftharpoons NADPH \qquad E^o = -0.32 \text{ v}$$

$$O_2 + 4H^+ + 4e^- \rightleftharpoons 2H_2O \qquad E^o = 1.23 \text{ v}$$

What potential is required by this photosynthetic process when the O_2 pressure is 0.21 atm and the electrolyte is 0.10 $\underline{M}$ in NADPH, 1.0×10^{-5} $\underline{M}$ in H^+, and 5.0×10^{-4} $\underline{M}$ in $NADP^+$?

*B.6. One step in the catabolism of glucose is characterized by the conversion of iso-citric acid to α-ketoglutaric acid. This reaction also contributes about 20% of the ATP molecules formed during glucose metabolism (problem B.4). The formation of α-ketoglutaric acid is a two-step process, beginning with a dehydrogenation reaction followed by an oxidative decarboxylation. The enzyme isocitric acid dehydrogenase catalyzes both steps with the aid of the metal activator Mg^{2+}. One of the two types of isocitrate dehydrogenase requires NAD^+ as electron acceptor, and the other requires $NADP^+$. The overall reactions catalyzed by the two types of isocitrate dehydrogenase are identical:

$$\text{isocitrate}^-_{(aq)} + NAD^+_{(aq)} \rightleftharpoons \text{α-ketoglutarate}^-_{(aq)} + CO_{2(g)} + NADH_{(aq)} + H^+_{(aq)}$$

$$\text{isocitrate}^-_{(aq)} + NADP^+_{(aq)} \rightleftharpoons \text{α-ketoglutarate}^-_{(aq)} + CO_{2(g)} + NADPH_{(aq)} + H^+_{(aq)}$$

for which the standard electrode potentials are:

$$\text{α-ketoglutarate}^- + CO_2 + 2H^+ + 2e^- \rightleftharpoons \text{isocitrate}^- \qquad E^o = -0.380 \text{ v}$$

$$NAD^+ + H^+ + 2e^- \rightleftharpoons NADH \qquad E^o = -0.320 \text{ v}$$

$$NADP^+ + H^+ + 2e^- \rightleftharpoons NADPH \qquad E^o = -0.324 \text{ v}$$

During the formation of α-ketoglutarate from isocitrate, the cell potential was found to be 0.109 volt. What was the approximate partial pressure of CO_2 in the system in which the electrolyte was 0.10 $\underline{M}$ in α-ketoglutarate, 0.10 $\underline{M}$ in NADH, 1.0×10^{-4} $\underline{M}$ each in isocitrate and NAD, and 1.0×10^{-7} $\underline{M}$ in H^+?

*B.7. The cytochromes are a very important group of substances involved in the oxidation-reduction reactions in living cells. These heme proteins are present in all oxygen-requiring (aerobic) organisms. Generally, the cytochrome content of tissues parallels their activity, the highest concentration being found in hard-working muscles such as the wing muscles of birds and insects. The cytochromes are a mixture of closely related compounds designated by the suffixes _a_, _a_3, _b_, and _c_. The best known is Cytochrome _c_, which is the only soluble member of the group. The others remain closely attached to the particulate matter of the cells, especially to the mito-chondria. Cytochrome _c_ was first isolated in a highly purified form from heart muscle, and was shown to contain a heme group similar to but not identical with that of hemoglobin. Unlike hemoglobin, <u>Cytochrome</u> <u>c</u> does not combine with oxygen or carbon monoxide, and its iron atom undergoes a cycle of oxidations and reductions from the "ferrous" (Fe^{2+}) to the "ferric" (Fe^{3+}) state as it performs its biochemical role. The oxidation of cytochrome c, Fe_c^{2+}, to the Fe_c^{3+} form can be formulated as:

$$O_{2(g)} + 4H^+_{(aq)} + 4Fe_c^{2+}{}_{(solid)} \rightleftharpoons 2H_2O_{(\ell)} + 4Fe_c^{3+}{}_{(solid)}$$

and the appropriate standard electrode potentials are tabulated as:

$$Fe_c^{3+} + e^- \rightleftharpoons Fe_c^{2+} \qquad\qquad E^o = 0.25 \text{ v}$$

$$O_2 + 4H^+ + 4e^- \rightleftharpoons 2H_2O \qquad\qquad E^o = 1.23 \text{ v}$$

The cytochrome system formulated has a measured cell potential of 0.55 volt and the P_{O_2} is 0.05 atm. What is the approximate pH of this system?

*B.8. <u>Cytochrome</u> a_3, also identified as <u>cytochrome oxidase</u>, is the terminal catalyst in the respiratory chain of the cell and reacts directly with the molecular oxygen provided by oxyhemoglobin. In this process, as in the case of <u>cytochrome</u> <u>c</u>, the iron ion oscillates between the "ferrous" and "ferric" states. <u>Cytochrome</u> a_3 is inhibited by very low concentrations of cyanide ion, which helps explain why cyanide is such a deadly poison to all aerobic organisms, including man. The overall reaction for the oxidation of <u>cytochrome</u> a_3 ($Fe_{a_3}^{2+}$) to <u>cytochrome</u> a_3 ($Fe_{a_3}^{3+}$) can be expressed by the following equation:

$$O_{2(g)} + 4Fe_{a_3}^{2+}{}_{(solid)} + 4H^+_{(aq)} \rightleftharpoons 4Fe_{a_3}^{3+}{}_{(solid)} + 2H_2O_{(\ell)}$$

for which the standard electrode potentials are:

$$O_2 + 4H^+ + 4e^- \rightleftharpoons 2H_2O \qquad\qquad E^o = 1.23 \text{ v}$$

$$Fe_{a_3}^{3+} + e^- \rightleftharpoons Fe_{a_3}^{2+} \qquad\qquad E^o = 0.385 \text{ v}$$

The <u>cytochrome</u> a_3 system formulated has a measured potential of 0.035 volt when the P_{O_2} is 1.1 atm. What is the approximate pH of this system?

*B.9. The ubiquinones are a group of coenzymes whose name reflects both structure and occurrence. These coenzymes are quinone derivatives (Unit 7, problem B.1) that are found not just in mitochondria but also in cell nuclei, microsomes, and elsewhere. Hence the contraction of the term "ubiquitous quinones" to ubiquinones. They are also identified by the synonym coenzyme Q (CoQ). In one particular respiratory reaction "CoQ" participates in the transfer of hydrogen from the reduced form of flavin adenine dinucleotide ($FADH_2$), a reaction which may be formulated as:

$$CoQ_{(s)} + FADH_{2(aq)} \rightleftharpoons CoQH_{2(s)} + FAD_{(aq)}$$

for which the standard electrode potentials are:

$$CoQ + 2H^+ + 2e^- \rightleftarrows CoQH_2 \qquad E^0 = 0.10 \text{ v}$$

$$FAD + 2H^+ + 2e^- \rightleftarrows FADH_2 \qquad E^0 = -0.22 \text{ v}$$

The coenzyme Q (CoQ) system formulated has a measured potential of 0.30 volt and is 1.2 $\underline{M}$ in FAD. What is the approximate concentration of $FADH_2$ in this system?

*B.10. Many enzymes, such as trypsin and ribonulease, consist entirely of protein. However, certain others contain in addition to the protein component (called the apoenzyme) another component, called the prosthetic group if firmly attached, or the coenzyme if easily separated by dialysis. As far as enzymatic activity is concerned, however, there is no significant difference between the function of a prosthetic group and a coenzyme. One role of the coenzyme may be to assist in the cleavage of the substrate by acting as an acceptor of the group or atoms being removed, or to act in the reverse way by contributing such fragments for the synthesis of molecules. For example, during the oxidation of carbohydrates, the hydrogen atoms and their electrons from the carbohydrate are removed in pairs. The two hydrogens and their electrons are "picked-up" by NAD^+, a coenzyme, and taken to the respiratory process as NADH + H^+. Subsequently, the hydrogens are "picked-up" by FAD, with the net result:

$$NADH + H^+ + FAD \rightleftarrows NAD^+ + FADH_2$$

In subsequent reactions, the hydrogens are transferred to oxygen to form water. The standard electrode reactions can be tabulated as:

$$FAD + 2H^+ + 2e^- \rightleftarrows FADH_2 \qquad E^0 = -0.22 \text{ v}$$

$$NAD + H^+ + 2e^- \rightleftarrows NADH \qquad E^0 \quad -0.32 \text{ v}$$

The coenzyme system formulated had a measured potential of 0.05 volt and is 1.0 $\underline{M}$ each in NADH and FAD, 8.0×10^{-4} $\underline{M}$ in NAD^+, and 9.0×10^{-4} $\underline{M}$ in $FADH_2$. What is the approximate pH of this system?

ANSWERS:

(B.1.) 0.15 v, (B.2.) 0.02 v, (B.3.) 1.46 v, (B.4.) 0.28 v, (B.5.) -1.46 v (The negative sign indicates the nonspontaneous nature of this energy-requiring process.), (B.6.) 0.2 atm, (B.7.) ∿ 7.0, (B.8.) 13.7, (B.9.) ∿ 0.25 $\underline{M}$, (B.10.) ∿ 7.8

RELEVANT

PROBLEMS

I.1. Electrochemical cells having a high degree of reversibility are broadly classified as storage cells. For prolonged commercial use, a storage battery should have a high electrical capacity per unit of weight. Adverse chemical effects that cause deterioration or loss of stored energy must be minimal. The transformation of electrical into chemical energy, as in recharging, and of chemical into electrical energy, in discharging, should be easily reversible. The ideal storage battery should have very low electrolyte resistance, simplicity and strength of construction, and low production cost. The battery best fitting these requirements at present is the lead storage battery. These batteries are manufactured in a wide range of sizes, from very small two-cell units up to massive batteries weighing several tons. The two electrodes in a lead storage cell consist of lead and lead dioxide, respectively, packed in grids immersed in a dilute sulfuric acid electrolyte. The standard electrode potentials are given by:

$$PbSO_{4(s)} + 2\ e^- \rightleftharpoons Pb_{(s)} + SO_{4(aq)}^{2-} \qquad E^o = -0.36 \text{ volt}$$

$$PbO_{2(s)} + SO_{4(aq)}^{2-} + 4\ H_{(aq)}^+ + 2\ e^- \rightleftharpoons PbSO_{4(s)} + 2\ H_2O_{(\ell)}$$
$$E^o = +1.68 \text{ volts}$$

Write the net equation for spontaneous discharge of a lead storage cell. What is the standard potential for this cell?

I.2. What is the potential of a lead storage cell (problem I.1) in which the concentration of sulfate is 0.05 $\underline{M}$ and the pH is 1.0?

I.3. The nickel-cadmium cell has come into wide use, especially in Europe. The positive electrode is a perforated steel plate impregnated with nickel(III) oxide and the negative electrode is a steel plate packed with finely divided cadminum. The typical commercial nickel-cadmium storage battery consists of six 60 mm x 70 mm x 2 mm positive plates and five negative plates separated by 0.8 mm spacers of polyvinyl chloride and assembled into a polystyrene case having nonspillable caps sealed by mercury on sintered glass. Nineteen such cells are assembled into a 24 volt battery. The electrolyte for the system is aqueous potassium hydroxide. Deterioration of the negative plate causes loss of capacity on standing fully charged, but the capacity can be recovered by slow discharge followed by rapid charge. Discharge at

50 amps for five minutes reduces the voltage to 1.0 volt per cell, thus the percentage reduction of capacity is the same as for the lead-sulfuric·acid storage battery. The alkaline-electrolyte cell compensates in part for the disadvantage of smaller potential by being far lighter, more durable, easier to recharge and immune to impairment effects of overcharging, to which the lead storage cell is susceptible. The overall cell reaction for the nickel-cadmium storage cell is given by the equation:

$$Cd_{(s)} + Ni_2O_{3(s)} + 3\ H_2O_{(\ell)} \rightleftarrows Cd(OH)_{2(s)} + 2\ Ni(OH)_{2(s)}$$

The role of the KOH electrolyte is best seen from the equations for the individual electrode processes:

$$Ni_2O_{3(s)} + 3H_2O_{(\ell)} + 2e^- \rightleftarrows 2Ni(OH)_{2(s)} + 2\ OH^-_{(aq)} \qquad E^o = 0.46\ v$$

$$Cd(OH)_{2(s)} + 2e^- \rightleftarrows Cd_{(s)} + 2\ OH^-_{(aq)} \qquad E^o = -0.80\ v$$

Calculate the individual electrode potentials of the "NICAD" cell (a) at pH 10.0 and (b) at pH 13.0.

I.4.   The voltage of the "NICAD" cell (problem I.3), unlike that of the lead storage battery (problem I.1), is essentially independent of pH. The maximum current flow is, however, a function of the electrolyte (KOH) concentration. Demonstrate the pH-independence of voltage for this cell by calculating the cell voltage from individual electrode potentials at (a) pH 10 and (b) pH 13, and (c) by calculating the value of the Q term in the Nernst equation based on the overall cell reaction.

I.5.   A fuel cell produces an electric current through the chemical consumption of two reactants. It generates the current as long as the reactants are fed to it. The fuel cell has the advantages of being light weight, quiet of operation and highly efficient. Fuel cells are galvanic cells, unique in that the electrode materials are generally in the form of gases. The gas-diffusion fuel cells require the diffusion of gases through a liquid electrolyte. Porous electrodes are used to provide a large number of contact areas between the gaseous fuel, liquid electrolyte, and solid electrode. These fuel cells require catalysts to activate the incoming gases. The most common catalysts are platinum, which is used at the fuel electrode, and silver at the oxidant electrode. Generally, the electrodes are made of porous carbon on which thin coatings of the catalyst are placed. Gas molecules are adsorbed on the catalytic surface, thereby speeding the electrochemical reaction. One of the most successful commercial cells uses hydrogen as the fuel, oxygen as the oxidant, and potassium hydroxide as the electrolyte. The overall electrochemical reaction in the fuel cell used to generate electrical power on the Apollo moon trips is described by

390

the equation:

$$2 \, H_{2(g)} + O_{2(g)} \rightarrow 2 \, H_2O_{(\ell)}$$

for which the individual electrode processes are:

$$2 \, H_2O_{(\ell)} + 2e^- \rightarrow H_{2(g)} + 2 \, OH^-_{(aq)} \qquad E^o = -0.828 \text{ v}$$

$$O_{2(g)} + 2 \, H_2O_{(\ell)} + 4e^- \rightarrow 4 \, OH^-_{(aq)} \qquad E^o = 0.401 \text{ v}$$

Calculate the voltage of this cell when the pressure of each gaseous reactant is 2.5 atm.

*I.6. The sulfuric acid used in manufacturing lead storage batteries (problem I.1) is prepared from pure concentrated sulfuric acid by dilution with <u>distilled</u> water. Ordinary "tap" water is not suitable, because the Ca^{2+} in moderately "hard" water can remove needed sulfate by precipitation of $CaSO_4$ (a problem often ignored by "service station" attendants). In a routine quality control test in a battery manufacturing plant, a series of batteries were all found to have a voltage of 1.85 v for each cell. Further examination revealed no apparent construction flaws, but a white sediment was observed in every electrolyte compartment. The electrolyte acidity was the same as that of a standard comparison cell, corresponding to 0.70 $\underline{M}$ H^+. If, as suspected, the problem was that the battery acid had accidentally been diluted with "hard" water so that some sulfate loss had occurred, what was the remaining sulfate concentration?

*I.7. A profitable "specialty" electronics industry has developed to serve the increasing technological demands of modern medicine. A particularly interesting area is concerned with the development of long-life batteries for heart "pacemakers". Conventional batteries require periodic replacement by relatively minor surgery, but a "biologically-fueled" cell already shown feasible with test animals shows promise of eliminating the need for surgical "battery replacement". By implanting a tiny magnesium anode in a proper subcutaneous location, the body's oxygen can serve as the cathode reactant to form a "lifetime" pacemaker cell with a potential of 3.28 v. The cell reaction may be formulated as:

$$2Mg_{(s)} + O_{2(g)} + 4H^+_{(aq)} \rightarrow 2Mg^{2+}_{(aq)} + 2H_2O_{(\ell)}$$

Under conditions corresponding to an oxygen gas pressure of 0.20 atm and a pH of 7.44, what is the concentration of Mg^{2+} for this cell? [For Standard Electrode Potentials, see Appendix E.]

*I.8. One of the most daring and extravagant ventures attempted by the petroleum industry is the Alaska pipeline, designed to help relieve some of the pressures of the "energy crunch". Corrosion is a problem with any pipeline operation, but it could be a

*Proficiency Level

massive problem with an operation as large as the Alaska pipeline. The corrosion of iron is believed to be essentially electrochemical, with the corroding area serving as the anode for initial oxidation of iron to Fe^{2+}. One method that can be used to protect buried iron or steel pipe from electrolytic corrosion is to bury chunks of magnesium metal at appropriate locations along the pipeline. The high reactivity of magnesium leads to its acceptance of the role of anode, with the iron then serving as a cathode to convert soil oxygen to water. The ions normally present in moist soil furnish the necessary electrolyte. The electrode processes appear to involve:

$$Mg_{(s)} \rightarrow Mg^{2+} + 2e^- \qquad \text{(at the anode)}$$

$$O_{2(g)} + 4H^+_{(aq)} + 4e^- \rightarrow 2H_2O_{(\ell)} \qquad \text{(at the iron cathode)}$$

What concentration of magnesium ion would correspond to a cell potential for this system of 3.44 v under conditions of 0.10 atm O_2 and a pH of 5.70?

*I.9. What soil moisture pH would correspond to a cell potential of 3.37 v for a magnesium-protected pipeline (problem I.8) under conditions of 3×10^{-8} M Mg^{2+} and 0.050 atm O_2?

*I.10. The direct combustion of hydrocarbon fuels is a very inefficient energy source and, invariably, a source of pollution problems. A private research corporation has developed a prototype minibus powered by propane-oxygen fuel cells. While the "power and speed" limitations of these cells are inappropriate for highway vehicles, the minibus shows promise for low speed mass transit applications. The electrode processes of the propane-oxygen fuel cell are represented as:

$$O_{2(g)} + 2H_2O_{(\ell)} + 4e^- \rightarrow 4\ OH^-_{(aq)} \qquad E^\circ = +0.40 \text{ v}$$

$$3CO_3^{2-}_{(aq)} + 17H_2O_{(\ell)} + 20e^- \rightarrow C_3H_{8(g)} + 26\ OH^-_{(aq)} \qquad E^\circ = -0.95 \text{ v}$$

What electrolyte concentration of KOH would correspond to a cell potential of 1.42 v when each gas is supplied at a pressure of 1.5 atm and the carbonate concentration is 1.2×10^{-6} M?

ANSWERS:

(I.1.) 2.04 v, (I.2.) 1.84 v, (I.3.) (a) [at Ni_2O_3] 0.70 v, [at Cd] −0.56 v (b) [at Ni_2O_3] 0.52 v, [at Cd] −0.74 v, (I.4.) (a) 1.26 v, (b) 1.26 v, (c) Q = 1 (The net process involves only solids and liquids, of unit activity.), (I.5.) 1.247 v, (I.6.) ∿ 1.3×10^{-3} M, (I.7.) 3.8×10^{-5} M, (I.8.) 3.2×10^{-7} M, (I.9.) 7.3, (I.10.) ∿ 6 M

RELEVANT
PROBLEMS

E.1. The chloride anion is essential for plant growth, being used by most crop plants in larger amounts than any other micronutrient except iron. Chloride is found normally in soils in the range of 10-1000 ppm, usually as NaCl or KCl. Since most chlorides are soluble, the ion is readily redistributed through flowing or percolating ground waters. In well-drained humid regions, most chloride is leached from the soil, but in poorly drained soils its concentration may reach the point of salt toxicity to plants. Nature maintains a chloride cycle by evaporation of salt waters near the ocean to provide submicroscopic salt dust to the atmosphere for return to the soil in snows and rains. Man has contributed a new source of chloride by the addition of chlorine as an antibacterial agent to effluent streams of industrial and domestic waste waters. The production of chloride by the chlorine oxidation of water is formulated as:

$$Cl_{2(aq)} + H_2O_{(\ell)} \rightleftharpoons HOCl_{(aq)} + H^+_{(aq)} + Cl^-_{(aq)}$$

The chlorine content of water can be determined by use of an electrochemical cell, for which the electrode processes (using inert platinum electrodes) are:

$$2\ HOCl_{(aq)} + 2H^+_{(aq)} + 2e^- \rightleftharpoons Cl_{2(aq)} + 2H_2O_{(\ell)} \qquad E^o = +1.63\ v$$

$$Cl_{2(aq)} + 2e^- \rightleftharpoons 2Cl^-_{(aq)} \qquad E^o = +1.36\ v$$

What cell potential would result from measurement of a sample of industrial waste effluent 0.10 $\underline{M}$ in Cl_2, 0.30 $\underline{M}$ in Cl^-, 0.025 $\underline{M}$ in HOCl, and 1.3×10^{-5} $\underline{M}$ in H^+? [The spontaneous cell discharge reaction produces chlorine.]

E.2. A municipal water supply uses the chlorine/hypochlorous acid/chloride cell (problem E.1) for monitoring chlorine concentration to ensure proper levels for bacteriocidal activity. At the desired level of 1.0 ppm Cl_2, what would be the cell potential, if the chloride concentration is 4.0×10^{-3} $\underline{M}$, the hypochlorous acid concentration is 1.2×10^{-5} $\underline{M}$, and the pH is 6.5? [Chlorine is consumed during cell discharge under these conditions.]

E.3. The formation of hydrogen sulfide is an undesirable side-effect of the sulfate ox- idation of organic wastes in ground water. Under normal circumstances the amount of H_2S produced is negligible, but it may become noticeable in cases of high sulfate contamination, as in the run-off from certain mining operations. What "biological potential" would correspond to a system for biodegradation of glucose at a concentration of 1.3×10^{-5} M, when the sulfate concentration is 1.8×10^{-3} M, the dissolved CO_2 and H_2S are each 4.3×10^{-4} M, and the pH is 5.7? The half-reactions for the process may be formulated as:

$$SO_{4(aq)}^{2-} + 10H_{(aq)}^{+} + 8e^{-} \rightleftharpoons H_2S_{(aq)} + 4H_2O_{(\ell)} \qquad E^{\circ} = -0.207 \text{ v}$$

$$6CO_{2(aq)} + 24H_{(aq)}^{+} + 24e^{-} \rightleftharpoons C_6H_{12}O_{6(aq)} + 6H_2O_{(\ell)} \qquad E^{\circ} = -0.479 \text{ v}$$

E.4. It has been suggested that "sewage fuel cells" might be employed in municipal waste water plants to utilize some of the energy from the oxidation of organic matter in sewage during aeration treatment. What voltage could be obtained from a cell designed to test this possibility, utilizing 0.10 M glucose and oxygen gas at 0.30 atm, under conditions of a CO_2 concentration of 0.034 M and pH 6.3? The cell discharge reaction is formulated as:

$$C_6H_{12}O_{6(aq)} + 6 \ O_{2(g)} \rightleftharpoons 6CO_{2(aq)} + 6H_2O_{(\ell)}$$

[For electrode potentials, see Appendix E and problem E.3]

E.5. How many glucose-oxygen fuel cells (problem E.4) would have to be connected in series to form a battery of approximately 24 volts, using 0.30 M glucose and oxygen gas at 0.20 atm, under conditions of 0.034 M CO_2 and pH 6.3?

*E.6. In a municipal water plant using the chlorine/hypochlorous acid/chloride cell for monitoring chlorine concentration (problem E.2), a test sample gave a potential reading of 0.45 v. Independent assay had shown the sample to contain 3.2×10^{-4} M Cl^- and 2.5×10^{-6} M HOCl, at a pH of 6.3. What was the chlorine concentration in the sample, in ppm Cl_2?

*E.7. The increasing concentrations of heavy metal ions in natural waters is a problem of major concern. Most such cations (e.g., Hg^{2+}, Cd^{2+}, Pb^{2+}) are toxic in moderate amounts. Although it is unlikely that toxic concentration levels would occur in drinking water supplies, the problem is more insidious than that. Many of these ions are cumulative poisons, so that ingestion of very tiny amounts over a prolonged period of time can lead to the gradual deterioration of certain vital biochemical processes, leading eventually to irreparable damage. Conventional wet-chemical methods cannot be used to detect metal ions in extremely dilute solutions, but electrochemical methods are available. A number of ion-specific electrodes, similar to those employed by a pH meter, have been developed that permit the assay of "trace" amounts of various heavy metal ions in water supplies. The electrode processes are relatively complex, and in some cases, the exact mechanisms of electrode functions are not yet understood (see: "Ion-Selective Electrodes", J. Chem. Ed., June 1974, pages 387-390). The utility of electrochemical assay of "trace pollutants" can be demonstrated, however, with simpler systems. Calculate the concentration of Cu^{2+} ion in a solution in contact with a copper cathode forming a cell with a Zn, 0.10 M Zn^{2+} anode system when the cell potential is 0.91 v. [See Appendix E for Standard Electrode Potentials.]

*E.8. Mercury is a particularly dangerous pollutant because of its conversion by certain aquatic organisms to organomercurials that are concentrated through various "food chains". Calculate the molar concentration of Hg^{2+} in an aqueous solution from a canned fish product and compare this with the official "safe" limit of 0.5 ppm. The sample was used as the cathode electrolyte in contact with mercury for a cell using a Zn, 0.10 M Zn^{2+} anode system. The cell potential was measured as 1.53 v. [For $Hg^{2+} + 2e^- \rightleftarrows Hg$, $E^o = +0.85$ v.]

*E.9. In an electrochemical assay of a water supply for heavy metal ions (problem E.7), analysis for Pb^{2+} used a water sample with a lead anode and a Ag, 0.10 M Ag^+ cathode system. Calculate the Pb^{2+} concentration, in ppm, corresponding to a cell potential of 1.01 v. Continued ingestion of Pb^{2+} at the rate of 1 mg per day is considered hazardous and there is considerable evidence that even lower levels may constitute a long-range health problem. [For Standard Electrode Potentials, see Appendix E.]

*E.10. Only within the last few years has cadmium been recognized as a really dangerous pollutant. Among other effects of Cd^{2+} accumulation are irreversible kidney damage and bone degeneration from loss of Ca^{2+}. Most cadmium taken into the body is excreted rather quickly, but about 2% is retained, with a biological half-life estimated at 10-25 years (compared to about two months for Hg^{2+}), so that Cd^{2+} is one of the more dangerous cumulative poisons. Cigarette smoke is particularly high in cadmium. An electrochemical assay for Cd^{2+} used a urine sample from a "heavy" smoker in the anode compartment with a cadmium electrode. The cathode system was silver in contact with 0.10 M Ag^+. What concentration of Cd^{2+} corresponded to a cell potential of 1.30 v? [See Appendix E for Standard Electrode Potentials.]

ANSWERS:

(E.1.) 0.09 v, (E.2.) 0.26 v, (E.3.) 0.23 v, (E.4.) 1.72 v, (E.5.) 14, (E.6.) 41 ppm, (E.7.) 3.8×10^{-8} M, (E.8.) $\sim 2 \times 10^{-4}$ M ($\sim$ 80 times the "safe" limit), (E.9.) $\sim$ 4 ppm, (E.10.) $\sim 4 \times 10^{-6}$ M

UNIT 15: ORGANIC CHEMISTRY REVISITED

Organic compounds may be involved in a number of different types of equilibrium systems.

Since most organic compounds are covalent, the solubility equilibrium represented by the saturated solution of an inorganic salt rarely finds a counterpart in organic chemistry. Somewhat analogous, however, and of considerable significance, is the equilibrium involving the "partitioning" of an organic compound between two immiscible solvents. This is the basis for separations by liquid/liquid extraction and by partition chromatography. The equilibrium is described by an equilibrium constant expression referring to a so-called "distribution coefficient", symbolized by K_D. The most common system involves distribution of the dissolved organic compound between an aqueous solution and an organic solvent, such as ether or chloroform. By convention, this equilibrium is represented by

$$\text{compound}_{(aq)} \rightleftarrows \text{compound}_{(org)}$$

for which:

$$K_D = \frac{C_{\text{compound(org)}}}{C_{\text{compound(aq)}}}$$

The distribution coefficient may be roughly approximated from measurement of the solubility of the compound in water and in the pure organic solvent used:

$$K_D \approx \frac{S_{\text{compound(org)}}}{S_{\text{compound(aq)}}}$$

This is only a rough approximation because the two liquids used in an extraction are each saturated with the other, and this changes somewhat the solvent properties of both.

The distribution coefficient is quite useful, even when it is only approximately known, in calculations involving extraction efficiency. Such calculations may be employed in planning extraction schemes for the isolation of a compound from a mixture or for determining optimum experimental conditions of relative solvent volumes or number of separate extractions to be employed. A mathematical expression has been developed for such calculations[1]:

$$f_{\text{extracted}} = 1 - \left[\frac{1}{K_D[V_{(org)}/V_{(aq)}] + 1} \right]^n$$

---

[1]If you are interested in the derivation of this equation, consult your instructor or see Unit 18 of Reference 1 (page 398).

This expression shows that the fraction (f) of a compound extracted from its aqueous solution is a function of three factors: the magnitude of the distribution coefficient (K_D), the relative volumes of organic solvent ($V_{(org)}$) and aqueous solution ($V_{(aq)}$) used, and the number of separate extractions (n) employed, each using "fresh" solvent. The extraction efficiency (expressed by $f_{extracted}$) will obviously be particularly favorable when multiple extractions are employed, using an organic solvent selected for a large K_D-value for the organic solution/aqueous solution equilibrium system.

Other types of equilibria of considerable importance include:

ACID/BASE EQUILIBRIA (aqueous[2] systems) [Compare Unit 11]

Examples

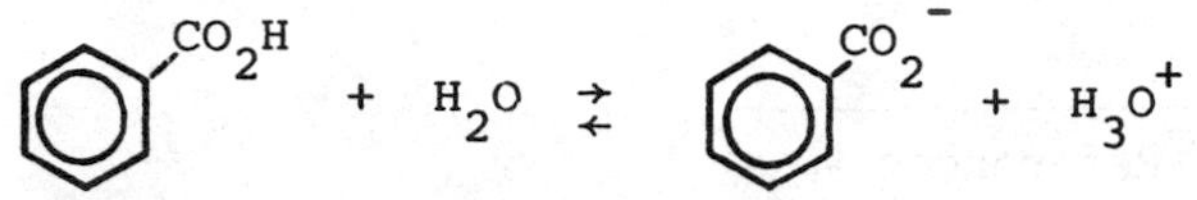

(benzoic acid in water, K_a = 6.6 x 10^{-5})

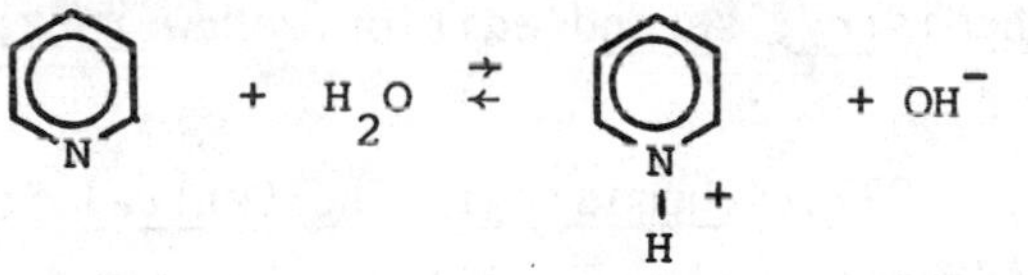

(pyridine in water, K_b = 2.0 x 10^{-9})

COMPLEX ION EQUILIBRIA [Compare Unit 12]

Example

$H_2NCH_2CH_2NH_2 + 2H_2O + Cu^{2+} \rightleftarrows$

(formation of an ethylenediamine complex)

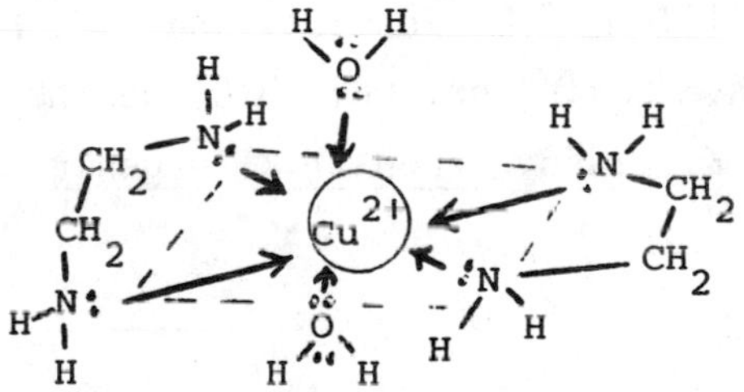

SIMULTANEOUS EQUILIBRIA [Compare Unit 13]

Example

$$CH_3\overset{O}{\overset{\|}{C}}OCH_2CH_3 + H_2O \rightleftarrows CH_3\overset{O}{\overset{\|}{C}}OH + HOCH_2CH_3$$

$$CH_3\overset{O}{\overset{\|}{C}}OH \rightleftarrows CH_3\overset{O}{\overset{\|}{C}}O^- + H^+$$

$$H_2O \rightleftarrows H^+ + OH^-$$

(ester hydrolysis, catalyzed by either acid or base)

---

[2]Similar equilibria are encountered with a number of solvent systems other than water, but the ionization constant varies appreciably with the solvent used.

<u>OXIDATION-REDUCTION PROCESSES</u> [Compare Unit 14]

Example

$$8CO_3^{2-} + 42H_2O + 50e^- \underset{\leftarrow}{\rightarrow} 66OH^- + CH_3\overset{\overset{\textstyle CH_3}{|}}{C}HCH_2CH_2CH_2CH_2CH_3, \quad E° = -0.72$$

$$O_2 + 2H_2O + 4e^- \underset{\leftarrow}{\rightarrow} 4OH^-, \quad E° = +0.40$$

(Standard Electrode Potentials for the isooctane/oxygen fuel cell, page 207)

We shall employ this final Unit as a review of some of the aspects of chemical equilibrium introduced in Units 10-14, using organic chemistry as the area of applications. You should probably review the use of "chemical shorthand" discussed in Unit 9 before beginning your work.

- References -

1. O'Connor, Rod. 1977. <u>Fundamentals of Chemistry</u>, second edition. New York: Harper & Row (Units 18, 28, 29, 31 and Excursion 4)

2. Brown, Theodore, and H. Eugene LeMay, Jr. 1977. <u>Chemistry: The Central Science</u>. Englewood Cliffs, N.J.: Prentice-Hall (Chapters 14, 17, 24)

3. Masterton, William, and Emil Slowinski. 1977. <u>Chemical Principles</u>, fourth edition. Philadelphia: W.B. Saunders (Chapters 10, 20, 21)

4. Nebergall, William, F.C. Schmidt, and H.F. Holtzclaw, Jr. 1976. <u>College Chemistry</u>, fifth edition. Lexington, MA: D.C. Heath (Chapters 17, 27, 28)

5. Stille, John K. 1968. <u>Industrial Organic Chemistry</u>. Englewood Cliffs, N.J.: Prentice-Hall

----

OBJECTIVES:

(1) *Be able to apply the <u>competency</u> level OBJECTIVES of Units 10-14 to processes involving organic compounds.*

*(2) *Be able to apply the <u>proficiency</u> level OBJECTIVES of Units 10-14 to processes involving organic compounds.*

- -

There is no PRE-TEST for this Unit. Rather, we will trace through a few examples and exercises to review earlier methods as applied to organic systems. The final Self-Test will let you determine how well you have accomplished your goals of "competency" or "proficiency" in the area of chemical EQUILIBRIUM.

- -

*Proficiency Level

EXAMPLE 1 (Heterogeneous Equilibria)

So-called "decaffeinated" tea or coffee preparations can be made by extracting the caffeine from an aqueous solution of the tea or coffee by chloroform. The recovered caffeine is itself a valuable commercial product, sold in various "stimulants", such as NO-DOZ, or in multi-ingredient analgesics, such as ANACIN.

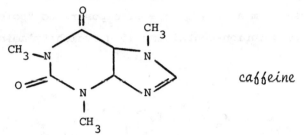

caffeine

The solubilities of caffeine in water and in chloroform, respectively, are 1.35 g per 100 ml (H_2O) and 14.2 g per 100 ml($CHCl_3$). What is the approximate distribution coefficient for caffeine in a water/chloroform system?

SOLUTION:

For the equilibrium system represented by:

$$CAFFEINE_{(aq)} \; \underset{\leftarrow}{\rightarrow} \; CAFFEINE_{(CHCl_3)}$$

the __approximate__ distribution coefficient is:

$$K_D \; \propto \; \frac{SOLUBILITY_{Caffeine\,(CHCl_3)}}{SOLUBILITY_{Caffeine\,(H_2O)}}$$

so that:

$$K_D \; \approx \; \frac{14.2 \text{ g(per 100 ml)}}{1.35 \text{ g(per 100 ml)}} \; \approx \; \underline{10.5}$$

- -

EXERCISE 1 (Acid/Base Equilibrium)

Artificial sweeteners have found extensive markets with persons requiring low-carbohydrate diets for health or cosmetic reasons. Two such sweeteners are widely known:

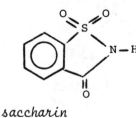

saccharin

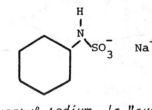

sucaryl sodium (a "cyclamate")

Cyclamates, formed from cyclohexylamine, were removed from the market a few years ago because of evidence suggesting a possible link between high cyclamate intake and cancer of the bladder. Since then, there has been some controversy over the validity of the tests employed and the issue is not yet fully resolved. The "parent" cyclohexylamine is a weak base with a pronounced "fishy" odor and, on contact with the skin, this amine can produce itching, watery blisters quite similar to those from a severe case of exposure to "poison ivy". What is the approximate pH of an aqueous solution containing 15 g cyclohexylamine per liter?

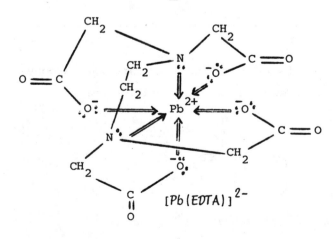

$$+ \; H_2O \; \rightleftarrows \qquad + \; OH^-, \quad K_b = 4 \times 10^{-4}$$

(answer, page 410)

EXERCISE 2 (Complex Ions)

Cases of lead poisoning in children have been reported in increasing numbers over the past few years, in spite of controls reducing the use of lead-based paints on items accessible to children. Recently, evidence has been reported which suggests that many cases of lead poisoning may be attributed to the inhalation of lead products from automobile emissions in the air near heavily used freeways. A treatment which is often effective for lead poisoning involves intramuscular injection of a solution of <u>versene</u>, more commonly known as <u>EDTA</u>. This reagent forms a very strong water-soluble complex with the Pb^{2+} ion, which is excreted in the urine. <u>Versene</u> is most effective at a pH above 7, since it is the 4- anion which acts best as a complexing agent. The instability constant for $[Pb(EDTA)]^{2-}$ is 9.1×10^{-19}. What is the molarity of free Pb^{2+} in a solution which is 2×10^{-4} <u>M</u> in $[Pb(EDTA)]^{2-}$ and 0.10 <u>M</u> in $EDTA^{4-}$?

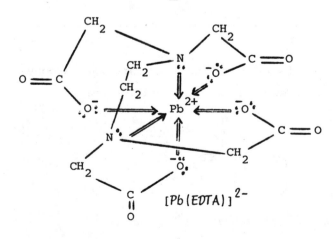

$[Pb(EDTA)]^{2-}$

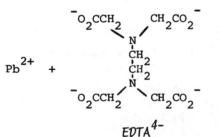

$EDTA^{4-}$

(ethylenediaminetetraacetate)

(answer, page 411)

400

EXERCISE 3 (Simultaneous Equilibria)

Acetylcholine is a "molecular cation" present in nerve cells in an inactive combination with protein. When the nerve cell receives a proper stimulus, acetylcholine is released in a cell-by-cell sequence along a nerve fiber to transmit the "nerve impulse". As each impulse passes, the acetylcholine must be destroyed by an enzyme-catalyzed hydrolysis:

$$CH_3\overset{O}{\overset{\|}{C}}OCH_2CH_2N(CH_3)_3^{+} + H_2O \xrightarrow{\text{(enzyme)}} CH_3CO_2H + HOCH_2CH_2N(CH_3)_3^{+}$$

(acetylcholine)   (choline)

If this hydrolysis does not occur properly, the accumulation of acetylcholine blocks further transmission of the nerve impulse and paralysis occurs. Most of the so-called "nerve-gases" developed as chemical warfare agents act by preventing the enzymatic hydrolysis of acetylcholine.

"Free" acetylcholine itself occurs in solution as a strong (ionic base):

$$CH_3\overset{O}{\overset{\|}{C}}OCH_2CH_2N(CH_3)_3^{+} \quad OH^{-}$$

Because of the high pH of such solutions, water used in biological experiments with free acetylcholine must be quite pure to avoid the precipitation of insoluble metal hydroxides.

What is the maximum concentration of Fe^{3+} which could exist in 0.10 $\underline{M}$ solution of "free" acetylcholine without danger of precipitation of $Fe(OH)_3$? [For K_{sp}-data, see Appendix D.]

(answer, page 411)

EXERCISE 4 (Oxidation-Reduction Processes)

The isooctane/oxygen fuel cell (pages 207, 398) is a more efficient way of "tapping the energy" of isooctane combustion than is a typical burning process. The fuel cell is, however, limited to rather slow rates of energy release, a problem inherent in the present "state of the art" of fuel cell technology. What voltage could be expected from such a cell under the conditions of injection of isooctane vapor and oxygen gas at 0.50 atm each through an electrolyte solution of pH 13.0 which is 0.50 $\underline{M}$ in carbonate ion? [For the cell equation and Standard Electrode Potentials, see page 398.]

(answer, page 411)

===

At this point you should try the competency level Self-Test questions on page 405.

===

*EXAMPLE 2 (Heterogeneous Equilibria)

Caffeine, for use as a stimulant in a variety of products ranging from "cola" beverages to nonprescription "pep pills", is obtained in large quantities as a by-product from the preparation of "decaffeinated" coffee. The coffee solution prepared from ground coffee beans may be extracted with chloroform to remove caffeine. Evaporation, or more commonly lyophilization ("freeze-drying"), of the aqueous solution produces so-called "instant coffee", while the caffeine is recovered by evaporation of the chloroform extract. The distribution coefficient for caffeine in a water/chloroform system is approximately 11. How many grams of caffeine could be extracted from 25 liters of an aqueous solution containing 12 g of caffeine per liter by 15 liters of chloroform, in a single extraction? [For the structural formula of caffeine, see EXERCISE 1.]

SOLUTION:

For the equilibrium system represented by:

$$CAFFEINE_{(aq)} \rightleftarrows CAFFEINE_{(CHCl_3)}$$

$$K_D = \frac{C_{caffeine(CHCl_3)}}{C_{caffeine(aq)}} = 11$$

and the efficiency of extraction (page 396) is:

$$f_{extracted} = 1 - \left[\frac{1}{K_D[V_{(org)}/V_{(aq)}] + 1}\right]^n$$

For a single extraction (n = 1) using 15 liters of $CHCl_3$ with 25 liters of aqueous solution:

$$f_{extracted} = 1 - \left[\frac{1}{11(15/25) + 1}\right]^1 = 1 - 0.13 = 0.87$$

Then, the amount of caffeine extracted is:

$$0.87 \times \frac{12 \text{ g}}{\text{liter}} \times 25 \text{ liters} = \underline{260 \text{ g}}$$

*Proficiency Level

*EXERCISE 5 (Acid/Base Equilibria)

Scopolamine, an ingredient of some of the nonpresciption "sleeping tablets", will be recognized by fans of World War II spy stories as the active principal of "truth serum". Intravenous injection of a buffered scopolamine solution apparently produces a semicomatose state in which the recipient may become sufficiently uninhibited to respond to questioning with information he would not consciously reveal. What would be the approximate pH of a solution prepared by adding 15 g of scopolamine to 250 ml of 0.10 M hydrobromic acid?

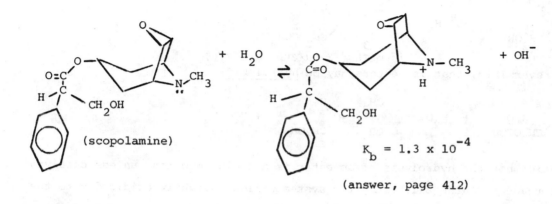

(scopolamine)   + H$_2$O   ⇌ $K_b = 1.3 \times 10^{-4}$ + OH$^-$

(answer, page 412)

*EXERCISE 6 (Complex Ions)

A few years ago an entire family was found to have severe lead poisoning from a most unusual source. The low-income family had found a free source of fuel, discarded empty battery cases. The lead compounds released into the air when the cases were burned were being inhaled on a regular basis, resulting in severe symptoms of lead poisoning. Once the problem had been identified, immediate EDTA therapy was started (see EXERCISE 2). This treatment is generally effective in quickly increasing the excretion of lead, as the EDTA complex, in the urine. To illustrate the very strong affinity of EDTA^{4-} for Pb^{2+}, estimate the molarity of "free" Pb^{2+} in a solution prepared by mixing equal volumes of 0.20 M EDTA^{4-} and 0.040 M Pb^{2+}. The formation constant for [Pb(EDTA)]$^{2-}$ is 1.1×10^{-18}.

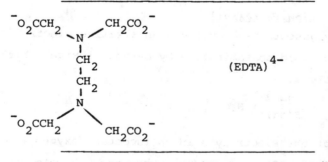

(EDTA)$^{4-}$

(answer, page 413)

---

*EXERCISE 7 (Simultaneous Equilibria)

Esters are, perhaps, the most pleasant of organic compounds in terms of odors. Many of the aromas of natural fruits and flowers, for example, are due to simple esters. Octyl acetate has the odor of oranges, while apricots owe their characteristic aroma to amyl butyrate. Many esters, both natural and synthetic, are used as flavoring agents. The unique aroma of JUICY FRUIT gum is due to an additive called 3-methyl-2-butenyl acetate. Esters may be formed by reaction between an alcohol and a carboxylic acid, for example:

$$\begin{array}{c}CH_3 \\ \end{array}C=C\begin{array}{c}H \\ \\ CH_2OH\end{array} \;+\; HO\overset{O}{\overset{\|}{C}}CH_3 \;\rightarrow\; \begin{array}{c}CH_3 \\ \end{array}C=C\begin{array}{c}H \\ \\ CH_2O\overset{O}{\overset{\|}{C}}CH_3\end{array} \;+\; H_2O$$

$$\text{(3-methyl-2-butenyl acetate)}$$

This reaction is reversible, that is, esters may be <u>hydrolyzed</u>:

$$\begin{array}{c}CH_3 \\ \end{array}C=C\begin{array}{c}H \\ \\ CH_2O\overset{O}{\overset{\|}{C}}OH_3\end{array} \;+\; H_2O \;\rightarrow\; \begin{array}{c}CH_3 \\ \end{array}C=C\begin{array}{c}H \\ \\ CH_2OH\end{array} \;+\; HO\overset{O}{\overset{\|}{C}}CH_3$$

Neither the formation nor the hydrolysis of an ester is a rapid reaction, unless catalyzed by an appropriate reagent. At equilibrium, the system will be slightly acidic due to the presence of the weak carboxylic acid in the mixture. What is the approximate pH of the equilibrium system resulting when a 13 g sample of 3-methyl-2-butenyl acetate is dissolved in 500 ml of water?

$$\begin{array}{c}CH_3 \\ \end{array}C=C\begin{array}{c}H \\ \\ CH_2O\overset{O}{\overset{\|}{C}}CH_3\end{array} \;+\; H_2O \;\rightleftarrows\; \begin{array}{c}CH_3 \\ \end{array}C=C\begin{array}{c}H \\ \\ CH_2OH\end{array} \;+\; CH_3CO_2H \qquad K_{eq} = \dfrac{C_{acid} \times C_{alcohol}}{C_{ester}}$$

$$= 0.17$$

$$CH_3CO_2H \;\rightleftarrows\; CH_3CO_2^- + H^+ \qquad\qquad K_a = 1.8 \times 10^{-5}$$

(answer, page 414)

---

*EXERCISE 8 (Oxidation-Reduction Processes)

Fuel cells which employ gaseous reactants and an alkaline electrolyte require careful purification of the gases to avoid contamination by carbon dioxide, which could cause rapid deterioration of the electrolyte:

$$2\,OH^-_{(aq)} + CO_{2(g)} \;\rightleftarrows\; CO_3^{2-}_{(aq)} + H_2O_{(\ell)}$$

What is the approximate pH of the electrolyte in an isooctane/oxygen fuel cell (page 398) when the cell potential is 1.06 volt, the gas pressures are 0.50 atm each, and the carbonate concentration is 1.0 $\underline{M}$? [For necessary information, see page 398.]

(answer, page 415)

= =

SELF-TEST (UNIT 15) [answers, page 415]

Americans consume more aspirins per capita than any other people on earth, which sometimes makes you wonder if too much prosperity might be "hazardous to your health". Although the analgesic (pain depressing) and antipyretic (fever reducing) properties of aspirin have been recognized for many years, we still have an incomplete understanding of how aspirin functions in the body. We do know that it can be poisonous in large dosages[3] (hence the "child-proof" bottle cap) and recent evidence suggests that excessive or long-term use may cause serious health problems. Aspirin is still, however, generally considered one of the safer nonprescription drugs.

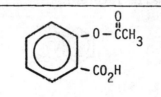

acetylsalicylic acid

(aspirin)

The _competency_ level Self-Test questions will deal with some aspects of the chemistry of aspirin in testing your achievement of the OBJECTIVES of Units 10-14, as applied to equilibrium systems involving organic compounds.

15.1. The sodium or calcium salt of acetylsalicylic acid is normally used in making "aspirin tablets", since the molecular acid is not very soluble in the aqueous fluid of the stomach. Advantage is taken of the low water solubility and high ether solubility of aspirin in extractive purifications employed by research groups seeking improved methods of aspirin synthesis and purification. If the solubility of aspirin is 0.35 g per 100 ml (H_2O) and 8.4 g per 100 ml (ether), what is the approximate distribution coefficient of acetylsalicylic acid in a water-ether system?

- -

[3]Single dose of 10-30 g usually fatal.

15.2. Acetylsalicylic acid itself is a weak acid, $K_a = 3.3 \times 10^{-4}$. What is the approximate pH of a saturated aqueous solution of aspirin? [solubility = 0.35 g per 100 ml (H_2O).]

15.3. Although we do not yet know the complete details of the biochemical function of aspirin, some aspects of its behavior are well established. We know, for example, that aspirin is fairly rapidly hydrolyzed[4]:

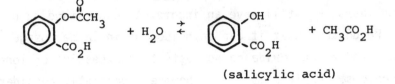

(salicylic acid)

Once the salicylic acid enters the bloodstream, which is buffered around pH 7.4, it exists mainly as the salicylate anion:

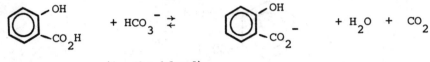

(in the blood)

This anion is capable of forming a complex with Cu^{2+} ion. It appears that one of the functions of aspirin is to furnish salicylate ion to "scavenge" Cu^{2+} from the blood, transfer it as a complex to various cells from which it was lost, and release it there from the complex:

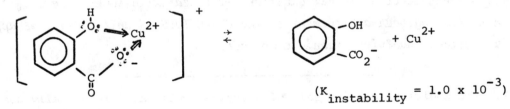

$(K_{instability} = 1.0 \times 10^{-3})$

What is the approximate concentration of "free" Cu^{2+} in a solution which is found to contain 1.6×10^{-5} moles per liter of the salicylate/Cu^{2+} complex and 2.5×10^{-2} moles per liter of the salicylate anion?

15.4. Acetylsalicylic acid itself is not very soluble in water (question 15.1), so most aspirin manufacturers sell it as either the sodium or calcium salt. Of the two, the calcium salt is the more soluble (17 g per 100 ml H_2O). Care must

---

[4]This hydrolysis may occur in bottles of aspirin which have been opened frequently in areas of high relative humidity. Open an "old" bottle of aspirin sometime and see if you notice the "vinegar" odor of acetic acid.

be exercised in the production of "soluble aspirin" (calcium acetylsalicylate) to avoid the contamination of the product from a number of sources. The hydrolysis of aspirin is particularly rapid at pH 8.5 or higher, so the reaction of calcium hydroxide with aspirin must be carried out very quickly and the product must be immediately dried to minimize hydrolysis. To avoid contamination of the product by inorganic salts, the water must be free of metal ions which would from insoluble hydroxides and CO_2 must be avoided because of its reaction with calcium hydroxide to form the insoluble $CaCO_3$. If a saturated solution of calcium hydroxide is to be used in the preparation of "soluble aspirin", what is the pH of that solution? [K_{sp} for $Ca(OH)_2$ is 4.0×10^{-6}]

- -

15.5. The measurement of very low concentrations of metal ion, such as that of "free" Cu^{2+} ion in a solution of the salicylate/Cu^{2+} complex, is difficult by conventional "wet chemical" methods. Precipitation or titration would be unsatisfactory because either would disturb the equilibrium system. However, the solution can be assayed by an electrochemical technique, since a voltage measurement can be made under conditions of negligible current flow so that, for all practical purposes, it is the <u>equilibrium</u> concentration of the free metal ion which is being measured. What would be the approximate voltage of an electrochemical cell in which one electrode is zinc immersed in a 0.10 <u>M</u> Zn^{2+} solution and the other is copper immersed in a solution of the salicylate/Cu^{2+} complex having a "free" Cu^{2+} concentration of 3.0×10^{-6} <u>M</u>?

Cell discharge equation:

$$Cu^{2+} + Zn \rightarrow Cu + Zn^{2+}$$

Standard Electrode Potentials:

$$Cu^{2+} + 2e^- \rightleftarrows Cu, \quad E° = +0.34$$
$$Zn^{2+} + 2e^- \rightleftarrows Zn, \quad E° = -0.76$$

==

* By the mid-1960's a heated controversy had developed between animal conservationists on the one hand and cattlemen on the other, concerning the use of <u>strychnine</u> in the control of coyote populations. From the point of view of many cattlemen and ranchers, the coyote was fast becoming a significant economic hazard. Loss of calves and other small livestock to coyotes (whose natural prey had been depleted by hunters and the extensive conversion of "wild territory" to farms, ranches, and metropolitan areas) was increasing every year. The wily coyote had learned to avoid traps and hunters, so some more drastic solution was

*Proficiency Level

needed. The answer to the problem appeared to be the use of "poisoned bait", generally employing strychnine or one of its salts. Conservationists and animal lovers soon discovered the results of this practice. Not only was the strychnine an "inhumane" poison, producing agonizing convulsions prior to death, but the bait was non-selective. Not only were coyotes being killed in increasing numbers (possibly to the extent of removing a major source of population control of rodents), but the bait was also being eaten by bears, mountain lions, hawks, and eagles. The problem has not yet been fully resolved and extensive research must continue to investigate ways of predator-control which are non-destructive to the natural ecosystem.

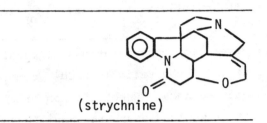

(strychnine)

The proficiency level Self-Test questions will deal with some aspects of the chemistry of strychnine in testing your achievement of the OBJECTIVES of Units 10-14, as applied to equilibrium systems involving organic compounds.

*15.6. Strychnine has been made synthetically [Woodward, et al, J. Am. Chem. Soc., 76, 4749(1954)], but the synthesis is not amenable to large-scale processes. As a result, the strychnine of commerce is obtained by extraction from seeds of *Strychnos nux-vomica L.*, or other species of *Strychnos*. The distribution coefficient of strychnine in a water/benzene system is 36. How many milligrams of strychnine could be extracted from 25 liters of an aqueous "seed extract" which contains 0.16 g (strychnine) per liter by a single extraction with 5.0 liters of benzene?

- -

*15.7. Because of the low water solubility of strychnine, it is usually prepared as a salt. These compounds are sold for poisons, but also for use in veterinary medicine in tonics and stimulants for debilitated animals. Caution must be exercised in strychnine therapy because of the small difference between a therapeutic and a toxic dosage (e.g., a 5 mg dosage could be used for a sheep, but a 35 mg dosage would prove fatal). What is the approximate pH of a pharmaceutical stock solution prepared by dissolving 600 mg of strychnine in 1.00 liter of 5.0×10^{-4} M nitric acid? (The basic character of strychnine is attributed primarily to only one of its nitrogen atoms.)

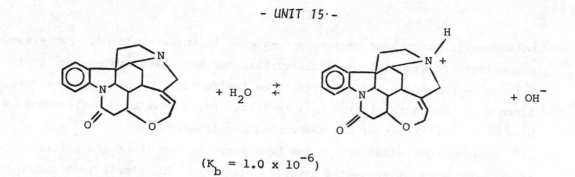

$$(K_b = 1.0 \times 10^{-6})$$

---

*15.8. The widespread use of strychnine in predator control has resulted in occasional cases of accidental strychnine poisoning in children. If the child is found with a labeled container, immediate medical attention (usually involving intravenous injection of one of the short-acting barbiturate drugs) may avoid a fatal reaction. If, however, there is some doubt as to the nature of the poison ingested, treatment may have to await the results of analytical tests. A slightly acidic solution of potassium hexacyanoferrate(II) may be used to precipitate strychnine from a solution of the free base or any of its soluble salts. The formation of the characteristic pale yellow needles of this precipitate can be used as a quick test on a solution suspected of containing strychnine, although some of the other plant alkaloids give similar results.

What is the approximate molarity of "free" Fe^{2+} in a solution of this reagent prepared by adding 5.0 ml of 0.10 $\underline{M}$ $FeSO_4$ to 15.0 ml of 2.86 $\underline{M}$ KCN? [The net formation constant for $[Fe(CN)_6]^{4-}$ is 1.0×10^{24}.]

---

*15.9. One of the more common strychnine compounds sold commercially is the sulfate salt:

$$(C_{21}H_{23}N_2O_2)_2SO_4 \cdot 5H_2O$$

A solution of this salt in water is slightly acidic due to the reaction:

$$C_{21}H_{23}N_2O_2^+ \rightleftharpoons H^+ + strychnine$$

The sulfate salt has a solubility product of 1.5×10^{-4}. What is the approximate pH of a saturated aqueous solution of the salt, assuming that hydrolysis of SO_4^{2-} is negligible? [For K_b for strychnine, see question 15.7.]

---

*15.10. The initial symptoms of strychnine poisoning may sometimes be mistaken for those of a number of other poisons often found on farms and ranches. It is important to identify the particular chemical involved in a case of accidental poisoning if the proper emergency treatment is to be employed. For arsenic poisoning, for example, the administration of the barbiturates used in treating strychnine poisoning would generally be ineffective. A specific antidote for arsenic poisoning is British Anti-Lewisite (BAL), developed during World War II because of the threat of war gas containing arsenic.

An electrochemical method has been proposed for the rapid determination of arsenic in cases of suspected arsenic poisoning. The sample is treated with zinc and hydrochloric acid to generate arsine gas. This is swept from the reaction vessel by a stream of helium, which then passes through an electrolyte containing an electrode made from a metallic form of arsenic. When the arsine/arsenic electrode is connected to another electrode of known potential, a cell voltage reading will permit a quantitative calculation of the arsenic content of the original sample.

What is the partial pressure of arsine gas in a cell consisting of the arsine/arsenic electrode buffered at pH 2.0 and a silver/0.10 $\underline{M}$ Ag^+ electrode when the cell voltage measured is 1.29 volts?

Cell discharge equation:

$$AsH_{3(g)} + 3Ag^+_{(aq)} \rightarrow As_{(s)} + 3Ag_{(s)} + 3H^+_{(aq)}$$

Standard Electrode Potentials:

$$As + 3H^+ + 3e^- \rightarrow AsH_3, \quad E° = -0.54$$
$$Ag^+ + e^- \rightarrow Ag, \quad E° = +0.80$$

ANSWERS to EXERCISES, Unit 15

1. (11.9) Solution: (following the method introduced in Unit 11.)
First, to simplify our work, we will "count atoms" and write molecular formulas as described in Unit 9:

$$C_6H_{13}N + H_2O \rightleftharpoons C_6H_{14}N^+ + OH^-$$

$$(K_b = 4 \times 10^{-4})$$

for which:

$$K_b = \frac{C_{C_6H_{14}N^+} \times C_{OH^-}}{C_{C_6H_{13}N}} \simeq \frac{C^2_{OH^-}}{\underline{M}(C_6H_{13}N)}$$

- UNIT 15 -

Using a mole/mass unity factor:

$$\underline{M}(C_6H_{15}N) = \frac{15\ g(C_6H_{13}N)}{1\ liter} \times \frac{1\ mole(C_6H_{13}N)}{99.2\ g(C_6H_{13}N)} = 0.15\ \underline{M}$$

so that:

$$4 \times 10^{-4} = \frac{C_{OH^-}^2}{0.15}$$

$$C_{OH^-} = \sqrt{(4 \times 10^{-4}) \times 0.15}$$

$$pOH = -\tfrac{1}{2} \log(6 \times 10^{-5}) = 2.1$$

$$pH = 14.00 - pOH = 14.00 - 2.1$$

----

2. $(1.8 \times 10^{-21}\ \underline{M})$ Solution: (following the method introduced in Unit 12.)
For the system represented by:

$$[Pb(EDTA)]^{2-} \rightleftarrows Pb^{2+} + EDTA^{4-}$$

$$K_{instability} = \frac{C_{Pb^{2+}} \times C_{EDTA^{4-}}}{C_{[Pb(EDTA)]^{2-}}}$$

From which, solving for $C_{Pb^{2+}}$ gives:

$$C_{Pb^{2+}} = \frac{K_{instability} \times C_{[Pb(EDTA)]^{2-}}}{C_{EDTA^{2-}}} = \frac{(9.1 \times 10^{-19}) \times (2 \times 10^{-4})}{0.10}$$

----

3. $[6.0 \times 10^{-35}\ \underline{M}\ (Fe^{3+})]$ Solution: (following the method introduced in Unit 13)
From Appendix D:

$$K_{sp} = 6.0 \times 10^{-38} = C_{Fe^{3+}} \times C_{OH^-}^3$$

for $Fe(OH)_{3(s)} \rightleftarrows Fe^{3+}_{(aq)} + 3OH^-_{(aq)}$

Since 0.10 $\underline{M}$ "free" acetylcholine would furnish 0.10 mole per liter of OH^-, precipitation of $Fe(OH)_3$ can be avoided only if the molarity of Fe^{3+} is equal to or less than:

$$C_{Fe^{3+}} = \frac{K_{sp}}{C_{OH^-}^3} = \frac{6.0 \times 10^{-38}}{(0.10)^3}$$

----

4. (~1.10 volt) Solution: (following the method introduced in Unit 14.)
Writing the molecular formula for isooctane (C_8H_{18}), the cell equation and

electrode potentials are expressed as:

$$2C_8H_{18(g)} + 25\ O_{2(g)} + 32\ OH^-_{(aq)} \rightleftharpoons 16\ CO_3^{2-}_{(aq)} + 34\ H_3O_{(\ell)}$$

$$8CO_3^{2-} + 42\ H_2O + 50e^- \rightleftharpoons C_8H_{18} + 66\ OH^-, \quad E° = -0.72$$

$$O_2 + 2\ H_2O + 4e^- \rightleftharpoons 4\ OH^-, \quad E° = +0.40$$

Since isooctane is converted to carbonate, the _reverse_ of that electrode equation:

$$E° = (+0.40) - (-0.72) = +1.12 \text{ volts}$$

The electron transfer for the balanced equation is:

$$n = 2 \times 50e^- = 100$$

$$(or\ 25 \times 4e^- = 100)$$

Using P to represent gas pressure (in atm) and C to represent the molar concentration of dissolved species:

$$Q = \frac{C^{16}_{CO_3^{2-}}}{P^2_{C_8H_{18}} \times P^{25}_{O_2} \times C^{32}_{OH^-}} = \frac{(0.50)^{16}}{(0.50)^2 \times (0.50)^{25} \times (0.10)^{32}}$$

(This expression will be easiest to evaluate by logarithms)

Substituting in the Nernst Equation:

$$E = E° - \frac{0.0592}{n} \log Q$$

$$E = 1.12 - \frac{0.0592}{100} \log\left[\frac{(0.50)^{16}}{(0.50)^{27} \times (0.10)^{32}}\right] = 1.12 - \frac{0.0592}{100} \log\left[\frac{(0.50)^{-11}}{10^{-32}}\right]$$

$$E = 1.12 - (5.92 \times 10^{-4})\ [-11 \log(0.50) - \log 10^{-32}]$$

$$E = 1.12 - (5.92 \times 10^{-4})\ [3.3 - (-32)]$$

$$E = 1.12 - [(5.92 \times 10^{-4}) \times 35.3] = 1.12 - 0.02$$

- -

*5. (~10.1) <u>Solution</u>: (following the method introduced in Unit 11)

To simplify the problem, we will "count atoms" (Unit 9) and use molecular formulas:

$$C_{17}H_{21}NO_4 + H_2O \rightleftharpoons C_{17}H_{22}NO_4^+ + OH^-$$

$$(\text{scopolamine}) \qquad [K_b = 1.3 \times 10^{-4}]$$

for which:

$$K_b = \frac{c_{C_{17}H_{22}NO_4^+} \times c_{OH^-}}{c_{C_{17}H_{21}NO_4}}$$

Now, to determine the composition of the final solution, we must first recognize that addition of scopolamine to hydrobromic acid will result in an acid/base reaction:

$$C_{17}H_{21}NO_4 + H^+ \rightarrow C_{17}H_{22}NO_4^+$$

We have available initially:

scopolamine -

$$\frac{15 \ g(C_{17}H_{21}NO_4)}{1} \times \frac{10^3 \ mg}{1 \ g} \times \frac{1 \ mmol(C_{17}H_{21}NO_4)}{303.4 \ mg(C_{17}H_{21}NO_4)} = 49 \ mmols$$

hydrobromic acid -

$$\frac{0.10 \ mmol(HBr)}{1 \ ml} \times \frac{250 \ ml}{1} = 25 \ mmols$$

We may assume, then, that at equilibrium:

$C_{17}H_{21}NO_4$ -

    49 mmols - 25 mmols = 24 mmols

    (original) (consumed)

$C_{17}H_{22}NO_4^+$ -

    25 mmols

    (newly formed)

This base/salt system corresponds to the characteristics of a conjugate-pair buffer, for which:

$$C_{OH^-} = K_b \times \frac{mmols(C_{17}H_{21}NO_4)}{mmols(C_{17}H_{22}NO_4^+)} = (1.3 \times 10^{-4}) \times \frac{24}{25} = 1.2 \times 10^{-4}$$

for which:

$$pOH = -\log C_{OH^-} = -\log(1.2 \times 10^{-4}) = 3.9$$

and:

$$pH = 14.00 - pOH = 14.00 - 3.9$$

- -

*6.   [2.3×10^{-19} $\underline{M}(Pb^{2+})$] Solution: (following the method introduced in Unit 12)

For the reaction represented by:

$$Pb^{2+} + EDTA^{4-} \rightleftharpoons [Pb(EDTA)]^{2-}$$

$$K_{formation} = 1.1 \times 10^{18} = \frac{C_{[Pb(EDTA)]^{2-}}}{C_{Pb^{2+}} \times C_{EDTA^{4-}}}$$

from which:

$$C_{Pb^{2+}} = \frac{C_{[Pb(EDTA)]^{2-}}}{(1.1 \times 10^{18}) \times C_{EDTA^{4-}}}$$

Assuming that the large excess of $EDTA^{4-}$ converts most of the original Pb^{2+} to the complex:

after mixing, before reaction -

$$C_{Pb^{2+}} = \frac{0.040 \text{ M}}{1} \times \frac{1}{2} = 0.020 \text{ M}$$

$$C_{EDTA^{4-}} = \frac{0.20 \text{ M}}{1} \times \frac{1}{2} = 0.10 \text{ M}$$

at equilibrium -

$$(Pb^{2+} + EDTA^{4-} \xrightarrow{\text{(mostly)}} [Pb(EDTA)]^{2-})$$

$$C_{[Pb(EDTA)]^{2-}} = 0.020 \text{ M}$$
$$\text{(newly formed)}$$

$$C_{EDTA^{4-}} = 0.10 \text{ M} - 0.020 \text{ M} = 0.08 \text{ M}$$
$$\text{(original) (consumed)}$$

Then:

$$C_{Pb^{2+}} = \frac{0.020}{(1.1 \times 10^{18}) \times (0.08)}$$

- -

*7. (~2.9) Solution:

First, we need to convert the structural formula of the ester to the more convenient molecular formula:

$$C_{17}H_{12}O_2$$

Then, using a mole/mass unity factor, we may find the initial concentration of the ester:

$$\frac{13 \text{ g}(C_7H_{12}O_2)}{0.500 \text{ liter}} \times \frac{1 \text{ mole}(C_7H_{12}O_2)}{128.2 \text{ g}(C_7H_{12}O_2)} = 0.10 \text{ M}$$

Since the acetic acid is only slightly ionized, we may approximate its concentration, represented by y:

at equilibrium

$$C_{CH_3CO_2H} = C_{alcohol} = y$$

$$C_{C_7H_{12}O_2} = 0.10 - y$$

$$K_{eq} \simeq \frac{y^2}{0.10 - y}$$

$$\frac{y^2}{0.10 - y} \simeq 0.17$$

from which:

$$y \simeq 0.070 \ \underline{M}$$

Then, from the ionization of acetic acid:

$$K_a = \frac{C_{H^+} \times C_{CH_3CO_2^-}}{C_{CH_3CO_2H}} \simeq \frac{C_{H^+}^2}{\underline{M} \ (CH_3CO_2H)}$$

from which:

$$C_{H^+} \simeq \sqrt{K_a \times \underline{M}(CH_3CO_2H)} \simeq \sqrt{(1.8 \times 10^{-5}) \times (0.07)}$$

$$pH \simeq -\tfrac{1}{2} \log(1.3 \times 10^{-6})$$

---

*8. (~11.1) <u>Solution</u>: (following the method introduced in Unit 14)

For the values of $\underline{E}^\circ$ *and* $\underline{n}$ *and the <u>form</u> of* $\underline{Q}$, *see the Solution to EXERCISE 4. Then:*

$$E = E^\circ - \frac{0.0592}{n} \log Q$$

$$1.06 = 1.12 - \frac{0.0592}{100} \log\left[\frac{(1.0)^{16}}{(0.50)^2 \times (0.50)^{25} \times C_{OH^-}^{32}}\right]$$

from which:

$$\log\left[\frac{1}{(0.50)^{27} \times C_{OH^-}^{32}}\right] = \frac{100(1.12 - 1.06)}{0.0592} = 101$$

$$-27 \log 0.50 - 32 \log OH^- = 101$$

$$-32 \log C_{OH^-} = 93$$

$$pOH = -\log C_{OH^-} = \frac{93}{32} = 2.9$$

and:

$$pH = 14.00 - pOH = 14.00 - 2.9$$

===

<u>ANSWERS to SELF-TEST, Unit 15</u>

15.1. (~24)

15.2. (~2.6)

15.3. [6.4×10^{-7} $\underline{M}$(Cu^{2+})]

15.4. (~12)

15.5. (~0.97 volt)

If you completed Self-Test questions 15.1-15.5 correctly, you have demonstrated a reasonable competency in chemical equilibrium. You may wish to try some of the RELEVANT PROBLEMS for Units 10-15 or to extend your skills by studying *proficiency* *level* sections of these Units. [If you missed them all you may wish to "take two aspirin and, if you're not better in the morning, call your instructor".]

---

*15.6. (~ 3500 mg)

*15.7. (~8.4)

*15.8. [~3.9×10^{-28} $\underline{M}$ (Fe^{2+})]

*15.9. (~4.6)

*15.10. (~3×10^{-6} atm)

If you completed Self-Test questions *15.6-*15.10 correctly, you have demonstrated proficiency in the area of chemical equilibrium. You may wish to try the RELEVANT PROBLEMS for Unit 15. If you missed any, you should review the corresponding Units.

This completes the instructional section of this text. We hope you have found it worthwhile.

A.1. The chelation of metal ions with organic ligands and their subsequent extraction into various solvents has long been used as an analytical technique. It is used for separation of metals from various matrices such as sea water, plant and animal extracts, soil solutions, and micronutrient solutions. An organic reagent that has found widespread application for extraction separations is diphenylthiocarbazone, usually called dithizone and formulated as:

$$C_6H_5-\overset{\overset{H}{|}}{N}-\overset{\overset{H}{|}}{N}$$
$$C_6H_5-N=N$$
$$C = S \qquad \text{(or simply, as: HDz)}$$

Both dithizone and its metal chelates are soluble in the widely used nonaqueous solvents, chloroform and carbon tetrachloride. Generally, the chelation reaction and partitioning equilibrium can be formulated as:

$$M^{2+}_{(aq)} + 2\ HDz_{(org)} \rightleftarrows [M(Dz)_2]_{(org)} + 2\ H^+_{(aq)}$$

If the solubility of dithizone is 2.0×10^{-7} $\underline{M}$ in H_2O and 8.0×10^{-2} $\underline{M}$ in chloroform ($CHCl_3$), what is the approximate distribution coefficient of dithizone in a water/chloroform system?

A.2. Dithizone itself is a weak acid, $K_a = 1.0 \times 10^{-5}$. What is the approximate pH of a saturated aqueous solution of dithizone (problem A.1)?

A.3. The concentration of nickel in sea water varies between 0.1 and 6.0 micrograms per liter. For precise measurements at this level, extraction and concentration methods are required. The organic compound 1-nitroso-2-naphthol (NIN) is frequently used to form the nickel complex which can easily be extracted into an organic solvent.

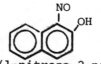

(1-nitroso-2-naphthol)

The chelation of Ni^{2+} ion by 1-nitroso-2-naphthol can be expressed in simple form by the equation:

$$Ni^{2+} + 2\ NIN^- \rightleftarrows [Ni(NIN)_2]$$
$$(K_{formation} = 3.2 \times 10^{11})$$

417

What is the concentration of "free" Ni^{2+} in a solution which is found to contain 1.8×10^{-5} mole per liter of the $[Ni(NIN)_2]$ complex and 3.5×10^{-3} mole per liter of the 1-nitroso-2-naphtholate anion?

A.4. To avoid precipitation of nickel(II) hydroxide, the nickel chelate, $[Ni(NIN)_2]$ should be formed in an acidic solution. What is the maximum allowable pH that could be maintained without precipitating $Ni(OH)_2$ at the concentration of "free" Ni^{2+} from problem A.3? [K_{sp} for $Ni(OH)_2$ is 1.6×10^{-14}.]

A.5. The measurement of very low concentrations of metal ion, such as that of "free" Ni^{2+} ion in a solution of the $[Ni(NIN)_2]$ complex (problem A.3) can be accomplished by an electrochemical technique. What would be the approximate voltage of an electrochemical cell in which one electrode is zinc immersed in a 0.20 $\underline{M}$ Zn^{2+} solution and the other is nickel immersed in a solution of the $[Ni(NIN)_2]$ complex having a "free" Ni^{2+} concentration of 4.0×10^{-5} $\underline{M}$? The cell discharge equation is:

$$Ni^{2+} + Zn \rightarrow Ni + Zn^{2+}$$

[See Appendix E for Standard Electrode Potentials.]

*A.6. The exact biochemical mechanism of aluminum toxicity in farm crops has not been established. However, distinct differences are recognized between aluminum-tolerant and aluminum-sensitive plants. The differential aluminum tolerance in plants appears to be associated with the plant's ability to alter soil pH in its root zone. For example, aluminum-sensitive wheat varieties induce lower pH values in their root zones than aluminum tolerant varieties. A lower pH increases the solubility of aluminum, and potential toxicity. Aluminum can be separated from soil solutions by complexing it with cupferron (CF), the ammonium salt of nitrosophenylhydroxylamine.

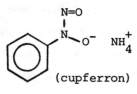

(cupferron)

The complex formation is represented by the equation:

$$Al^{3+} + 3\ CF^- \rightleftharpoons [Al(CF)_3]$$

How many grams of aluminum cupferrate could be extracted from 30.0 liters of aqueous soil extract which contains 0.005 g of aluminum cupferrate per liter by a single extraction with 10.0 liters of chloroform? [K_D of $[Al(CF)_3]$ in a water/chloroform system is 1.5×10^5.]

*A.7. The ammonium salt of nitrosophenylhydroxylamine (problem A.6) is freely soluble in water and fairly stable in neutral or alkaline solutions. In the presence of strong acids the "free" acid precipitates and rapidly decomposes. What is the approximate pH of a solution of the "free" acid prepared by dissolving 3.1 g of nitrosophenyl-hydroxylamine in 1.00 liter of water? [K_a for the monoprotic acid is 4.0×10^{-5}.]

*A.8. It has often been suggested that garbage should be buried, rather than burned, to reduce air pollution and recycle organic wastes as plant nutrients. This was, in fact, common practice many years ago and is still employed in a number of simple agrarian societies. However, the nature of our "garbage" has been changed by modern technology, presenting some rather complex problems. Most plastics, for example, have very long "half-lives", since they are relatively inert to chemical and biological degradation in soil systems. The phthalate esters used as additives in soft, flexible plastics present an added problem. As these "plasticizers" undergo hydrolysis, the resulting phthalate ions can complex with various metal ions in soils to transport the ions into plant systems, often in amounts toxic to the plants. What concentration of "free" Co^{2+} ion would result from the mixing of equal volumes of 0.050 $\underline{M}$ phthalate and 2.0×10^{-4} $\underline{M}$ Co^{2+} solutions?

$$(K_{formation} = 3.2 \times 10^4)$$

*A.9. Organic matter from plants is a very complex mixture containing carbohydrates, lignins, tannins, pigments, proteins, fats, oils, and waxes. Some of the organic matter supplies energy and nutrients for various forms of life in the soil. Other forms of the organic matter help control the water and micronutrient levels in the soil, while still other forms may react with various soil components to control the acidity or alkalinity in the soil solution. The simple aliphatic acids are very important components of the soil solution because they are effective solubilizers of mineral matter. The amounts of acids in the soil solution are very significant. With a soil moisture content of 20% the concentration of acetic acid in the aqueous phase is often as high as 4.0×10^{-3} $\underline{M}$. Other acids include formic, malonic, malic, tartaric, succinic and butyric acids. What would be the approximate pH of a solution prepared by adding 20 g of butyric acid ($CH_3CH_2CH_2CO_2H$) to 250 ml of 0.20 $\underline{M}$ sodium hydroxide? [For butyric acid, $K_a = 1.5 \times 10^{-5}$.]

*A.10. A rancher in West Texas discovered that his home had been built over a pocket of methane gas. Unfortunately, the amount of methane present was too small for commercial exploitation, but the leakage rate was more than that needed for the ranch's fuel purposes, so the surplus had to be wastefully burned. It was suggested that

some of the surplus methane be used in methane-oxygen fuel cells for a 24 volt light-
ing system. The electrode processes for this fuel cell are formulated as:

$$CO_{3(aq)}^{2-} + 7\ H_2O_{(\ell)} + 8e^- \rightleftarrows CH_{4(g)} + 10\ OH_{(aq)}^- \qquad E^{\circ} = -1.06\ v$$

$$O_{2(g)} + 2\ H_2O_{(\ell)} + 4e^- \rightleftarrows 4\ OH_{(aq)}^- \qquad E^{\circ} = 0.40\ v$$

What electrolyte pH would give a proper potential so that a series of seventeen equal
cells would provide 24.0 volts, using methane and oxygen each at 1.0 atm with a carbonate concentration of 0.010 $\underline{M}$?

ANSWERS:

(A.1.) 4×10^5, (A.2.) ~5.8, (A.3.) 4.6×10^{-12} $\underline{M}$, (A.4.) 12.8, (A.5.) 0.40 v, (A.6.)
~0.15 g, (A.7.) 3.0, (A.8.) 5.1×10^{-6} $\underline{M}$, (A.9.) 4.3, (A.10.) ~9.7

RELEVANT PROBLEMS

Unit 15: *Organic Chemistry Revisited*

Set B: *Biological & Medical Sciences*

B.1. Salicylic acid, a natural product, was first extracted from the bark of willow trees
by Dr. Thomas Sydenham, a London physician, in 1650. Willow trees, members of the
family Salicaceae, are widely distributed through Europe, Asia, and the Americas. During the extraction of salicylic acid from an aqueous willow bark extract, its solubility was determined to be 0.22 g per 100 ml (H_2O) and 2.4 g per 100 ml ($CHCl_3$).
What is the approximate distribution coefficient of salicylic acid in a water/chloro-
form system?

B.2. Salicylic acid (problem B.1) was first synthesized in 1852 and processes amenable to
its large-scale production were developed in 1874. The synthetic material soon became widely used in medicine, especially in treating rheumatic fever. However, undesirable side-effects were soon recognized, some of which resulted from its acidic
properties. What is the approximate pH of a saturated aqueous solution of salicylic
acid?

(salicylic acid) [$K_a = 1.05 \times 10^{-3}$]

[Dissociation of the phenolic group is
negligible at low pH.]

B.3. The biochemical function of salicylic acid has not been entirely defined. However, it is known that the salicylate ion is capable of forming complexes with several biochemically important cations. Thus, it appears to function, in part, by stabilizing certain metal ions and facilitating their transport through the circulatory system. The salicylate ion forms a zinc complex according to the equation:

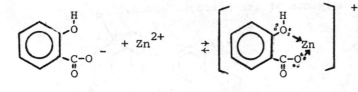

$$(K_{formation} = 1.0 \times 10^4)$$

What is the approximate concentration of "free" Zn^{2+} in a solution which is found to contain 2.6×10^{-6} moles per liter of the salicylate/Zn^{2+} complex and 3.0×10^{-2} moles per liter of salicylate anion?

B.4. Several salts of salicylic acid are used in medicine. For example, magnesium salicylate is used as an intestinal antiseptic and calcium salicylate is used in treatment of diarrhea and gastroenteritis. A saturated solution of calcium hydroxide is used to prepare calcium salicylate. What is the pH of the saturated calcium hydroxide solution used? [See Appendix D for K_{sp}-data.]

B.5. What would be the approximate voltage of an electrochemical cell in which one electrode is manganese immersed in a $0.15 \underline{M} \ Mn^{2+}$ solution and the other is copper immersed in a solution of the salicylate/Cu^{2+} complex having a "free" Cu^{2+} concentration of $4.5 \times 10^{-5} \underline{M}$? [See Appendix E for Standard Electrode Potentials.] The cell discharge can be formulated as:

$$Cu^{2+} + Mn \rightarrow Cu + Mn^{2+}$$

*B.6. Brucine, an important alkaloid, can be produced synthetically, but the synthesis is not amenable to large-scale processes. As a result, brucine is obtained commercially by extraction from seeds of Strychnos nux-vomica Loganiaceae. Trees of the family Loganiaceae (found in India, Ceylon, Burma, Thailand, and Australia) are small-to-moderate in size, having short, thick, often crooked trunks. The leaves are smooth, with three to five veins, and ovate. The flower is a small light green tubule. The hard orange-size fruit has a gelatinous pulp containing from one to five disc-shaped seeds. The seeds contain several powerful alkaloids, including brucine and strychnine. The distribution coefficient of brucine in a chloroform/water system is 256. How many grams of brucine could be extracted from 20.0 liters of an aqueous Strychnos seed extract which contains 0.78 g of brucine per liter, by a single extraction with 5.0 liters of chloroform?

*Proficiency Level

*B.7. Brucine is a milder poison than strychnine and has been used in a therapeutic capacity in veterinary medicine, as a central nervous system stimulant. It is also widely used in denaturing alcohol and for separating racemic mixtures in analytical chemistry. What is the approximate pH of an alcohol-denaturing stock solution prepared by dissolving 7.8 g of brucine in 1.0 liter of water? (The basic character of brucine is attributed to only one of its nitrogen atoms.)

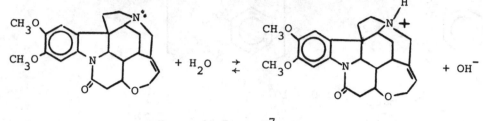

$$(K_b = 9.2 \times 10^{-7})$$

*B.8. Brucine and strychnine, related alkaloids, can be precipitated using an acidic solution of potassium hexacyanoferrate(II). Both of these alkaloids form characteristic pale yellow needles when precipitated by potassium hexacyanoferrate(II), thus providing an easy test for solutions suspected of containing brucine or strychnine. What is the approximate molarity of "free" Fe^{2+} in a solution of this reagent prepared by adding 5.0 ml of 0.10 $\underline{M}$ $FeSO_4$ to 25.0 ml of 3.50 $\underline{M}$ KCN? [The net formation constant for $[Fe(CN)_6]^{4-}$ is 1.0 x 10^{24}.]

*B.9. A common derivative of brucine sold commercially is the sulfate salt:

$$(C_{23}H_{27}N_2O_4)_2SO_4$$

A solution of this salt in water is slightly acidic due to the reaction:

$$C_{23}H_{27}N_2O_4^+ \rightleftharpoons H^+ + Brucine$$

The sulfate salt has a solubiltiy product of 1.1 x 10^{-5}. What is the approximate pH of a saturated aqueous solution of the salt, assuming that the hydrolysis of SO_4^{2-} is negligible? [K_b for brucine is 9.2 x 10^{-7}.]

*B.10. The C=C bond in brucine (problem B.7) is vulnerable to attack by oxidizing agents such as hydrogen peroxide or potassium permanganate. Aqueous solutions of $KMnO_4$ are not entirely stable because of the tendency of MnO_4^- ions to oxidize water. This process may be formulated as:

$$4\ MnO_4^-{}_{(aq)} + 2\ H_2O_{(\ell)} \rightarrow 4\ MnO_2{}_{(s)} + 4\ OH^-{}_{(aq)} + 3\ O_2{}_{(g)}$$

for which the Standard Electrode Potentials are:

$$MnO_4^-{}_{(aq)} + 2\ H_2O_{(\ell)} + 3e^- \rightarrow MnO_2{}_{(s)} + 4\ OH^-{}_{(aq)} \qquad E^o = +0.57\ v$$

$$O_{2(g)} + 2 H_2O_{(\ell)} + 4e^- \overset{\rightarrow}{\leftarrow} 4 OH^-_{(aq)} \qquad\qquad E^o = +0.40 \text{ v}$$

A permanganate electrochemical cell maintained with an oxygen pressure of 1.0 atm registered a potential of 0.26 volt. What was the approximate pH of the cell, in which the electrolyte was 0.10 $\underline{M}$ in MnO_4^-?

ANSWERS:

(B.1.) ~11, (B.2.) ~2.4, (B.3.) 8.7×10^{-9} $\underline{M}$, (B.4.) 12.3, (B.5.) 1.42 v, (B.6.) 15 g, (B.7.) 10.1, (B.8.) 3.3×10^{-29} $\underline{M}$, (B.9.) ~4.8, (B.10.) ~8.4

RELEVANT
PROBLEMS

Unit 15: *Organic Chemistry Revisited*

Set I: *Industrial Chemistry*

I.1. The compound phenacetin has been used in a number of common nonprescription drugs such as ANACIN, EMPIRIN, and EXCEDRIN. Phenacetin has about the same analgesic and antipyretic activity as aspirin, but has now been implicated in possible blood and kidney damage. As a result, most pharmaceutical companies have discontinued its use. ANACIN, for example, no longer contains phenacetin. The "preference" of the drug for absorption into body lipids can be illustrated by its facile extraction by ether (a good solvent for lipid-like materials) from the aqueous solution. Calculate the distribution coefficient for phenacetin in an ether-water system.

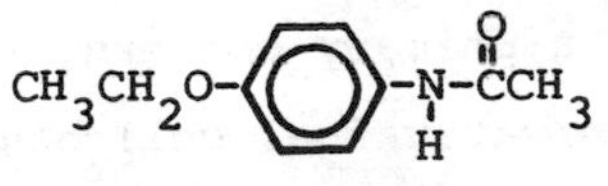

(phenacetin)

water solubility: 0.70 g $(100 \text{ ml})^{-1}$
ether solubility: 1.8 g $(100 \text{ ml})^{-1}$

I.2. Acetaminophen has largely replaced phenacetin in nonprescription drugs, such as EXCEDRIN, so the industrial production of acetaminophen has found a vastly increased market. Unlike phenacetin, the acetaminophen is acidic and this presents some minor problems in work up procedures and corrosion control. What is the approximate pH of a solution 8.5 g liter^{-1} in acetaminophen?

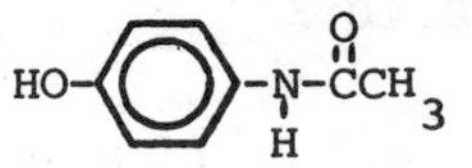

$(K_a = 3 \times 10^{-9})$

I.3. For a few brief months the production of nitrilotriacetic acid (NTA) showed a fantastic profit potential, with the probability of its use in detergents as a replacement for phosphates. A number of chemical industries scaled up their NTA production capabilities and began to stock large amounts of the substance. However, before any really major market could develop, NTA was banned as a detergent additive because of concerns over its possible health hazard in drinking water. As a result, a potential profit became a very real economic loss. Today there is only a limited market for NTA, for such purposes as removal of boiler scale in steam plants. What would be the approximate concentration of "free" Ca^{2+} in a system containing 12 g liter^{-1} of NTA^{3-} and 0.050 $\underline{M}$ $[Ca(NTA)_2]^{4-}$?

$$2\ N(CH_2CO_2)_3^{3-} + Ca^{2+} \rightleftarrows [Ca(NTA)_2]^{4-}$$

$$[NTA^{3-}] \qquad [K_{formation} = 4.1\times 10^{11}]$$

I.4. Many products of the pharmaceuticals industry, developed for specific useful purposes, have been misused to such an extent that their continued production represents "bad public relations policy". Methamphetamine is such a drug, developed as an appetite depressant for persons required to diet for medical reasons, but now identified in the public mind with the drug culture as "speed". The drug is frequently used as the hydrochloride salt, which is much more water soluble than the free base. What is the approximate pH of a solution containing 25 g liter^{-1} of methamphetamine hydrochloride?

$$(K_b = 1.6\times 10^{-4})$$

I.5. Fuel cell technology has found a ready market in a number of areas, such as the aerospace industries, in which energy efficiency is a primary concern, rather than high power output. A methanol/formaldehyde - oxygen/water fuel cell has been studied because of its capability of producing energy along with a valuable chemical product from relatively inexpensive "wood alcohol". The cell discharge reaction is formulated as:

$$2\ CH_3OH_{(aq)} + O_{2(aq)} \rightarrow 2\ \underset{H}{\overset{H}{\diagdown}}C = O_{(aq)} + 2\ H_2O_{(\ell)}$$

for which the electrode processes are:

$$O_{2(g)} + 2\ H_2O_{(\ell)} + 4e^- \rightleftarrows 4\ OH^-_{(aq)} \qquad E^o = +0.40\ v$$

$$H_2CO_{(aq)} + 2\ H_2O_{(\ell)} + 2e^- \rightleftarrows CH_3OH_{(aq)} + 2\ OH^-_{(aq)} \qquad E^o = -0.59\ v$$

What voltage would this cell produce with an oxygen pressure of 2.5 atm and an electrolyte pH of 12.0, with 5.0 $\underline{M}$ CH_3OH and 0.25 $\underline{M}$ H_2CO?

*I.6.   In the purification of phenacetin (problem I.1), what mass of the compound could be extracted from 250 liters of its aqueous solution, initially containing 0.65 g liter^{-1}, by two successive extractions with fresh 50 liter portions of ether?

*I.7.   A commercially valuable buffer, used expecially in medical and biochemical labora-tories in protein separation by electrophorsis, involves tris (hydroxymethyl)amino-methane, usually referred to just as "TRIS". A buffer made by mixing 50.0 ml of 0.10 $\underline{M}$ "TRIS", 42.0 ml of 0.10 $\underline{M}$ HCl, and sufficient water to form 100.0 ml of solu-tion has a pH of 7.36 (approximating that of human blood). What pH change would occur on addition of 5.0 ml of 0.30 $\underline{M}$ NaOH to 95 ml of this buffer?

$$\begin{array}{c} CH_2OH \\ | \\ HOCH_2-C-NH_2 \\ | \\ CH_2OH \end{array} \qquad (K_b = 1.2 \times 10^{-6})$$

("TRIS")

*I.8.   An industrial operation requiring the circulation of a large volume of coolant water through heat exchangers must often employ some additive to prevent the formation of "hard water deposits" in lines and valves. If a plant uses well water with a "hard-ness" of 120 ppm Ca^{2+}, (a) What minimum volume of 0.30 $\underline{M}$ NTA^{3-} solution must be added to 250,000 liters of coolant to reduce the equilibrium concentration of "free" Ca^{2+} to an acceptable level of 10 ppm? (b) What is the equilibrium concentration of NTA^{3-}? [See problem I.3 for additional data. Assume no significant concentration of the intermediate complex.]

*I.9.   What mole ratio of "TRIS" to "TRIS-hydrochloride" (problem I.7) would provide a buffer of maximum pH for a solution 0.12 $\underline{M}$ in Mg^{2+} if $Mg(OH)_2$ precipitation must be avoided?

*I.10. For the recovery of formaldehyde from a methanol-oxygen fuel cell operation (problem I.5), the anode compartment must be drained periodically for isolation of the pro-duct. Ideally, chemical recovery would be most profitable from the highest possible methanol-formaldehyde conversion, but a minimum voltage must be maintained to make the fuel cell aspect worthwhile. Thus, some compromise is necessary in selecting the optimum conditions for anode compartment draining. What concentration of formaldehyde would be found in the anode compartment of this cell when the potential had dropped to 0.98 v, the minimum acceptable operating voltage, with an oxygen pressure of 2.5 atm and a methanol concentration of 0.50 $\underline{M}$?

ANSWERS:

(I.1.) 2.6, (I.2.) 4.9, (I.3.) 3×10^{-11} $\underline{M}$, (I.4.) 5.5, (I.5.) 1.03 v, (I.6.) 92 g, (I.7.) increase by 0.68 pH unit, (I.8.) ~4600 liters, 5.2×10^{-6} $\underline{M}$, (I.9.) ["TRIS"]:["TRIS·HCl"] $\simeq$ 8.3 : 1, (I.10.) ~4 $\underline{M}$

RELEVANT

PROBLEMS

Unit 15: *Organic Chemistry Revisited*

Set E: *Environmental Sciences*

E.1. Many of the chlorinated organic compounds used as insecticides, such as DDT, have now been banned or restricted to carefully controlled use. Many of these compounds, however, are extremely resistant to natural degradation processes and, as a result, their accumulation from past use or their spread from current "localized" applications still represents an environmental problem. Industrial plants manufacturing these compounds have a special obligation to avoid dumping contaminated wastes into rivers and lakes. A few years ago, several million fish were killed in the Mississippi River from the insecticide ENDRIN, apparently dumped by a chemical plant. Even today, measurable concentrations of several insecticides have been found in several municipal water supplies. The long range effects of drinking water containing parts-per-billion levels of such compounds are not known, but many insecticides are accumulated in fatty tissues because of their preferential solubility in "lipid-like" solvents. What is the approximate distribution coefficient for ENDRIN in a water-ether system?

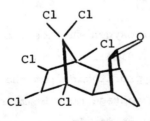

(ENDRIN)

water solubility: ~0.005 g $(100\,ml)^{-1}$

ether solubility: ~12 g $(100\ ml)^{-1}$

E.2. The compound β-naphthylamine has been used extensively as an intermediate for making certain dyes and as an optical bleaching agent. Within the last few years, it was found that workers continually exposed to β-naphthylamine had an unusually high incidence of cancer of the bladder. Further tests have now resulted in the classification of β-naphthylamine as a "known potent carcinogen", and its use is now restricted to carefully controlled conditions. What is the approximate pH of a saturated aqueous solution of β-naphthylamine?

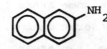

(β-naphthylamine)

water solubility: 0.44 g $(100$ ml$)^{-1}$

$K_b = 1.3 \times 10^{-10}$

E.3. Esters of phthalic acid are produced in enormous quantities for the polymer industries as plasticizers, to make plastics softer and more flexible. The indiscriminate dumping of waste phthalic acid, or phthalate salts, into rivers and streams represents a particular environmental problem because of the action of phthalate ion in complexing with certain heavy metal cations. These complexes maintain heavy metal ions in solutions from which they would otherwise precipitate as hydroxides and thus help to spread these toxic ions through natural aquatic systems. What is the approximate concentration of "free" Co^{2+} ion in a solution that is 0.050 $\underline{M}$ in phthalate ion and 1.5×10^{-5} $\underline{M}$ in $[Co(PHTH)_2]^{2-}$?

$$2 \underset{CO_2^-}{\overset{CO_2^-}{\bigcirc}} + Co^{2+} \rightleftarrows [Co(PHTH)_2]^{2-}$$

(phthalate) $(K_{formation} = 3.2 \times 10^4)$

E.4. Almost all the synthetic detergents now used in the United States are biodegradable alkylbenzenesulfonates. Microorganisms can "digest" the alkyl groups of these detergents, unlike those of the "nondegradable" detergents used a few years ago. However, recent evidence suggests that the "benzene" portion of the detergent may not be fully degraded (to $CO_2 + H_2O$). Instead, this part of the molecule may be converted to phenol, a toxic substance whose increasing concentration could present a "new" environmental problem. In aquatic regions of relatively high pH, highly soluble salts of phenol could accumulate. What is the approximate pH of a solution containing 0.20 g liter^{-1} of the sodium salt of phenol?

$$\underset{OH}{\bigcirc} \rightleftarrows \underset{O^-}{\bigcirc} + H^+$$

$(K_a = 1.0 \times 10^{-10})$

E.5. Although fuel cells could theoretically replace internal combustion engines, with vastly improved fuel economy and essentially zero air pollution, there are valid reasons why these cells will never successfully compete with all uses of internal combustion engines. Fuel cells have a very limited rate of energy production, making them unsuitable for high speed or high power engine requirements. The problems of

427

providing adequate storage tanks for gaseous fuels and of periodically replacing electrolyte solutions add the further limitation of short distance use of fuel cell powered vehicles. Even with these limitations, there are some areas in which conversion to fuel cells could reduce pollution and help conserve energy, as in small vehicles (minibuses) for mass transit in downtown metropolitan areas. How many methane-oxygen fuel cells would be required for a 24 volt minibus engine, with each cell in the series using methane and oxygen at 3.0 atm each and an electrolyte 0.20 $\underline{M}$ in KOH and 1.0×10^{-5} $\underline{M}$ in CO_3^{2-}? The electrode processes are formulated as:

$$CO_{3(aq)}^{2-} + 7\ H_2O_{(l)} + 8e^- \rightleftharpoons CH_{4(g)} + 10\ OH_{(aq)}^- \qquad E^o = -1.06\ v$$

$$O_{2(g)} + 2\ H_2O_{(l)} + 4e^- \rightleftharpoons 4\ OH_{(aq)}^- \qquad E^o = +0.40\ v$$

*E.6. To illustrate the problem of "lipid-like" solvent accumulation in fatty tissues of chlorinated organic insecticides, calculate the percentage of ENDRIN (problem E.1) that would be extracted from 2.5 liters of its saturated aqueous solution by two successive extractions with fresh 50 ml portions of ether.

*E.7. In certain apple products industries, the apples are soaked in a sodium hydroxide solution of pH 12.0 to loosen the peels. During use, this solution gradually decreases in pH and must be replenished. However, the waste solution is still too alkaline to be discarded into the effluent stream and must first have enough HCl added to reduce the pH to below 11.0. Typical pH meters cannot be used for routine monitoring of highly alkaline solutions because of alkali corrosion of the glass electrodes. A special pH indicator is available for the pH 11-12 range that permits an adequate monitoring of "soak" and "effluent" solutions by addition of a few drops of the indicator to periodically withdrawn samples. As with most indicators, the two different colored species coexist as a conjugate-pair buffer at intermediate pH. Below pH 11.0, the yellow p-nitrophenylhydrazone of benzaldehyde (p-NPB) is the principal species, while the intense purple color above pH 12 is due to the anion (p-NPB)$^-$. The intermediate system is tan.

(p-NPB)   [$K_a = 3.2 \times 10^{-12}$]   (p-NPB)$^-$

What <u>change</u> in the mole ratio of (p-NPB) to (P-NPB)$^-$ occurs when the solution pH is changed from 11.5 to 10.8, an acceptable waste effluent pH?

*E.8. What equilibrium concentration of "free" Co^{2+} would result from the addition of 250 liters of 0.15 $\underline{M}$ phthalate solution (problem E.3) to 2500 liters of 1.3×10^{-5} $\underline{M}$ Co^{2+} solution?

*E.9. The consequences of phthalate complexing of heavy metal ions (problems E.3 and E.8) can be avoided by maintaining a low enough pH in the effluent stream to adequately reduce the available phthalate concentration, while keeping the pH high enough to permit precipitation of the metal hydroxide. What mole ratio of hydrogen phthalate to phthalate would provide the minimum pH to initiate precipitation of $Co(OH)_2$ from 1.3×10^{-3} Co^{2+} solution, assuming complex formation does not occur?

$$(K_a = 3.9 \times 10^{-6})$$

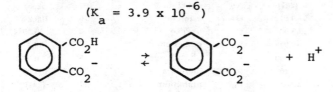

[hydrogen phthalate]   [phthalate]

[For K_{sp}-data, see Appendix D.]

*E.10. What would be the electrolyte pH of a partially discharged methane-oxygen fuel cell (problem E.5) when the cell potential is 1.30 v, the methane and oxygen pressures are 1.0 atm each, and the carbonate concentration is 0.10 $\underline{M}$?

ANSWERS:

(E.1.) ~2.4×10^3, (E.2.) 8.3, (E.3.) 1.9×10^{-7} $\underline{M}$, (E.4.) 10.7, (E.5.) 16 cells (E.6.) "99.96%" [100% within limits of significant figures], (E.7.) [p-NPB]:[p-NPB]⁻ changed from 1 : 1 to 5 : 1, (E.8.) ~2.0×10^{-6} $\underline{M}$, (E.9.) [hydrogen phthalate]:[phthalate] ≈ 6.5×10^{-3} : 1, (E.10.) ~2.7

APPENDIX A: THE CHEMICAL ELEMENTS

Symbol	Name*	Atomic number	Atomic weight†	Symbol	Name*	Atomic number	Atomic weight†
Ac	actinium	89	(227)	Hg	mercury (hydrargyrum)	80	200.59
Al	aluminum	13	26.9815	Mo	molybdenum	42	95.94
Am	americium	95	(243)	Nd	neodymium	60	144.24
Sb	antimony (stibium)	51	121.75	Ne	neon	10	20.18
Ar	argon	18	39.948	Np	neptunium	93	237.0
As	arsenic	33	74.9216	Ni	nickel	28	58.71
At	astatine	85	(210)	Nb	niobium	41	92.91
Ba	barium	56	137.34	N	nitrogen	7	14.0067
Bk	berkelium	97	(247)	No	nobelium§	102	(255)
Be	beryllium	4	9.012	Os	osmium	76	190.2
Bi	bismuth	83	208.98	O	oxygen	8	15.9994
B	boron	5	10.811	Pd	palladium	46	106.4
Br	bromine	35	79.909	P	phosphorus	15	30.974
Cd	cadmium	48	112.40	Pt	platinum	78	195.09
Ca	calcium	20	40.08	Pu	plutonium	94	(244)
Cf	californium	98	(251)	Po	polonium	84	(210)
C	carbon	6	12.0115	K	potassium (kalium)	19	39.102
Ce	cerium	58	140.12	Pr	praseodymium	59	140.91
Cs	cesium	55	132.91	Pm	promethium	61	(147)
Cl	chlorine	17	35.453	Pa	protactinium	91	231.0
Cr	chromium	24	51.996	Ra	radium	88	(226)
Co	cobalt	27	58.933	Rn	radon	86	(222)
Cu	copper (cuprum)	29	63.546	Re	rhenium	75	186.2
Cm	curium	96	(247)	Rh	rhodium	45	102.91
Dy	dysprosium	66	162.50	Rb	rubidium	37	85.47
Es	einsteinium	99	(254)	Ru	ruthenium	44	101.07
Er	erbium	68	167.26	Rf	rutherfordium**	104	(261)
Eu	europium	63	151.96	Sm	samarium	62	150.4
Fm	fermium	100	(257)	Sc	scandium	21	44.96
F	fluorine	9	18.9984	Se	selenium	34	78.96
Fr	francium	87	(223)	Si	silicon	14	28.086
Gd	gadolinium	64	157.25	Ag	silver (argentum)	47	107.868
Ga	gallium	31	69.72	Na	sodium (natrium)	11	22.9898
Ge	germanium	32	72.59	Sr	strontium	38	87.62
Au	gold (aurum)	79	196.97	S	sulfur	16	32.06
Hf	hafnium	72	178.49	Ta	tantalum	73	180.95
Ha	hahnium‡	105	(260)	Tc	technetium	43	(99)
He	helium	2	4.0026	Te	tellurium	52	127.60
Ho	holmium	67	164.93	Tb	terbium	65	158.9
H	hydrogen	1	1.0080	Tl	thallium	81	204.37
In	indium	49	114.82	Th	thorium	90	232.0
I	iodine	53	126.904	Tm	thulium	69	168.93
Ir	iridium	77	192.22	Sn	tin (stannum)	50	118.69
Fe	iron (ferrum)	26	55.847	Ti	titanium	22	47.90
Kr	krypton	36	83.80	W	tungsten (wolfram)	74	183.85
La	lanthanum	57	138.91	U	uranium	92	238.03
Lr	lawrencium	103	(256)	V	vanadium	23	50.94
Pb	lead (plumbum)	82	207.2	Xe	xenon	54	131.30
Li	lithium	3	6.94	Yb	ytterbium	70	173.04
Lu	lutetium	71	174.97	Y	yttrium	39	88.91
Mg	magnesium	12	24.31	Zn	zinc	30	65.37
Mn	manganese	25	54.938	Zr	zirconium	40	91.22
Md	mendelevium	101	(258)				

*The names in parentheses are the Latin forms used in complex formation: e.g., gold (*aurum*); $[AuCl_4]^-$ is tetrachloroaurate(III). (Exception is wolfram, which has a German derivation.)

†Atomic weights in parentheses are those of the most stable radioisotope.

‡This name has been suggested by American researchers and has not yet been approved by IUPAC. Russian researchers have suggested the name *nielsbohrium*.

§Although the name nobelium has official IUPAC sanction, some Russian researchers use the name *joliotium*.

**This name has been suggested by American researchers and has not yet been approved by IUPAC. Russian researchers have suggested the name *kurchatovium*.

APPENDIX B: MATHEMATICAL OPERATIONS

(Taken from FUNDAMENTALS OF CHEMISTRY: A LEARNING

SYSTEMS APPROACH, Rod O'Connor, Harper & Row, 1974)

The following discussions represent simple pragmatic approaches to the types of mathematical operations most frequently encountered in an introductory chemistry course. They are not intended as mathematical theory, but rather as a review or introduction of the how-to-do-it of solving simple problems.

Section A: Algebraic Manipulations

The key to solving simple algebraic equations containing a single unknown (e.g., $x + 6 = 10$) is to realize that the equation represents a statement that two quantities are equal. Thus any mathematical operation can be applied to one side of the equation if it is also applied to the other. It is then necessary to determine what operations are required for the side containing the unknown (to "isolate" the unknown) and to apply these operations to both sides of the equation. This principle is illustrated by the following solutions to the problems in part A of the Pre-Test.

EXAMPLE 1

$$x + 6 = 10$$

SOLUTION

To isolate x, it is necessary to subtract 6:

$$(x + 6) - 6 = (10) - 6$$
$$x = 4$$

EXAMPLE 2

$$7 - x = 24$$

SOLUTION

To isolate x, it is necessary to subtract 7, then multiply by -1:

$$(7 - x) - 7 = (24) - 7$$
$$-x = 17$$
$$(-x)(-1) = (17)(-1)$$
$$x = -17$$

EXAMPLE 3

$$2x + 8 = 4$$

SOLUTION

To isolate x, it is necessary to subtract 8, then divide by 2:

$$(2x + 8) - 8 = (4) - 8$$
$$2x = -4$$
$$\frac{(2x)}{2} = \frac{(-4)}{2}$$
$$x = -2$$

EXAMPLE 4

$$\frac{x}{6} = \frac{1}{4}$$

SOLUTION

To isolate x, it is necessary to multiply by 6:

$$\left(\frac{x}{6}\right)(6) = \frac{1}{4}(6)$$
$$x = \frac{6}{4}$$
$$x = 1.5$$

EXAMPLE 5

$$\frac{3}{x} = \frac{2}{7}$$

SOLUTION

First, convert the unknown term from a denominator to a numerator position (by multiplying by x):

$$\left(\frac{3}{x}\right)x = \left(\frac{2}{7}\right)x$$

$$3 = \frac{2x}{7}$$

Then, to isolate x it is necessary to multiply by 7 and divide by 2; that is, multiply by $\frac{7}{2}$:

$$(3)\left(\frac{7}{2}\right) = \left(\frac{2x}{7}\right)\left(\frac{7}{2}\right)$$

$$\frac{21}{2} = x$$

$$x = 10.5$$

EXAMPLE 6

$$\frac{5x}{12} = \frac{9}{16}$$

SOLUTION

To isolate x, it is necessary to multiply by 12 and divide by 5 (i.e., multiply by $\frac{12}{5}$):

$$\left(\frac{5x}{12}\right)\left(\frac{12}{5}\right) = \left(\frac{9}{16}\right)\left(\frac{12}{5}\right)$$

$$x = \frac{108}{80}$$

Section B: Dimensional Conversions

Most scientific measurements are expressed in metric system units.* Until these units are universally accepted, it is necessary to convert from one measuring system to another. A large number of conversion factors have been tabulated in such sources as the *Handbook of Chemistry and Physics* (Chemical Rubber Company), but most conversions required in an introductory study of chemistry can be made by using the simple factors given in Table 1 and the metric system prefixes in Table 2. Conversions among the Celsius (centigrade), Fahrenheit, and Kelvin ("absolute") temperature scales are given in Table 3.

Table 1 Common conversion factors

Approximate English-metric equivalents
1.00 pound (lb) = 454 grams (g)
1.00 quart (qt) = 0.946 liters (l)

Exactly defined equivalents

English
1 mile = 5280 feet (ft)
1 yard = 3 feet (ft)
1 foot (ft) = 12 inches (in.)
1 ton = 2000 pounds (lb)
1 pound (lb) = 16 ounces (oz)
1 gallon (gal) = 4 quarts (qt)
1 quart (qt) = 2 pints (pt)
1 pint (pt) = 16 ounces (fluid)(oz)

Metric
1 cubic centimeter (cm^3) = 1 milliliter (ml)
1 angstrom unit (Å) = 1×10^{-8} centimeter (cm)
1 micron (μ) = 1×10^{-6} meter (m)

English-metric equivalents
1 inch (in.) = 2.54 centimeters (cm) (exactly)

Table 2 Common metric system prefixes

mega = million (10^6)
kilo = thousand (10^3)
deci = tenth (10^{-1})
centi = hundredth (10^{-2})
milli = thousandth (10^{-3})

Table 3 Temperature scale conversions

$°F = \left(\frac{9}{5}\right)(°C) + 32°$
$°C = \left(\frac{5}{9}\right)(°F - 32°)$
$°K = °C + 273°$

*For information concerning the new "SI units," consult your instructor.

Unity Factors

Any quantity can be multiplied by unity (one) without changing its value. A unity factor employs this to change the dimensions of a quantity without changing the value of the quantity, although the numbers used to express the value may be changed. If someone asked you the number of inches in 2 yards, you would reply, "72." The value of the length was not changed; that is, it is the same length whether you call it 2 yards or 72 inches (or 6 feet). You probably made this conversion almost intuitively. All the unity factor method does is to provide a systematic procedure for making such conversions in cases a bit too complicated for the "intuitive approach."

Since any fraction whose numerator and denominator are equivalent has the value of unity, you can make unity factors from any dimensional equivalents, such as those given in Table 1 or implied by the prefix meanings of Table 2, for example,

Equivalents	Corresponding unity factors	
1 in. = 2.54 cm	$\dfrac{1 \text{ in.}}{2.54 \text{ cm}}$ or	$\dfrac{2.54 \text{ cm}}{1 \text{ in.}}$
1 mile = 5280 ft	$\dfrac{1 \text{ mile}}{5280 \text{ ft}}$ or	$\dfrac{5280 \text{ ft}}{1 \text{ mile}}$
1 kg = 1000 g (kilo = thousand)	$\dfrac{1 \text{ kg}}{1000 \text{ g}}$ or	$\dfrac{1000 \text{ g}}{1 \text{ kg}}$

Unity factors are valuable in problems of dimensional conversion and, as we shall see later, in a variety of other types of problem situations. The method simply requires setting up the problem in fractional form, using dimensions for all quantities involved. The quantity to be converted is multiplied by appropriate unity factors until all dimensions can be canceled except for those desired in the answer. In multistep conversions it is well to have a systematic sequence; typically, numerator dimensions are converted first, then denominator units.

EXAMPLE 7

Convert 60 miles per hour (60 miles hr^{-1}) to feet per second. (Remember that "per" means "divide by.")

STEPWISE SOLUTION

$$\frac{60 \text{ miles}}{1 \text{ hr}} = \underline{\hspace{2cm}} \text{ ft sec}^{-1}$$

First, convert the numerator unit (miles ⟶ feet), using a unity factor expressing the feet-miles equivalence, in a form such that the "miles" will cancel:

$$\frac{60 \text{ miles}}{1 \text{ hr}} \times \frac{5280 \text{ ft}}{1 \text{ mile}} = \frac{(60 \times 5280) \text{ ft}}{1 \text{ hr}}$$

Next, convert the denominator unit (hour ⟶ seconds), using a set of unity factors expressing hour ⟶ minutes ⟶ second equivalences, in proper form for dimensional cancellation:

$$\frac{(60 \times 5280) \text{ ft}}{1 \text{ hr}} \times \frac{1 \text{ hr}}{60 \text{ min}} \times \frac{1 \text{ min}}{60 \text{ sec}} = \frac{(60 \times 5280) \text{ ft}}{(60 \times 60) \text{ sec}}$$

The combined solution is

$$\frac{60 \text{ miles}}{\text{hr}} \times \frac{5280 \text{ ft}}{\text{mile}} \times \frac{1 \text{ hr}}{60 \text{ min}} \times \frac{1 \text{ min}}{60 \text{ sec}} = \frac{60 \times 5280 \text{ ft}}{60 \times 60 \text{ sec}} = 88 \text{ ft sec}^{-1}$$

Note that if any factors had been omitted or included "upside down," the necessary cancellation of units would not have worked.

Once the dimensions desired for the answer are obtained, the numbers are placed in scientific notation for calculation of the final answer. It is important that an "estimation" step be included as a check on the reasonableness of an answer. The estimation step involves rounding all numerical quantities to the nearest whole number and performing indicated operations to determine the order of magnitude expected for the final answer.

EXAMPLE 8

Convert 62.4 pounds per cubic foot (62.4 lb ft^{-3}) to units of grams per cubic centimeter.

SOLUTION

First, convert all dimensions:

$$\frac{62.4\ lb}{1\ ft^3} \times \frac{454\ g}{1.00\ lb} \times \frac{1\ ft^3}{(12\ in.)^3} \times \frac{(1\ in.)^3}{(2.54\ cm)^3} = \frac{(62.4 \times 454)\ g}{[(12)^3 \times (2.54^3]\ cm^3}$$

Second, convert to scientific notation (see Section E):

$$\frac{(6.24 \times 10^1 \times 4.54 \times 10^2)\ g}{[(1.2 \times 10^1)^3 \times (2.54)^3]\ cm^3} = \frac{(6.24 \times 4.54 \times 10^3)\ g}{[(1.2)^3 \times (2.54)^3 \times 10^3]\ cm^3}$$

Third, estimate the order of magnitude:

$6.25 \approx 6$
$4.54 \approx 5$
$1.2\ \ \approx 1$
$2.54 \approx 3$

$$\frac{6 \times 5 \times 10^3}{(1)^3 \times (3)^3 \times 10^3} = \frac{30 \times 10^3}{27 \times 10^3} = \sim 1$$

Hence, the final answer should be closer to 1 than to, for example, 0.1 or 10.

Finally, perform the calculation (e.g., using a slide rule):

$$\frac{(6.24 \times 4.54 \times 10^3)\ g}{[(1.2)^3 \times (2.54)^3 \times 10^3]\ cm^3} = 1.00\ g\ cm^{-3} \qquad \text{(to three-place accuracy)}$$

Section C: Significant Figures

The numerical value of a measured quantity is limited by both the precision and the accuracy of the measurements involved. Precision is concerned with how close together the results of a series of measurements of the same quantity are, whereas accuracy is determined by how close a measured value is to the "true" value (usually compared to an exactly defined standard).

For example, suppose someone wishes to know the time at which an event took place. He might consult his wristwatch. However, if for some reason he wants to be more certain of the time, he might check his watch against his alarm clock or even use the telephone to dial "time and temperature." The chances are that he would arrive at three different values, and he might choose to average these values. However, if the three values were widely different (poor precision), he might wish to consult a number of other timepieces. If he can obtain several independent readings that differ only slightly, then he feels a certain confidence in accepting the average value. If the knowledge of the time were really important, then he might telephone the National Bureau of Standards and set ("calibrate") his watch against an accepted standard. This procedure would guarantee an improvement in accuracy, whereas any improvement in precision alone would simply increase the confidence in the measured value. Ten very close readings from ten different electric clocks during a power failure would give good precision, but poor accuracy.

Measurements in science should, whenever possible, use measuring devices that have been calibrated against accepted standards and repetitive determinations whose precision indicates minimized errors in the use of the devices. There is still another limitation on the accuracy of a reported value; this involves the calculations made with measured quantities. Although pure numbers (dimensionless) and exactly defined whole-number conversion factors (e.g., 1 ft = 12 in., or 1 in. = 2.54 cm) do not limit the number of meaningful figures in a computed quantity, other factors do. Common sense suggests that one cannot improve the accuracy of a measured value simply by performing mathematical operations. The accuracy of a computed value (number of significant figures reported) cannot exceed that of the least accurately known quantity used in the calculation. In most problem situations at the introductory level, it is routine to use "slide rule accuracy"; that is, answers are reported with the number of significant figures that can be read from a simple slide rule. One must, however, be aware of the limitations described above and apply them when necessary. (See also Sections D and E for exponential manipulations.)

EXAMPLE 9

The diameter of a cylinder is estimated, using a yard-stick, to be 1.1 in. The diameter of a second cylinder is measured carefully with a calibrated micrometer and found to be 0.8417 ± 0.0002 in. What value can meaningfully be reported as the sum of the diameters of the two cylinders?

SOLUTION

> 1.1
> 0.8417
> ‾‾‾‾‾‾
> 1.9417

But the less accurately known quantity is measured only to the nearest 0.1 in., so the sum can be reported only as approximately 1.9 in.

EXAMPLE 10

How many samples of 25.0 mg each could be prepared from approximately 2 lb of a crystalline salt?

SOLUTION

$$\frac{1 \text{ sample}}{25.0 \text{ mg}} \times \frac{10^3 \text{ mg}}{1 \text{ g}} \times \frac{454 \text{ g}}{1 \text{ lb}} \times \frac{2 \text{ lb}}{1} = \underline{\hspace{1cm}} \text{ samples}$$

Estimating

$$\frac{10^3 \times 5 \times 10^2 \times 2}{2.5 \times 10^1} = 4 \times 10^4$$

$$\frac{10^3 \times 4.54 \times 10^2 \times 2}{2.50 \times 10^1} = 3.63 \times 10^4$$

However, the number of significant figures is limited by the measured quantity "approximately 2 lb." (Note: not by the exact figures "1 sample," "10^3 mg," "1 g," or "1 lb.") Thus the answer can be reported only as "approximately 4×10^4 samples."

The proper use of scientific notation clearly indicates the number of significant figures.

Suggestion: To convince yourself that the above solution is correct, remember that "approximately 2 lb" would generally indicate a weight between 1.5 and 2.5 lb (i.e., values in this range would "round off" to 2 lb). Calculate the number of samples if 1.5 lb were available and if 2.5 lb were available. Then explain why rounding to one digit was reasonable for this example.

Section D: Manipulation of Exponents

Scientific calculations are frequently handled most conveniently by expressing quantities in scientific notation (see Section E). Such operations require simple manipulation of exponents, usually exponents of 10. When the same base (e.g., 10) is involved, the following rules apply:

1. When the operation involves multiplication, add the exponents algebraically; for example,

$$10^3 \times 10^2 = 10^{(3+2)} = 10^5$$
$$10^4 \times 10^8 \times 10^{-5} = 10^{[4+8+(-5)]} = 10^7$$

2. When the operation involves division, subtract the divisor exponent from the numerator exponent; for example,

$$\frac{10^5}{10^3} = 10^{(5-3)} = 10^2$$

$$\frac{10^7}{10^{11}} = 10^{(7-11)} = 10^{-4}$$

$$\frac{10^8}{10^{-2}} = 10^{[8-(-2)]} = 10^{10}$$

$$\frac{10^6 \times 10^4}{10^3 \times 10^9} = 10^{[6+4-(3+9)]} = 10^{-2}$$

435

3. When the operation involves roots or powers, divide the exponent by the root number or multiply the exponent by the power number, respectively; for example,

$$\sqrt{10^4} = (10^4)^{1/2} = 10^{4/2} = 10^2$$
$$\sqrt[5]{10^{20}} = (10^{20})^{1/5} = 10^{20/5} = 10^4$$
$$(10^2)^3 = 10^{2 \times 3} = 10^6$$
$$(10^{-7})^4 = 10^{-7 \times 4} = 10^{-28}$$

[Note: To take "roots" for cases in which the exponent is not a multiple of the "root number," use logarithms (see Section F).]

The operations described apply only to cases in which the same base number is involved. To illustrate this point, show that $2^3 \times 10^2$ is not equal to $(2 \times 10)^5$.

Section E: Scientific Notation

It is inconvenient to work with the standard form of very large or very small numbers. An exponential notation (see also Section D) is commonly employed to simplify calculations involving such numbers. The procedure for converting from standard form to scientific (exponential) notation* follows.

Step 1: Reset the decimal point so that there is a single digit (not zero) to its left, for example,

$$0.00044 \longrightarrow 0.0004\,4 \longrightarrow 0.0004_{\circ}4$$

Step 2: Count the number of digits between the original and new decimal point positions to determine the magnitude of the exponent of 10, for example,

$$10^4$$

Step 3: Determine the sign for the exponent of 10 by the direction from the new to the original decimal point position (left is negative, right is positive), for example,

$$0.0004_{\circ}4$$
$$\longleftarrow$$
$$10^{-4}$$

Thus,

$$0.00044 = 4.4 \times 10^{-4}$$

EXAMPLE 11

Assume that the approximate diameter of the solar system, to the nearest million miles, is 7,300,000,000. Write this in scientific notation.

SOLUTION

Step 1:

(understood location of decimal point)

$$7,\!.300,\!000,\!000_{\circ}$$

Step 2:

$$7_{\circ}300,\!000,\!000.$$
$$10^9$$

---

*In scientific notation, the only zeros recorded are those that represent actual measured quantities. Zeros used in standard form merely to indicate numerical magnitude are omitted.

Step 3:

7.300,000,000.

(Direction indicates positive exponent.)

Since the measured quantity is known only to the nearest million, the last six zeros (indicating order of magnitude) are omitted. Thus 7,300,000,000 miles (to the nearest million) is written as 7.300×10^9 miles.

EXAMPLE 12

Write 0.00020 in. in scientific notation.

SOLUTION

Step 1:

0.0002₀0

Step 2:

0.0002₀0

10^4

Step 3:

0.0002₀0

(Direction indicates negative exponent.)

The four zeros preceding the 2 are not measured quantities but only represent order of magnitude. The final zero does indicate a measurement to the nearest 0.00001 in. Thus 0.00020 in. is written as 2.0×10^{-4} in.

Section F: Logarithms

Chemical calculations normally use base 10 logarithms. These logarithms are exponents of 10 and, as such, are manipulated as described in Section D. Logarithms consist of a whole number (characteristic) followed by a decimal number (mantissa) or zero. The mantissa is found from a logarithm table.

1. To find the logarithm of a number:
   a. Write the number in scientific notation (Section E), for example,

   $$343 = 3.43 \times 10^2$$

   b. Look up the logarithm (mantissa) of the number preceding the power of 10 in a log table. (Note: All such mantissas are decimal numbers, although most log tables do not show the decimal point.) For example,

   $$\log 3.43 = 0.5353$$

   c. Add this to the exponent of 10. For example,

   $$\log 343 = 0.5353 + 2 = 2.5353$$

In some cases the log will be a negative number. This is perfectly legitimate, although other forms are more frequently used. Operational uses of logs in many chemical calculations are simpler using negative logs. For example,

$$\log 0.0200 = ?$$

Step a: $0.0200 = 2.00 \times 10^{-2}$
Step b: $\log 2.00 = 0.3010$
Step c: $\log 0.0200 = 0.3010 + (-2) = -1.6990$

(Entire logarithm is negative)

2. To find a number from its logarithm:*

    a. Be sure the log is written in a form such that the decimal part (mantissa) is positive. (That is how the logs are conventionally tabulated.) For example, what number has the logarithm -4.3010? Add the log to the next largest whole number, then subtract that number. For example,

$$(-4.3010 + 5) - 5 = 0.6990 - 5$$

    b. Look up the mantissa in the logarithm part of the table and find the number to which it most closely corresponds. For example, the number 5.00 corresponds to 0.6990 in the mantissa part of the table.

    c. Write this number as the first part of a scientific notation. Then use the whole number in the log as the exponent of 10. For example,

from log table $\left[\begin{array}{l} 0.6990\ (-5) \\ 5.00 \times 10^{-5} \end{array}\right.$

If the original log is positive, follow steps **b** and **c** above. For example, what number has the logarithm 11.0828?

from log table   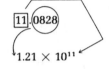

                      1.21×10^{11}

3. To multiply or divide, using logs:

    a. Write the operation in fractional form. (Use scientific notation.) For example,

$$\frac{(7.1 \times 10^{-3}) \times (1.48 \times 10^2) \times 2.73}{8.1 \times 6.02 \times 10^{23}}$$

    b. Combine all powers of 10 into a single power of 10 to appear in the numerator (see Section D). For example,

$$\frac{7.1 \times 1.48 \times 2.73}{8.1 \times 6.02} \times 10^{(-3+2-23)} = \frac{7.1 \times 1.48 \times 2.73}{8.1 \times 6.02} \times 10^{-24}$$

    c. Obtain the sum of the logs of all numbers in the numerator. Subtract from this the sum of the logs of all numbers in the denominator. For example,

Numerator	Denominator	
log 7.1 = 0.8513	log 8.1 = 0.9085	1.4578
log 1.48 = 0.1703	log 6.02 = 0.7796	−1.6881
log 2.73 = 0.4362	sum: 1.6881	−0.2303
sum: 1.4578		

    d. Find the number that has this logarithm (Section F, 2). Combine it with the power of 10 found in step **b** above. For example,

$$(-.2303 + 1) - 1 = .7697\ (-1)$$

from log table $.7697 - 1$

                5.88×10^{-1}

---

* The number is calculated by raising 10 to the power of the logarithm.

$$\text{number} = 10^{\text{LOG}}$$
$$y^x = 10^{-4.3010} = 5.00 \times 10^{-5}$$
$$y^x = 10^{11.0828} = 1.21 \times 10^{11}$$

Then

(step **b**)

$5.88 \times 10^{-1} \times 10^{-24} = 5.88 \times 10^{-25}$

Note that if any of the numbers in the calculation are measured quantities or other than exactly defined conversion units, round off the final answer to the proper number of significant figures (see Section C).

4. To find roots and powers, using logs: Since logs are exponents of 10, roots and powers are found in the same way as in any other exponential situation (Section D).

Roots

$$\sqrt[3]{1728} = ?$$

a. Find the log of the number.

$$\log(1728) = \log(1.728 \times 10^3) = 3.2375$$

b. Divide the log by the "root number."

$$\log \sqrt[3]{1728} = \tfrac{1}{3}(3.2375) = 1.0792$$

c. Find the antilog (the number that has this logarithm).

$$\sqrt[3]{1728} = \text{antilog } 1.0792 = 1.2 \times 10^1$$

Powers

$$(25)^2 = ?$$

a. Find the log of the number.

$$\log(25) = \log(2.5 \times 10^1) = 1.3979$$

b. Multiply the log by the power number.

$$\log(25)^2 = 2 \times 1.3979 = 2.7958$$

c. Find the antilog.

$$(25)^2 = \text{antilog } 2.7958 = 6.25 \times 10^2$$

LOGARITHM TABLE

Natural numbers	0	1	2	3	4	5	6	7	8	9	Proportional parts								
											1	2	3	4	5	6	7	8	9
10	0000	0043	0086	0128	0170	0212	0253	0294	0334	0374	4	8	12	17	21	25	29	33	37
11	0414	0453	0492	0531	0569	0607	0645	0682	0719	0755	4	8	11	15	19	23	26	30	34
12	0792	0828	0864	0899	0934	0969	1004	1038	1072	1106	3	7	10	14	17	21	24	28	31
13	1139	1173	1206	1239	1271	1303	1335	1367	1399	1430	3	6	10	13	16	19	23	26	29
14	1461	1492	1523	1553	1584	1614	1644	1673	1703	1732	3	6	9	12	15	18	21	24	27
15	1761	1790	1818	1847	1875	1903	1931	1959	1987	2014	3	6	8	11	14	17	20	22	25
16	2041	2068	2095	2122	2148	2175	2201	2227	2253	2279	3	5	8	11	13	16	18	21	24
17	2304	2330	2355	2380	2405	2430	2455	2480	2504	2529	2	5	7	10	12	15	17	20	22
18	2553	2577	2601	2625	2648	2672	2695	2718	2742	2765	2	5	7	9	12	14	16	19	21
19	2788	2810	2833	2856	2878	2900	2923	2945	2967	2989	2	4	7	9	11	13	16	18	20
20	3010	3032	3054	3075	3096	3118	3139	3160	3181	3201	2	4	6	8	11	13	15	17	19
21	3222	3243	3263	3284	3304	3324	3345	3365	3385	3404	2	4	6	8	10	12	14	16	18
22	3424	3444	3464	3483	3502	3522	3541	3560	3579	3598	2	4	6	8	10	12	14	15	17
23	3617	3636	3655	3674	3692	3711	3729	3747	3766	3784	2	4	6	7	9	11	13	15	17
24	3802	3820	3838	3856	3874	3892	3909	3927	3945	3962	2	4	5	7	9	11	12	14	16
25	3979	3997	4014	4031	4048	4065	4082	4099	4116	4133	2	3	5	7	9	10	12	14	15
26	4150	4166	4183	4200	4216	4232	4249	4265	4281	4298	2	3	5	7	8	10	11	13	15
27	4314	4330	4346	4362	4378	4393	4409	4425	4440	4456	2	3	5	6	8	9	11	13	14
28	4472	4487	4502	4518	4533	4548	4564	4579	4594	4609	2	3	5	6	8	9	11	12	14
29	4624	4639	4654	4669	4683	4698	4713	4728	4742	4757	1	3	4	6	7	9	10	12	13
30	4771	4786	4800	4814	4829	4843	4857	4871	4886	4900	1	3	4	6	7	9	10	11	13
31	4914	4928	4942	4955	4969	4983	4997	5011	5024	5038	1	3	4	6	7	8	10	11	12
32	5051	5065	5079	5092	5105	5119	5132	5145	5159	5172	1	3	4	5	7	8	9	11	12
33	5185	5198	5211	5224	5237	5250	5263	5276	5289	5302	1	3	4	5	6	8	9	10	12
34	5315	5328	5340	5353	5366	5378	5391	5403	5416	5428	1	3	4	5	6	8	9	10	11
35	5441	5453	5465	5478	5490	5502	5514	5527	5539	5551	1	2	4	5	6	7	9	10	11
36	5563	5575	5587	5599	5611	5623	5635	5647	5658	5670	1	2	4	5	6	7	8	10	11
37	5682	5694	5705	5717	5729	5740	5752	5763	5775	5786	1	2	3	5	6	7	8	9	10
38	5798	5809	5821	5832	5843	5855	5866	5877	5888	5899	1	2	3	5	6	7	8	9	10
39	5911	5922	5933	5944	5955	5966	5977	5988	5999	6010	1	2	3	4	5	7	8	9	10
40	6021	6031	6042	6053	6064	6075	6085	6096	6107	6117	1	2	3	4	5	6	8	9	10
41	6128	6138	6149	6160	6170	6180	6191	6201	6212	6222	1	2	3	4	5	6	7	8	9
42	6232	6243	6253	6263	6274	6284	6294	6304	6314	6325	1	2	3	4	5	6	7	8	9
43	6335	6345	6355	6365	6375	6385	6395	6405	6415	6425	1	2	3	4	5	6	7	8	9
44	6435	6444	6454	6464	6474	6484	6493	6503	6513	6522	1	2	3	4	5	6	7	8	9
45	6532	6542	6551	6561	6571	6580	6590	6599	6609	6618	1	2	3	4	5	6	7	8	9
46	6628	6637	6646	6656	6665	6675	6684	6693	6702	6712	1	2	3	4	5	6	7	7	8
47	6721	6730	6739	6749	6758	6767	6776	6785	6794	6803	1	2	3	4	5	5	6	7	8
48	6812	6821	6830	6839	6848	6857	6866	6875	6884	6893	1	2	3	4	4	5	6	7	8
49	6902	6911	6920	6928	6937	6946	6955	6964	6972	6981	1	2	3	4	4	5	6	7	8
50	6990	6998	7007	7016	7024	7033	7042	7050	7059	7067	1	2	3	3	4	5	6	7	8
51	7076	7084	7093	7101	7110	7118	7126	7135	7143	7152	1	2	3	3	4	5	6	7	8
52	7160	7168	7177	7185	7193	7202	7210	7218	7226	7235	1	2	2	3	4	5	6	7	7
53	7243	7251	7259	7267	7275	7284	7292	7300	7308	7316	1	2	2	3	4	5	6	6	7
54	7324	7332	7340	7348	7356	7364	7372	7380	7388	7396	1	2	2	3	4	5	6	6	7

Natural numbers	0	1	2	3	4	5	6	7	8	9	Proportional parts								
											1	2	3	4	5	6	7	8	9
55	7404	7412	7419	7427	7435	7443	7451	7459	7466	7474	1	2	2	3	4	5	5	6	7
56	7482	7490	7497	7505	7513	7520	7528	7536	7543	7551	1	2	2	3	4	5	5	6	7
57	7559	7566	7574	7582	7589	7597	7604	7612	7619	7627	1	2	2	3	4	5	5	6	7
58	7634	7642	7649	7657	7664	7672	7679	7686	7694	7701	1	1	2	3	4	4	5	6	7
59	7709	7716	7723	7731	7738	7745	7752	7760	7767	7774	1	1	2	3	4	4	5	6	7
60	7782	7789	7796	7803	7810	7818	7025	7832	7839	7846	1	1	2	3	4	4	5	6	6
61	7853	7860	7868	7875	7882	7889	7896	7903	7910	7917	1	1	2	3	4	4	5	6	6
62	7924	7931	7938	7945	7952	7959	7966	7973	7930	7987	1	1	2	3	3	4	5	6	6
63	7993	8000	8007	8014	8021	8028	8035	8041	8048	8055	1	1	2	3	3	4	5	5	6
64	8062	8069	8075	8082	8089	8096	8102	8109	8116	8122	1	1	2	3	3	4	5	5	6
65	8129	8136	8142	8149	8156	8162	8169	8176	8182	8189	1	1	2	3	3	4	5	5	6
66	8195	8202	8209	8215	8222	8228	8235	8241	8248	8254	1	1	2	3	3	4	5	5	6
67	8261	8267	8274	8280	8287	8293	8299	8306	8312	8319	1	1	2	3	3	4	4	5	6
68	8325	8331	8338	8344	8351	8357	8363	8370	8376	8382	1	1	2	3	3	4	4	5	6
69	8388	8395	8401	8407	8414	8420	8426	8432	8439	8445	1	1	2	2	3	4	4	5	6
70	8451	8457	8463	8470	8476	8482	8408	8494	8500	8506	1	1	2	2	3	4	4	5	6
71	8513	8519	8525	8531	8537	8543	8549	8555	8561	8567	1	1	2	2	3	4	4	5	5
72	8573	8579	8585	8591	8597	8603	8609	8615	8621	8627	1	1	2	2	3	4	4	5	5
73	8633	8639	8645	8651	8657	8663	8669	8675	8681	8686	1	1	2	2	3	4	4	5	5
74	8692	8698	8704	8710	8716	8722	8727	8733	8739	8745	1	1	2	2	3	4	4	5	5
75	8751	8756	8762	8768	8774	8779	8785	8791	8797	8802	1	1	2	2	3	3	4	5	5
76	8808	8814	8820	8825	8831	8837	8842	8848	8854	8859	1	1	2	2	3	3	4	5	5
77	8865	8871	8876	8882	8887	8893	8899	8904	8910	8915	1	1	2	2	3	3	4	4	5
78	8921	8927	8932	8938	8943	8949	8954	8960	8965	8971	1	1	2	2	3	3	4	4	5
79	8976	8982	8987	8993	8998	9004	9009	9015	9020	9025	1	1	2	2	3	3	4	4	5
80	9031	9036	9042	9047	9053	9058	9063	9069	9074	9079	1	1	2	2	3	3	4	4	5
81	9085	9090	9096	9191	9106	9112	9117	9122	9128	9133	1	1	2	2	3	3	4	4	5
82	9138	9143	9149	9154	9159	9165	9170	9175	9180	9186	1	1	2	2	3	3	4	4	5
83	9191	9196	9201	9206	9212	9217	9222	9227	9232	9238	1	1	2	2	3	3	4	4	5
84	9243	9248	9253	9258	9203	9269	9274	9279	9284	9289	1	1	2	2	3	3	4	4	5
85	9294	9299	9304	9309	9315	9320	9325	9330	9335	9340	1	1	2	2	3	3	4	4	5
86	9345	9350	9355	9360	9365	9370	9375	9380	9385	9390	1	1	2	2	3	3	4	4	5
87	9395	9400	9405	9410	9415	9420	9425	9430	9435	9440	0	1	1	2	2	3	3	4	4
88	9445	9450	9455	9460	9465	9469	9474	9479	9484	9489	0	1	1	2	2	3	3	4	4
89	9494	9499	9504	9509	9513	9518	9523	9528	9533	9538	0	1	1	2	2	3	3	4	4
90	9542	9547	9552	9557	9562	9566	9571	9576	9581	9586	0	1	1	2	2	3	3	4	4
91	9590	9595	9600	9605	9609	9614	9619	9624	9628	9633	0	1	1	2	2	3	3	4	4
92	9638	9643	9647	9652	9657	9661	9666	9671	9675	9680	0	1	1	2	2	3	3	4	4
93	9685	9689	9694	9699	9703	9708	9713	9717	9722	9727	0	1	1	2	2	3	3	4	4
94	9731	9736	9741	9745	9750	9754	9759	9763	9768	9773	0	1	1	2	2	3	3	4	4
95	9777	9782	9786	9791	9795	9800	9805	9809	9814	9818	0	1	1	2	2	3	3	4	4
96	9823	9827	9832	9836	9841	9845	9850	9854	9859	9863	0	1	1	2	2	3	3	4	4
97	9868	9872	9877	9881	9886	9890	9894	9899	9903	9908	0	1	1	2	2	3	3	4	4
98	9912	9917	9921	9926	9930	9934	9939	9943	9948	9952	0	1	1	2	2	3	3	4	4
99	9956	9961	9965	9969	9974	9978	9983	9987	9991	9996	0	1	1	2	2	3	3	3	4

APPENDIX C: CONSTANTS AND CONVERSION FACTORS

1. Metric System Prefixes[1]

giga (G) = 10^9
mega (M) = 10^6
kilo (k) = 10^3
deci (d) = 10^{-1}
centi (c) = 10^{-2}
milli (m) = 10^{-3}
micro (μ) = 10^{-6}
nano (n) = 10^{-9}
pico (p) = 10^{-12}

2. Metric/English Conversions

a. *Length*
2.54 cm = 1 in. (exactly defined)
1.0000 m = 39.370 in.
1.0000 km = 0.62137 mile

b. *Volume*
(Metric "bulk"/volume)
1.0000 liter = 1.0567 qt (liquid)
1.0000 m^3 = 35.314 ft^3

c. *Mass and weight* (Compared at Standard Earth Gravity)
1.0000 kg = 2.2046 lb (avoirdupois)
1.0000 ton (av.) = 907.18 kg
1.0000 lb = 453.59 g

d. *Pressure*
1.0000 atmosphere (atm) = 760.00 torr = 760.00 mm Hg

e. *Temperature*
°C (centigrade or Celsius) = $\frac{5}{9}$(°F − 32)
°K (absolute or Kelvin) = °C + 273.15°

3. International System of Units (SI)

In 1960, an agreement was reached on a system of international units (the *Système International*, SI). In many respects, the new system is based on the metric units, first used in 1790. Although many scientific journals and textbooks have adopted the SI units, they have not yet gained universal acceptance. Throughout this book, we have used, for the most part, units consistent with SI, with three major exceptions. In referring to very small dimensions, we have sometimes employed the *angstrom* (Å) = 10^{-10} meter, in deference to common practice in the bulk of the chemical literature prior to 1960. For pressures, we have employed either *atmospheres* or *torr*, which are more convenient than is the SI unit of *pascal* (Pa) = newton meter^{-2}. Finally, we have chosen to use *calories* as the basic unit of energy, rather than the SI unit *joule* (J). This particular choice is, perhaps, the most difficult to defend because it is the most strongly contested by those scientists concerned with precision of terminology. The choice was made on the basis of the fact that the majority of the currently available chemical literature employs the calorie as an energy unit. There is, beyond question, a definite trend toward the SI units, and you should be aware of at least the most common of these.[2]

Some Basic SI Units

Physical quantity	SI unit (symbol)[3]
length	meter (m)
mass	kilogram (kg)
time	second (s)
electric current	ampere (A)
thermodynamic temperature	Kelvin (K)

Some Derived SI Units

Physical quantity	SI unit (symbol)[3]	Definition of unit[3]
energy	joule (J)	kg m^2 s^{-2}
force	newton (N)	kg m s^{-2}
electric charge	coulomb (C)	A s
electric potential difference	volt (V)	kg m^2 s^{-3} A^{-1}
frequency	hertz (Hz)	cycles s^{-1}
pressure	pascal (Pa)	N m^{-2}
temperature (customary)	degree Celcius (°C)	K − 273.15°

Some Units Accepted by SI (but not recommended)

Physical quantity	Unit (symbol)	SI relationship
volume	liter (l)	10^{-3} m^3
energy	electron volt (eV)	1.6021×10^{-19} J
mass	tonne [metric ton](t)	10^3 kg

Some Units Not Accepted by SI (but employed in this textbook)

Physical quantity	Unit (symbol)	SI equivalent
length	angstrom (Å)	10^{-10} m
volume	milliliter (ml)	cm^3 (10^{-6} m^3)
force	dyne (dyn)	10^{-5} N
pressure	atmosphere (atm)	101.325 kN m^{-2}
	torr (torr)	133.322 N m^{-2}
energy	erg (erg)	10^{-7} J
	calorie (cal)	4.184 J

4. Mathematical Formulas

area of a rectangle = length × width
area of a triangle = $\frac{1}{2}$(length of base × height)
area of a circle = π × (radius)^2
surface area of a sphere = π × (diameter)^2
volume of an orthoganal box = length × width × height
volume of a sphere = $\frac{4}{3}\pi$ (radius)^3

5. Logarithmic Manipulation (base 10)

$\log(m \times n) = \log m + \log n$
$\log(m/n) = \log m - \log n$
$\log(1/m) = -\log m$
$\log(m)^x = x \log m$
$\log(\sqrt[x]{m}) = (\log m)/x$

6. Physical constants

absolute zero = 0°K = −273.15°C
atomic mass unit (amu) = 1.6602×10^{-24} g
Avogadro's number = 6.0225×10^{23} unit particles mole^{-1}
Boltzmann's constant = 1.3805×10^{-16} erg °K^{-1} per molecule
charge on the electron = 1.6021×10^{-19} C
Faraday's constant (F) = 9.6487×10^4 C mole^{-1} (electrons)
gravitational acceleration = 980.66 cm sec^{-2}
molar gas constant (R) = 0.08206 liter atm mole^{-1} °K^{-1}
 = 1.987 cal mole^{-1} °K^{-1}
molar volume = 22.414 liters mole^{-1} (at STP)
Planck's constant (h) = 6.6257×10^{-27} erg sec
 (1.00 erg = 2.39×10^{-8} cal)
speed of light (c) = 2.9979×10^{10} cm sec^{-1}
standard temperature and pressure (STP) = 273.15°K, 1.000 atm

---

[1] Compound prefixes are to be avoided. For example, 10^{-8} meter could be expressed as 10 nanometers, but not as 10 millimicrometers. Note also that attaching a prefix to a unit constitutes a *new* unit. For example, 1 mm^3 = (10^{-3} m)^3 = 10^{-9} m^3, and not 1 m (m)^3 = 10^{-3} m^3.

[2] For further details, see "International System of Units (SI)," by Martin A. Paul in *Chemistry*, Vol. 45, p. 14 (1972) and "The International System of Units (SI)," *National Bureau of Standards Special Publication 330*, 1972, U.S. Government Printing Office, Washington, D.C.

[3] Note that we have used more generally employed abbreviations for time in seconds (sec) and current in amperes (amp) throughout this book, and we have used deg K (or °K), rather than just K alone.

Table 1 ACID DISSOCIATION CONSTANTS

ACID	K_a
acetic (CH_3CO_2H)	1.8×10^{-5}
benzoic ($C_6H_5CO_2H$)	6.6×10^{-5}
boric (ortho) (H_3BO_3)	6.0×10^{-10} (K_{a_1})
carbonic ($H_2O + CO_2$)	4.2×10^{-7} (K_{a_1})
	4.8×10^{-11} (K_{a_2})
formic (HCO_2H)	2.0×10^{-4}
hydrobromic (HBr)	$\sim 10^0$
hydrochloric (HCl)	$\sim 10^6$
hydrocyanic (HCN)	4.8×10^{-10}
hydrofluoric (HF)	6.9×10^{-4}
hydroiodic (HI)	$\sim 10^9$
hydrosulfuric (H_2S)	1.0×10^{-7} (K_{a_1})
	1.3×10^{-13} (K_{a_2})
hypochlorous (HOCl)	3.2×10^{-8}
nitric (HNO_3)	~ 30
nitrous (HNO_2)	4.5×10^{-4}
oxalic ($H_2C_2O_4$)	6.3×10^{-2} (K_{a_1})
	6.3×10^{-5} (K_{a_2})
perchloric ($HClO_4$)	$\sim 10^{10}$
phosphoric (ortho) (H_3PO_4)	7.5×10^{-3} (K_{a_1})
	6.2×10^{-8} (K_{a_2})
	2.0×10^{-13} (K_{a_3})
phosphorous (H_3PO_3)	1.6×10^{-2} (K_{a_1})
	6.9×10^{-7} (K_{a_2})
propionic ($CH_3CH_2CO_2H$)	1.4×10^{-5}
sulfuric (H_2SO_4)	$\sim 10^3$ (K_{a_1})
	1.3×10^{-2} (K_{a_2})
sulfurous (H_2SO_3)	1.6×10^{-2} (K_{a_1})
	1.3×10^{-7} (K_{a_2})

Table 2 BASE IONIZATION CONSTANTS

BASE	K_b
ammonia (NH_3)	1.8×10^{-5}
aniline ($C_6H_5NH_2$)	3.8×10^{-10}
dimethylamine [$(CH_3)_2NH$]	5.1×10^{-4}
ethylamine ($CH_3CH_2NH_2$)	5.6×10^{-4}
methoxide ion (CH_3O^-)	$\sim 10^3$
methylamine (CH_3NH_2)	5.0×10^{-4}
trimethylamine [$(CH_3)_3N$]	5.3×10^{-5}

Table 3 SOLUBILITY PRODUCTS

(Most simple salts not listed in this table are soluble in water.)

SALT	K_{sp}	SALT	K_{sp}
Acetates		**Hydroxides**	
$Ag(CH_3CO_2)$	4.1×10^{-3}	(for amphoteric hydroxides, see Table 5)	
$Hg_2(CH_3CO_2)_2$	3.6×10^{-10}	$Ca(OH)_2$	4.0×10^{-6}
Arsenates		$Cd(OH)_2$	2.0×10^{-14}
Ag_3AsO_4	1.0×10^{-22}	$Co(OH)_2$	2.0×10^{-16}
Bromates		$Cu(OH)_2$	1.8×10^{-19}
$AgBrO_3$	6.0×10^{-5}	$Fe(OH)_2$	1.8×10^{-15}
$Ba(BrO_3)_2$	5.6×10^{-6}	$Fe(OH)_3$	6.0×10^{-38}
Bromides		$Hg(OH)_2$	3.2×10^{-26}
$AgBr$	5.0×10^{-13}	$Mg(OH)_2$	1.2×10^{-11}
Hg_2Br_2	5.0×10^{-23}	$Mn(OH)_2$	2.0×10^{-13}
$PbBr_2$	1.0×10^{-6}	**Iodides**	
Carbonates		AgI	8.5×10^{-17}
Ag_2CO_3	8.2×10^{-12}	Hg_2I_2	4.5×10^{-29}
$BaCO_3$	1.6×10^{-9}	HgI_2	2.5×10^{-26}
$CaCO_3$	4.7×10^{-9}	PbI_2	8.3×10^{-9}
$CuCO_3$	2.5×10^{-10}	**Phosphates**	
$FeCO_3$	2.0×10^{-11}	Ag_3PO_4	1.6×10^{-18}
$MgCO_3$	4.0×10^{-5}	$Ba_3(PO_4)_2$	3.2×10^{-23}
$NiCO_3$	1.4×10^{-7}	$Ca_3(PO_4)_2$	1.3×10^{-32}
$PbCO_3$	1.5×10^{-13}	Li_3PO_4	3.2×10^{-13}
$SrCO_3$	7.0×10^{-10}	$Sr_3(PO_4)_2$	4.0×10^{-28}
$ZnCO_3$	3.0×10^{-11}	**Sulfates**	
Chlorides		Ag_2SO_4	6.4×10^{-5}
$AgCl$	2.4×10^{-10}	$BaSO_4$	7.9×10^{-11}
Hg_2Cl_2	1.2×10^{-18}	$CaSO_4$	2.5×10^{-5}
$PbCl_2$	1.6×10^{-5}	$PbSO_4$	1.3×10^{-8}
Chromates		$SrSO_4$	3.2×10^{-7}
Ag_2CrO_4	1.9×10^{-12}	**Sulfides**	
$BaCrO_4$	8.5×10^{-11}	Ag_2S	5.5×10^{-51}
$CaCrO_4$	7.1×10^{-4}	CdS	1.0×10^{-28}
$PbCrO_4$	1.8×10^{-14}	CoS	5.0×10^{-22}
$SrCrO_4$	5.7×10^{-5}	CuS	8.0×10^{-37}
Fluorides		FeS	4.0×10^{-19}
BaF_2	1.0×10^{-7}	Fe_2S_3	$\sim 10^{-88}$
CaF_2	1.7×10^{-10}	HgS	1.6×10^{-54}
LiF	$\sim 10^{-2}$	MnS	7.0×10^{-16}
MgF_2	6.8×10^{-9}	NiS	3.0×10^{-21}
SrF_2	2.5×10^{-9}	PbS	7.0×10^{-29}
		SnS	1.1×10^{-26}
		ZnS	1.6×10^{-23}

Table 4 FORMATION CONSTANTS OF COMPLEXES

$$(K_{net} = K_1 \times K_2 \times K_3 \cdots)$$

EQUILIBRIUM	K
Ammine complexes	
$[Ag(H_2O)_2]^+ + NH_3 \rightleftharpoons [Ag(H_2O)(NH_3)]^+ + H_2O$	2.0×10^3 (K_1)
$[Ag(H_2O)(NH_3)]^+ + NH_3 \rightleftharpoons [Ag(NH_3)_2]^+ + H_2O$	8.0×10^3 (K_2)
$[Cd(H_2O)_6]^{2+} + NH_3 \rightleftharpoons [Cd(H_2O)_5(NH_3)]^{2+} + H_2O$	4.5×10^2 (K_1)
$[Cd(H_2O)_5(NH_3)]^{2+} + NH_3 \rightleftharpoons [Cd(H_2O)_4(NH_3)_2]^{2+} + H_2O$	1.3×10^2 (K_2)
$[Cd(H_2O)_4(NH_3)_2]^{2+} + NH_3 \rightleftharpoons [Cd(H_2O)_3(NH_3)_3]^{2+} + H_2O$	2.8×10^1 (K_3)
$[Cd(H_2O)_3(NH_3)_3]^{2+} + NH_3 \rightleftharpoons [Cd(H_2O)_2(NH_3)_4]^{2+} + H_2O$	8.5×10^0 (K_4)
$[Co(H_2O)_6]^{2+} + NH_3 \rightleftharpoons [Co(H_2O)_5(NH_3)]^{2+} + H_2O$	4.5×10^2 (K_1)
$[Co(H_2O)_5(NH_3)]^{2+} + NH_3 \rightleftharpoons [Co(H_2O)_4(NH_3)_2]^{2+} + H_2O$	3.0×10^1 (K_2)
$[Co(H_2O)_4(NH_3)_2]^{2+} + NH_3 \rightleftharpoons [Co(H_2O)_3(NH_3)_3]^{2+} + H_2O$	8.0×10^0 (K_3)
$[Co(H_2O)_3(NH_3)_3]^{2+} + NH_3 \rightleftharpoons [Co(H_2O)_2(NH_3)_4]^{2+} + H_2O$	4.0×10^0 (K_4)
$[Co(H_2O)_2(NH_3)_4]^{2+} + NH_3 \rightleftharpoons [Co(H_2O)(NH_3)_5]^{2+} + H_2O$	1.3×10^0 (K_5)
$[Co(H_2O)(NH_3)_5]^{2+} + NH_3 \rightleftharpoons [Co(NH_3)_6]^{2+} + H_2O$	2.0×10^{-1} (K_6)
$[Cu(H_2O)_6]^{2+} + NH_3 \rightleftharpoons [Cu(H_2O)_5(NH_3)]^{2+} + H_2O$	1.4×10^4 (K_1)
$[Cu(H_2O)_5(NH_3)]^{2+} + NH_3 \rightleftharpoons [Cu(H_2O)_4(NH_3)_2]^{2+} + H_2O$	3.2×10^3 (K_2)
$[Cu(H_2O)_4(NH_3)_2]^{2+} + NH_3 \rightleftharpoons [Cu(H_2O)_3(NH_3)_3]^{2+} + H_2O$	7.8×10^2 (K_3)
$[Cu(H_2O)_3(NH_3)_3]^{2+} + NH_3 \rightleftharpoons [Cu(H_2O)_2(NH_3)_4]^{2+} + H_2O$	1.4×10^2 (K_4)
net $[Zn(H_2O)_4]^{2+} + 4NH_3 \rightleftharpoons [Zn(NH_3)_4]^{2+} + 4H_2O$	2.8×10^9 (K_{net})

TABLE 4 (Continued) $(K_{net} = K_1 \times K_2 \times K_3 \cdots)$

Equilibrium	K
Chloro complexes	
$[Cd(H_2O)_6]^{2+} + Cl^- \rightleftharpoons [CdCl(H_2O)_5]^+ + H_2O$	1.0×10^2 (K_1)
$[CdCl(H_2O)_5]^+ + Cl^- \rightleftharpoons [CdCl_2(H_2O)_4] + H_2O$	5.0×10^0 (K_2)
$[CdCl_2(H_2O)_4] + Cl^- \rightleftharpoons [CdCl_3(H_2O)_3]^- + H_2O$	4.0×10^{-1} (K_3)
$[CdCl_3(H_2O)_3]^- + Cl^- \rightleftharpoons [CdCl_4(H_2O)_2]^{2-} + H_2O$	2.0×10^{-1} (K_4)
$[Hg(H_2O)_4]^{2+} + Cl^- \rightleftharpoons [HgCl(H_2O)_3]^+ + H_2O$	5.5×10^6 (K_1)
$[HgCl(H_2O)_3]^+ + Cl^- \rightleftharpoons [HgCl_2(H_2O)_2] + H_2O$	3.0×10^6 (K_2)
$[HgCl_2(H_2O)_2] + Cl^- \rightleftharpoons [HgCl_3(H_2O)]^- + H_2O$	7.1×10^0 (K_3)
$[HgCl_3(H_2O)]^- + Cl^- \rightleftharpoons [HgCl_4]^{2-} + H_2O$	1.0×10^1 (K_4)
net $\quad [Pb(H_2O)_4]^{2+} + 3Cl^- \rightleftharpoons [PbCl_3(H_2O)]^- + 3H_2O$	2.5×10^1 (K_{net})
net $\quad [Sn(H_2O)_4]^{2+} + 4Cl^- \rightleftharpoons [SnCl_4]^{2-} + 4H_2O$	1.8×10^1 (K_{net})
net $\quad [Sn(H_2O)_6]^{4+} + 6Cl^- \rightleftharpoons [SnCl_6]^{2-} + 6H_2O$	6.6×10^0 (K_{net})
Cyano complexes	
$[Cd(H_2O)_6]^{2+} + CN^- \rightleftharpoons [Cd(CN)(H_2O)_5]^+ + H_2O$	3.0×10^5 (K_1)
$[Cd(CN)(H_2O)_5]^+ + CN^- \rightleftharpoons [Cd(CN)_2(H_2O)_4] + H_2O$	1.3×10^5 (K_2)
$[Cd(CN)_2(H_2O)_4] + CN^- \rightleftharpoons [Cd(CN)_3(H_2O)_3]^- + H_2O$	4.3×10^4 (K_3)
$[Cd(CN)_3(H_2O)_3]^- + CN^- \rightleftharpoons [Cd(CN)_4(H_2O)_2]^{2-} + H_2O$	3.5×10^3 (K_4)
net $\quad [Fe(H_2O)_6]^{2+} + 6CN^- \rightleftharpoons [Fe(CN)_6]^{4-} + 6H_2O$	1.0×10^{24} (K_{net})
net $\quad [Fe(H_2O)_6]^{3+} + 6CN^- \rightleftharpoons [Fe(CN)_6]^{3-} + 6H_2O$	1.0×10^{31} (K_{net})
Fluoro complexes	
net $\quad [Al(H_2O)_6]^{3+} + 6F^- \rightleftharpoons [AlF_6]^{3-} + 6H_2O$	6.1×10^{19} (K_{net})
net $\quad [Fe(H_2O)_6]^{3+} + 5F^- \rightleftharpoons [FeF_5(H_2O)]^{2-} + 5H_2O$	2.0×10^{15} (K_{net})
Thiocyanato complexes	
$[Co(H_2O)_6]^{2+} + NCS^- \rightleftharpoons [Co(NCS)(H_2O)_5]^+ + H_2O$	1.0×10^3 (K_1)
$[Co(NCS)(H_2O)_5]^+ + NCS^- \rightleftharpoons [Co(NCS)_2(H_2O)_4] + H_2O$	1.0×10^0 (K_2)
$[Co(NCS)_2(H_2O)_4] + NCS^- \rightleftharpoons [Co(NCS)_3(H_2O)_3]^- + H_2O$	2.0×10^{-1} (K_3)
$[Co(NCS)_3(H_2O)_3]^- + NCS^- \rightleftharpoons [Co(NCS)_4(H_2O)_2]^{2-} + H_2O$	1.0×10^0 (K_4)
$[Fe(H_2O)_6]^{3+} + NCS^- \rightleftharpoons [Fe(NCS)(H_2O)_5]^{2+} + H_2O$	1.1×10^3 (K_1)
Thiosulfato complexes	
net $\quad [Ag(H_2O)_2]^+ + 2S_2O_3^{2-} \rightleftharpoons [Ag(S_2O_3)_2]^{3-} + 2H_2O$	2.9×10^{13} (K_{net})

Table 5 SOME COMMON AMPHOTERIC HYDROXIDES

EQUILIBRIUM	K
Aluminum	
$[Al(H_2O)_6]^{3+}(aq) + H_2O(l) \rightleftharpoons [Al(OH)(H_2O)_5]^{2+}(aq) + H_3O^+(aq)$	$K_{a_1} = 1.4 \times 10^{-5}$
net $\quad Al(OH)_3(s) \rightleftharpoons Al^{3+}(aq) + 3OH^-(aq)$	$K_{sp} = 5.0 \times 10^{-33}$
$Al(OH)_3(s) + OH^-(aq) + 2H_2O(l) \rightleftharpoons [Al(OH)_4(H_2O)_2]^-(aq)$	$K_4 = 4.0 \times 10$
Chromium	
$[Cr(H_2O)_6]^{3+}(aq) + H_2O(l) \rightleftharpoons [Cr(OH)H_2O)_5]^{2+}(aq) + H_3O^+(aq)$	$K_{a_1} = 1.0 \times 10^{-10}$
net $\quad Cr(OH)_3(s) \rightleftharpoons Cr^{3+}(aq) + 3OH^-(aq)$	$K_{sp} = 7.0 \times 10^{-31}$
$Cr(OH)_3(s) + OH^-(aq) + 2H_2O(l) \rightleftharpoons [Cr(OH)_4(H_2O)_2]^-(aq)$	$K_4 = 9.0 \times 10^4$
Zinc	
$[Zn(H_2O)_4]^{2+}(aq) + H_2O(l) \rightleftharpoons [Zn(OH)(H_2O)_3]^+(aq) + H_3O^+(aq)$	$K_{a_1} = 2.5 \times 10^{-10}$
net $\quad Zn(OH)_2(s) \rightleftharpoons Zn^{2+}(aq) + 2OH^-(aq)$	$K_{sp} = 4.5 \times 10^{-17}$
net $\quad Zn(OH)_2(s) + 2OH^-(aq) \rightleftharpoons [Zn(OH)_4]^{2-}(aq)$	$K_{3,4} = 1.0 \times 10^{-15}$

APPENDIX E: STANDARD ELECTRODE POTENTIALS

Temperature = 298°K
All solids in most stable form
All gases at 1.00 atm pressure
All solutes at unit activity (approx. 1.0 M)

COUPLE	$\mathcal{E}°$ (volts)	COUPLE	$\mathcal{E}°$ (volts)
A. Solutions of pH $\leq$ 7.0		$Cu^+(aq) + e^- \rightleftharpoons Cu(s)$	+0.52
$Li^+(aq) + e^- \rightleftharpoons Li(s)$	−3.05	$I_2(aq) + 2e^- \rightleftharpoons 2I^-(aq)$	+0.54
$K^+(aq) + e^- \rightleftharpoons K(s)$	−2.93	$MnO_4^-(aq) + e^- \rightleftharpoons MnO_4^{2-}(aq)$	+0.56
$Cs^+(aq) + e^- \rightleftharpoons Cs(s)$	−2.92	$O_2(g) + 2H^+(aq) + 2e^- \rightleftharpoons H_2O_2(aq)$	+0.68
$Ra^{2+}(aq) + 2e^- \rightleftharpoons Ra(s)$	−2.92	$Fe^{3+}(aq) + e^- \rightleftharpoons Fe^{2+}(aq)$	+0.77
$Ba^{2+}(aq) + 2e^- \rightleftharpoons Ba(s)$	−2.90	$Hg_2^{2+}(aq) + 2e^- \rightleftharpoons 2Hg(l)$	+0.79
$Sr^{2+}(aq) + 2e^- \rightleftharpoons Sr(s)$	−2.89	$Ag^+(aq) + e^- \rightleftharpoons Ag(s)$	+0.80
$Ca^{2+}(aq) + 2e^- \rightleftharpoons Ca(s)$	−2.87	$2Hg^{2+}(aq) + 2e^- \rightleftharpoons Hg_2^{2+}(aq)$	+0.92
$Na^+(aq) + e^- \rightleftharpoons Na(s)$	−2.71	$HNO_2(aq) + H^+(aq) + e^- \rightleftharpoons NO(g) + H_2O(l)$	+1.00
$Mg^{2+}(aq) + 2e^- \rightleftharpoons Mg(s)$	−2.37	$Br_2(l) + 2e^- \rightleftharpoons 2Br^-(aq)$	+1.07
$Be^{2+}(aq) + 2e^- \rightleftharpoons Be(s)$	−1.87	$ClO_4^-(aq) + 2H^+(aq) + 2e^- \rightleftharpoons ClO_3^-(aq)$	
$Al^{3+}(aq) + 3e^- \rightleftharpoons Al(s)$	−1.66	$\quad + H_2O(l)$	+1.19
$Ti^{2+}(aq) + 2e^- \rightleftharpoons Ti(s)$	−1.63	$ClO_3^-(aq) + 3H^+(aq) + 2e^- \rightleftharpoons HClO_2(aq)$	
$Mn^{2+}(aq) + 2e^- \rightleftharpoons Mn(s)$	−1.18	$\quad + H_2O(l)$	+1.21
$V^{2+}(aq) + 2e^- \rightleftharpoons V(s)$	−1.16	$O_2(g) + 4H^+(aq) + 4e^- \rightleftharpoons 2H_2O(l)$	+1.23
$TiO^{2+}(aq) + 2H^+(aq) + 4e^- \rightleftharpoons Ti(s) + H_2O(l)$	−0.89	$MnO_2(s) + 4H^+(aq) + 2e^- \rightleftharpoons Mn^{2+}(aq)$	
$B(OH)_3(aq) + 3H^+(aq) + 3e^- \rightleftharpoons B(s) + 3H_2O(l)$	−0.87	$\quad + 2H_2O(l)$	+1.23
$(SiO_2)_n(s) + 4nH^+(aq) + 4ne^- \rightleftharpoons nSi(s) + 2nH_2O(l)$	−0.00	$Cr_2O_7^{2-}(aq) + 14H^+(aq) + 6e^- \rightleftharpoons 2Cr^{3+}(aq)$	
$Zn^{2+}(aq) + 2e^- \rightleftharpoons Zn(s)$	−0.76	$\quad + 7H_2O(l)$	+1.33
$Cr^{3+} + 3e^- \rightleftharpoons Cr(s)$	−0.74	$Cl_2(aq) + 2e^- \rightleftharpoons 2Cl^-(aq)$	+1.36
$As(s) + 3H^+(aq) + 3e^- \rightleftharpoons AsH_3(g)$	−0.60	$Au^{3+}(aq) + 3e^- \rightleftharpoons Au(s)$	+1.50
$Ga^{3+}(aq) + 3e^- \rightleftharpoons Ga(s)$	−0.53	$MnO_4^-(aq) + 8H^+(aq) + 5e^- \rightleftharpoons Mn^{2+}(aq)$	
$H_3PO_2(aq) + H^+(aq) + e^- \rightleftharpoons P(s) + 2H_2O(l)$	−0.51	$\quad + 4H_2O(l)$	+1.51
$H_3PO_3(aq) + 2H^+(aq) + 2e^- \rightleftharpoons H_3PO_2(aq)$		$2HOCl(aq) + 2H^+(aq) + 2e^- \rightleftharpoons Cl_2(aq) + 2H_2O(l)$	+1.63
$\quad + H_2O(l)$	−0.50	$HClO_2(aq) + 2H^+(aq) + 2e^- \rightleftharpoons HOCl(aq)$	
$Fe^{2+}(aq) + 2e^- \rightleftharpoons Fe(s)$	−0.44	$\quad + H_2O(l)$	+1.64
$Cr^{3+}(aq) + e^- \rightleftharpoons Cr^{2+}(aq)$	−0.41	$PbO_2(s) + SO_4^{2-}(aq)$	
$Cd^{2+}(aq) + 2e^- \rightleftharpoons Cd(s)$	−0.40	$\quad + 4H^+(aq) + 2e^- \rightleftharpoons PbSO_4(s)$	
$Se(s) + 2H^+(aq) + 2e^- \rightleftharpoons H_2Se(g)$	−0.40	$\quad + 2H_2O(l)$	+1.68
$Ti^{3+}(aq) + e^- \rightleftharpoons Ti^{2+}(aq)$	−0.37	$H_2O_2(aq) + 2H^+(aq) + 2e^- \rightleftharpoons 2H_2O(l)$	+1.77
$PbSO_4(s) + 2e^- \rightleftharpoons Pb(s) + SO_4^{2-}(aq)$	−0.36	$Co^{3+}(aq) + e^- \rightleftharpoons Co^{2+}(aq)$	+1.82
$Co^{2+}(aq) + 2e^- \rightleftharpoons Co(s)$	−0.28	$O_3(g) + 2H^+(aq) + 2e^- \rightleftharpoons O_2(g) + H_2O(l)$	+2.07
$H_3PO_4(aq) + 2H^+ + 2e^- \rightleftharpoons H_3PO_3(aq)$		$F_2(g) + 2e^- \rightleftharpoons 2F^-(aq)$	+2.87
$\quad + H_2O(l)$	−0.28		
$V^{3+}(aq) + e^- \rightleftharpoons V^{2+}(aq)$	−0.26	**B. Solutions of pH $>$ 7.0**	
$Ni^{2+}(aq) + 2e^- \rightleftharpoons Ni(s)$	−0.25	$Ca(OH)_2(s) + 2e^- \rightleftharpoons Ca(s) + 2OH^-(aq)$	−3.03
$Sn^{2+}(aq) + 2e^- \rightleftharpoons Sn(s)$	−0.14	$Mg(OH)_2(s) + 2e^- \rightleftharpoons Mg(s) + 2OH^-(aq)$	−2.69
$Pb^{2+}(aq) + 2e^- \rightleftharpoons Pb(s)$	−0.13	$[Al(OH)_4(H_2O)_2]^-(aq) + 3e^- \rightleftharpoons Al(s) + 4OH^-(aq)$	
$2H^+(aq) + 2e^- \rightleftharpoons H_2(g)$	ZERO (Reference Standard)	$\quad + 2H_2O(l)$	−2.35
		$Zn(OH)_2(s) + 2e^- \rightleftharpoons Zn(s) + 2OH^-(aq)$	−1.24
$P(s) + 3H^+(aq) + 3e^- \rightleftharpoons PH_3(g)$	+0.06	$Se(s) + 2e^- \rightleftharpoons Se^{2-}$	−0.93
$Sn^{4+} + 2e^- \rightleftharpoons Sn^{2+}(aq)$	+0.15	$Fe(OH)_2(s) + 2e^- \rightleftharpoons Fe(s) + 2OH^-(aq)$	−0.88
$Cu^{2+}(aq) + e^- \rightleftharpoons Cu^+(aq)$	+0.15	$S(s) + 2e^- \rightleftharpoons S^{2-}$	−0.48
$SO_4^{2-}(aq) + 3H^+(aq) + 2e^- \rightleftharpoons HSO_3^-(aq)$		$NO_3^-(aq) + H_2O(l) + 2e^- \rightleftharpoons NO_2^-(aq)$	
$\quad + H_2O(l)$	+0.17	$\quad + 2OH^-(aq)$	+0.01
$Cu^{2+}(aq) + 2e^- \rightleftharpoons Cu(s)$	+0.34	$IO_3^-(aq) + 3H_2O(l) + 6e^- \rightleftharpoons I^-(aq) + 6OH^-(aq)$	+0.26
		$ClO_2(g) + e^- \rightleftharpoons ClO_2^-(aq)$	+1.16

APPENDIX F: SELECTED THERMODYNAMIC DATA

Some standard[a] heats of formation.

Compound	ΔH°_f (kcal mole^{-1})
acetic acid(l)	-116.4
ammonium nitrate (s)	- 87.4
ammonia (g)	- 11.0
benzene (l)	+ 11.7
carbon dioxide (g)	- 94.0
carbon monoxide (g)	- 26.4
carbon tetrachloride (l)	- 32.4
ethane (g)	- 20.2
ethanol (l)	- 66.4
hydrogen bromide (g)	- 8.7
hydrogen chloride (g)	- 22.1
hydrogen fluoride (g)	- 64.8
hydrogen iodide (g)	+ 6.2
hydrogen sulfide (g)	- 4.8
methane (g)	- 17.9
methanol (l)	- 57.0
nitric acid (l)	- 41.6
nitric oxide (g)	+ 21.6
nitrous oxide (g)	+ 19.6
propane (g)	- 24.8
sodium bicarbonate (s)	-226.5
sodium bromide (s)	- 86.0
sodium carbonate (s)	-270.3
sodium chloride (s)	- 98.4
sodium hydroxide (s)	-102.0
sodium fluoride (s)	-136.5
sulfur dioxide (g)	- 70.9
sulfur trioxide (g)	- 94.6
sulfuric acid (l)	-194.6
water (l)	- 68.3
water (g)[b]	- 57.8

[a]To facilitate calculations, values are tabulated for experimental quantities corrected to conditions of standard states (ΔH°). These are defined as follows: temperature = 25°C, all gases at 1.00 atm pressure, and all solids in their most stable form. A negative sign indicates an exothermic process (heat evolved). A positive sign indicates an endothermic process (heat required).

[b]The value reported is "corrected" for differences between water vapor and the standard state (liquid) at 25°C.

INDEX

453

454

456

The Periodic System of the Elements

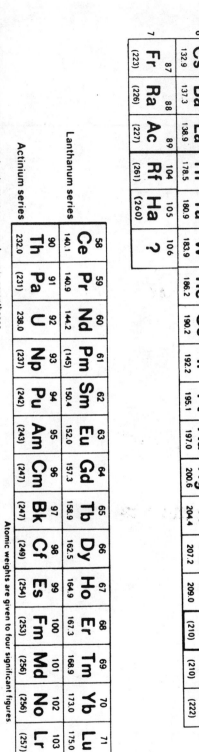
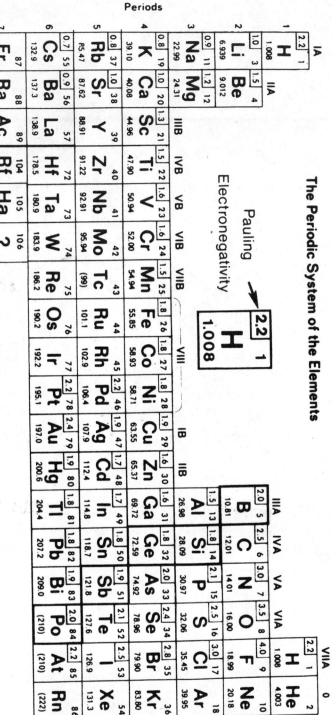